Ecological Processes and Cumulative Impacts

Illustrated by Bottomland Hardwood Wetland Ecosystems

Edited by

James G. Gosselink
Center for Wetland Resources
Louisiana State University
Baton Rouge, Louisiana 70803

Lyndon C. Lee
L. C. Lee Associates, Inc.
3016 W. Elmore Street
Seattle, Washington 98199

Thomas A. Muir
Division of Endangered Species and Habitat Conservation
U.S. Fish and Wildlife Service
Washington, D.C. 20240

Library of Congress Cataloging-in-Publication Data

Ecological processes and cumulative impacts : illustrated by bottomland hardwood wetland ecosystems / edited by James G. Gosselink, Lyndon C. Lee, Thomas A. Muir.

p. cm.

Includes bibliographical references and index.

Summary: Summary of reports from three workshops sponsored by the U.S. Environment Protection Agency.

ISBN 0-87371-339-7

1. Forest ecology—Southern States—Congresses. 2. Floodplain ecology—Southern States—Congresses. 3. Wetland ecology—Southern States—Congresses. 4. Hardwoods—Southern States—Ecology—Congresses. I. Gosselink, James G. II. Lee, Lyndon C. III. Muir, Thomas A. IV. United States. Environmental Protection Agency.

QH104.5.S59E35 1990 90-6360

574.5′2642′0975—dc20 CIP

LEWIS PUBLISHERS, INC.
121 South Main Street, Chelsea, Michigan 48118

PRINTED IN THE UNITED STATES OF AMERICA

PREFACE

By the late 1970s and early 1980s the rapid decimation of the most extensive wetlands - bottomland forests - in the United States was becoming clear to anyone concerned with wetland conservation. Compared to coastal salt marshes inland freshwater wetlands had received little study, and practically no regulatory attention. Suddenly it was realized that over 50 percent of the nation's bottomland hardwood forests were gone, most cleared for a booming, subsidized, agricultural industry. The clearing itself was often made economically feasible by large publicly-financed flood control projects.

To begin to address the implication sof these bottomland forest losses, and the appropriate regulatory and non-regulatory responses, a number of federal agencies and private organizations sponsored and financed two workshops on the ecology of bottomland and riparian ecosystems, the first in 1978 at Callaway Gardens, Georgia (Johnson and McCormick 1979), the second in 1980 at Lake Lanier, Georgia (Clark and Benforado 1981). Beginning in 1984 the U.S. Environmental Protection Agency sponsored 3 workshops to describe ecological processes in bottomland hardwood forest ecosystems, and to relate these processes to human activities that affect them. These workshops are the subject of this volume. The chapters range from reports of original research into ecological processes, through the impacts of human activities on flood plain ecosystems, to human perceptions of bottomland values and management strategies for ecosystem restoration and conservation. A general recognition in the second workshop, of the importance of cumulative impacts in bottomland hardwood forest ecosystems, changed the emphasis in the third workshop to issues of cumulative impact assessment and management, with a consequent scale shift from local sites and ecoystems to processes in large scale landscapes such as watersheds and hydrologic basins.

The editors appreciate the work and cooperative spirit of the contributors to this volume. The authors of the workgroup reports, especially, deserve a strong vote of thanks. Their job was to merge, simplify, and make structurally coherent three workgroup reports, each report written a year apart, usually by a panel composed of different people. This was no mean feat! The lag in time between the last workshop in 1986 and the publication of this volume posed an additional problem. The chapter authors responded to the rapid expansion of the scientific literature during this period, and updated their individual chapters accordingly with references to new reserach findings, major conferences, and major policy changes.

We thank John Meagher, Susan Sarason, and Allan Hirsch, U.S. Environmental Protection Agency, Washington, D.C., for their support and direction in the workshops; and James Roelle, Gregor Auble, David Hamilton, Richard Johnson, Charles Segelquist, and Gerald Horak, of the National Ecology Research Center, U.S. Fish and Wildlife Service, Fort Collins, CO, who were facilitators for the workshops, and who prepared the reports of the individual workshops. All the participants worked diligently on various workgroup panels. The results of their deliberations are contained in Chapters 11-18.

We owe a special debt to Susan Hamilton, who edited the manuscript and coordinated the production of the camera-ready copy; and to Kathy Joiner and Tiny Sikkes, who spent hundreds of hours on the computer producing figures and formatting the individual chapters.

The preparation of this volume was supported by the Office of Wetland Protection, U.S. Environmental Agency, Washington, D.C. (Project No. CR814646-01-0), and by the Center for Wetland Resources, Louisiana State University, Baton Rouge, Louisiana. The views expressed in this report are those of the authors and are not offical policy of the U.S. Environmental Protection Agency.

REFERENCES CITED

Clark, J. R., and J. Benforado, eds. 1981. Wetlands of bottomland hardwood forests. Elsevier Scientific Publ. Co., New York. 401 pp.

Johnson, R. R., and J. F. McCormick, Tech. Coords. 1979. Strategies for protection and management of floodplain wetlands and other riparian ecosystems. U.S. Department of Agriculture, Forest Service, Washington, D.C. General tech. Rep. WO-12.410 pp.

JAMES G. GOSSELINK
LYNDON C. LEE
THOMAS A. MUIR

CONTENTS

CONTRIBUTORS AND PARTICIPANTS

Gregor T. Auble
National Ecology Research Center,
U.S, Fish and Wildlife Service
4512 McMurray Avenue
Fort Collins, CO 80525

Paul Adamus
NSI Technology Services Corporation
USEPA Environmental Research Laboratory
200 SW 35th Street
Corvallis, OR 97333

Roger Banks
U.S. Fish and Wildlife Service
P. O. Box 12559
Charleston, SC 29412

Charles Belin
U.S. Army Corps of Engineers
P. O. Box 889
Savannah, GA 31402-0889

Jay Benforado
U.S. Environmental Protection Agency
Office of Research and Development (RD-682)
401 M Street, SW
Washington, DC 20460

Rex Boner
The Nature Conservancy
3179 Maple Drive, N.E. Room 8
Atlanta, GA 30305

Roel Boumans
Center for Wetland Resources
Louisiana State University
Baton Rouge, LA 70803

Mark Brinson
Biology Department
East Carolina University
Greenville, NC 27834

James Brown
U.S. Fish and Wildlife Service
75 Spring Street, S.W., Suite 1276
Atlanta GA 30303

C. Frederick Bryan
Louisiana Cooperative Fishery Research Unit
Louisiana State University
Baton Rouge, LA 70803

John Cairns
Center for Environmental Studies
Virginia Polytech. Instit. and State Univ.
Blacksburg, VA 24061

Milady A. Cardamone
Maasstraat 14,
8226 LP Lelystad,
The Netherlands

Thomas Cavinder
U.S. Environmental Protection Agency
Environmental Services Division
College Station Road
Athens, GA 30613

Jack Chowning
Department of the Army
Office of the Chief of Engineers
20 Massachusetts Avenue
Washington, DC 20314-1000

Thomas G. Ciravolo
University of Georgia
Savannah River Ecology Laboratory, Drawer E
Aiken, SC 29801

Ellis Clairain
U. S. Army Engineer
Waterways Experiment Station
Environmental Laboratory,
P.O.Box 631
Vicksburg, MS 39180

William Conner
Baruch Forest Science Institute
Box 596, Georgetown, SC
29442

Robert Davis
Office of Policy Analysis
U.S. Department of the Interior
Washington, DC 20240

John Day
Center for Wetland Resources
Louisiana State University
Baton Rouge, LA 70803

Joseph Dowhan
U.S. Fish and Wildlife Service
Division of Ecological Services
18th and C Streets, NW
Washington, DC 20240

Fred Dunham
Louisiana Department of Wildlife
and Fisheries
P. O. Box 15570
Baton Rouge, LA 70895

John F. Elder
U.S. Geological Survey
227 N. Borough Street, #3015
Tallahassee, FL 32301

Jack S. Erwin
University of Georgia,
Savannah River Ecology
Laboratory, Drawer E
Aiken SC 29801

Beverly Ethridge
Environmental Assessment
Branch
U.S. Environmental Protection
Agency
345 Courtland Street, N.E.
Atlanta, GA 30365

Stephen Faulkner
4417 Regis Avenue
Druham, NC 27705

Stephen Forsythe
U.S. Fish and Wildlife Service
222 South Houston
Suite A, Tulsa, OK 74127

Tom Glatzel
Dredge and Fill Section
U.S. Environmental Protection
Agency
230 S. Dearborn Street
Chicago, IL 60604

Linda Glenboski
U.S. Army Corps of Engineers
P. O. Box 60267
New Orleans, LA 70160-0267

James Gosselink
Center for Wetland Resources
Louisiana State University
Baton Rouge, LA 70803

Dale Hall
U.S. Fish and Wildlife Service
18th and C Street, NW, Room
3245
Washington, DC 20240

David Hamilton
National Ecology Research
Center
U.S. Fish and Wildlife Service
4512 McMurray Ave,
Fort Collins, CO 80525

William R. Harms
Southeastern Experiment Station
U.S. Forest Service
2730 Savannah Highway
Charleston, SC 29407

Larry Harris
Department of Wildlife and Range Sciences
118 Newins-Ziegler Hall
University of Florida
Gainesville, FL 32611

Susan Harrison
U.S. Army Corps of Engineers
Lower Mississippi Valley Division
P. O. Box-Attn: LMVCO-N
Vicksburg, MS 39180-0080

Presley Hatcher
U.S. Environmental Protection Agency
1201 Elm Street
Dallas, TX 75270

Robert Heeren
Coordinator, Hardwood Management
Union Camp, Woodlands Division
Franklin, VA 23851

John Hefner
U.S. Fish and Wildlife Service
75 Spring Street, SW
Atlanta, GA 30303

Mr. Kelly Hendricks
Regulatory Branch
U. S. Army Corps of Engineers
P. O. Box 889
Savannah, GA 31402

Delbert Hicks
U.S. Environmental Protection Agency
Environmental Services Division
College Station Road
Athens, GA 30613

Allan Hirsch
Office of Federal Activities104)
U.S. Environmental Protection Agency
401 M Street, S.W.
Washington, DC 20460

Luther Holloway
Holloway Environmental Services
101 Willow Glen Drive
Vicksburg, MS 39180

Hoke Howard
U.S. Environmental Protection Agency
Environmental Services Division
College Station Road
Athen, GA 30613

Robert Terry Huffman
Huffman Technologies Company
69 Aztec Street
San Francisco, CA 94110

Eric Hughes
U.S. Environmental Protection Agency
Environmental Assessment Branch
345 Courtland Street, NE
Atlanta, GA 30365

Joseph Jacob, Jr.
Southeastern Regional Director
Preserve Selection and Design
The Nature Conservancy
P. O. Box 270
Chapel Hill, NC 27514

E. S. Jemison
Tishomingo National Wildlife Refuge,
U.S. Fish and Wildlife Service
Route 1, Box 151
Tiohomingo, OK 73460

Richard L. Johnson
National Ecology Research Center
U.S. Fish and Wildlife Service
4512 McMurray Ave
Fort Collins, CO 80525

Robert L. Johnson
U.S. Forest Service
Southern Hardwood Laboratory
P. O. Box 227
Stoneville, MS 38776

Phillip H. Jones
U.S. Army Engineers, Waterways Experiment Station
P. O. Box 631
Vicksburg, MS 39180

Barbara Keeler
U.S. Environmental Protection Agency
1201 Elm Street
Dallas, TX 75270

Clyde Kiker
Food and Resource Economics Dept.
McCarty Hall, University of Florida
Gainesville, FL 32611

Wiley Kitchens
University of Florida,
117 Newins Ziegler Hall,
Gainesville, FL 32611

John Kittelson
Environmental Resource Branch, St. Paul District
U. S. Army Corps of Engineers
P.O. Box 1125
St. Paul, Mn 55101

Barbara Kleiss
U.S. Waterways Experiment Station,
P.O. Box 631
Vicksburg, MS 39180

Albert Korgi
U.S. Environmental Protection Agency
Environmental Assessment Branch
345 Courtland Street, NE
Atlanta, GA 30365

David Kovacic
Department of Landscape Architecture
University of Illinois,
Urbana, IL 61801

William Kruczynski
U.S. Environmental Protection Agency
Environmental Assessment Branch
345 Courtland Street, NE
Atlanta, GA 30365

Victor Lambou
U.S. Environmental Protection Agency
P. O. Box 93478
Las Vegas, Nevada 89193-3478

Russell Lea
Hardwood Research Cooperative
North Carolina State University
Box 8002, Raleigh, NC 27695

Lyndon C. Lee
L.C. Lee Associates, Inc.
3016 W. Elmore Street
Seattle, WA 981907

David Lofton
Regulatory Branch
U.S. Army Corps of Engineers
P. O. Box 60
Vicksburg, MS 39180

Allen Lucas
U.S. Environmental Protection Agency
Environmental Assessment Branch
345 Courtland Street, NE
Atlanta, GA 30365

Kenneth W. McLeod
University of Georgia,
Savannah River Ecology Laboratory,
Drawer E
Aiken SC 29801

Mr. John Meagher
Office of Wetland Protection
U.S. Environmental Protection Agency
401 M Street, SW
Washington, DC 20460

Dennis Mengel
Hardwood Research Cooperative
North Carolina State University
Raleigh, Nc 27695-8008

Edwin Miller
Weyerhauser Corporation
Southern Forestry Research Center
P. O. Box 1060
Hot Springs, AR 71901

William J. Mitsch
School of Natural Resources,
Ohio State University,
Columbus, OH 43210-1085

Thomas Muir
U.S. Fish Wildl Serv.
Div. of Endangered Species and Habitat Conservation
Office of Special Projects
Arlington Square, Room 400
Washington, D.C. 20240

James Neal
U. S. Fish and Wildlife Service
P. O. Box 4655 SFA Station
Nacogdoches, TX 75962

Lawrence Neville
Office of Regional Counsel
U.S. Environmental Protection Agency, Region IV
345 Courtland Street, NE
Atlanta, GA 30365

Charles Newling
U.S. Army Engineers, Waterways Experiment Station
Environmental Laboratory
P. O. Box 631
Vicksburg, MS 39180

Richard Novitzki
NSI Technology Services Corporation
USEPA Environmental Research Laboratory
200 SW 35th Street
Corvallis, OR 97333

William Patrick
Laboratory for Wetland Soils and Sediments
Center for Wetland Resources
Louisiana State University
Baton Rouge, LA 70803

Sam Patterson
Department of Environmental Science
Clark Hall, University of Virginia
Charlottesville, VA 22903

Leonard Pearlstine
National Wetlands Research Center
U.S. Fish and Wildlife Service
1010 Gause Blvd.
Slidell, LA 70458

John Pomponio
U.S. Environmental Protection Agency
Curtis Building, 6th and Walnut Street
Philadelphia, PA 19106

Thomas Pullen
U. S. Army Corps of Engineers
Lower Mississippi Valley Division
P. O. Box 80
Vicksburg, MS 39180-0080

Susan Ray
U.S. Environmental Protection Agency
Environmental Review Branch
726 Minnesota Avenue
Kansas City, KS 64101

Max Reed
U. S. Army Corps of Engineers
Lower Mississippi Valley Division
P. O. Box 80
Vicksburg, MS 39180

James E. Roelle
National Ecology Research Center,
U.S. Fish and Wildlife Service,
4512 McMurray Ave,
Fort Collins, CO 80525

Charles Rhodes
U.S. Environmental Protection Agency
Environmental Impact and Marine Policy Branch
841 Chestnut Building
Philadelphia, PA 19107

John M. Safley
Ecological Sciences Division
USDA Soil Conservation Service
P. O. Box 2890
Washington, DC 20013

Dana Sanders
U. S. Army Engineers, Waterways Experiment Station
Environmental Laboratory
P. O. Box 631
Vicksburg, MS 39180

Susan Sarason
Office of Wetland Protection
U.S. Environmental Protection Agency
401 M Street, SW
Washington, DC 20460

J. Henry Sather
103 Oakland Lane
Macomb, Il 61455

Michael Scott
U.S. Fish and Wildlife Service
National Ecology Research Center
4512 McMurray Ave
Fort Collins, CO 80525

Rebecca R. Sharitz
Department of Botany and Savannah River Ecology Laboratory
University of Georgia
Drawer E
Aiken, SC 29802

Rebecca L. Schneider
Ecology and Systematics
Cornell University
Ithaca, NY 14853

C. A. Segelquist
U.S. Fish and Wildlife Service,
National Ecology Research
Center
4512 McMurray Ave,
Fort Collins, CO 80525

Richard Smardon
School of Landscape
Architecture
College of Environmental
Science and Forestry
State University of New York
Syracuse, NY 13210

Edward Ray Smith
USDA, Soil Conservation
Service
South National Technical Center
P. O. Box 6567
Fort Worth, TX 76115

John Stierna
Economics Division
USDA, Soil Conservation
Service
P. O. Box 2890
Washington, DC 20013

Jan R. Taylor
West Virginia Water Resources
Board,
1260 Greenbriar street,
Charleston, WV 25311

Lee Tebo
U.S. Environmental Protection
Agency
Environmental Services Division
College Station Road
Athens, GA 30613

Russell Theriot
U. S. Army Engineers,
Waterways Experiment
Station
P. O. Box 631
Vicksburg, MS 39180

William Tomlinson, Jr.
Anderson-Tulley Company
P. O. Box 38
Vicksburg, MS 39180

B. Arville Touchet
USDA, Soil Conservation
Service
3737 Government Street
Alexandria, LA 71302

Johannes van Beek
Coastal Environments, Inc.
1260 Main Street
Baton Rouge, LA 70802

Don Walker
Kentucky Department of
Environmental Protection
Division of Water
18 Reilly Road
Franklin, KY 40601

Thomas Welborn
Environmental Assessment
Branch
U.S. Environmental Protection
Agency
345 Courtland Street, NE
Atlanta, GA 30365

Jean Wooten
University of Southern
Mississippi
Dept. of Biological Sciences
Box 5018
Hattiesburg, MS 39401

Ecological Processes and Cumulative Impacts

Illustrated by Bottomland Hardwood Wetland Ecosystems

1. INTRODUCTION

James E. Roelle
Gregor T. Auble
National Ecology Research Center, U.S. Fish and Wildlife Service, Fort Collins, CO 80525

James G. Gosselink
Center for Wetland Resources, Louisiana State University, Baton Rouge, LA 70803

Bottomland hardwood forests occupying floodplains of the southeastern United States represent a valuable ecological resource that is being rapidly destroyed. While exact historical data are difficult to obtain, the lower Mississippi Valley probably contained 9 to 10 million ha of floodplain forests prior to settlement by European immigrants (Fredrickson 1979; Harris 1984; Turner et al. 1981). By 1937 only about half of this area of hardwood forest remained and an additional 2.7 million ha were lost between 1937 and 1978 (MacDonald et al. 1979). For the entire southeastern United States, the average rate of loss between about 1960 and 1975 was approximately 175,000 ha annually. In 1978, about 80% of the converted forest in the Mississippi River alluvial floodplain was in agriculture (Turner et al. 1981).

Under current federal regulations, most floodplain forests are wetlands, which are defined as "...those areas that are inundated or saturated by surface or groundwater at a frequency and duration sufficient to support, and that under normal circumstances do support, a prevalence of vegetation typically adapted for life in saturated soil conditions" [33 CFR 323.2(c)]. Many activities on sites meeting this definition are regulated under Section 404 of the Clean Water Act of 1977 (CWA; 33 USC 1344), which prohibits discharges of dredged or fill material into waters of the United States, including wetlands, except by permit from the U.S. Army Corps of Engineers (ACE). While the ACE issues such permits, the U.S. Environmental Protection Agency (EPA) is intimately involved in the administration of the CWA in general and Section 404 in particular. With respect to Section 404, EPA responsibilities include

---

Ecological Processes and Cumulative Impacts: Illustrated by Bottomland Hardwood Wetland Ecosystems. Edited by James G. Gosselink, Lyndon C. Lee, and Thomas A. Muir. Chelsea, MI 48118. Printed in USA.

issuance (in conjunction with ACE) of the Section 404 (b)(1) guidelines, which establish criteria to be used in evaluating permit applications; designation, under Section 404(c), of areas where discharge of dredged or fill material will not be permitted due to unacceptable impacts on municipal water supplies, shellfish beds, fishery areas, wildlife, or recreational areas; enforcement actions for unauthorized discharges; and delineation of wetland boundaries consistent with the above definition.

The EPA is concerned about the loss of forested wetlands because of its responsibilities under the CWA. In an effort to better fulfill its responsibilities, the EPA sought a synthesis of the best current scientific information pertaining to bottomland forest ecosystems, with particular reference to:

1) Structure. Bottomland communities are diverse and complex. Accurate characterization of structure is important both to understand the resource and to make rational and objective determinations of wetland boundaries. Delineation of the wetland portions of bottomland ecosystems has been particularly troublesome, because these ecosystems occupy a continuum of sites ranging from nearly permanently inundated to rarely flooded.
2) Function. Wetlands are regulated under the CWA because of the functions (e.g., water quality enhancement) they perform and the value of those functions to society. However, performance of these functions varies from one bottomland site to another. Regulation under Section 404 should take such variation into account.
3) Impacts. Bottomland forest ecosystems are affected by a variety of human activities. An understanding of these activities and their impacts is essential to determine which activities fall under the jurisdiction of Section 404 (i.e., constitute a deposition of dredged or fill material), and to assess how these activities affect the functions that the CWA is intended to protect.

The EPA convened a series of three workshops to solicit expert advice on bottomland forest ecosystems. These workshops were attended by scientists and regulators familiar with bottomland issues, and facilitated by a team from the National Ecology Research Center, U.S. Fish and

Wildlife Service, Fort Collins, CO. Each of the workshops consisted of a set of contributed papers and workgroup discussions concerning specific questions relevant to EPA regulatory concerns.

WORKSHOP I

The first workshop in the series was held December 3-7, 1984, in St. Francisville, Louisiana. The general objective of this workshop was to examine ways in which the structure and function of forested bottomland ecosystems can be characterized and, in particular, to investigate the utility of the conceptual framework described in Clark and Benforado (1981). In this framework, the transition from aquatic to upland habitats through a floodplain is divided into six zones based on concomitant variations in the soil-moisture regime and associated vegetation (Larson et al. 1981). This zonation concept was of interest to the EPA from at least two perspectives. First, as a framework for organizing information, recognizable zones with identifiable functions might form a useful cataloguing system to facilitate tasks such as assessing the impacts of a particular site-specific activity. Second, the zonation concept has potential utility for identifying the wetland portions of bottomland communities. If the zones can be recognized in the field, and if one or more of them can be consistently shown to have wetland characteristics while others do not, then zones might be used to identify areas that should fall under Section 404 jurisdiction.

To examine these issues systematically, six workgroups, each dealing with a single subject area (hydrology, soils, vegetation, fish, wildlife, and ecosystem processes), were asked to consider three basic questions.

1) How well does the zonation concept describe the structure of bottomland communities?
2) What are the functions performed by bottomland ecosystems?
3) How does performance of these functions vary across bottomland zones?

WORKSHOP II

One of the conclusions of the first workshop was that the zonation concept does not provide a particularly good framework for representing the functions performed by forested bottomland ecosystems. Other characteristics are likely to be at least as important as the zones in determining function. Based in part on this conclusion, groups at the second workshop were assigned the following tasks.

1) Using the results of the first workshop as a starting point, develop a revised list of the functions performed by forested bottomland ecosystems.
2) Identify human activities that affect the major functions.
3) Identify a set of characteristics or indices that can be used to evaluate the performance of each function at any site. Describe the relationship of each characteristic to the function.
4) Develop, with supporting evidence where possible, an analysis of the impact of each activity (Task 2) on each characteristic (Task 3) and on each function as a whole.

Table 1 is a list of all of the human activities identified by the participants as having significant impacts in bottomland hardwood forests. A set of seven activities (e.g., conversion to soybeans) and a number of specific actions associated with each (e.g., land clearing) were derived from this list for analysis by the workgroups (Table 2). These activities were selected on the basis of their perceived importance in forested bottomland ecosystems and their interest from the perspective of EPA.

The second workshop did not have separate vegetation and soils workgroups. Water Quality and Cultural/Recreational/Economic workgroups were added to the list from the first workshop.

WORKSHOP III

Certain limitations of the second workshop approach influenced the focus of the third workshop, held January 13-17, 1986, in Savannah, Georgia. First, site-specific analyses such as those conducted at the second workshop do not allow for adequate consideration of cumulative

Table 1. Activities identified as having significant impacts in bottomland hardwoods.

Landclearing	Clearcutting
Leveling	Thinning
Channelization	Selective harvesting
Tiling	Depredation control
Ditching	Impoundment construction
Pumping	Bank stabilization
Dredging and filling	Diversion construction
Fertilizing	Highway construction
Applying pesticides	Utility right- of-way clearing
Chemical conditioning	Refuse disposal
Discing	Sewage disposal
Plowing	Drilling mud disposal
Planting	Hazardous waste disposal
Farm road construction	Mining
Irrigation	Oil and gas exploration and production

impacts in the regulatory process. Second, contextual variables (i.e., position of a bottomland site with respect to surrounding landscape features) are often as important as on-site variables in determining the extent to which functions are performed. Finally, while there is a need for a consistent, defensible functional assessment methodology, such a methodology will require an approach that does more than just "add up" impacts on individual characteristics to obtain the overall impact on a function.

Considering these conclusions from the second workshop, the objectives of the third workshop were to: 1) develop an approach for considering cumulative impacts on bottomland forest functions in the regulatory process, and 2) develop an approach for field personnel to assess the functions performed by forested bottomland sites, including the important contextual variables. To achieve these objectives, workgroups were asked to complete the following tasks.

Table 2. General activities and associated specific actions for analysis by all workgroups.

Activity	Actions	
Conversion to rice	Landclearing Leveling Levee construction Flooding	Drainage or drying Fertilization Pesticide application Seed bed preparation
Conversion to soybeans	Landclearing Leveling Ditching	Fertilization Pesticide application Seed bed preparation
Impoundment construction Upstream of site On-site Downstream of site	Landclearing (on-site)	Filling Dredging

1) Review and refine the functions developed at the first two workshops.
2) Refine the list of site characteristics that determine function, and identify site variables that might be used as indices of functional performance.
3) Summarize existing information showing the relationships between the characteristics and bottomland forest functions.
4) Identify target values or management goals for these characteristics.
5) Develop a preliminary approach or "assessment model" that might be used to evaluate forested bottomland sites with respect to these functions.

Four workgroups -- hydrology, water quality, fisheries, and wildlife -- were asked to approach these tasks from the site-specific perspective of particular subject areas. An ecosystem workgroup was asked to approach the tasks from the perspective of cumulative impacts.

This book presents the results of the three workshops. It is organized in three sections. Sections I and II, are based on invited "theme" presentations at the workshops. The subject of Section I is the ecology of bottomland forests and Section II concerns cumulative impacts in bottomland forests. Section III contains summaries of the workgroup reports. Three reports (Roelle et al. 1987a,b,c) contain the full record of the workgroup discussions in each of the three workshops. In this book these reports are summarized in individual chapters on hydrology, soils, water quality, vegetation, fisheries, wildlife, ecosystem processes and cumulative impacts, and culture/recreation/economics.

REFERENCES CITED

Clark, J. R., and J. Benforado, eds. 1981. Wetlands of bottomland hardwood forests. Elsevier, Amsterdam. 401 pp.

Fredrickson, L. G. 1979. Lowland hardwood wetlands: current status and habitat values for wildlife. Pages 296-306 *in* P. E. Greeson, J. R. Clark, and J. E. Clark, eds. Wetland Functions and Values: the State of Our Understanding. Proceedings of the National Symposium on Wetlands. November 7-10, 1978. American Water Resources Assoc., Minneapolis, Minnesota.

Harris, L. D. 1984. Bottomland hardwoods: valuable, vanishing, vulnerable. Cooperative Extension Service, Institute of Food and Agricultural Sciences, University of Florida, Gainesville, Florida. 18 pp.

Larson, J. S., M. S. Bedinger, C. F. Bryan, S. Brown, R. T. Huffman, E. L. Miller, D. G. Rhodes, and B. A. Touchet. 1981. Transition from wetlands to uplands in southeastern bottomland hardwood forests. Pages 225-273 *in* J. R. Clark and J. Benforado, eds. Wetlands of bottomland hardwood forests. Elsevier, Amsterdam.

MacDonald, P. O., W. E. Frayer, and J. K. Clauser. 1979. Documentation, chronology, and future projections of bottomland hardwood habitat losses in the lower Mississippi alluvial plain. U.S. Department of Interior, Fish and Wildlife Service, Washington, D. C.

Roelle, J. E., G. T. Auble, D. B. Hamilton, R. L. Johnson and C. A. Segelquist. 1987a. Results of a workshop concerning ecological zonation in bottomland hardwoods. U.S. Fish Wildl. Serv., National Ecology Center, Fort Collins, Colorado. NEC-87/14. 141 pp.

Roelle, J. E., G. T. Auble, D. B. Hamilton, G. C. Horak, R. L. Johnson and C. A. Segelquist. 1987b. Results of a workshop concerning impacts of various activities on the functions of bottomland hardwoods. U.S. Fish Wildl. Serv., National Ecology Center, Fort Collins, Colorado. NEC-87/15. 171 pp.

______________________________, R. L. Johnson and C. A. Segelquist. 1987c. Results of a workshop concerning assessment of the functions of bottomland hardwoods. U.S. Fish Wildl. Serv., National Ecology Center, Fort Collins, Colorado. NEC-87/16. 173 pp.

Turner, R. E., S. W. Forsythe and N. J. Craig. 1981. Bottomland hardwood forest land resources of the southeastern United States. Pages 13-28 *in* J. R. Clark, and J. Benforado, eds. Wetlands of bottomland hardwood forests. Elsevier, Amsterdam.

SECTION I. THE ECOLOGY OF BOTTOMLAND HARDWOOD FORESTS

PROLOGUE TO SECTION I

Chapters 1- 6 describe the ecology of bottomland forests. We start with a general overview by Taylor, Cardamone and Mitsch of ecological processes in forested bottomlands, and their values to human society. Mengel and Lea describe studies at the Hardwood Research Cooperative, North Carolina State University, to quantify the productivity and standing biomass of hardwood forests across the southeast. Biotic processes in bottomland forests, particularly primary production, are intimately tied to nutrient dynamics. Kovacic, Ciravolo, McLeod and Erwin examine the potential for soil nitrate and ammonium losses across the floodplain. Lambou presents a detailed analysis of the role of bottomland forests in fish and crayfish production in the Atchafalaya Basin, Louisiana. Finally, Sharitz, Schneider and Lee describe plant community structure in a disturbed palustrine wetland in South Carolina and the role of hydrology in seed availability, seed distribution, and seedling survivorship, as they relate to the forest regeneration process.

2. BOTTOMLAND HARDWOOD FORESTS: THEIR FUNCTIONS AND VALUES

Jan R. Taylor
West Virginia Water Resources Board, 1260 Greenbriar Street, Charleston, WV 25311

Milady A. Cardamone
Maasstraat 14, 8226 LP Lelystaad, The Netherlands

William J. Mitsch
School of Natural Resources, The Ohio State University, Columbus, OH 43210-1085

ABSTRACT

Bottomland hardwood forests of the southeastern United States result from the physical forces, particularly the hydrologic and geomorphologic conditions, that interact with, and provide an energy and nutrient subsidy to, the solar-powered forest. The ecological functions of bottomland hardwood forests - community dynamics, physio-chemical processes, surface water storage, and ground water storage - result in valuable services to humans: biomass production, food chain support, fish and wildlife habitat, erosion control, water quality protection, flood storage and control, low flow augmentation, and deep aquifer recharge. Conversion of these forests to agricultural production results in a number of functional changes to the ecosystem that, in general, reduce the values derived from natural forested wetlands.

INTRODUCTION

Historically, the term "bottomland hardwood forests" has been used to describe the floodplain forests found throughout the southeastern United States (Figure 1). The definition has been in some cases

---

Ecological Processes and Cumulative Impacts: Illustrated by Bottomland Hardwood Wetland Ecosystems. Edited by James G. Gosselink, Lyndon C. Lee, and Thomas A. Muir. Chelsea, MI 48118. Printed in USA.

broadened to include floodplain forests in the eastern and central United States as well. Huffman and Forsythe (1981) described several characteristics of bottomland hardwood forests:

1) the habitat is inundated or saturated by surface or groundwater periodically during the growing season;
2) the soils within the root zone become saturated periodically during the growing season; and
3) the prevalent woody plant species associated with a given habitat have demonstrated the ability, because of morphological and/or physiological adaptation(s), to survive, achieve maturity, and reproduce in a habitat where the soils within the root zone may become anaerobic for varying periods during the growing season.

There is concern that bottomland hardwood forests are one of the most rapidly diminishing wetland ecosystems in the United States. An estimated 4.8 million ha of bottomland hardwood wetlands in the Mississippi Alluvial Plain in 1937 had been reduced to 2.1 million ha by 1977 and were expected to be only 1.6 million ha by 1995 (MacDonald et al. 1979, Clark and Benforado 1981). Turner et al. (1981) found that an average of 174,400 ha of bottomland hardwoods were being lost each year from 12 southeastern states from 1960 to 1975. Much of this loss of bottomland hardwood forests in the southeastern United States is through clearing and drainage for agriculture.

Bottomland hardwood forests are important ecosystems for maintenance of water quality, provision of a habitat for a variety of fish and wildlife, and regulation of flooding and stream recharge. If these contributions are significant to regional ecological balances, and if these wetlands continue to be drained and cleared for agricultural development, a serious loss of values to society will result. There also remain questions about how bottomland hardwood forests fit into the definition of wetlands, whether they are covered by wetland regulations such as Section 404 of the Clean Water Act, and how the functions and values provided by these ecosystems are affected by their disruption.

This paper will describe the spatial patterns of certain functions of bottomland hardwood forests and the values that accrue to humans as a result of those functions. A general discussion of the cumulative impacts of large-scale land clearing of bottomland hardwood forests will also be included. A proper understanding of these functions, values, and impacts will lead to more effective management of these important ecosystems.

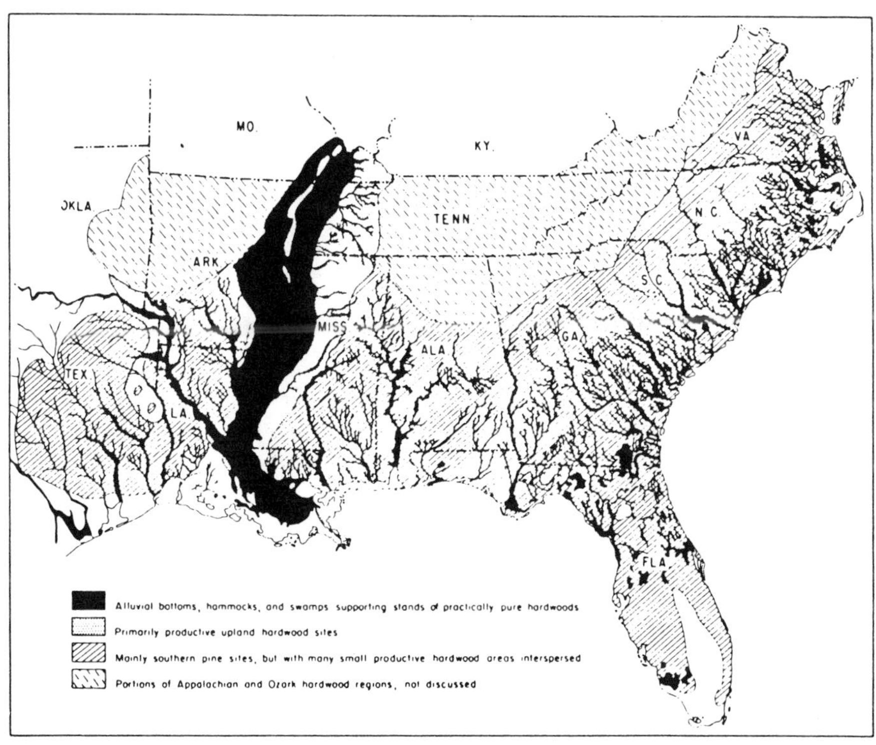

Figure 1. Extent of bottomland and hardwood forests of southern United Stated (from Patrick et al. 1981, p. 277; Copyright © by Elsevier Scientific Publishing Co., reprinted with permission).

The Physical Setting

Bottomland hardwood forests are a result of physical forcing functions, particularly the hydrologic and geomorphologic conditions where water, energy, and materials from upstream areas converge in the relatively narrow floodplain to subsidize the solar-powered forest (Figure 2). The bottomland forest, in turn, contributes to the physiochemical and biological characteristics of the river through water and nutrient exchange and organic export. In order to understand the bottomland hardwood ecosystem, it is useful first to understand these physical conditions.

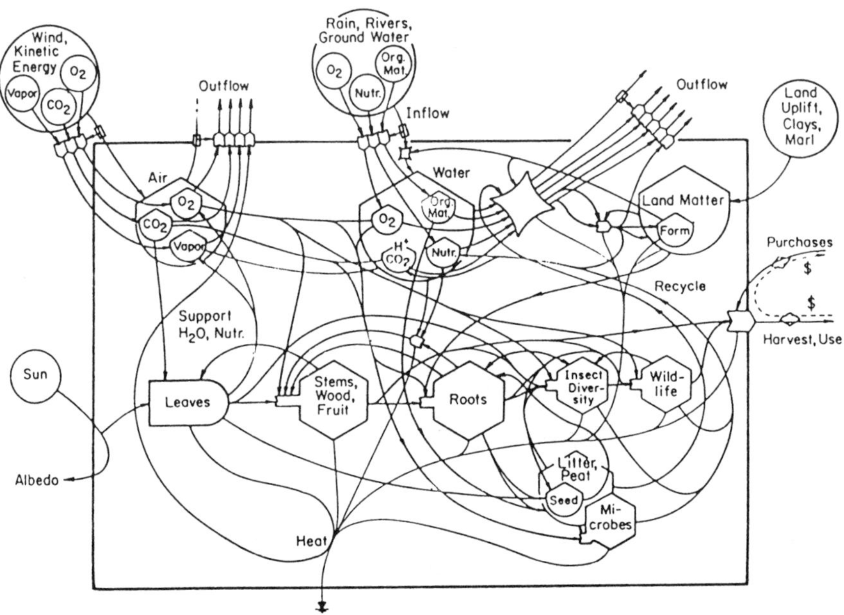

Figure 2. Energy flow diagram of a floodplain wetland (from Odum 1983, p. 433; Copyright © 1983 by John Wiley & Sons, Inc., reprinted with permission).

Floodplains and their rivers are in a continual dynamic balance between the building of structure through the deposition of alluvial materials (aggradation) and the removal of structure through the downcutting of surface geology (degradation). There are two major aggradation processes that are responsible for the formation of most floodplains--deposition of sediments on the inside curves, or point bars, of rivers and deposition from overbank flooding. In the gently sloping coastal plain rivers of the southeastern United States, conditions are ideal for the development of meandering rivers through wide floodplains. Major characteristics of southern bottomland floodplains, all of which influence the zonation and function of the bottomland hardwood forest community, include the natural levee, meander scrolls, oxbows and backswamps, point bars, and sloughs (Figure 3). The streams are not straight nor is the floodplain surface uniform from stream to upland in the bottomland hardwood forest.

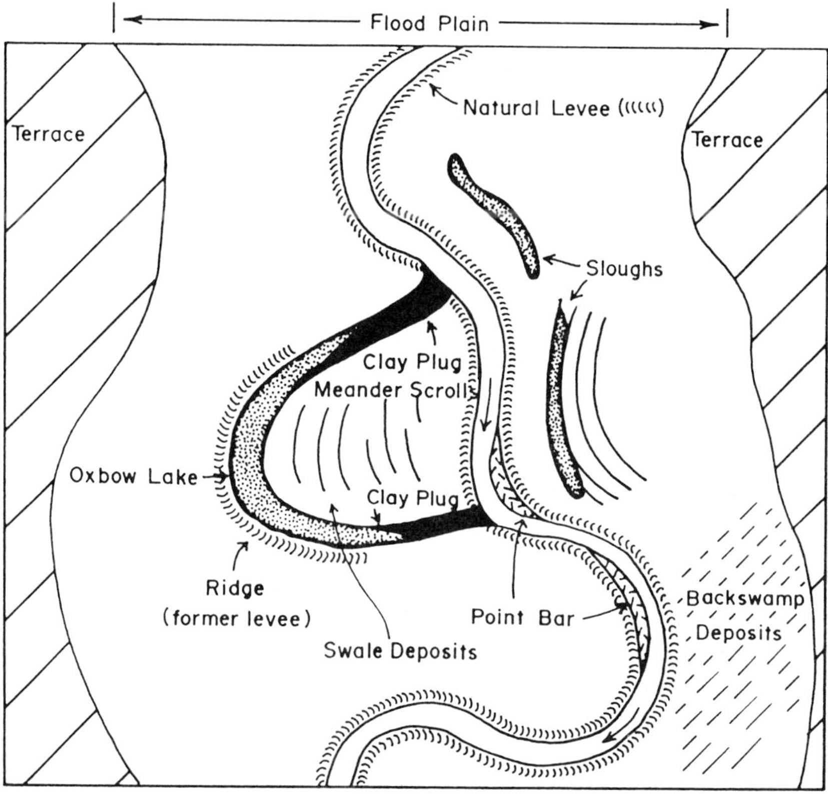

Figure 3. Major geomorphological features of a floodplain (from Mitsch and Gosselink 1986, p. 359; Copyright © by Van Nostrand Reinhold, reprinted with permission).

The frequency and duration of flooding and the subsequent soil moisture are among the main determinants in the functioning of the bottomland hardwood forest. Flooding waters also bring nutrient-rich sediments to the forest and carry away organic and inorganic materials. The timing and length of flooding depend on several factors, including precipitation patterns in local and upstream areas, the floodplain level itself, the drainage area upstream of the bottomlands in question, the channel slope, and the soil composition. Nevertheless there is remarkable consistency in the average flood recurrence intervals of floodplain forests. Leopold et al. (1964) estimated that most stream and river floodplains have recurrence intervals between 1 and 2 years, with an average of 1.5 years; on the average, streams and rivers overflow their banks in 3 out of 4 years (Figure 4a). Bedinger (1981), in describing the frequency and duration of flooding of bottomlands in the Ouachita and White Rivers, demonstrated that there is an inverse relationship between the duration of flooding and the recurrence interval (Figure 4b). Bottomland sites that were flooded annually remained flooded from 6% to 40% of the time, while floodplains with recurrence intervals of 3 to 10 years generally were flooded for less than 1% of the time.

Zonation of Bottomland Hardwood Forests

Ecologists, foresters, and hydrologists long have noted the gradient of tree species and physical conditions as one proceeded from the wettest to the driest conditions in a bottomland hardwood forest. Based on the hydrologic conditions and on previously defined vegetation associations, a zonation of bottomland hardwood forests was developed at an interdisciplinary scientific workshop held at Lake Lanier Georgia in June 1980 (Clark and Benforado 1981). Six zones were described ranging from zone I, which is the permanently wet stream or river itself, to zone VI, which is a transition zone between the floodplain and the uplands and is rarely flooded. A summary of the characteristics of the zones is given in Figure 5 and described below.

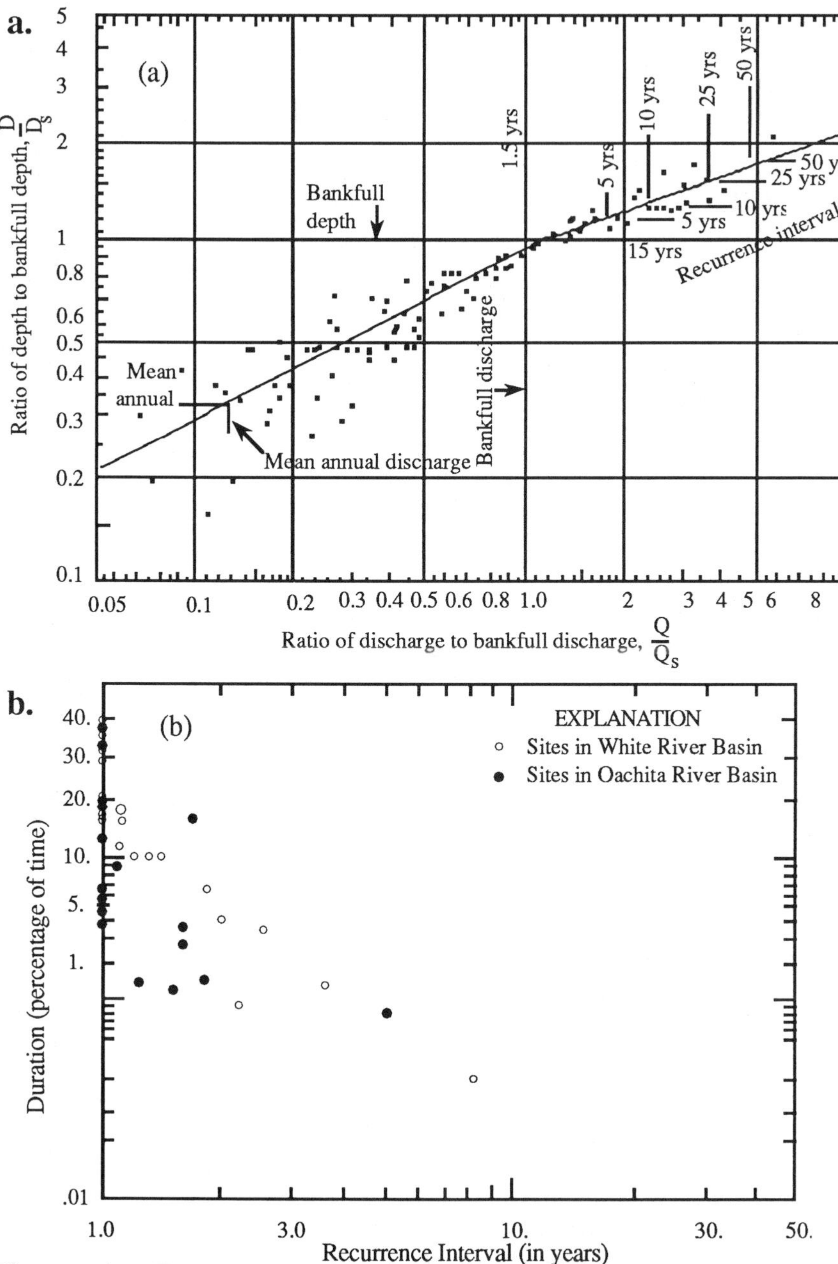

Figure 4. **Bottomland forest flooding. a) recurrence interval for bankfull flooding (after Leopold et al. 1964, p.219; Copyright © by W. H. Freeman and Co., reprinted with permission), and b) relationship between flooding duration and recurrence interval (from Bedinger 1971).**

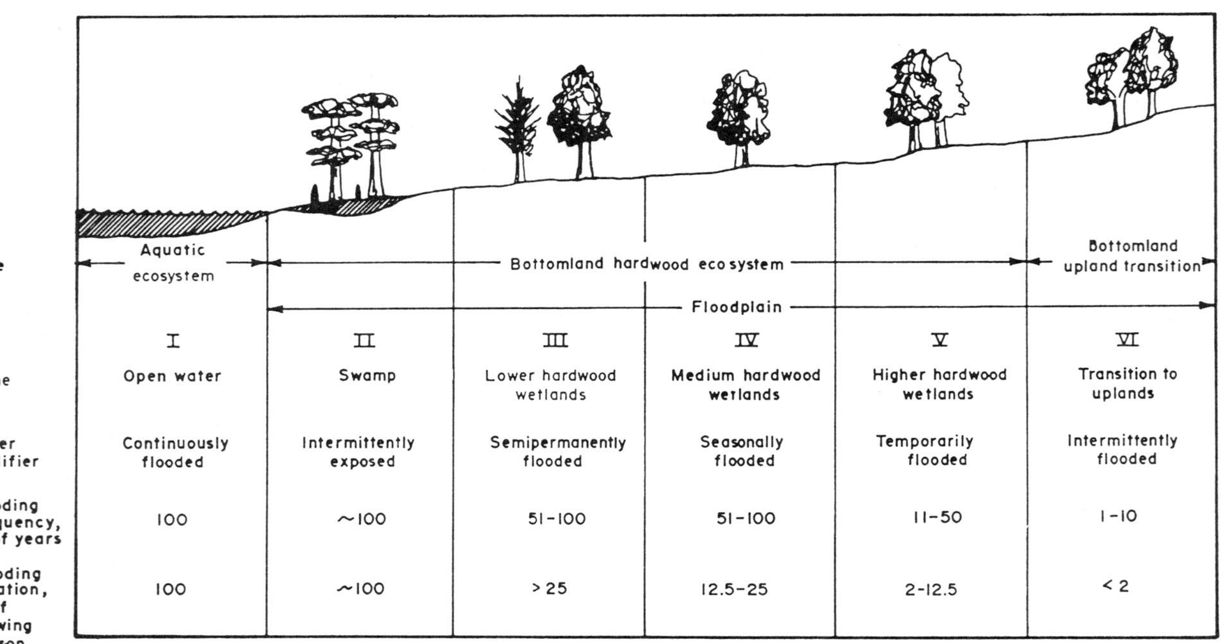

Figure 5. Zonal classification of bottomland hardwood forest wetlands (from Mitsch and Gosselink 1986, p. 364; Copyright © by Van Nostrand Reinhold, reprinted with permission).

Hydrologic Zonation

The following water regimes are used to define the zonation of bottomland hardwood forests (from Larson et al. 1981 and Cowardin et al. 1979):

Zone I - Continuously flooded. This is the permanently flooded zone which in most cases is the stream or river itself. This is not part of the bottomland hardwood forest.

Zone II- Intermittently exposed. Surface water is present throughout the year except in years of extreme drought. The probability of annual flooding is nearly 100% and vegetation is in saturated or flooded soil for the entire growing season.

Zone III- Semipermanently flooded. Surface water or soil saturation persists for a major portion of the growing season in most years. Flooding frequency ranges from 51 to 100 years per 100 years. Flooding duration typically exceeds 25% of the growing season.

Zone IV - Seasonally flooded. Surface water or saturated soil is present for extended periods, especially early in the growing season, but is absent by the end of the season in most years. Flooding frequency ranges from 51 to 100 years per 100 years and flooding duration is typically 12.5% to 25% of the growing season.

Zone V - Temporarily flooded. Surface water or soil saturation is present for brief periods during the growing season, but the water table usually lies well below the soil surface for most of the season. A typical frequency of flooding is 11 to 50 years out of 100. Typical duration is 2% to 12.5% of the growing season.

Zone VI - Intermittently flooded. Soil inundation or saturation rarely occurs, but surface water may be present for variable periods without detectable seasonal periodicity. Flood frequency typically ranges from 1 to 10 years per 100 years. Total duration of flood events is typically less than 2% of the growing season.

It should not be assumed that these flooding conditions or zones occur in sequence as one proceeds from the river's edge to the uplands as implied in Figure 5. The floodplain normally does not rise uniformly from the river but is transected with levees, meander scrolls, sloughs, and oxbow lakes that allow for different zones to appear in almost any location on the floodplain.

Soil Zonation

The zonation in hydrologic conditions leads to a differentiation in soil texture, composition, oxygenation, and color, and differences in available nutrients. Many of these distinctions are summarized in Table 1 and are discussed below.

Zone II - Soil texture is dominated by clays with a high percentage of organic matter. Anaerobic conditions dominate, causing gleyed soils that are gray to olive gray. Soil chemistry shows relatively high levels of nutrients, although nutrient availability is significantly altered from continuous flooding.

Zone III - Wharton et al. (1982) have suggested that 5% organic matter is a good dividing line between the almost permanently flooded zone II and the periodically flooded zones. Zone III is at this division. Soils are gray with mottled or olive color, but flooding is not so permanent as to cause gleying. Clays dominate the soil, and lack of oxygen prevails for much of the year.

Zone IV - Soils are alternatively aerobic and anaerobic with soil chemistry dependent on the prevailing condition. Clay soils dominate, but coarser material is possible; soil is generally gray with brownish-gray and grayish-brown mottles. Nutrients may be more available due to prolonged dry periods than in more frequently flooded zones.

Zone V - Because soils are typically flooded for only short periods in the growing season, soil-air-water are most favorable for root respiration. Clay and sandy loams

Table 1. Physiochemical characteristics of floodplain soils by zones (after Wharton et al. 1982).

	Zone	
Characteristic	II	III
Soil Texture	Dominated by silty clays or sands	Dominated by dense clays
Sand:silt:clay (% composition)		
Blackwater	69:20:12	-
Alluvial	29:23:48	34:22:44
Organic matter, %		
Blackwater	18.0	-
Alluvial	4.5	3.4
Oxygenation	Moving water aerobic; stagnant water anaerobic	Anaerobic for portions of the year
Soil color	Gray to olive gray with greenish gray, bluish gray, grayish green mottles	Gray with olive mottles
pH[a]		
Blackwater	5.0	-
Alluvial	5.0	5.3
Phosphorus (ppm)		
Blackwater	11.2	-
Alluvial	9.1	6.3
Calcium (ppm)		
Blackwater	607	-
Alluvial	1,079	752
Magnesium (ppm)		
Blackwater	98	-
Alluvial	154	140
Sodium (ppm)		
Blackwater	46	-
Alluvial	94	31
Potassium (ppm)		
Blackwater	48	-
Alluvial	51	28

[a]Range includes drought years.

Table 1. (Concluded).

Characteristic	Zone IV	Zone V	Zone VI
Soil Texture	Clays dominate surface; some coarser fractions (sands) increase with depth	Clay and sandy loams dominate; sandy soils frequent	Sands to clay
Sand:silt:clay (% composition)			
Blackwater	74:14:12	-	-
Alluvial	34:20:45	71:16:14	-
Organic matter, %			
Blackwater	7.9	-	-
Alluvial	2.8	3.8	-
Oxygenation	Alternating anaerobic and aerobic conditions	Alternating: mostly aerobic, occasionally anaerobic	Aerobic year-round
Soil color	Dominantly gray on blackwater floodplains and reddish on alluvial with brownish gray and grayish brown mottles	Dominantly gray or grayish brown with brown, yellowish brown and reddish brown mottles	Dominantly red, brown, reddish brown, yellowish brown, with a wide range of mottle color
pH[a]			
Blackwater	5.1	-	-
Alluvial	5.5	5.6	-
Phosphorus (ppm)			
Blackwater	9.8	-	-
Alluvial	8.1	4.8	-
Calcium (ppm)			
Blackwater	346	-	-
Alluvial	669	186	-
Magnesium (ppm)			
Blackwater	36	-	-
Alluvial	145	39	-
Sodium (ppm)			
Blackwater	31	-	-
Alluvial	28	23	-
Potassium (ppm)			
Blackwater	29	-	-
Alluvial	32	20	-

[a]Range included drought years.

dominate, although sandy soils are frequent. Soil color is gray or grayish-brown with mottles of brown, yellowish-brown, or reddish-brown, all typical of alternating reduced and oxidized conditions.

Zone VI - Soils are generally well-drained and texture can be anything from sands to clays. Soils are generally well-structured, and have colors typical of well-drained soils, such as red, brown, and yellow. Because these soils are least frequently-flooded, available nutrients may be somewhat less than zone IV or V, although there are little data on the subject.

Vegetation Zonation

The vegetation of bottomland hardwood forests is dominated by a high diversity of trees that are adapted to the wide variety of environmental conditions described above. The most important environmental condition is the hydrology, which determines the "moisture gradient," or as Wharton et al. (1982) prefer, the "anaerobic gradient" which varies in time and space across the floodplain. Table 2 presents a list of tree and shrub species common to bottomland hardwood forests from wettest (zone II) to driest (zone VI) conditions.

Zone II - Vegetation in this zone is adapted to almost continuous flooding. Bald cypress (*Taxodium distichum*) and water tupelo (*Nyssa aquatica*) usually dominate the canopy of southern bottomlands. Major associates include buttonbush (*Cephalanthus occidentalis*) and water elm (*Planera aquatica*). Pond cypress (*Taxodium distichum* var. *nutans*) and black gum (*Nyssa sylvatica* var. *biflora*) are also found in some blackwater floodplains in Florida and Georgia.

Zone III - This zone supports vegetation such as black willow (*Salix nigra*), silver maple (*Acer saccharinum*), and sometimes cottonwood (*Populus deltoides*) in the pioneer stage. A more common association in this zone includes overcup oak (*Quercus lyrata*) and water

Table 2. Selected tree and shrub species for bottomland hardwood forests in southeastern United States (after Larson et al. 1981).

Species	Ecological Zone				
	II	III	IV	V	VI
Taxodium distichum (bald cypress)	X	X			
Nyssa aquatica (water tupelo)	X	X			
Cephalanthus occidentalis (buttonbush)	X	X			
Salix nigra (black willow)	X	X			
Planera aquatica (water elm)	X	X			
Forestiera acuminata (swamp privet)	X	X			
Acer rubrum (red maple)		X	X	X	X
Fraxinus caroliniana (water ash)		X	X		
Itea virginica (Virginia willow)		X			
Ulmus americana var. *floridana* (Florida elm)		X	X		
Quercus laurifolia (laurel oak)		X	X	X	
Carya aquatica (bitter pecan)		X	X		
Quercus lyrata (overcup oak)		X	X		
Styrax americana (smooth styrax)		X			
Gleditsia aquatica (water locust)		X	X		
Fraxinus pennsylvanica (green ash)		X	X		
Diospyros virginiana (persimmon)		X	X	X	X
Nyssa sylvatica var. *biflora* (swamp tupelo)		X			
Amorpha fruticosa (lead plant)		X	X		
Betula nigra (river birch)		X	X		
Populus deltoides (eastern cottonwood)		X	X		
Baccharis glomeruliflora (groundsel)			X	X	X
Cornus foemina (stiff dogwood)		X	X		
Viburnum obovatum (black haw)			X		
Celtis laevigata (sugarberry)			X	X	X
Liquidambar styraciflua (sweetgum)			X	X	
Acer negundo (box elder)			X	X	
Sabal minor (dwarf palmetto)			X	X	
Gleditsia triacanthos (honey locust)			X	X	X
Ilex decidua (possum haw)			X	X	X
Crataequs viridis (green hawthorn)			X		
Quercus phellos (willow oak)			X	X	X
Platanus occidentalis (sycamore)			X	X	X

Table 2. (Continued).

Species	Ecological Zone				
	II	III	IV	V	VI
Alnus serrulata (common alder)			X		
Ulmus crassifolia (cedar elm)			X		
Ulmus alata (winged elm)			X	X	X
Ulmus americana (American elm)			X	X	X
Quercus nuttallii (nuttall oak)			X		
Quercus virginiana (live oak)			X	X	X
Schinus terebinthifolius (Brazilian peppertree)			X	X	
Ascyrum hypericoides (St. Andrews cross)			X	X	X
Bumelia reclinata (bumelia)			X	X	
Carya illinoensis (pecan)			X	X	X
Carpinus caroliniana (blue beach)			X	X	
Myrica cerifera (wax myrtle)			X	X	X
Psychotria sulzneri (wild coffee)			X	X	
Psychotria nervosa (wild coffee)			X	X	
Zanthoxylum fragara (wild lime)			X	X	
Morus rubra (red mulberry)			X	X	X
Ximenia americana (hog plum)			X	X	X
Sambucus canadenis (elderberry)			X	X	X
Magnolia virginiana (sweet bay)			X		
Sabal palmetto (cabbage palm)			X	X	
Ligustrum sinense (privet)			X	X	X
Crataegus marshallii (parsley haw)			X	X	
Quercus nigra (water oak)			X	X	X
Quercus michauxii (cow oak)			X	X	
Quercus falcata var. *pagodaefolia* (cherrybark oak)				X	X
Nyssa sylvatica (black gum)				X	X
Pinus taeda (loblolly pine)				X	X
Carya ovata (shagbark hickory)				X	X
Juniperus virginiana (eastern red cedar)				X	X
Callicarpa americana (American beautyberry)				X	X
Asimina triloba (paw paw)				X	
Ilex opaca (American holly)				X	X
Serenoa repens (saw palmetto)				X	X
Prunus serotina (black cherry)				X	X

Table 2. (Concluded).

Species	II	III	IV	V	VI
Fagus grandifolia (American beech)				X	X
Magnolia grandiflora (southern magnolia)				X	X
Ostrya virginiana (eastern hop-hornbeam)				X	X
Sassafras albidum (sassafras)				X	X
Sargeretia minutiflora (sargeretia)				X	X
Quercus alba (white oak)				X	X
Cornus florida (flowering dogwood)				X	X
Tilia caroliniana (basswood)				X	X
Asimina parviflora (dwarf paw paw)				X	X
Euonymus americana (strawberry bush)				X	X
Carya glabra (pignut hickory)				X	X
Ptelea trifoliata (water ash)				X	X

hickory (*Carya aquatica*) which often occur in relatively small depressions in the floodplain. Several other species of ash, maple, and birch can be found in this zone. New point bars that form in river channels are often in this hydrologic zone and are colonized by monospecific stands of willow, maple, birch, or cottonwood.

Zone IV - This zone supports a wider array of trees and shrubs. Common species in southeastern bottomlands include laurel oak (*Quercus laurifolia*), green ash (*Fraxinus pennsylvanica*), American elm (*Ulmus americana*), and sweetgum (*Liquidambar styraciflua*). This zone can also support several other species of oaks such as willow oak (*Quercus phellos*), Nuttall oak (*Quercus nuttallii*) and pin oak (*Quercus palustris*).

Zone V - Oaks (*Quercus* spp.) and hickories (*Carya* spp.) usually dominate the associations at this level in the floodplain. This zone is particularly difficult to delineate in the field due to the overlap of species with zone IV. Some species of pine, for example loblolly pine (*Pinus taeda*), can occur in this zone.

Zone VI - This zone is no longer considered part of the "wetlands" of the bottomland hardwood forest, but is a transition zone to the dry upland. Several species of oak, ash, and hickory are found in these generally aerobic soils. In fact, trees that are intolerant of soil saturation first appear in this zone.

FUNCTIONS OF BOTTOMLAND HARDWOOD FORESTS

In much of the current literature concerning wetlands, functions and values are often discussed in the same context. In this way, the human worth of a wetland is implied as being identical to its biological or physical functioning. In reality, functions are characteristics of the bottomlands in the absence of any consideration of their importance to humans. Values, on the other hand, refer to characteristics of bottomlands that happen to provide benefit to current human needs.

The relationships between ecosystem functions of bottomland hardwoods, discussed in this section, and subsequent values to humans, discussed in Section III, are illustrated in Figure 6. In bottomland hardwoods, the biological community responds to the physical conditions created by solar energy, hydrology, and geomorphology of the watershed. Subsequently, an interdependent system exists that operates in several ways to maintain itself and to interface with other ecological systems. The functions can be described in terms of biological or physical operations as follows: 1) community dynamics; 2) physio-chemical processes that include the deposition of sediments, retention of nutrients and toxins in the system, and biochemical transformations; 3) surface water storage through floodplain structure, soil types, and vegetative cover; and 4) groundwater storage that results in groundwater discharge and recharge.

Community Dynamics

The biotic community of southeastern bottomland hardwood forests is diverse and contains a variety of woody and herbaceous plants,

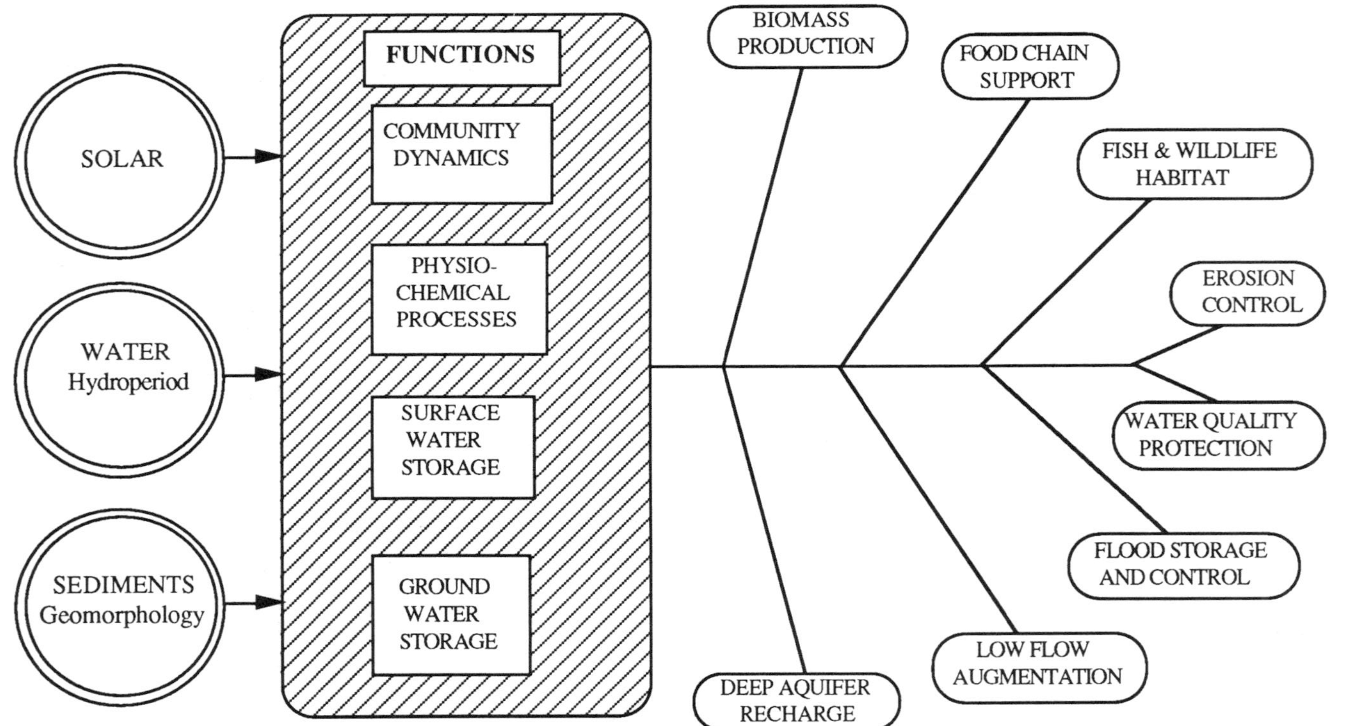

Figure 6. Relationships of functions and values of bottomland hardwood wetlands.

terrestrial and aquatic animals, and microorganisms. The composition of the community changes across the floodplain with different species associations dominating each zone. Each association is uniquely suited to the physical forces shaping its habitat and, likewise, has specific effects on its environment. Figure 7 shows four major functional characteristics of the biotic community within the bottomland hardwood ecosystem. The curves illustrate the relative importance of the zones in performing these functions.

Primary Productivity

Bottomland hardwood forests are subsidized by the adjacent streams during periodic flooding. The inputs include water, particulate and dissolved organic matter, and nutrients in dissolved, particulate, and sediment-adsorbed forms. The patterns of primary productivity across the bottomland zones reflect these energy subsidies (Figure 7a). Primary productivity reaches its highest level in floodplain forests with a pulsing hydroperiod by using the additional nutrients and cycling them within the ecosystem (Figure 8).

Primary productivity may be considered a gauge of the "health" of an ecosystem. In this case, natural bottomland hardwood forests are among the healthiest ecosystems known (Conner and Day 1976). Floodplain forests with unaltered hydroperiods generally have aboveground net primary productivity in excess of 1000 $g/m^2/yr$. Forested wetlands with stagnant or sluggish waters are usually less productive (Table 3). Flowing water systems maintain a more oxygenated root zone than those with stagnant waters. Nutrients are imported and conserved to increase primary productivity.

In developing ecosystems, primary productivity is invested in the accumulation of plant and animal biomass. This results in a more complex ecosystem structure as well as food chain support. Structural diversity and food availability are important for secondary productivity of fish and wildlife resources. Mature bottomland hardwood forests are equally as productive, but the energy flow is used in maintenance, rather than in accrual, of existing structure and biomass. The high natural productivity of bottomland hardwood wetlands provides a variety of natural products harvested by humans, including timber, fish, and wildlife (Section III).

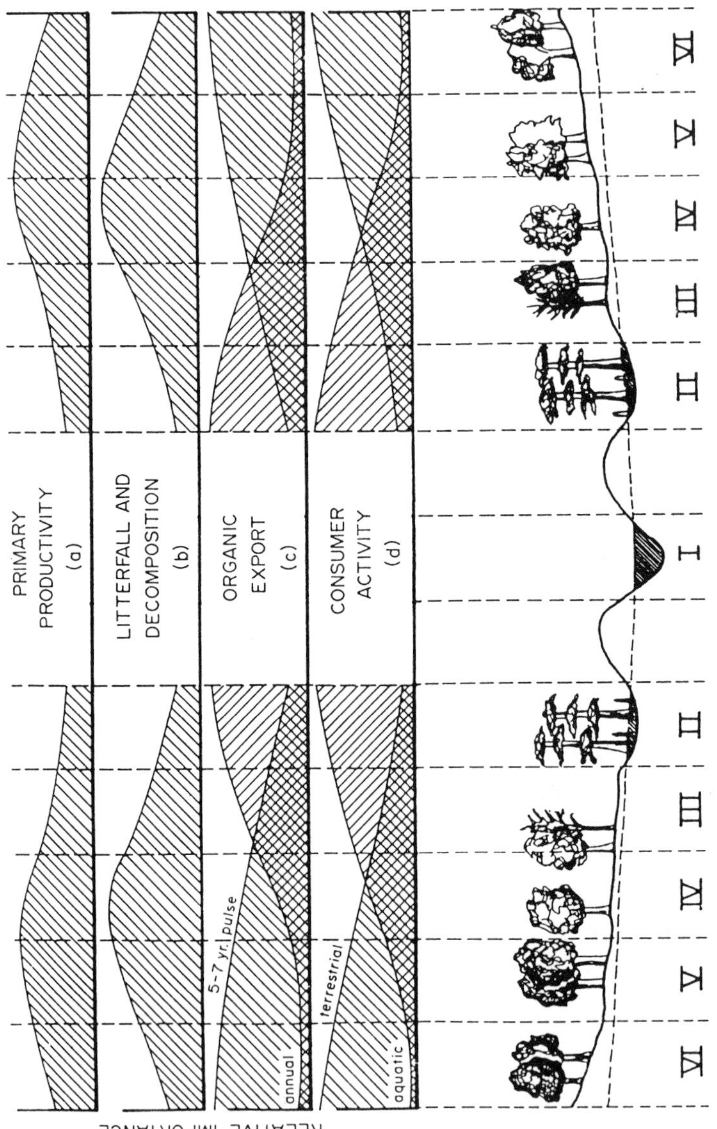

Figure 7. Patterns of community dynamics in bottomland hardwood wetlands (from Mitsch and Gosselink 1986, p. 379; Copyright © by Van Nostrand Reinhold, reprinted with permission).

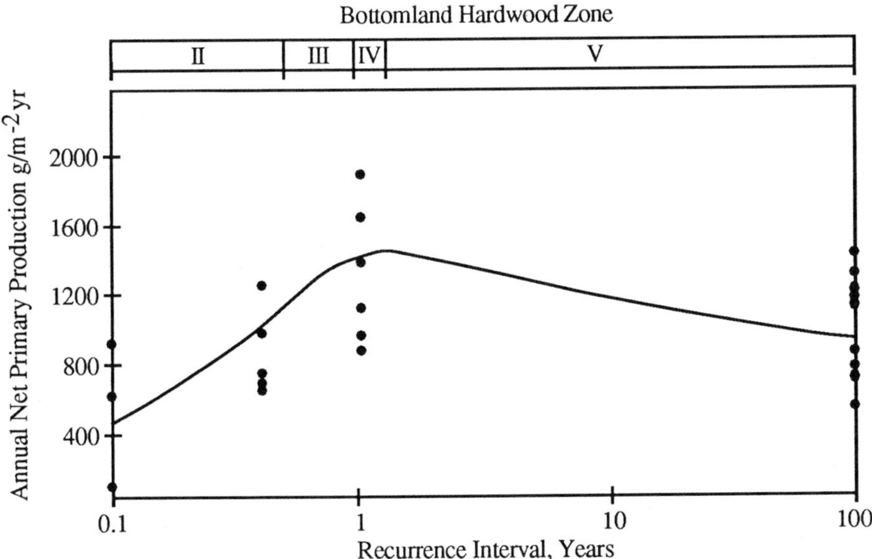

Figure 8. Relationship of bottomland hardwood wetland productivity to flooding conditions (after Gosselink et al. 1981).

Table 3. Net aboveground productivity of forested wetlands in g/m^2/yr (from Brinson et al. 1981a).

	Total tree prod.	Total aboveground prod.	Reference
Still water:			
Okefenokee Swamp, GA	595	712	Schlesinger 1978
Cypress dome, low fertility, FL	750	---	Brown 1981
Cypress dome, enriched, FL	1794	---	Brown 1981
Slow flowing water:			
Cypress strand, FL	1111	1363	Burns 1978
Cypress strand, enriched, FL	1290	---	Nessel 1978
Cypress strand, drained, FL	681	816	Burns 1978
Flowing water:			
Floodplain forest, FL	1607	---	Brown 1981
Bottomland hardwood, LA	1374	1574	Conner & Day 1976
Cypress-tupelo, LA	1120	1140	Conner & Day 1976
Cypress-tupelo, IL	678	1780	Mitsch et al. 1979

Litterfall and Decomposition

Both litterfall and organic decomposition have similar patterns throughout the bottomland hardwood forest (Figure 7b). Most forested wetlands are more productive than adjacent uplands (Odum 1978) and generally produce more litterfall except for some permanently inundated swamps (Elder and Cairns 1982). The ecological value of this high productivity and litterfall depends on its decomposition and entry into aquatic food chains.

The process of decomposition includes leaching of soluble materials, mechanical fragmentation, and biological decay (de la Cruz 1979). Leaching of organic carbon occurs quite rapidly in southeastern forested bottomlands with the majority of early weight reduction occurring in this way (Brinson 1977). Concurrently, weathering and abrasion from wave action, wind, or other mechanical means break the litter into smaller pieces. Detritivores also aid in fragmentation. Microorganisms colonize detrital fragments and fecal material. Overall, particle size is decreased and chemical structure is simplified.

The effect of hydroperiod on decomposition rates has been studied by several researchers. On a Michigan floodplain, Merritt and Lawson (1978) found that less than 5% of the leaf litter remained at a point bar site (probably zone III) after nine months. About 40% of the litter remained at their upland site which probably contained zones V and VI. The terrace site (zone VI) was intermediate. They attributed the difference in rates to soil moisture and the diversity and abundance of litter macroinvertebrates. A southern bottomland hardwood forest in North Carolina was studied by Brinson (1977) where he found that the decomposition rate of cellulose sheets was greatest in flowing water (zone I), intermediate in a floodplain swamp (zone II or III), and lowest on a natural, well-drained levee (probably zone V). On the other hand, Duever et al. (1975) discovered that decomposition was slower in sites that were flooded from 16% to 50% of the time than in dry sites of Corkscrew Swamp in Florida. It seems likely that, in general, the rate of decomposition is greatest in aerobic bottomland zones with sufficient soil moisture. Dry sites probably have a slightly lower rate with permanently anaerobic zones the slowest (Brinson et al. 1981b).

Organic Export

Bottomland hardwood forests provide food chain support to aquatic ecosystems through the export of detritus (Figure 7c). Food chain support refers to the direct or indirect use of exported organic materials by aquatic organisms outside the immediate wetland environment. The major characteristics of the bottomland hardwood forest that determine its ability to export organic material are (de la Cruz 1979):

1) Primary productivity and nutrient cycling - Highly productive ecosystems generally "leak" more nutrients overall than ecosystems with low productivity even if relative "leakiness" is comparable.
2) Litterfall and decomposition - The amount and type of litterfall and its decomposition characteristics will determine, in large part, the amount and type of organic materials available for export.
3) Hydroperiod - The hydrologic character of the ecosystem controls the timing and rate of detrital dispersal as well as affecting the decomposition process.

All river basins export organic material to some extent. Annual organic carbon export and runoff exhibit a linear relationship in both upland and wetland watersheds. There is evidence, however, that watersheds containing wetlands export more organic carbon than watersheds without wetlands (Mulholland and Kuenzler 1979). This is demonstrated in Figure 9. Streams in upland watersheds receive leaf litter primarily from overhanging riparian vegetation since floodplains are narrow or absent. Wetland watersheds, on the other hand, have broad floodplains where floods may transport additional organic material to the stream ecosystem.

The organic material available for export is in both dissolved and particulate forms. Particulate carbon is generally a small percentage of the total carbon in most streams, but has a disproportionately large importance as a high-quality food source for certain organisms (Brinson et al. 1981b). Studies have shown that many species of fish and invertebrates feed preferentially or entirely on particulate detritus (de la Cruz and Kawanabe 1967, Odum 1970). Dissolved organic carbon is utilized by microorganisms, converted to microbial biomass, and made available to estuarine and river filter feeders (Correll 1978).

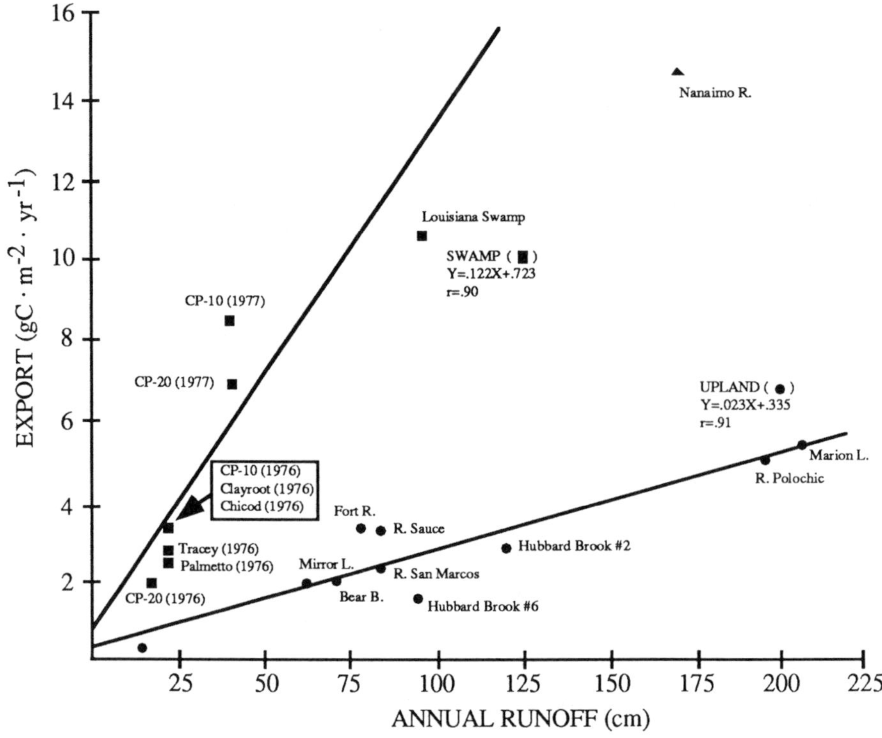

Figure 9. Organic export from wetland-dominated and non-wetland watersheds (from Mulholland and Kuenzler 1979, p. 964; Copyright © 1979 by the American Society of Limnology and Oceanography, reprinted with permission).

The physical export of organic materials depends on the flooding cycle. When the stream overflows into the bottomland, a large surface area of litter and detritus is exposed to the water, often for a long time. During this time, significant leaching and fragmentation occur, and both dissolved and particulate organic material are removed from the floodplain. Export of organic matter, therefore, follows seasonal, annual, or even less-frequent hydrologic pulses.

Consumer Activity

The bottomland hardwood wetland provides habitat, both temporary and permanent, for a wide range of aquatic and terrestrial consumers

(Figure 7d), leading to a relatively high abundance and diversity of consumer species. This is due primarily to the abundance of preferred food items and to the structural heterogeneity of the ecosystem. Brinson et al. (1981b) state that, with few exceptions, "riparian ecosystems are among the most productive areas for wildlife in the U.S.A." The high secondary productivity is not only a function of bottomland-dependent species, it is also related to those upland and aquatic animals that use wetlands opportunistically.

As transitional ecosystems between streams and uplands, bottomland hardwood wetlands have two distinctly different ecotones. The aquatic/wetland ecotone supports species from both ecosystems as well as those endemic to that area. Likewise, the upland/wetland ecotone supports its own characteristic fauna. The linear nature of floodplain forests provides a relatively large ecotone to wetland ratio.

The "edge effect" of higher species abundance and diversity is illustrated especially by bird and amphibian populations. In South Carolina, Hair et al. (1978) found that riparian wetlands and beaver (*Castor canadensis*) ponds had 1.5 to 2 times the number of birds found in upland areas. Dickson (1978) reported similar findings for Louisiana and east Texas; the densities of breeding birds in bottomland hardwoods were 2 to 4 times higher than in upland pine and pine/hardwood forests. Blem and Blem (1975) noted that both birds and small mammals of floodplains had higher density, biomass, and standard metabolism than similar populations in upland areas.

Many amphibians are restricted to wetland environments since aquatic habitats are required for breeding and overwintering (Wharton 1978, Fredrickson 1979). These animals are able to exploit seasonally-flooded ecosystems such as bottomland hardwood wetlands. Southern river swamps serve as breeding areas for large numbers of amphibians which may be permanent residents or migrants to temporarily-flooded zones. Wharton (1978) reported that 1600 salamanders and 3800 frogs and toads annually used a small tupelo pond in Georgia for breeding. Species richness is high in zones II and III of southern bottomlands; 15 species of frogs and toads were collected from one Florida tupelo pond and 9-10 species are commonly found in cypress-tupelo swamps across the Southeast (Wharton 1978).

In summary, bottomland hardwoods appear to have higher species diversity than uplands for birds, amphibians, small and large mammals, and turtles; lizards and snakes are not as diverse. Density of amphibians, some small mammals, and breeding/wintering birds is greater in floodplain ecosystems than in adjacent uplands. Detailed descriptions of faunal utilization of bottomland hardwood forest zones are given by Fredrickson (1979), Wharton et al. (1981, 1982), and Brinson et al. (1981b).

Physiochemical Processes

Sediments and water in wetland systems exist in dynamic states. Periodic import, retention, and export of sediments and water occur in bottomland hardwood forests; various transformations within the soil and at the sediment-water interface also play a major functional role in these ecosystems. Some of the major functional attributes of physio-chemical processes are displayed in Figure 10. Both organic and inorganic suspended solids may be carried into bottomland hardwood forests with overbank flooding. Nutrients and other materials are sorbed onto sediment particles. The amount of sediment moving into and out of the wetland depends on the volume of water moving through the forest, the water velocity in the channel and in the bottomland itself, and the availability of erodible material in upstream ecosystems (Gosselink et al. 1981). The characteristics of the floodwater while in the wetland and when it leaves are primarily a result of biochemical and physical reactions at the sediment/water interface.

Sediment Deposition

Deposition of sediments occurs when streamwater leaves its channel and spreads across the floodplain. By utilizing the broad floodplain as a shallow channel, water velocity decreases and particles in the water settle out. The bottomland hardwood vegetation also obstructs water flow, and the frictional resistance enhances the entrapment of sediments within the wetland.

Maximum flow reduction and water retention yields maximum deposition. Gosselink et al. (1981) reported that an extreme flood in 1979

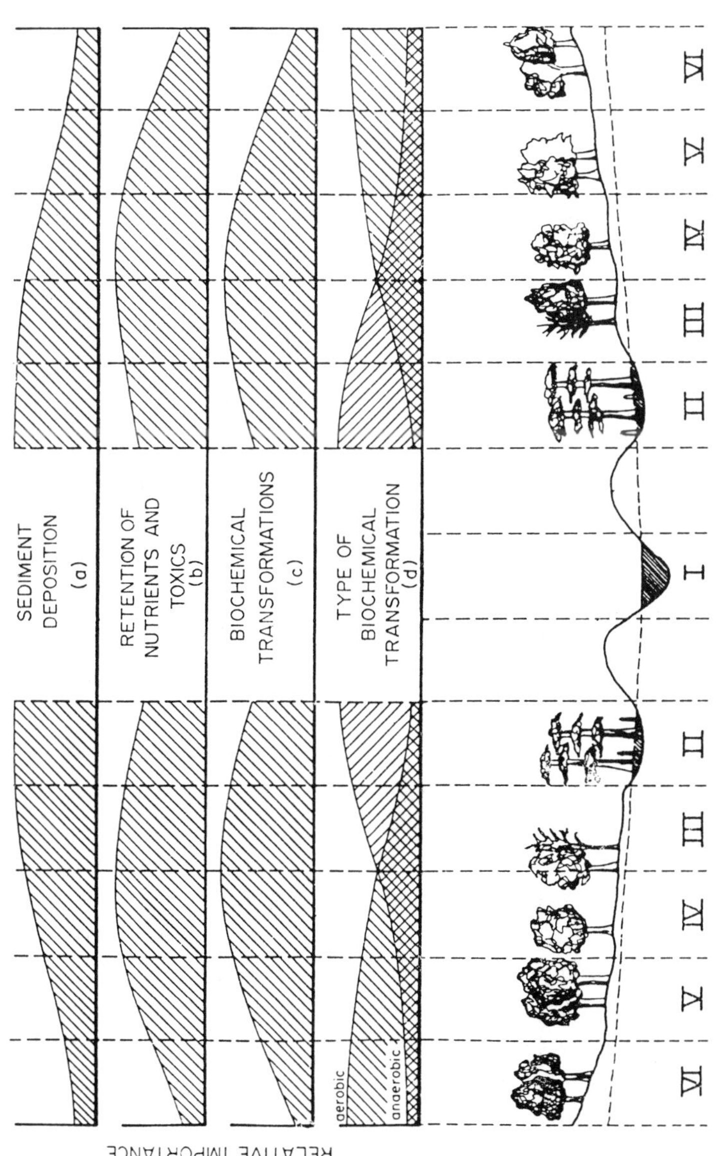

Figure 10. Patterns of physiochemical processes in bottomland hardwood wetlands.

left up to 45.7 cm of new sediments on the lower Atchafalaya River floodplain. Conversely, small floods of short duration may deposit only a thin film. Figure 10a illustrates that lower bottomland hardwood zones that are flooded for a greater time period have greater deposition than higher, less frequently flooded zones. Novitski (1979) reported that a Wisconsin watershed with 40% wetland and lake areal coverage had stream sediment loads that were 90% lower than in a comparable watershed with no wetlands. A large annual flood that inundated a zone II and III bottomland in Southern Illinois deposited about 3-6% (450-780 g/m^2) of its sediment load in the wetland over a 3-day period (Mitsch et al. 1977). Kuenzler et al. (1980) reported annual sediment deposition rates for Creeping Swamp, North Carolina. In low areas (zone II or III), deposition was 305 $g/m^2/yr$, while in zone IV and zone V, about 148 and 39 $g/m^2/yr$ were deposited, respectively. Sediment export may exceed import during extremely high flows in some bottomlands (Adamus and Stockwell 1983).

Retention of Nutrients and Toxins

Both nutrients and toxic materials are thought to be retained, at least temporarily, in the bottomland hardwood forest. Several studies have documented the nutrient dynamics, but there are few studies on toxic retention. Figure 10b suggests a possible spatial pattern.

Nutrient cycling in bottomland forests certainly influences the character of water in adjacent streams. During overbank flooding, the floodplain ecosystem is coupled with both upstream and downstream ecosystems. As water flows slowly through the wetland, there is opportunity for nutrient uptake and transformation. Table 4 shows the exchange of nutrients between floodwaters and forested wetlands as reported in a number of recent mass balance studies.

Bottomland hardwood trees derive their nutrients primarily from the sediments (Farnsworth et al. 1979). Plant uptake of nutrients from the sediments may not affect water quality directly, but may upset the water/sediment equilibrium creating movement of nutrients from the water to the sediments (Boyt et al. 1977, Klopatek 1978). Even if the nutrients

Table 4. Input-output nutrient studies in forested wetlands ($g/m^2/yr$).

	Input	Output	Net Uptake (%)
Cypress Swamp, IL (Mitsch et al. 1979)			
Total P	80.420	76.940	3.480 (4.3)
Cypress swamp, LA (Day et al. 1977)			
Total P	0.374	0.200	0.174 (46)
Total N	2.690	1.360	1.330 (49)
Cypress-tupelo, enriched, LA (Kemp & Day 1981)			
PO_4-P	0.730	0.830	-0.100
Dissolved organic P	0.350	0.440	-0.090
Partial P	3.120	1.240	1.880 (60)
Total P	4.200	2.510	1.690 (40)
NO_3+NO_2-N	0.690	0.260	0.430 (62)
NH_4-N	0.620	0.250	0.370 (60)
Dissolved organic N	3.400	4.130	-0.730
Partial N	10.830	7.030	3.800 (35)
Total N	15.540	11.620	3.920 (25)
Okefenokee cypress swamp, GA (Blood 1980)			
PO_4-P	0.041	0.003	0.038 (93)
Total P	0.570	0.144	0.426 (75)
NO_3-N	0.403	0.008	0.395 (98)
NO_2-N	0.044	0.000	0.042 (95)
NH_4-N	0.349	0.014	0.335 (96)
Total N	1.753	1.070	0.683 (39)
Tar River Swamp, NC (Brinson et al. 1981a)			
Total P	43.700	18.600	25.100 (57)
Total N	51.300	32.100	19.200 (37)
Cornish Creek Swamp, GA (Wharton & Hopkins 1980)			
Total P	4.942	4.497	0.445 (9)

Creeping Swamp, NC (Kuenzler et al. 1980)	Input Year 1	Input Year 2	Output Year 1	Output Year 2	Net Uptake Year 1	Net Uptake Year 2
PO_4-P	0.663	0.372	0.382	0.077	0.281 (42)	0.295 (79)
Dissolved organic P	0.061	0.185	0.072	0.201	-0.011	-0.016
Partial P	0.274	0.668	0.281	0.262	-0.007	0.406 (61)
Total P	1.045	1.290	0.735	0.560	0.310 (30)	0.730 (57)

are stored only temporarily in non-perennial tissues, stream nutrient concentrations are reduced during the growing season which may lessen the possibility of downstream eutrophication.

Nutrient cycling studies in bottomland hardwood forests have concentrated primarily on phosphorus dynamics since phosphorus is believed to be the limiting nutrient in freshwater wetlands. The largest flux of phosphorus occurs at the soil-water interface (Yarbro 1979). Bottomland hardwood plants take part in this process in several ways. Most bottomland trees have shallow root systems that can take advantage of phosphorus deposition in sediments. The adventitious roots which develop as a response to flooding in some wetland trees are capable of phosphorus uptake from soil surface, litter layer, and water (Hook et al. 1970, 1971).

Phosphorus concentration and annual increments of phosphorus in stem wood are low when compared to the concentration in litterfall and the other flows in the ecosystem (Brinson et al. 1981b). However, total standing crop of phosphorus in perennial tissues is one of the major phosphorus pools in the ecosystem (Yarbro 1979). This implies that ecosystems with net annual production accumulate phosphorus. Studies by Vitousek and Reiners (1975) and Vitousek (1977) showed that successional communities with accumulating biomass tended to retain nutrients, while mature communities tended to be in equilibrium with both biomass and nutrients. Kuenzler et al. (1980) supported this conclusion in their studies of Creeping Swamp, a zone III swamp forest in North Carolina. The forest community is about 40-50 years old and is adding biomass; the study reports a net accumulation of phosphorus and tight recycling of vegetative phosphorus.

Although sediment accretion accounts for a large proportion of nutrient removal, studies imply that increased vegetative growth is also important. Hartland-Rowe and Wright (1975) suggested that the 97% decrease in phosphorus of sewage effluent applied to a Canadian floodplain swamp was due primarily to increased primary production. A hardwood swamp in Florida exposed to sewage effluent reduced phosphorus concentrations in its output to a lower level than seen in the waters leaving a similar unexposed swamp nearby (Boyt 1976). The fate of the retained phosphorus was not reported, but there was no evidence of increased phosphorus buildup in the sediments of the sewage-treated swamp (Boyt et al. 1977). For a Florida cypress strand, stem wood uptake of phosphorus increased by three times when treated sewage effluent was applied to the system (Nessel 1978). Brown and Lugo (1982) have presented a correlation between phosphorus input and primary production in their comparison of freshwater forested wetlands (Figure 11).

Nitrogen cycling has also been studied in wetland ecosystems. Nitrate is usually the most abundant form of nitrogen in surface waters and can create water-quality problems when occurring in high concentrations. Wetland trees can take up nitrate and accumulate small amounts of nitrogen in their woody tissues; nitrogen cycling in forested wetlands is quite efficient (Brinson et al. 1981b). Permanent nitrogen removal is accomplished only by denitrification as discussed below.

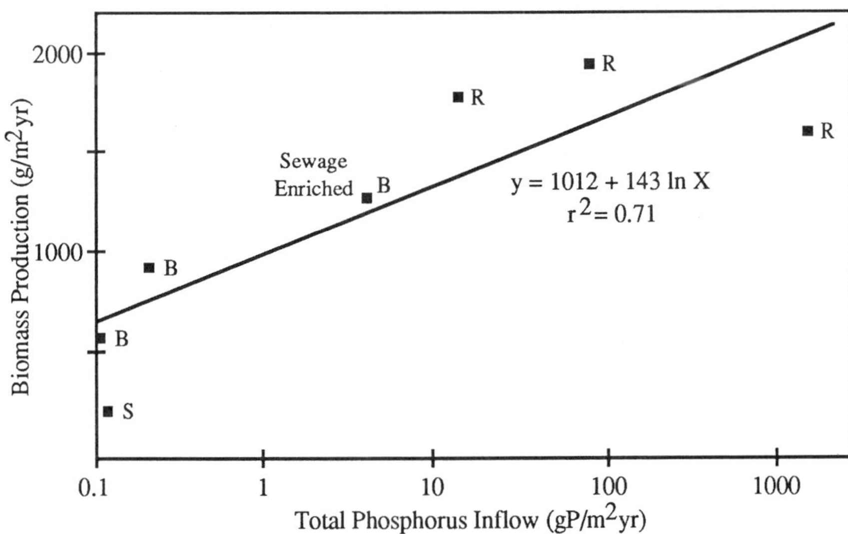

Figure 11. Relationship between biomass production and total phosphorus inflow for freshwater forested wetlands. S = scrub forests, B = basin forests, and R = riverine forests (from Brown and Lugo 1982, p.120; Copyright © 1982 by National Institute of Ecology and International Scientific Publications, Jaipir, India, reprinted with permission).

Biochemical Transformations

There are a myriad of biochemical transformations, both anaerobic and aerobic, that occur in the standing water and sediments of bottomland hardwood forests (Figure 10c and d). When water fills the interstitial space in soils and sediments, the rate of oxygen diffusion is greatly reduced. Low diffusion rates lead to anaerobic soil conditions within a few hours or a few days. Ambient temperature and the amount of organics available for microbial respiration are primarily responsible for determining the rate of oxygen depletion (Gambrell and Patrick 1978). The anaerobic or reduced soil condition strongly affects the availability of nutrients and toxic materials.

In bottomland hardwood forests, oxygen is usually not totally lacking in sediments. At the sediment-water interface, a thin layer of aerobic sediments may be present. With the exception of some permanently inundated zone II bottomlands, flooded wetland forests have currents which help mix oxygen into floodwaters. This oxygenated layer may also result from oxygen production by photosynthetic algae and bacteria or wind mixing of surface water (Gambrell and Patrick 1978). Although deeper layers are still in a reduced condition, the aerobic interface layer can be very important in determining the degree of sediment-water exchange of nutrients and toxic materials.

The redox potential of saturated soils drops as organic materials are oxidized. As a result, predictable biochemical transformations occur. Nitrate reduction is one of the first transformations. Several microbiological processes are involved, some causing nutrient availability to decrease. In most wetland sediments, ammonium is the dominant inorganic form of nitrogen (Mohanty and Dash 1982). Much nitrogen can also be bound in organic form in highly organic soils. The oxidized layer is important for several of the reactions that take place. The reactions that can occur, in order, are mineralization of organic nitrogen, diffusion of ammonium toward the surface, nitrification, diffusion of nitrate away from the surface, and denitrification.

Mineralization of organic nitrogen via ammonification yields the ammonium form. The ammonium ion can be absorbed by microorganisms or plants and converted to organic matter. Immobilization can occur through ion exchange onto negatively-charged sediment particles; further oxidation is restricted in anaerobic conditions. The upward diffusion of ammonium toward the oxidized layer reduces the chance of excessive buildup. At the aerobic layer, ammonium is oxidized to nitrate in a two-step, microbially mediated nitrification process. Since nitrate is negatively charged, it cannot be immobilized by soil particles with a negative charge (Atlas and Bartha 1981). Nitrate can be assimilated by plants and microorganisms, lost into groundwater, or undergo dissimilatory nitrogenous oxide reduction. This reduction process may be to ammonia or, more commonly, to gaseous nitrogen by denitrification.

Denitrification has been reported by several studies as a major process in the reduction of nitrogen concentration in wetlands. Most of the studies have been in salt marsh ecosystems, but Brinson et al. (1981b) reported the loss of 29 g N/m^2/yr from an alluvial cypress swamp on the Tar River in North Carolina. This study estimated denitrification from the uptake of labelled nitrogen. Evidence of denitrification was reported by Kitchens et al. (1975) for the Santee River Swamp in South Carolina. Nitrate concentrations decreased from the river across the floodplain to the swamp backwaters. Increased contact time with the forest floor was suggested as the reason for the concentration decrease. The mechanism was presumably denitrification, but it was not measured directly. Although none of the studies measured production of nitrogen gas, it is probable that denitrification is significant in wetlands, and that it takes place in the sediments rather than in the water column (Patrick et al. 1976).

The rate of denitrification in bottomland hardwood forests is probably comparable to that of other freshwater systems because of the alternating wet-dry cycle allowing oxidation of ammonium to nitrate (Brinson et al. 1981a) and the availability of organic carbon as an energy supply (Graetz et al. 1980). The denitrification rate appears to be related to the amount of nitrogen enrichment as reported by Odum and Ewel (1978) in Florida cypress domes. Farnsworth et al. (1979) noted that high concentrations of nitrate and ammonium inhibit the rate of nitrogen fixation. These studies suggest that wetlands have a substantial value in water quality protection especially where stream nutrient concentrations are highest.

Phosphorus has been described as a major limiting nutrient in many freshwater wetlands including bogs (Heilman 1968), marshes (Klopatek 1978), and southern cypress swamps (Mitsch et al. 1979, Brown 1981).

It occurs in both organic and inorganic forms as soluble and insoluble complexes in wetland soils. Orthophosphate is the primary inorganic form with the specific form (PO_4^{-3}, $HPO_4^{=}$,$H_2PO_4^{-}$) dependent on pH. Although phosphorus is not directly affected by changes in redox potential as is nitrogen, it is indirectly affected by its associations with other elements that are altered (Mohanty and Dash 1982). Phosphorus can be relatively unavailable due to (Mitsch and Gosselink, 1986):

1) precipitation of insoluble phosphates with ferric iron, calcium, and aluminum under aerobic conditions;
2) adsorption of phosphate onto sediment particles, particularly clay, peat, and ferric and aluminum hydroxides and oxides; and
3) incorporation into living tissues.

Adsorption of phosphorus onto clay particles involves both chemical bonding of negatively charged phosphates to positively charged areas of the clay particle and phosphate/silicate exchange (Stumm and Morgan 1970). The clay-phosphorus complex is important in bottomland hardwood forests because most of the phosphorus introduced to these wetlands is in floodwaters carrying clay sediments (Mitsch et al. 1979, Kuenzler et al. 1980).

In zone II bottomland hardwoods, the sediment-water interface may become anoxic. Under these conditions, the reduction of ferric phosphate compounds to ferrous compounds releases the phosphorus into the water column. Hydrolysis of ferric and aluminum phosphates and anion exchange of clay-adsorbed phosphorus may also release phosphorus (Mortimer 1941). In the other bottomland hardwood zones, release of phosphorus is less likely due to the oxidizing layer which is usually present in flowing-water wetlands. This layer helps trap phosphorus in the interstitial water within the sediments (Darnell et al. 1976, Farnsworth et al. 1979). The breakdown of the aerobic layer has been observed to result in phosphorus release (Teskey and Hinckley 1977, Sloey et al. 1978, Crow and McDonald 1979). Kuenzler et al. (1977) reported that phosphorus concentrations in the water column increased under low dissolved oxygen conditions in isolated pools on a North Carolina swamp floodplain.

Heavy metals are also associated with the sediments carried in floodwaters. After deposition and burial in wetland sediments, elements such as iron and manganese are found mostly in their reduced and more soluble forms. The anoxic conditions create the remobilization of these metals and their subsequent diffusion to the sediment surface. Absence of an oxidized layer allows the soluble metals to enter the water column. If the sediment-water interface is aerobic, the metals may be reoxidized and precipitated ore readsorbed by the sediments (Atlas and Bartha 1981). Ferrous iron can impair primary productivity by immobilizing phosphorus and by coating the roots of plants creating a barrier to nutrient uptake (Gambrell and Patrick 1978). Iron and manganese can accumulate to toxic levels in wetland soils. Kuenzler et al. (1980) reported an annual deposition rate of 53 mg/m^2/yr for manganese in Creeping Swamp, North Carolina. Some metals are cycled by plants (e. g. mercury, cadmium) and may be transferred through the food chain (Kadlec and Kadlec 1979). Damman (1979) reported that a red maple swamp acted as a sink for copper, nickel, manganese, iron, sodium, and potassium, but did not remove lead or zinc. A report by Tchobanoglous and Culp (1980), as cited by the Office of Technology Assessment (OTA 1984), noted that heavy metals were removed from water by wetlands at 20-100% efficiencies depending on the type of metal and wetland characteristics. Chlorinated hydrocarbons are also taken up by some plants, but specific studies in bottomland hardwoods were not found. Many of these synthetic chemicals degrade easily in sediments, which could make forested wetlands potentially efficient sites of removal (OTA 1984). Studies concerning the fate of metals and other toxic materials in wetland sediments are sparse, and the removal of metals and synthetic organics from floodwaters by bottomland hardwood forests has not been studied in detail.

Surface Water Storage

The pulses of high water through bottomland hardwood systems create the specific biotic environment of the wetland. Consequently, that environment becomes dependent on the recurrence of water pulses. Surface water storage in bottomland hardwood forests suits the needs of the ecosystem by providing periods of inundation and saturation along with the transport of nutrients and sediments as part of the vital support required to maintain a wetland system. The surface water storage function is dependent on: 1) geomorphological characteristics of the drainage basin upstream of the wetland which affect duration of flooding, and 2) capacity of the wetland floodplain. The spatial characteristics of surface water storage in bottomland hardwood wetlands is shown in Figure 12.

Duration of Flooding

The size, topography, and geology of the upstream areas influence both the biotic and structural nature of floodplain wetlands. This is specifically represented in terms of flood duration. Flood duration is a measure of the length of time floodwaters are held within the basin and is directly related to basin drainage area and runoff characteristics. Flood duration patterns across the floodplain zones are illustrated in Figure 12a, suggesting that the greater duration, and therefore the greater storage capability, occurs within zone II and III. In a study of the Ouachita and White River Basins in Arkansas, Bedinger (1981) showed that flood duration increases with upstream increased drainage area. Bedinger also showed that those areas with up to 40% duration tended toward greater wetland storage.

Characteristics of the drainage basin that affect runoff include vegetative cover, soil types, and topographic structure. Duration is influenced by the rate of runoff. Watersheds with steep slopes and "smoother" surfaces will produce floods of shorter duration, passing the water downstream in higher peaks. In bottomland hardwood wetlands, runoff rates are reduced by expansion of surface area and high vegetation densities.

Bottomland hardwoods in the southeastern United States can be found on small floodplains along tributary streams with rather small drainage areas and on large floodplains along major river basins. Both tributary storage and mainstem wetlands contribute to overall basin flood storage capacity. Floodplain storage in the tributaries acts as a flood desynchronization process, delaying and offsetting peak flows on the mainstem. Mainstem wetlands may have significant storage capabilities, depending on areal extent of wetlands. Since water storage is a function of the nature of the wetland, the larger the wetland the greater the storage in most cases.

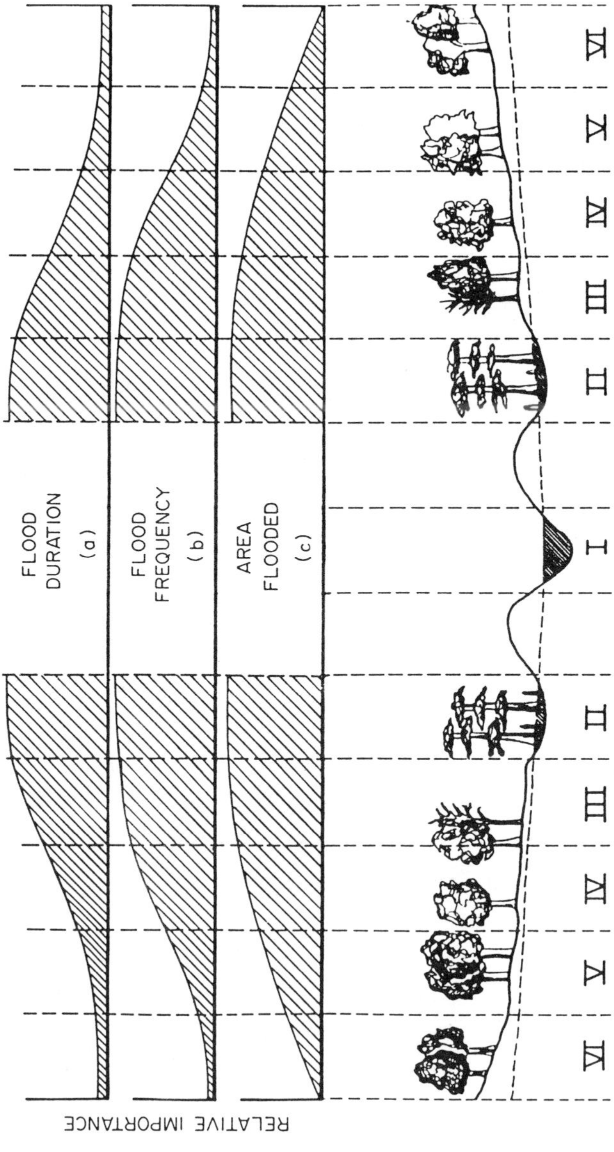

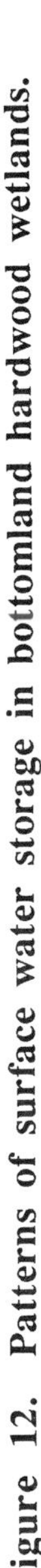

Figure 12. Patterns of surface water storage in bottomland hardwood wetlands.

Flooding Frequency

The lower elevations of the floodplain wetland are the most active storage systems. Generally, the frequency of flooding and the depth of flooding above the bankfull stage are inversely proportional to the floodplain elevation (Brinson et al. 1981b). Figure 12b illustrates the general distribution of flood frequency across the zones of a typical floodplain. The frequency curve can also represent the relative depth of flooding across the zones.

Area Flooded

The more frequent floods occupy a greater portion of the floodplain area than the less frequent floods (Brinston et al. 1981b). Figure 12c shows the relative percentage of floodplain area that is flooded by any major flood. Zones II, III, and IV are likely to occupy the greatest area of the floodplain.

The type of vegetative cover plays a significant role in the extent to which wetlands in the river basin contribute to surface water storage. The extent and density of wetland vegetation can act to slow floodwaters (Carter et al. 1979) and may reduce flow through high rates of evapotranspiration during growing seasons. Zonal distribution of vegetation density is wetland specific and must be considered in terms of herbaceous, understory, and canopy densities.

Typically, floodplains that contain backwater swamps provide greater storage capacity than those without swamps (Wharton et al. 1982). The depressions of backwater swamps, sloughs, and oxbows can act as storage tanks. Also, the highly organic soils in these areas can function as "sponges" if they are not already fully saturated (see Groundwater Storage, next section). Soil saturation levels vary according to season and thereby cause seasonal variation in the water storage capacity of forested wetlands.

Groundwater Storage

Storage of groundwater supports bottomland hardwood forests during dry seasons and can aid in retarding surface water flow during flood periods. Groundwater storage in bottomland hardwood forests depends on watershed characteristics, water table levels, proximity to streambeds, and most importantly, the characteristics and extent of surface and subsurface soils. A general budget of groundwater flows is shown in Figure 13.

In floodplain regions, groundwater inflows (recharge) consist of:

1) infiltration from precipitation - Surface water infiltration is usually distributed within the upper layer of soil and does not contribute to groundwater storage unless the soils are highly permeable or the zone of aeration very thin (Linsley and Franzini 1979) as in swamp lands;
2) percolation from water bodies (surface depressions, oxbows, and streams); and
3) overbank flows stored in the floodplain (overbank storage) which may percolate into the groundwater, recharging water table aquifers (Mundorff 1950 in Carter et al. 1979) depending on soils, flood duration, and water table levels.

Outflows of groundwater (discharge) include:

1) effluent streams, springs, and seepage resulting from intersections of the water table with the surface;
2) base flow, a discharge into the stream from groundwater and differentiated from stream discharge of surface water runoff by the time of arrival (Linsley and Franzini 1979);
3) direct evaporation where the water table is close to the surface; and
4) evapotranspiration through vegetation. The floodplain capacity to store water is responsive to the soil profile and seasonal water table fluctuations (Carter et al. 1979).

Soil porosity and water table levels control the saturation condition of the floodplain. Soils with higher organic content tend to be more porous and can hold more water. As the soils become more saturated, runoff volumes increase. Soil porosity is site specific, but it is likely that more organic soils are found in the lower zones than toward the upland zones. Figure 14a shows highest organic content in those soils of zone II based on information from selected wetland and upland soil types (after Wharton et al. 1982). In some wetland areas, surficial aquifer recharge can act as a flow reduction function (Wharton et al. 1982), especially in

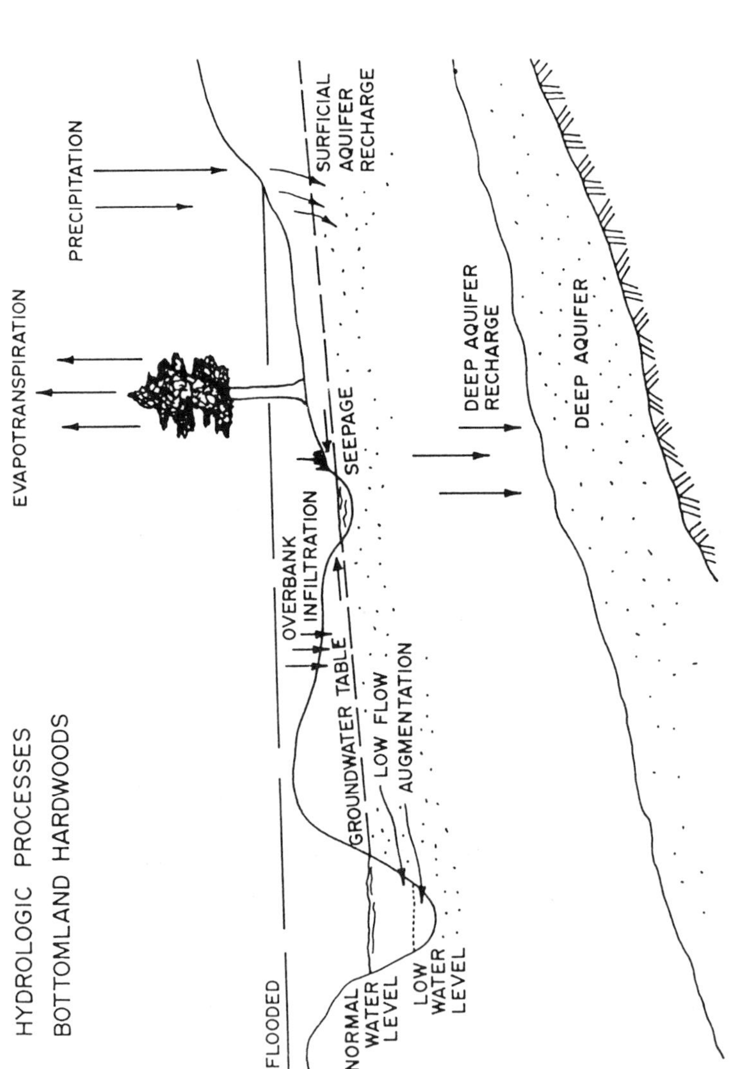

Figure 13. Groundwater hydrologic processes in bottomland hardwood wetlands.

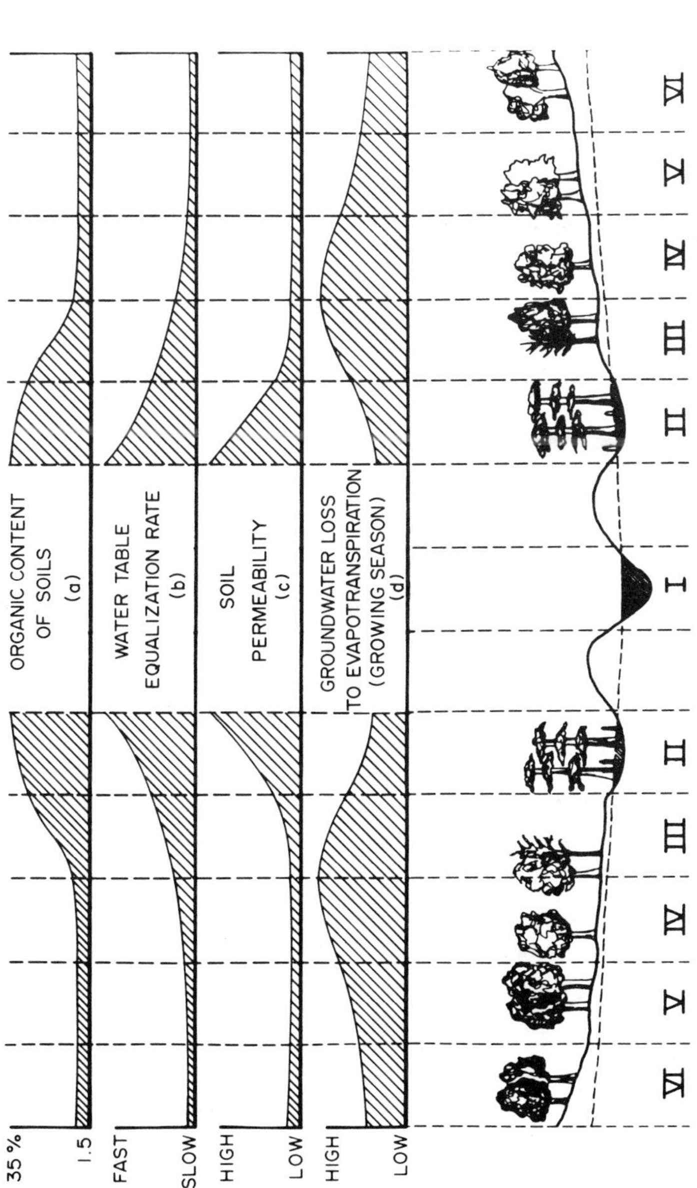

Figure 14. Patterns of groundwater storage in bottomland hardwood wetlands.

alluvial soils found in natural levees and point bar formations. Alluvial soils with high inflow rates can be found throughout southern bottomland hardwood wetlands.

Some wetlands may act as recharge systems, but most are discharge systems (Carter et al. 1979). Identifying hydraulic gradients in the wetland area can be used to determine if the wetland recharges or discharges.

Groundwater Recharge

Groundwater storage occurs at two levels: surficial aquifer storage and deep aquifer storage. Floodplain systems have been identified as surficial aquifer recharge systems, but little information is available on their contribution to deep aquifer recharge. Adamus and Stockwell (1983) completed an extensive review of the literature and concluded that, from the little information available, as recharge systems "adjacent undeveloped uplands are usually, but not always, more important (than wetlands), and no inherent characteristic makes wetlands distinctly important for recharge." However, they indicated that some seasonally flooded wetlands that have been scoured of impermeable sediments may be more important recharge systems than upland areas. It is clear that recharge potential is dependent on soil conditions and must be determined on a site-specific basis.

The hydraulic gradient in wetland systems identifies water levels and the direction of water flow and is measured as a sum of the elevation head, pressure head, and velocity head. Floodplain systems have low elevation heads due to their flat terrain. Large floodplain areas may have high pressure heads and, when combined with long retention times, could contribute to increased recharge capability. Velocity head is influenced by sediment porosity, transmissibility, permeability, and level of saturation. Soils of sand and gravel allow greater movement of groundwater and increased velocity head (Adamus and Stockwell 1983). Granneman and Sharp (1979) noted that distance from the river channel affects groundwater levels in that hydraulic gradient equalizes more slowly farther from the river. Figure 14b illustrates this rate of change across the floodplain zones.

Generalizing zonal distribution of floodplain recharge potential is not attempted here. Soil permeability, evapotranspiration rates, and floodplain size can directly influence recharge capabilities. Estimated distribution of relative soil permeability and evapotranspiration rates across the floodplain zones are presented in Figure 14c and 14d. In alluvial soils, recharge is more likely to occur. Bedinger (1981) suggests that the well-drained sandy and silty soils of natural levees and point bar formations make them important as areas of groundwater recharge. Also, the areal extent of these lands influences their significance as recharge areas. Brinson et al. (1981b) relate aquifer storage potential with the size of the alluvial field. If the floodplain is narrow, it is less likely that alluvial aquifer storage will be of significance. If the floodplain is large, the potential for groundwater storage is large and may have an influence on surface water flows by either receiving surface water during high flows or acting as a source during low flow periods. The highly productive vegetation of bottomland hardwood forests can reduce the wetland's ability to act as a recharge system because of water lost through evapotranspiration and inhibition of the downward percolation of water (Carter et al. 1979).

Groundwater Discharge

It is believed that floodplain wetlands store water during flooded times and discharge water during low flow periods (Wharton 1970 in Connor and Day 1982). In Barataria Swamp in Louisiana, Hopkinson and Day (1980) determined that under natural conditions upland runoff percolates slowly through the swamp and is released slowly to downstream areas. However, there are questions as to when, where, and how much water is discharged to the stream.

The discharge of groundwater from forested floodplain areas is affected by evapotranspiration rates and the hydraulic gradient of the wetland. When evapotranspiration rates are low and soils are saturated (spring), wetlands will release water to streams quickly (Carter et al. 1979). When evapotranspiration rates are high and soil saturation is low, groundwater discharges to the stream may be significantly reduced.

In a study of bog-type wetlands, Bay (1966) indicated that evapotranspiration was dependent on tree species, vegetative cover, climate, vegetation density, stage of vegetation development including seasonal stages in vegetation, and stand age (Carter et al. 1979). It is likely that floodplain evapotranspiration rates are functions of these same criteria, but with varying degrees of dependency. There is indication that wetland plants transpire more rapidly than those less tolerant to flooding, yet evapotranspiration rates of bottomland hardwoods vary and are not well defined (Carter et al. 1979). In a study of cypress and other forested wetlands, Brown (1981) determined that the ratio of transpiration rate to the standard pan evaporation (pan ratio) of a floodplain forest was higher than that of open water. Wharton (1980) determined that a Mississippi floodplain forest lost twice as much water by evapotranspiration as would be evaporated by a proposed reservoir. At these high rates, seasonal evapotranspiration may act in regulating the highwater fluctuations of the river.

VALUES OF BOTTOMLAND HARDWOOD FORESTS

Values are those benefits derived from the functions of bottomland hardwood wetlands. Usually, these benefits are those that can be useful to the human elements of the greater system in which the bottomland hardwood wetland ecosystem lies. The values of bottomland hardwood forests are: biomass production, including timber and fish and wildlife harvest; downstream food chain support through the export of organics; fish and wildlife habitats that provide recreation and aesthetic enjoyment; water quality protection through retention of nutrients/toxins and sediments; erosion control; flood storage and control; lowflow augmentation; and deep aquifer recharge.

Biomass Production

Timber Harvest

About 33 million ha of commercially valuable forested wetlands exist in the continental United States (Johnson 1979). Eastern bottomland hardwood forests comprise about 23 million ha of this total with 55% of the bottomlands located in the Southeast (Turner et al. 1981). These forests provide timber for construction, furniture, pulp, and firewood. The value of southeastern bottomland hardwoods has been estimated from

$3 billion (Turner et al. 1981) to $8 billion (Johnson 1979). Harvesting timber need not cause serious habitat deterioration and with proper management can complement other uses of the resource.

The most economically important bottomland hardwood forests fall into two categories. Oak-gum-cypress forests prevail in the Southeast, while elm-ash-cottonwood forests occupy a large portion of north-central floodplains (U.S. Forest Service 1974, Shifley and Brown 1978). Harvest practices include selective cutting and clearcutting. High-grading (i.e., cutting only the high-quality trees leaving the less desirable species) has resulted in deterioration of many southeastern hardwood stands in the past (Langdon et al. 1981). Selective cutting of bald cypress in zone II across the Southeast has shifted community composition to water tupelo and green ash associations (Wharton et al. 1982). Clearcutting is preferred by foresters (Putnam et al. 1960) and can enhance wildlife habitat if practiced in small (0.8-1.2 ha) plots scattered throughout the bottomland resource (Johnson 1979). Natural regeneration of clearcuts in some areas results in a highly desirable species mix (Bowling and Kellison 1983) but may also lead to forests with entirely different canopy constituents than the natural ecosystem. Wharton et al. (1982) noted that extensive logging of zone IV Nuttall oak-willow oak stands led to the development of a community of sugarberry-American elm-green ash. However, attention to regeneration requirements of desirable species can allow timber harvesting and other uses to be compatible.

Fish and Wildlife Harvest

Southeastern bottomland hardwood forests are among the most productive habitats for game species. Zone II sloughs provide opportunities for fishing and waterfowl hunting. Waterfowl are also abundant in zones III and IV especially when flooded and managed as greentree reservoirs. Large bottomland tracts, such as the Delta national Forest in Mississippi, are used extensively for sport hunting of deer (*Odocoileus virginianus*), turkeys (*Meleagris gallopavo*), and several small mammals (Ellis Clairain, Jr., U.S. Army Corps of Engineers [USACE], Waterways Experiment Station, Vicksburg, Mississippi, personal communication). Fur-bearing animals are almost exclusively wetland inhabitants, and the bottomlands of Louisiana yield the largest harvest of furs in the United States (Tiner 1984). Forested wetlands (zones II, III, IV) had higher populations of minks (*Mustela vison*), raccoons (*Procyon lotor*), and river otters (*Lutra canadensis*) than other wetland types. Muskrats (*Ondatra zibethicus*) and nutria (*Myocastor coypus*) were also harvested in these zones (Chabreck 1979). The commercially-important red swamp crawfish (*Procambarus clarkii*) is taken from zone II or from swamp forests with artificially manipulated hydroperiods which are common in the Lower Mississippi River basin (Bryan et al. 1975). Estimates of the annual crawfish harvest are given as 11.3 million kg or about $11 million in revenue (OTA 1984).

Food Chain Support

Seasonal shifts in food choice have been reported for some aquatic organisms (Nixon and Oviatt 1973, Haines and Montague 1979). Mulholland (1981) noted a shift to detrital feeding in fall and winter when other food sources were less available. Annual patterns of organic export from a Louisiana alluvial swamp to Barataria Bay are timed with feeding and spawning of migrant fishes in that estuary (Day et al. 1977). Rarely flooded bottomland hardwoods (zone V) may be just as important in food chain support as frequently-flooded zones. Unless organics are lost by burial or by peat accumulation, large amounts of organic material can be transported with less-frequent floods (Lugo 1981). Livingston et al. (1976) correlated peak commercial fishing and cyclic productivity in the Apalachicola Bay with the major pulse of organics received with 5-7 year floods. Since bottomland hardwood forests can regulate the rate of organic export to some degree, wild fluctuations of food availability in downstream ecosystems are less likely to occur where wetlands are coupled with both upland and riverine systems (Adamus and Stockwell 1983).

Fish and Wildlife Habitat

Bottomland hardwood wetlands are important to many species of fish and wildlife. They provide food, cover, and water and are vegetatively diverse. As an interface ecosystem between aquatic and terrestrial systems, riparian forested wetlands contain a large proportion of ecotone. This area illustrates the classic "edge effect" with high species abundance and diversity. Brinson et al. (1981b) discussed four ecological attributes of riparian systems that are important to fish and wildlife:

1) Predominance of woody plant communities - In regions where agricultural conversions are heavy (e.g. the Lower Mississippi River basin), remaining tracts of bottomland hardwoods are especially important as refugia for many species. Living trees and shrubs provide food and shelter for birds and other animals, and standing dead timber and snags are also important as habitat for many terrestrial and aquatic animals. Stream shading, bank stabilization, and input of allochthonous organics are all important to the aquatic community.
2) Presence of surface water and abundant soil moisture - The stream that defines the bottomland hardwood forest ecosystem provides food, protection, and breeding areas. During flooding, the bottomland itself becomes a feeding and spawning area for fish and a staging area for migratory waterfowl.
3) Diversity and interspersion of habitat features - Bottomland hardwoods provide a diversity of habitats related primarily to hydrologic characteristics. A soil moisture gradient from permanently inundated in zone II to infrequently flooded in zone VI creates numerous microhabitats for a variety of fish and wildlife.
4) Corridors for dispersal and migration - Because bottomland forests line streams, they are essentially linear in nature. This increases the relative proportion of ecotone and intensifies the edge effect. The vegetative cover provides protection for animal movement up and down stream.

Fredrickson (1979), Brinson et al. (1981b), and Wharton et al. (1981, 1982) reviewed in detail the fauna of forested riparian wetlands from which the following information is taken. Bottomland hardwood forests have relatively high plant productivity which is reflected in their high carrying capacity for several species. Probably because of an abundance of food, forested floodplains have been found to support twice the whitetail deer as an equivalent amount of upland forest (Zwank et al. 1979). Zones IV and V are generally most important as sources of browse for birds and mammals. Two mammals that are almost exclusively found in zones IV and V are *Sylvilagus aquaticus* and *S. palustris*, the swamp and marsh rabbits (Wharton et al. 1982). Bird density in riparian forests has also been reported as double that in uplands of many areas (Brinson et al. 1981b). The spatial stratification of the upper zones provides a variety of niches for birds. Generally, the upper zones that take in the upland/wetland ecotone support the highest faunal diversity. The greatest diversity of breeding birds, small mammals, amphibians, and reptiles is seen in zones IV and V. These zones contain an abundance of invertebrate food sources and mast-producing trees and shrubs which attract many upland species. They also serve as refugia for non-aquatic bottomland inhabitants during high water.

Zones II and III would naturally have more importance to fish, waterfowl, and amphibious mammals. Flooded bottomland hardwood zones are used extensively by migrating waterfowl as staging and overwintering areas. Wooded swamps are extremely important for wood duck (*Aix sponsa*) nesting. A number of mammals live or feed preferentially in these zones. Mink, for example, are inhabitants of backwater swamps and cypress-tupelo wetlands in particular. They feed heavily on crawfish, fish, and frogs which inhabit zones II and III. Up to 53 species of fish have been reported to spawn or feed in flooded bottomland forests. Most common are members of the catfish (*Ictaluridae*), sunfish (*Centrarchidae*), gar (*Lepisosteidea*), perch (*Aphredoderidae*), and sucker (*Catostomidae*) families. Many clupeids use oxbows and sloughs as spawning areas. Twigs and debris on the floodplain floor provide sites for egg-attachment so that eggs will not be swept away. Sport fish such as largemouth bass (*Micropterus salmoides*) also spawn in flooded bottomland forests. Zones II and III provide an abundant supply of invertebrates which are necessary during the period after the eggs hatch.

Without bottomland hardwood forests, many species would disappear or become rare. According to Brinson et al. (1981b), eighty, or 29%, of the 276 species listed as threatened or endangered by the U.S. Fish and Wildlife Service (FWS) in 1980 were at least partially dependent on riparian habitats. Based on these reviews and other available information, bottomland hardwood ecosystems can be regarded as highly valuable fish and wildlife habitat across all the zones.

Water Quality Protection

Wetlands are believed to be effective in the removal of excess nutrients, sediments, metals, and other pollutants from surface waters. Bottomland hardwood forests perform this ecological service in different ways with varying degrees of success. Water-quality protection depends not only on the amount of vegetative uptake, biochemical transformation, and sediment deposition but also on the hydrologic character of the watershed. Nutrient retention may be seasonal with storage in spring/summer and release in fall/winter, or it may be temporary over a period of years with storage in woody biomass. The materials released by the wetland are often converted to less harmful forms by plants and microorganisms. Essentially permanent retention of nutrients occurs when sediments are buried.

The transformation of nutrients in wetlands is a potentially important process in water-quality protection. Inorganic nutrients are most important in eutrophication. Forested wetlands are quite effective in converting inorganic nitrogen and phosphorus to organic forms. A nutrient budget of Okefenokee Swamp (Blood 1980) showed that inorganic nitrogen made up 48% of the input, but only 2% of the output implying preferential removal or transformation by the swamp. In the same study, phosphate was 7% of the total phosphorus input and 2% of the output. In Creeping Swamp, North Carolina, Kuenzler et al. (1980) had similar results. Over a two-year period, inputs of phosphorus were 63% and 29% phosphate, while outputs were 52% and 14%, respectively.

Toxic materials such as heavy metals and chlorinated hydrocarbons may be removed from water by wetlands. The primary means of removal is by adsorption onto suspended solids. Metals are mobilized by anaerobic conditions but will remain trapped in the sediment if an aerobic surface layer is present (Mortimer 1941, 1942). A report by Tchobanoglous and Culp (1980), as cited by OTA (1984), noted that heavy metals were removed by wetlands at 20-100% efficiencies depending on the type of metal and wetland characteristics. On the whole, very little research has been done with bottomland hardwood forests and their role in metal removal. Chlorinated hydrocarbons are also taken up by some plants, but specific studies in bottomland hardwoods were not found. Many of these synthetic chemicals degrade easily in sediments which could make forested wetlands potentially efficient sites for removal (OTA 1984).

Erosion Control

Reduction of water velocities within the watershed controls streambank erosion, sheet erosion in the wetland, and erosion of downstream systems. Erosion control is experienced at various points within any inland watershed: upland (slope) erosion control, floodplain erosion control, and streambank erosion control. Surface water storage in forested wetlands, by reducing throughflow velocities, acts to provide a valuable means of stabilizing land masses and protecting downstream embankments.

The forces of soil erosion and deposition play a significant role in floodplain dynamics. The flat, broad nature of floodplains interacts with these forces causing meandering, migrating stream patterns across the floodplain. The decrease of in-channel flow velocities is a function of these meanders. Overbank flows begin to utilize the increased volume of the floodplain terraces and again velocities are diminished. Annual flooding is less likely to create major changes in the watercourse than flooding caused by higher flows (Wharton 1980).

The value of floodplain areas acting as water storage systems and thereby reducing flow velocities and diminishing downstream erosion has not been determined directly. The information provided by hydrology and hydraulics assures us that, if water is allowed to occupy a greater volume, flows will slow down. Inland wetland activities that reduce peak flows downstream have been studied in terms of stormflow modification and flood storage (see below) but have not addressed the question of reduced erosion of downstream areas. Most studies that have been done on erosion control capabilities of wetlands have centered around how wetland vegetation acts to hold soils.

The biotic community of bottomland hardwood wetlands helps control erosion in the following ways (Adamus and Stockwell 1983): 1) wave energy and current velocity are reduced by friction against the plant structure, and 2) root systems bind the sediments and stabilize the shoreline. The degree to which erosive forces are dissipated increases with vegetation density, rigidity of obstructions, and height above waves or currents (Adamus and Stockwell 1983). Bottomland hardwoods, then, should be among the most effective wetlands in absorbing wave and current energy. Reduced current velocity allows the deposition of sediments which enhances shoreline anchoring. The wide, shallow root mats of floodplain trees anchor the soil and prevent loss by scouring (Wharton et al. 1982). A literature review of erosion control by wetland vegetation (Allen 1979) presented substantial evidence that wetland plants do bind soil and stabilize banks but have inherent limitation. Riparian vegetation may be undermined by water currents, damaged by ice and floating debris, or covered by silt (Carter et al. 1979). However, trees are excellent stabilizers; some riverbanks have remained essentially uneroded for 100-200 years due to the presence of riparian trees (Sigafoos 1964).

Bottomland hardwood wetlands are not only resistant to erosion but also protect erodible uplands adjacent to the wetland (OTA 1984). The same forces that reduce erosion of the shoreline are effective at higher elevations. Less water with lower current velocity reaches the upland areas and has less ability to erode. On the other hand, Turner (1980) suggested that some wetlands may contribute occasionally to downstream erosion by filtering out sediment which, in turn, increases the sediment-carrying capacity of the water.

Flood Storage and Control

The ability of a floodplain wetland to slow and hold water is valuable in that downstream flood control and stormflow modification are provided. These values help to maintain human use and control of downstream areas.

Three processes within bottomland hardwood wetlands act together to provide flood storage and stormflow modification: 1) temporary storage of surface water through physical volume of the wetland; 2) recharge of surficial aquifers; and 3) the reduction of floodwater velocities (Carter et al. 1979).

The best example of the value of wetlands as storage systems is still the Charles River wetland retention program in Massachusetts. The USACE showed that the floodplain wetlands of the Charles River could provide suitable flood protection and that purchase of these lands was less expensive than constructing flood control facilities (USACE 1971).

Wharton (1980) showed that the discharge volumes of the Oconee River in Georgia cease to rise, even with a significant increase in drainage area, as floodwaters begin to flow through the wide floodplain area. Flood peaks are smoothed due to the storage action of floodplain areas. Leitman (1978) showed significant lowering of hydrograph peaks recorded on the Apalachicola River upstream and downstream of an increase in floodplain width. Overbank flooding of Heron Pond on the Cache River, Illinois was shown to smooth the hydrograph, indicating that without swamp storage of flood levels, peak discharges would be higher (Mitsch et al. 1977). Novitzki (1979) showed that watersheds in Wisconsin with wetlands and lake areas held floodwaters longer than those without. In North Dakota, Brun et al. (1981) compared surface water flows before and after the drainage of wetland areas which suggested that wetlands play a significant role in desynchronizing peak flows (Adamus and Stockwell 1983).

There are unknowns involved in determination of the potential value of a specific bottomland hardwood as a flood storage and control facility. There are two methods available for determining capacity: 1)actual measurement through detailed field studies of inflows and outflows; and 2) developing mathematical models with realistic coefficients for various wetland types which could then be applied to specific cases of similar wetlands. Calculations of runoff volume have been attempted using a modified Manning's equation which requires a subjective definition of the "n" roughness coefficient (Carter et al. 1979). The development of close-to-real roughness coefficients is dependent on field experiments. Therefore, field measurement is a vital component of modeling methodologies.

The value of a wetland for flood control and protection of downstream property must be determined on an individual basis and in conjunction with other values it may provide. In some watersheds, the wetland may be one of the most significant means of flood control, while in others the flood protection value would not override the value of different landuses on the floodplain. There is some claim that wetlands provide greater flood peak attenuation for more frequent floods of lesser volume, and that their value diminishes as flood volumes increase. These greater floods (50 to 100 year storms) usually cause most of the downstream damage (Adamus and Stockwell 1983).

Low Flow Augmentation

There is significant disagreement as to the extent wetlands actually contribute to baseflow, suggesting that during dry periods and the concurrent growing season, the evapotranspiration rates increase and virtually use up available groundwater leaving little for baseflow returns to the streams. In a study done by Bell and Johnson (1974) on the Sangamon River, Illinois, the groundwater loss to evapotranspiration in the middle areas of the floodplain was greater than river infiltration and upland drainage, resulting in little or no low flow augmentation. Carter et al. (1979) cite several studies that showed <u>reduction</u> in flows caused by wetlands as opposed to augmentation of low flows. In Wisconsin basins with much lake and wetland area, springtime streamflow is greater and fall baseflow is lower than in basins with no lakes or wetlands (Novitzki 1979).

Opposing views are equally justified. The importance of wetlands areas as groundwater storage systems and low flow augmentation systems has been shown in several studies. A study of the Missouri River showed that during periods of sustained high river stages, lateral seepage into the groundwater maintained high groundwater levels. During low flows, the hydraulic gradient was reversed, resulting in groundwater discharge to the river as baseflow (Granneman and Sharp 1979 in Brinson et al. 1981b). Wharton et al. (1982) describe baseflow as an important component of the discharge of blackwater streams in the southeastern United States. They cite a water budget study done by Winner and Simmons (1977) on a small stream in Creeping Swamp, North Carolina which showed the baseflow runoff to be 20%, indicating a high stream dependency on groundwater discharge. The Alcovy and Yellow Rivers in Georgia have low water flows that are close in volume, although the Yellow River basin is 36% larger than the Alcovy. The Alcovy has much more swamp land, suggesting that the swamp lands influence baseflow (Wharton 19700. A Gila River study done by Gatewood et al. (1950) showed that in this arid region of the country where the demands on groundwater supplies are extremely competitive, groundwater outflows were extremely important in maintaining surface water flows. In Florida, Littlejohn (1977) studied the effects of swamp areas on aquifer storage and discharge of Naples Bay. He showed that the swamplands helped to stabilize flows even though seasonal precipitation patterns are sharp and frequent (Conner and Day 1982).

Deep Aquifer Recharge

There is little information available that validates the value of bottomland hardwood forests as deep aquifer recharge systems. The large, alluvial levee formations on the Mississippi River are potential water supply areas. A floodplain wetland overlying saturated sand could be a source of deep aquifer recharge and water supply. However, the wetland could also reduce groundwater levels through evapotranspiration and by inhibiting percolation of water (Carter et al. 1979).

IMPACTS OF AGRICULTURAL CONVERSION ON BOTTOMLAND HARDWOOD FOREST FUNCTIONS AND VALUES

The loss of bottomland hardwood area described in the Introduction has been due, in large part, to the conversion of floodplain forests to agricultural land. Turner et al. (1981) estimated that 80% of lost bottomlands in the Southeast have been converted to agricultural use. Agricultural conversion usually requires water management as well as removal of natural vegetation. In some cases, the channelization of streams reduces flooding sufficiently to encourage clearing of the floodplain. In other cases, levees, ditches, or other water diversion methods are necessary to produce arable land for crops. Although land clearing eliminates the bottomland hardwood forest as such, natural succession will replace the ecosystem if sufficient propagules are available and the land is taken out of crop production. On the other hand, hydrologic alterations modify the driving force of the wetland and change the natural functional patterns.

Bottomland hardwood forests are intimately tied to the hydrologic patterns of the watershed. In fact, any consideration of the structure and function of floodplain wetlands must include the stream because of their inseparable coupling. Water is the single most important element in the development and maintenance of bottomland hardwood ecosystems.

Bottomland hardwood forests are highly productive ecosystems largely as a result of the pulsing inputs of floodwater, sediments, and nutrients. The fertile floodplain soils are also attractive to farmers for crop production. Soybeans, cotton, and corn grown in southeastern bottomlands produce large harvests and yield at least twice as much profit to the farmer as silviculture (MacDonald et al. 1979). The sections below will discuss the impacts of biomass removal and hydrologic modifications as employed in the conversion of these wetlands to agricultural systems.

Biomass Removal

Although intensive agricultural practices create agroecosystems with high net primary productivity, biomass in croplands is only a fraction of that in natural bottomland hardwood forests. For example, Odum (1971) noted that the biomass of a corn crop was about 0.4 kg/m^2 at its peak and near zero after harvest. In comparison, natural forested wetlands often exceed 20 kg/m^2 throughout the year (Conner and Day 1982). Removal of natural vegetation not only results in biomass reduction but also in direct and indirect effects upon the intrinsic functions and values of the unaltered bottomland hardwood forest.

Repeated harvest of crops artificially maintains an early successional stage. Energy is almost totally devoted to the accumulation of biomass, and very little structural diversity is developed in agricultural monocultures. Bottomland hardwood forests, through the process of natural succession, have achieved a high degree of spatial heterogeneity and species diversity as well as a large standing stock of biomass. Diversity in plant species is directly related to niche availability which, in turn, determines the type and abundance of consumers present in the ecosystem (Gosselink et al. 1981).

Organisms with life cycles intimately tied to bottomland hardwood forests are most likely to be seriously affected by agricultural conversion. For example, the prothonotary warbler (*Protonotaria citrea*) nests only in trees standing in or leaning over water and feeds on the abundant insects in the forested wetland. Wood ducks may feed in flooded agricultural fields, yet, for successful breeding, cavities in hardwood trees must be available for nest sites (Maki et al. 1980). At the very least, animals that use bottomland forests will be forced to migrate to other areas or will feed on crops and be a nuisance to farmers.

Another direct effect of biomass removal is loss of timber resources. The traditional use of bottomland hardwood forests for timber production is being replaced by crop production in many areas. If present conversion trends continue, bottomland hardwood timber will become scarce, particularly in the Lower Mississippi River Valley. The demand for wood products will be met from other timber resources with concomitant habitat deterioration. Timber harvesting would probably increase in the remaining bottomland hardwood forests as well as in upland forests.

Any alteration of floodplain vegetation will have effects on the associated stream. Floodplain export of organic materials is a major source of energy for aquatic organisms in the stream and in estuaries. Clearing of floodplain trees deprives the aquatic ecosystem of its primary detrital input. Livingston (1978) reported that clearcutting in the Apalachicola Delta severely damaged productivity in the East Bay of the estuary. Other studies have shown low abundance and diversity of aquatic invertebrates in streams that flow through cleared floodplains (Minshall 1968). In North Carolina, Kuenzler et al. (1977) showed that organic carbon in bottomland hardwood streams was significantly higher than in streams flowing through non-forested floodplains. Likewise, organic export from watersheds containing forested wetlands exceeded that from upland watersheds (Mulholland and Kuenzler 1979).

Riparian trees have also been shown to act as an effective buffer against temperature extremes. Greene (1950) compared the temperature of a farm stream that once flowed through a hardwood forest with that of a stream draining a mature hardwood forest. He found that temperatures in the farm stream were significantly higher (6-9 C) in summer and lower (3-5 C) in winter than in the woodland stream. Similar results were recorded by Gray and Edington (1969) from a stream before and after riparian clearing and by Karr and Gormann (1975) comparing shaded and unshaded stream segments.

Stream temperature is very important in regulating biotic communities and water quality. As temperature increases with reduced shading, the oxygen-holding capacity of the water decreases. Since the decomposition of organic matter uses oxygen, the ability of the stream to assimilate organic material without oxygen depletion is severely reduced.

Even more important with respect to eutrophication is the effect of temperature on the release of nutrients from sediment. Sommers et al. (1975) found that the release of phosphorus increases exponentially with an increase in temperature; slight temperature increases above 15 C produced substantial increases in phosphorus release.

Nutrient concentration in streams has also been correlated with landuse patterns. Streams draining agricultural watersheds have, on the average, considerably higher nutrient concentrations than those draining forested watersheds. In general, inorganic forms of nitrogen prevail in streams draining cropland. Inorganic nitrogen comprised 18% of total nitrogen in forest streams and almost 80% in agricultural streams (Kuhner 1980). Mean concentrations of both total nitrogen and total phosphorus were nearly nine times greater in streams draining agricultural lands than in streams flowing through forested floodplains.

Peterjohn and Correll (1984) studied the nutrient dynamics of both cultivated and natural riparian ecosystems in Maryland. They determined mass balances for nitrogen and phosphorus. Nitrogen retention by the cropland was estimated to be only 8%, while the riparian forest was calculated to retain 89% of the nitrogen input. The cropland and the bottomland forest retained 41% and 80% of the phosphorus inputs, respectively. Although most of the nitrogen and phosphorus loss by the cropland was through the harvest of corn, about 30% of the nitrogen and 20% of the phosphorus was exported via surface and groundwater flows.

Most of the surface waters in the United States contain chlorinated hydrocarbons originating from insecticides and herbicides (Reese et al. 1972). Pesticides are of special concern in streams due to their toxicity to aquatic organisms and their potential hazard to human health. Most of the pesticides in streams result from storm runoff and overland flow (Spencer 1971). Agricultural lands receive approximately 51% of all pesticides used and, consequently, contribute greatly to stream loading.

Soluble pesticides may enter streams dissolved in surface runoff. However, most of the pesticides probably enter receiving waters sorbed onto particulates carried by runoff. Row crops have been shown to significantly reduce the amount of sediment carried by overland flow as compared to that from bare soil (Karr and Schlosser 1977, Schlosser and Karr 1981). On the other hand, Hewlett and Nutter (1970) have reported that overland flow rarely occurs in naturally vegetated watersheds. When forested watersheds are compared to areas under cultivation, runoff may be 40 times greater from agricultural watersheds (Hornbeck et al. 1970).

Hydrologic Modifications

Although some floodplains can be cleared and farmed without drainage, by far most agricultural conversions of bottomland hardwood forests require water management. Water management practices include channelization, drainage canals, and other water diversion techniques. All of these practices reduce or eliminate flooding. Not only are the normal inputs to the floodplain ecosystems eliminated, but outputs from the systems are also altered.

Channelization of rivers and streams involves removal of natural meanders, streambank clearing, increases in channel depth and width, and disposal of dredged materials. The purposes of channelization include navigation improvement, flood control, and production of arable land. By straightening and deepening the stream, the movement of water is improved. Consequently, flooding in the bottomland adjacent to a channelized stream segment is reduced or eliminated. This severely decreases the water, sediment, and nutrient subsidies that make the bottomland hardwood forest so productive. On the other hand, the flood storage capabilities of the bottomland hardwood forest have largely been bypassed, and flooding in unchannelized downstream segments will be much more severe. Deepening the channel also encourages the discharge of groundwater from the bottomland to the stream. This creates a reduction in soil moisture which, along with reduced nutrient input, stresses the flood-adapted vegetation and decreases primary productivity.

Maki et al. (1980) studied four channelized and seven unchannelized streams in eastern North Carolina. In the unaltered watersheds, winter flooding and summer drying prevailed due to precipitation and evapotranspiration patterns. In channelized watersheds, flooding occurred in winter only when prolonged, intense storms caused the streams to exceed spoil bank height. Usually, soils were saturated but without standing water. During summer, the unchannelized streams generally became intermittent during dry periods while channelized streams maintained perennial flow. They concluded that the deepened channels intercepted the groundwater table and depleted soil moisture in the adjacent bottomlands. The water table in channelized basins remained at least 40 cm below the level found in natural watersheds regardless of landuse.

Channelization also affects the fish and wildlife resources of the bottomland ecosystem. For some species, channelization is detrimental, while others species may benefit. Generally, the more water-dependent species are adversely affected. Fish are no longer able to use the floodplain for spawning and feeding areas. Other animals adversely affected include birds such as prothonotary warblers, wood ducks, and herons (*Ardeidae*); reptiles such as eastern cottonmouths (*Agkistrodon piscivorus leucostoma*); mammals such as otter, mink, and other furbearers; and, particularly, amphibians (Maki et al. 1980). More terrestrial species may replace these species in the drier conditions of channelized bottomlands. Several reports including those by Arner et al. (1976), Fredrickson (1979), and Klimas et al. (1981) detail the response of wildlife to altered hydrologic conditions.

Water quality in natural streams differs from that of channelized streams. Three natural streams and four channelized streams in the North Carolina Coastal Plain were studied by Kuenzler et al. (1977). Concentrations of nitrate in channelized streams were ten times greater than those in natural streams. Much of the increase was due to the agricultural conversion of parts of the channelized watershed. Not only were there nutrient inputs from livestock and agricultural activities, but the absence of natural swamp forests also eliminated the nutrient filtering and transformation capabilities of the floodplains. Phosphorus concentrations (filterable reactive-P and particulate-P) were also several times higher in channelized streams without point sources than in unpolluted natural streams. Monitoring during two flood periods showed that the flux rates of nutrients from one of the channelized streams were up to ten times higher than from one of the natural streams.

Other drainage methods are often used in conjunction with channelization. Many agricultural floodplains employ drainage canals for surface drainage and/or tile fields for subsurface drainage. Artificial drainage and conversion of forested wetlands may increase total runoff. Daniel (1981) reported that total annual runoff as a percentage of rainfall may be greater from agriculturally-developed areas of coastal North Carolina than from forested floodplains. The suggested explanation was reduced interception by vegetation and reduced residence time leading to less evapotranspiration. Of more importance, perhaps, are the changes in timing, duration, and rate of discharge of floodwaters that occur as a result of drainage and agricultural conversion. Discharge records from North Carolina showed that peak discharges were higher and baseflows were lower in the intensively farmed Albemarle Canal drainage than in the naturally forested Van Swamp watershed even though annual runoff totals were not significantly different.

Kirby-Smith and Barber (1979) investigated the effects of drainage and agricultural conversion of forested floodplains on estuarine systems in North Carolina. The major water-quality changes seen in the estuary were decreased surface water salinity, increased turbidity, and increased concentrations of phosphate, nitrate, and ammonia. The extent and duration of the changes were correlated with rainfall and runoff from the watershed. Although no significant biological changes were observed in the South River estuary, the authors noted that the water quality changes could only serve to reduce the future assimilation capacity of the estuary.

Three drained and three undrained sites in the North Carolina Tidewater region were compared by Skaggs et al. (1980). Because the rich, organic soils of the region hold nitrate well, low levels of inorganic nitrogen are found in drainage waters from agricultural fields. However, these concentrations are still significantly higher than in exports from natural floodplains in the area. The reduced conditions of saturated soils in undrained bottomlands favor the removal of nitrogen from groundwater via denitrification. Nitrogen in well-drained soils is more likely to be lost to the stream since denitrification is uncommon in dry, aerated soils. The exported nitrogen may be transported by surface flow in drainage canals, or it may be lost by artificial or natural subsurface drainage.

Summary and Conclusions

Individual bottomland hardwood forests are unique ecosystems with individual responses to agricultural impacts. However, these wetlands have many biological, chemical, and physical processes in common and should respond similarly to environmental changes though with varying degrees of response. In summary, conversion to intensive agriculture may generally have the following effects on the functions of bottomland hardwood forests:

1) Primary productivity - The primary productivity of agroecosystems may equal or exceed that of the bottomland forest, but biomass and structural diversity will be much lower.
2) Litterfall and decomposition - Harvest of crops removes most of the crop biomass with a corresponding decrease in litter available for decomposition. Decomposition rates will vary with soil type and moisture.
3) Organic export - Removal of riparian trees deprives the aquatic system of much of its detrital food base. Flood diversion eliminates the transport mechanism.
4) Consumer activity - Agricultural monocultures provide few niches for consumers. Original bottomland inhabitants will likely be replaced by small, upland species. Water management shifts species composition from flood-adapted assemblages to more terrestrial communities.
5) Sediment deposition - When flooding is eliminated or greatly reduced, much of the source of sediment is removed. Drainage canals in the fields may increase sediment export to the stream.
6) Retention of nutrients and toxins - Agricultural ecosystems retain some nutrients and toxins but are less efficient than bottomland hardwood forests.
7) Biochemical transformations - Water management practices eliminate anaerobic soil conditions during most of the year and decrease anaerobic biochemical processes such as denitrification. Aerobic processes will likely be increased.
8) Flooding dynamics - The frequency and duration of floods and the area flooded in the bottomland are all reduced by most agricultural water-management techniques. However, downstream ecosystems will experience an increase in all phases of flooding.
9) Groundwater dynamics - The water table is significantly lower in most agricultural bottomlands than in the natural forest. Any decreases in evapotranspiration are often exceeded by groundwater loss via discharge into channelized streams and drainage systems.

REFERENCES CITED

Adamus, R. R., and L. T. Stockwell. 1983. A method for wetland functional assessment, Volume I. U.S. Department of Transportation, Federal Highway Administration Report No. FHWA-IP-82-23, Washington, D.C. 176 pp.

Allen, H. H. 1979. Role of wetland plants in erosion control of riparian shorelines. Pages 403-414 *in* P. E. Greeson, J. R. Clark, and J. E. Clark, eds. Wetland functions and values: the state of our understanding. American Water Resources Association, Minneapolis, Minnesota.

Arner, D. H., H. R. Robinette, J. E. Frasier, and M. H. Gray. 1976. Effects of channelization of the Luxapalila River on fish, aquatic invertebrates, and furbearers. U.S. Fish and Wildlife Service, Washington, D.C. FWS/OBS-76-08.

Atlas, R. M., and R. Bartha. 1981. Microbial ecology: fundamentals and applications. Addison-Wesley Publishing Co., New York. 560 pp.

Bay, R. R. 1966. Factors influencing soil-moisture relationships in undrained bogs. Pages 335-343 *in* International symposium on forest hydrology. Proceedings of the National Science Foundation Advances in Science Seminar, Washington, D. C.

Bedinger, M. S. 1971. Forests species as indicators of flooding in the lower White River Valley, Arkansas. U.S. Geological Survey, Washington, D. C. Professional Paper 750-C, pp. 248-253.

Bell, D. T., and F. L. Johnson. 1974. Groundwater level in the floodplain and adjacent uplands of the Sangamon River. Trans. Ill. St. Acad. Sci. 67:376-383.

Blem, C. R., and L. B. Blem. 1975. Density, biomass, and energetics of the bird and mammal populations of an Illinois deciduous forest. Trans. Ill. Acad. Sci. 68:156-164.

Blood, E. R. 1980. Surface water hydrology and biogeochemistry of the Okefenokee Swamp watershed. Ph.D. Dissertation, University of Georgia, Athens. 194 pp.

Bowling, D. R., and R. C. Kellison. 1983. Bottomland hardwood stand development following clearcutting. Southern J. of Applied Forestry 7:110-116.

Boyt, F. L. 1976. A mixed hardwood swamp as an alternative to tertiary wastewater treatment. M.S. Thesis, University of Florida, Gainesville. 99 pp.

__________, S. E. Bayley, and J. Zoltek, Jr. 1977. Removal of nutrients from treated municipal wastewater by wetland vegetation. Water Pollut. Cont. Fed. J. 49:789-799.

Brinson, M. M. 1977. Decomposition and nutrient exchange of litter in an alluvial swamp forest. Ecology 58:601-609.

__________, A. E. Lugo, and S. Brown. 1981a. Primary productivity, decomposition and consumer activity in freshwater wetlands. Annual Review Ecol. and Systematics. 12:123-161.

__________, B. L. Swift, R. C. Plantico, and J. S. Barclay. 1981b. Riparian ecosystems: their ecology and status. U.S. Fish and Wildlife Service, Kearneysville, West Virginia. FWS/OBS-81/17.

Brown, S. 1981. A comparison of the structure, primary productivity, and transpiration of cypress ecosystems in Florida. Ecological Monographs 51:403-427.

_________, and A. E. Lugo. 1982. A comparison of structural and functional characteristics of saltwater and freshwater forested wetlands. Pages 109-130 *in* B. Gopal, R. E. Turner, R. G. Wetzel, and D. F. Whigham, eds. Wetlands: ecology and management. National Institute of Ecology and International Scientific Publications, Jaipur, India.

Brun, L. J., J. L. Richardson, J. W. Enz, and J. K. Larsen. 1981. Streamflow changes in the Southern Red River Valley of North Dakota. N. Dak. Farm Res. 38:11-14.

Bryan, C. F., F. M. Truesdale, and D. S. Sabins. 1975. A limnological survey of the Atchafalaya Basin. Annual report of the Louisiana Cooperative Fisheries Unit, Louisiana State University, Baton Rouge. 203 pp.

Burns, L. A. 1978. Productivity, biomass, and water relations in a Florida cypress forest. Ph.D. Dissertation. University of North Carolina, Chapel Hill. 170 pp.

Carter, V., M. S. Bedinger, R. P. Novitski, and W. O. Wilen. 1979. Water resources and wetlands. Pages 344-376 *in* P. E. Greeson, J. R. Clark, and J. E. Clark, eds. Wetland functions and values: the state of our understanding. American Water Resources Association, Minneapolis, Minnesota.

Chabreck, R. H. 1979. Wildlife harvest in wetlands of the United States. Pages 618-631 *in* P. E. Greeson, J. R. Clark, and J. E. Clark, eds. Wetland functions and values: the state of our understanding. American Water Resources Association, Minneapolis, Minnesota.

Clark, J. R., and J. Benforado, eds. 1981. Wetlands of bottomland hardwood forests. Elsevier Science Publishing Co., Amsterdam. 401 pp.

Conner, W. H., and J. W. Day, Jr. 1976. Productivity and composition of a bald cypress-water tupelo site and a bottomland hardwood site in a Louisiana swamp. Amer. J. Bot. 63:1354-1364.

______________________________. 1982. The ecology of forested wetlands in the southeastern United States. Pages 69-87 *in* B. Gopal, R. E. Turner, R. G. Wetzel, and D. F. Whigham, eds. Wetlands: ecology and management. National Institute of Scientific Publications, Jaipur, India.

Correll, D. L. 1978. Estuarine productivity. Bioscience 28:646-650.

Cowardin, L. M., V. Carter, F. C. Golet, and E. T. LaRoe. 1979. Classification of wetlands and deepwater habitats of the United States. U.S. Fish and Wildlife Service, Washington, D.C. FWS/OBS-81/17.

Crow, J. H., and K. B. MacDonald. 1979. Wetland values: secondary production. Pages 146-161 *in* P. E. Greeson, J. R. Clark, and J. E. Clark, eds. Wetland functions and values: the state of our understanding. American Water Resources Association, Minneapolis, Minnesota.

Damman, A. W. H. 1979. Mobilization and accumulation of heavy metals in freshwater wetlands. Institute of Water Resources, Connecticut University, Storrs. 17 pp.

Darnell, R. M., W. E. Pequegnat, B. M. James, F. J. Benson, And R. E. Defenbaugh. 1976. Impacts of construction activities in the wetlands of the United States. U.S. Environmental Protection Agency, Ecological Research Services, Washington, D. C. EPA-600/3-76-45.

Day, J. W., Jr., T. J. Butler, and W. H. Conner. 1977. Production and nutrient export studies in a cypress swamp and lake system in Louisiana. Pages 255-269 *in* M. Wiley, ed. Estuarine processes, Vol. 2. Academic Press, New York.

de la Cruz, A. 1979. Production and transport of detritus in wetlands. Pages 162-174 *in* P. E. Greeson, J. R. Clark, and J. E. Clark, eds. Wetland functions and values: the state of our understanding. American Water Resources Association, Minneapolis, Minnesota.

__________, and H. Kawanabe. 1967. The population and food habits of fish in a small estuary. Nat. Appl. Sci. Bull. 20:473-477.

Dickson, J. G. 1978. Forest bird communities of the bottomland hardwoods. Pages 66-73 *in* R. M. DeGraaf, ed. Proceedings of the workshop: management of southern forests for nongame birds. U.S. Department of Agriculture, Forest Service, Atlanta, Georgia. GTR/SE-14.

Duever, M. J., J. E. Carlson, and L. A. Riopelle. 1975. Ecosystem analysis of Corkscrew Swamp. Pages 627-725 *in* H. T. Odum and K. C. Ewel, eds. Cypress wetlands for water management, recycling, and conservation, University of Florida, Gainesville.

Elder, J. F., and J. Cairns. 1982. Production and decomposition of forest litter fall on the Apalachicola River floodplain, Florida. U.S. Geological Survey, Washington, D. C. Water Supply Paper 2196-B.

Farnsworth, E. G., M. C. Nichols, C. N. Vann, L. G. Wolfson, R. W. Bosserman, P. R. Hendrix, F. B. Golley, and J. L. Cooley. 1979. Impacts of sediments and nutrients on biota in surface waters of the United States. Environmental Research Laboratory, U.S. Environmental Protection Agency, Athens, Georgia. EPA-600/3-79-105.

Fredrickson, L. H. 1979. Lowland hardwood wetlands: current status and habitat values for wildlife. Pages 296-306 *in* P. E. Greeson, J. R. Clark, and J. E. Clark, eds. Wetland functions and values: the state of our understanding. American Water Resources Association, Minneapolis, Minnesota.

Gambrell, R. P., and W. H. Patrick. 1978. Chemical and microbiological properties of anaerobic soils and sediments. Pages 375-423 *in* D. D. Hook and R. M. Crawford, eds. Plant life in anaerobic environments. Ann Arbor Science, Ann Arbor, Michigan.

Gatewood, J. S., J. W. Robinson, B. R. Colby, J. D. Hem, and L. C. Halpenny. 1950 Use of water by bottomland vegetation in Lower Safford Valley, Arizona. U.S. Geological Survey, Washington, D. C. Water Supply Paper 1103.

Gosselink, J. G., S. E. Bayley, W. H. Conner, and R. E. Turner. 1981. Ecological factors in the determination of riparian wetland boundaries. Pages 197-219 *in* J. R. Clark and J. Benforado, eds. Wetlands of bottomland hardwood forests. Elsevier Science Publishing Co., Amsterdam.

Graetz, D. A., P. A. Krottje, N. L. Erickson, J. Fiskell, and D. F. Rothwell. 1980. Denitrification in wetlands as a means of water quality improvement. Florida Water Resources Research Center Publ. No. 48, University of Florida, Gainesville. 75 pp.

Granneman, N. G., and J. M. Sharp, Jr. 1979. Alluvial hydrogeology of the lower Missouri River Valley. J. Hydrology 40:85-99.

Gray, J. R., and J. M. Eddington. 1969. Effect of woodland clearance on stream temperature. J. Fisheries Res. Bd. Canada 26:399-403.

Greene, G. F. 1950. Land use and trout streams. J. Soil Water Conservation 5:125-126.

Haines, E. B., and G. L. Montague. 1979. Food sources of estuarine invertebrates analyzed using 13C/12C ratios. Ecology 60:48-56.

Hair, J. D., G. T. Hepp, L. M. Luckett, K. P. Reese, and D. K. Woodward. 1978. Beaver pond ecosystems and their relationships to multi-use natural resource management. Pages 80-92 *in* R. R. Johnson and J. F. McCormick, eds. Strategies for protection and management of floodplain wetlands and other riparian ecosystems. U.S. Department of Agriculture, Forest Service, Washington, D.C. GTR-WO-12.

Hartland-Rowe, R., and P. P. Wright. 1975. Effects of sewage effluent on a swampland stream. Verh. Internat. Verin. Limnol. 19:1575-1583.

Heilman, P. E. 1968. Relationship of availability of phosphorus and cations to forest succession and bog formation in interior Alaska. Ecology 49:331-336.

Hewlett, J. D., and W. L. Nutter. 1970. The varying source area for streamflow from upland basins. Pages 65-83 *in* Proc. symposium on interdisciplinary aspects of watershed management. American Society of Civil Engineering, New York.

Hook, D. D., C. L. Brown, and P. P. Kormanik. 1970. Lenticel and water root development of swamp tupelo under various flooding conditions. Bot. Gaz. 131:217-224.

__________. 1971. Inductive flood tolerance in swamp tupelo (*Nyssa sylvatica* var. *biflora*). J. Exp. Bot. 22:78-89.

Hopkinson, C. S., and J. W. Day, Jr. 1980. Modeling hydrology and eutrophication in a Louisiana swamp forest ecosystem. Env. Management 4:325-335.

Hornbeck, J. W., R. S. Price, and C. A. Federer. 1970. Streamflow changes after forest clearing in New England. Water Resources Research 6:1124-1132.

Huffman, R. T., and S. W. Forsythe. 1981. Bottomland hardwood forest communities and their relation to anaerobic soil conditions. Pages 187-196 *in* J. R. Clark and J. Benforado, eds. Wetlands of bottomland hardwood forests. Elsevier Science Publishing Co., Amsterdam.

Johnson, R. L. 1979. Timber harvests from wetlands. Pages 598-605 *in* P. E. Greeson, J. R. Clark, and J. E. Clark, eds. Wetland functions and values: the state of our understanding. American Water Resources Association, Minneapolis, Minnesota.

Kadlec, R. H., and J. A. Kadlec. 1979. Wetlands and water quality. Pages 436-456 *in* P. E. Greeson, J. R. Clark, and J. E. Clark, eds. Wetland functions and values: the state of our understanding. American Water Resources Association, Minneapolis, Minnesota.

Karr, J. R., and O. T. Gorman. 1975. Effects of land treatment on the aquatic environment. Pages 120-150 *in* Non-point source pollution seminar. U.S. Environmental Protection Agency, Chicago, Illinois. EPA 905/9-75-007.

Karr, J. R., and I. J. Schlosser. 1977. Impact of nearstream vegetation and stream morphology on water quality and stream biota. Environmental Research Laboratory, U.S. Environmental Protection Agency, Athens, Georgia. EPA-600/3-77-097.

Kemp, G. P., and J. W. Day. 1981. Floodwater nutrient processing in a Louisiana swamp forest receiving agricultural runoff. Water Resources Research Institute, Louisiana State University, Baton Rouge. Report No. A-043-LA.

Kirby-Smith, W. W., and R. T. Barber. 1979. The water quality ramifications in estuaries of converting forest to intensive agriculture. Water Resources Research Institute, University of North Carolina, Raleigh. Report No. 148.

Kitchens, W. M., Jr., J. M. Dean, L. H. Stevenson, and J. H. Cooper. 1975. The Santee swamp as a nutrient sink. Pages 349-366 *in* F. G. Howell, J. B. Gentry, and M. H. Smith, eds. Mineral cycling in southeastern ecosystems. National Technical Information Service, Washington, DC. ERDA Conf-740513.

Klimas, C. V., C. O. Martin, and J. W. Teaford. 1981. Impacts of flooding regime modification on wildlife habitats of bottomland hardwood forests in the Lower Mississippi Valley. U.S. Army Engineer Waterways Experiment Station, Vicksburg, Miss. Technical Report EL-81-13.

Klopatek, J. M. 1978. Nutrient dynamics of freshwater riverine marshes and the role of emergent macrophytes. Pages 195-216 *in* R. E. Good, D. F. Whigham, and R. L. Simpson, eds. Freshwater Wetlands. Academic Press, New York.

Kuenzler, E. J., P. J. Mulholland, L. A. Ruley, and R. P. Sniffen. 1977. Water quality in North Carolina coastal plain streams and effects of channelization. Water Resources Research Institute of the University of North Carolina, Raleigh. Report No. 127.

______________, P. J. Mulholland, L. A. Yarbro, and L. A. Smock. 1980. Distributions and budgets of carbon, phosphorus, iron, and manganese in a floodplain swamp ecosystem. Water Resources Research Institute of the University of North Carolina, Raleigh. Report No. 157.

Kuhner, J. 1980. Agricultural land use water quality interaction: problem abatement, project monitoring, and monitoring strategies. U.S. Environmental Protection Agency, Rural Nonpoint Section, Washington, D.C.

Langdon, O. G., J. P. McClure, D. D. Hook, J. M. Crockett, and R. Hunt. 1981. Extent, condition, management, and research needs of bottomland hardwood-cypress forests in the southeastern United States. Pages 71-85 *in* J. R. Clark and J. Benforado, eds. Wetlands of bottomland hardwood forests. Elsevier Science Publishing Co. Amsterdam.

Larson, J. S., M. S. Bedinger, C. F. Bryan, S. Brown, R. T. Huffman, E. L. Miller, D. G. Rhodes, and B. A. Touchet. 1981. Transition from wetlands to uplands in southeastern bottomland hardwood forests. Pages 225-273 *in* J. R. Clark and J. Benforado, eds. Wetlands of bottomland hardwood forests, Elsevier Science Publishing Co., Amsterdam.

Leitman, H. M. 1978. Correlation of Apalachicola river floodplain tree communities with water levels, elevation and soils. M. S. Thesis, Florida State University, Tallahassee. 56 pp.

Leopold, L. B., M. G. Wolman, and J. P. Miller. 1964. Fluvial processes in geomorphology. W. H. Freeman and Co., San Francisco.

Linsley, R. K., and J. B. Franzini. 1979. Water-resources engineering. McGraw-Hill Book Co., New York. 682 pp.

Littlejohn, C. B. 1977. An analysis of the role of natural wetlands in regional water management. Pages 451-476 *in* C.A.S. Hall and J. W. Day, Jr., eds. Ecosystem modeling in theory and practice, John Wiley and Sons, New York.

Livingston, R. J. 1978. Short and long term effects of forestry operations on water quality and the biota of the Apalachicola estuary. Florida Sea Grant Program, University of Florida, Gainesville. Tech. Paper No. 5.

_______________, R. L. Iverson, and D. C. White. 1976. Energy relationships and the productivity of Apalachicola Bay. Final Report. Florida Sea Grant College.

MacDonald, P. O., W. E. Frayer, and J. K. Clauser. 1979. Documentation, chronology, and future predictions of bottomland hardwood habitat losses in the Lower Mississippi Alluvial Plain. Vols. 1 and 2. U.S. Fish and Wildlife Service. Jackson, Miss. 133 pp. and 295 pp.

Maki, T. E., A. J. Weber, D. W. Hazel, S. C. Hunter, B. T. Hyberg, D. M. Flinchum, J. P. Lollis, J. B. Kognstad, and J. D. Gregory. 1980. Effects of stream channelization on bottomland and swamp forest ecosystems. Report No. 147. Water Resources Research Institute, University of North Carolina, Raleigh.

Merritt, R. W., and D. L. Lawson. 1978. Leaf litter processing in floodplain and stream communities. Pages 93-105 *in* R. R. Johnson and J. F. McCormick, eds. Strategies for protection and management of floodplain wetlands and other riparian ecosystems. U.S. Department of Agriculture, Forest Service, Washington, D.C. GTR-WO-12.

Minshall, G. W. 1968. Community dynamics in a woodland spring brook. Hydrobiologia 32:305-339.

Mitsch, W. J. 1979. Interactions between a riparian swamp and a river in southern Illinois. Pages 63-72 *in* R. R. Johnson and F. McCormick, tech. coord. Strategies for protection and management of floodplain wetlands and other riparian ecosystems. U.S. Department of Agriculture, Forest Service, Washington, DC. GTR-WO-12.

__________, C. L. Dorge, and J. R. Wiemhoff. 1977. Forested wetlands for water resource management in southern Illinois. Illinois Water Resources Center Research Report 132. Urbana. 275 pp.

__________, C. L. Dorge, and J. R. Wiemhoff. 1979. Ecosystem dynamics and a phosphorus budget of an alluvial cypress swamp in southern Illinois. Ecology 60:1116-1124.

__________, and J. G. Gosselink. 1986. Wetlands. Van Nostrand Reinhold Co., New York 539 pp.

Mohanty, S. K., and R. N. Dash. 1982. The chemistry of waterlogged soils. Pages 389-396 *in* B. Gopal, R. E. Turner, R. G. Wetzel, and D. F. Whigham, eds. Wetlands: ecology and management. National Institute of Ecology and International Scientific Publications, Jaipur, India.

Mortimer, C. H. 1941. The exchange of dissolved substances between mud and water in lakes. J. Ecol. 29:280-329.

_____________. 1942. The exchange of dissolved substances between mud and water in lakes. J. Ecol. 30:147-201.

Mulholland, P. J. 1981. Organic carbon flow in a swamp-stream ecosystem. Ecological Monographs 51:307-322.

______________, and E. J. Kuenzler. 1979. Organic carbon export from upland and forested wetland watersheds. Limnol. Oceanogr. 24:960-966.

Mundorff, M. J. 1950. Floodplain deposits of North Carolina piedmont and mountain streams as a possible source of groundwater supply. North Carolina Division of Mineral Resources. North Carolina Department of Conservation and Development, Raleigh. Bulletin 59.

Nessel, J. 1978. Distribution and dynamics of organic matter and phosphorus in a sewage enriched cypress swamp. Thesis. University of Florida, Gainesville, Florida. 159 pp.

Nixon, S. W., and C. A. Oviatt. 1973. Ecology of a New England salt marsh. Ecol. Monogr. 43:463-498.

Novitski, R. P. 1979. Hydrologic characteristics of Wisconsin's wetlands and their influence on floods, streamflow, and sediment. Pages 377-388 *in* P. E. Greeson, J. R. Clark, and J. E. Clark, eds. Wetland functions and values: the state of our understanding, American Water Resources Association, Minneapolis, Minnesota.

Odum, E. P. 1978. Ecological importance of the riparian zone. Pages 2-4 *in* R. R. Johnson and J. F. McCormick, eds. Strategies for protection and management of floodplain wetlands and other riparian ecosystems. U.S. Department of Agriculture, Forest Service, Washington, D.C. GTR-WO-12.

Odum, H. T. 1983. Systems ecology. John Wiley, New York. 644 pp.

_________, and K. C. Ewel. eds. 1978. Cypress wetlands for water management, recycling and conservation. 4th Annual Report. Center for Wetlands, University of Florida, Gainesville, Florida. 945 pp.

Odum, W. E. 1970. Pathways of energy flow in South Florida estuary. Ph.D. Dissertation, University of Miami, Miami, Florida. 161 pp.

Office of Technology Assessment. 1984. Wetlands: their use and regulation. Office of Technology Assessment, U.S. Congress, Washington, D.C. OTA-O-206.

Patrick, W. H., Jr., R. D. DeLaune, R. M. Engler, and S. Gotoh. 1976. Nitrate removal from water at the water-mud interface in wetlands. Environmental Research Laboratory, U.S. Environmental Protection Agency, Corvallis, Oregon. EPA 600/3-76-042.

____________, G. Dissmeyer, D. D. Hook, V. W. Lambou, H. M. Leitman, and C. H. Wharton. 1981. Characteristics of wetland ecosystems of southeastern bottomland hardwood forests. Pages 276-300 *in* J. R. Clark and J. Benforado, eds. Wetlands of bottomland hardwood forests. Elsevier Science Publishing Co., Amsterdam.

Putnam, J. A., G. M. Furnival, and J. S. McKnight. 1960. Management and inventory of southern hardwoods. Agricultural Handbook 181. U.S. Department of Agriculture.

Reese, C. D., I. W. Dodson, V. Ulrich, D. L. Becker, and C. J. Kempter. 1972. Pesticides in the aquatic environment. Office of Water Programs, U.S. Environmental Protection Agency, Washington, D.C.

Schlesinger, W. H. 1978. Community structure, dynamics and nutrient cycling in the Okefenokee cypress swamp-forest. Ecological Monographs 48:43-65.

Schlosser, I. J., and J. R. Karr. 1981. Riparian vegetation and channel morphology impacts on spatial patterns of water quality in agricultural watersheds. Environ. Management 5:233-243.

Shifley, S. R., and K. M. Brown. 1978. Elm-ash-cottonwood forest type bibliography. U.S. Department of Agriculture, Forest Service , Washington, D.C. GTR-NC-42.

Sigafoos, R. S. 1964. Botanical evidence of floods and floodplain deposition. U.S. Geological Survey, Washington, D. C. Professional Paper 485-A.

Skaggs, R. W., J. W. Gilliam, T. J. Sheets, and J. S. Barnes. 1980. Effect of agricultural land development on drainage waters in the North Carolina tidewater region. Water Resources Research Institute, University of North Carolina, Raleigh. Report No. 159.

Sloey, W. E., F. L. Spangler, and C. W. Felter, Jr. 1978. Management of freshwater wetlands for nutrient assimilation. Pages 321-340 *in* R. E. Good, D. F. Whigham, and R. L. Simpson, eds. Freshwater wetlands. Academic Press, New York City, New York.

Sommers, L. E., D. W. Nelson, and D. B. Kaminsky. 1975. Nutrient contributors to the Maumee River. Pages 105-119 *in* Non-point source pollution seminar. U.S. Environmental Protection Agency, Chicago, Illinois. EPA 905/9-75-1007.

Spencer, D. A. 1971. Trends in pesticide use. Env. Science and Tech. 4:478-481.

Stumm, W., and J. J. Morgan. 1970. Aquatic chemistry. Wiley Interscience, New York City, New York.

Tchobanoglous, G., and G. L. Culp. 1980. Wetland systems for wastewater treatment. *In* S. C. Reed and R. K. Bastian, eds. Aquaculture systems for wastewater treatment: an engineering assessment. U.S. Environmental Protection Agency, Washington, D.C. EPA 430/9-80-007.

Teskey, P. O., and T. M. Hinckley. 1977. Impact of water level changes on woody riparian and wetland communities. Vol. I: Plant and soil responses to flooding. U.S. Fish and Wildlife Service, Columbia, Missouri. FWS/OBS-77/58.

Tiner, R. W., Jr. 1984. Wetlands of the United States and recent trends. U.S. Fish and Wildlife Service, National Wetlands Inventory, Washington, D.C. 59 pp.

Turner, R. E., S. W. Forsythe, and N. J. Craig. 1981. Bottomland hardwood forest land resources of the southeastern United States. Pages 13-28 *in* J. R. Clark and J. Benforado, eds. Wetlands of bottomland hardwood forests. Elsevier Science Publishing Co., Amsterdam.

U. S. Army Corps of Engineers. 1971. Charles River, Massachusetts: Main report and attachments. New England Division, Waltham, Massachusetts 76 pp.

U. S. Forest Service. 1973. The outlook for timber in the United States. Forest Resources Report 20. U.S. Department of Agriculture, Washington, D.C. 367 pp.

Vitousek, P. M. 1977. The regulation of element concentrations in mountain streams in the northeastern United States. Ecol. Monogr. 47:65-87.

______________, and W. A. Reiners. 1975. Ecosystem succession and nutrient retention: a hypothesis. Bioscience 25:376-381.

Wharton, C. H. 1970. The southern river swamp - a multiple-use environment. Georgia State University, Atlanta. 48 pp.

______________. 1978. The natural environments of Georgia. Georgia Department of Natural Resources, Atlanta. 227 pp.

______________. 1980. Values and functions of bottomland hardwoods. Pages 341-353 *in* Transactions of the 45th North American Wildlife and Natural Resources Conference. Wildlife Management Institute, Washington, D.C.

______________, and H. P. Hopkins. 1980. In-situ evaluation of the filtering function of a piedmont creek swamp. Environmental Resources Center, Georgia Institute of Technology, Atlanta. Report ERC-02-80.

______________, W. M. Kitchens, E. C. Pendleton, and T. W. Sipe. 1982. The ecology of bottomland hardwood swamps of the Southeast: a community profile. U.S. Fish and Wildlife Service, Washington, D.C. FWS/OBS-81/37.

______________, V. W. Lambou, J. Newsom, P. V. Winger, L. L. Gaddy, and R. Mancke. 1981. The fauna of bottomland hardwoods in southeastern United States. Pages 87-160 *in* J. R. Clark and J. Benforado, eds. Wetlands of bottomland hardwood forests. Elsevier Science Publishing Co., Amsterdam.

Winner, M. D., Jr., and C. E. Simmons. 1977. Hydrology of the Creeping Swamp watershed, North Carolina, with reference to potential effects of stream channelization. U.S. Geological Survey Water Resources Investigations 77-26.

Yarbro, L. A. 1979. Phosphorus cycling in the Creeping Swamp floodplain ecosystem and exports from the Creeping Swamp watershed. Ph.D. dissertation, University of North Carolina, Chapel Hill. 231 pp.

Zwank, P. J., R. D. Sparrowe, W. R. Porath, and O. Tergerson. 1979. Utilization of threatened bottomland habitats by white-tailed deer. Wildl. Soc. Bull. 7:226-232.

3. GROWTH AND YIELD, NUTRIENT CONTENT, AND ENERGETICS OF SOUTHERN BOTTOMLAND HARDWOOD FORESTS

Dennis Mengel
Russell Lea
Hardwood Research Cooperative, North Carolina State University,
Raleigh, NC 27695-8008

ABSTRACT

Coastal Plain bottomland hardwood forests comprise approximately 4,396,000 ha in the southeastern United States. The extent of the resource and continued expansion of the hardwood forest products market necessitate methods to assess the productivity of these forest types. The North Carolina State University/Industry-sponsored Hardwood Research Cooperative is coordinating efforts to provide information on bottomland hardwood forest productivity. Research is directed towards three broad site-types: Bottomland forests, Wet flat forests, and Swamp forests. Data from permanent-growth plots has been collected since 1969, resulting in the recent release of a software package to project diameter distribution in five-year increments. Biomass, nutrient content, and energetics data have been reported in numerous publications. Research exploring the harvesting impacts on productivity of bottomland hardwood forests continues to expand as forest managers require more detailed information on ecosystem fluxes in response to their activities.

Ecological Processes and Cumulative Impacts: Illustrated by Bottomland Hardwood Wetland Ecosystems. Edited by James G. Gosselink, Lyndon C. Lee, and Thomas A. Muir. Chelsea, MI 48118. Printed in USA.

INTRODUCTION

Southern bottomland hardwood forests are located on the Atlantic and Gulf Coastal Plains and the Mississippi River floodplain. According to U.S. Forest Inventory and Analysis survey reports, hardwood forest types account for more than 50% of the 20,216,500 ha of forested land in the Southeastern Coastal Plain (Tansey and Kellison 1985). Of that 50%, approximately 42%, or 4,395,600 ha, are in bottomland hardwood types. (Bottomland forest communities are broadly classified by the U.S. Forest Service as either an oak-gum-cypress [*Quercus-Nyssa-Taxodium*] or an elm-ash-cottonwood [*Ulmus-Fraxinus-Populus*] forest type.) The survey reports also show that water tupelo (*Nyssa aquatica* L.), blackgum (*Nyssa sylvatica* var. *biflora* L.), and sweetgum (*Liquidambar styraciflua* L.) make up 39% of the total, merchantable hardwood volume in the Coastal Plain. Oaks (*Quercus* spp.), some of which are found on bottomlands, account for another 29%. Management of bottomland hardwood forests for fiber and solid wood products continues to expand, as does the demand for accurate quantification of the productivity, nutrient content, and energetics of these complex forest systems.

Based on applied and ecological needs to quantify and better understand the growth and yield, nutrient content, and energetics of bottomland hardwood forests, the North Carolina State University/Industry-sponsored Hardwood Research Cooperative decided to compile related information. The Hardwood Cooperative intends to continue to strengthen its research efforts in bottomland hardwood ecosystems because a majority of its cooperators own and manage lands subject to high water tables. Availability of data to support current and future management plans is a major concern. The data required to construct specific models that would serve to identify the effects of forest management on wetland forests have been collected and analyzed for tidally influenced wetlands in Alabama (Aust et al. 1989, Mader et al. 1989) and are currently being collected on South Carolina river bottoms (Perison 1989). These efforts reflect abiotic, forest, and wildlife responses to management. The framework guiding our research and development is provided in Figure 1.

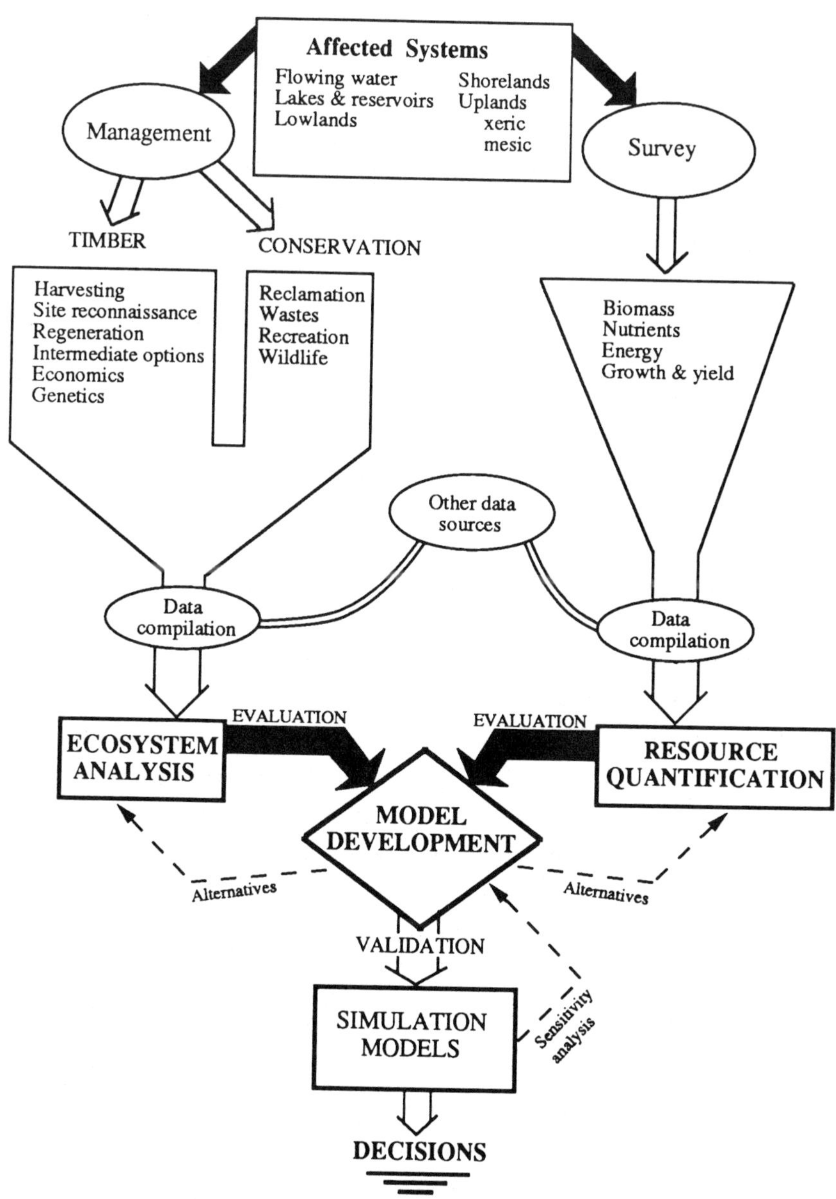

Figure 1. Conceptual framework for data assembly and model development for the North Carolina State Hardwood Research Cooperative.

In our data framework, the hardwood ecotypes are analyzed for management options or surveyed for certain attributes (Figure 1). Management topics can be analyzed separately or in combination to provide an understanding of ecosystem function. Likewise, certain attributes of a hardwood site can be inventoried and combined to quantify the resource. Once the resource is quantified and certain ecosystem functions are understood, scientists can provide preliminary models or, more specifically, management models for the hardwood types. The ultimate use of the models is to provide managers of land and water resources a means to make decisions. This paper presents overviews of the types of information that currently are available from the Hardwood Research Cooperative for southern bottomland hardwood forests.

STUDY SITE LOCATIONS

The Hardwood Research Cooperative recognizes three broad bottomland hardwood forest site types--Bottomland forests (red and black river bottoms), Wet Flat forests, and Swamp forests--because they support the largest proportion of the commercial hardwood volume in the southern Atlantic and Gulf Coastal Plains. Red and black river bottom forests occupy floodplains of major drainage systems originating in the mountain or Piedmont and Coastal Plain physiographic provinces, respectively. Drainage of red river bottoms is moderately rapid, and the soils are sandy loam to silt loam. Black river bottoms drain slowly and are characterized by sandy to clay loam soils with surface organic matter accumulation. Wet Flat forests generally occupy broad interstream areas and are strongly influenced by water table versus overbank flooding (Figure 2). Drainage of Wet Flats is intermediate between river bottoms and swamps; the soils are nonalluvial in origin, typically have restricted infiltration, and contain some organic matter. Swamp forests occupy areas such as old oxbows; sloughs between stream bottoms; and large, poorly-drained inter-river areas. These areas are characterized by an extended hydroperiod for much of the growing season. Soils range from silt loam to clay and have large accumulations of raw organic matter. These broad site classifications were used for all projects on bottomland hardwood forests recognized by our cooperative supporters.

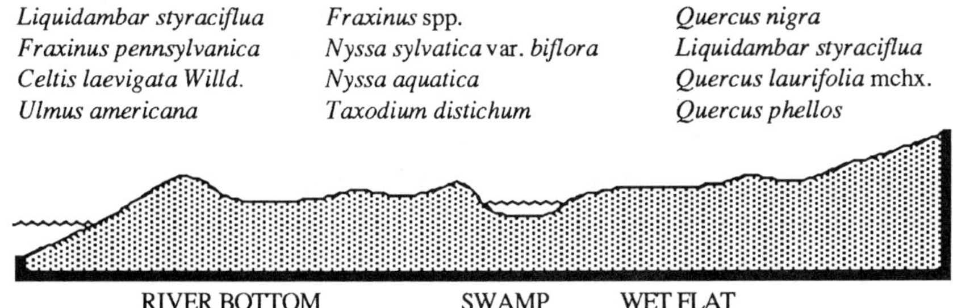

Figure 2. Schematic diagram of the physiographic location of Atlantic and Gulf Coastal Plain River Bottom, Wet Flat, and Swamp forests.

GROWTH AND YIELD PROJECT

To fulfill the need of forest managers for growth and yield information, a natural stand growth and yield study was initiated in 1969. Six hundred and forty-one, 0.08-ha plots were established throughout the South (Delaware to Texas). Yield equations were developed and published (Smith et al. 1975). A refined version of that report (Gardner et al. 1982) emphasized alternative merchandising options, including biomass and utilization of residual stand material. A later report (Roeder and Gardner 1984) using re-measurement data provided growth estimates compatible with previous yield publications. Roeder and Gardner (1984) presented our first attempt to actually project stand growth of the bottomland hardwood types. Given the co-dominant stand height and initial merchantable stand basal area, tables were presented to project the expected five-year percentage change in merchantable basal area and height. Percentage changes in total gross volume, pole volume, pulpwood volume, sawtimber volume, board-foot volume, and green and dry-weight biomass could then be determined. Published yield tables (Gardner et al. 1982) could then be used to determine current and projected yields. Our most recent effort in growth and yield modeling has resulted in the development of a software package to project five-year diameter growth of bottomland hardwood stands. The Bottomland Yield Projection System (BYPS) is an interactive, user-friendly software package written in Turbo Pascal 4.0 and is compatible with any DOS-based computer system. Input data can be entered interactively or input from data files. The model was developed from data collected on 43 permanent-growth plots representing 95 re-measurement periods. A diameter-class matrix model is used to project the present diameter distribution five years into the future. Equation 1 shows the three regression equations that simultaneously predict the probability of a tree staying in the same 2.54-cm diameter class (P_{0t}), moving up one diameter class (P_{1t}), or moving up two or more diameter classes (P_{2t}). The probability of mortality (Pmt) is calculated as the difference between 1 and the sum of the three diameter-class movement probabilities, because all four probabilities must equal 1 for each diameter class. Predictions are based on diameter class, diameter class squared, stand basal area, and trees per hectare.

$$
\begin{aligned}
P_{i0t} &= b_{00} + b_{10}D_i + b_{20}D_i^2 + b_{30}BA + b_{40}QMD \qquad (1)\\
P_{i1t} &= b_{01} + b_{11}D_i + b_{21}D_i^2 + b_{31}BA + b_{41}QMD\\
P_{i2t} &= b_{02} + b_{12}D_i + b_{22}D_i^2 + b_{32}BA + b_{42}QMD\\
P_{imt} &= 1 - (P_{i0t} + P_{i1t} + P_{i2t})
\end{aligned}
$$

where

P_{ikt} = Proportion of trees within diameter class that stay in the same class, move up one or two classes, or die at time t

D_i = Midpoint of diameter class (cm)

D_i^2 = Square of D_i (cm^2)

BA = Stand basal area (m^2 / ha)

QMD = Quadratic mean diameter

b_{jk} = Regression coefficient, j, for movement of k diameter classes

k $\in$ {0,1,2,m}

Submerchantable ingrowth, added to the lower diameter classes, is a constant based on the mean number of ingrowth trees per plot in the data. Merchantable ingrowth, used when the observed distribution does not contain submerchantable trees and added to the 15.24-cm diameter class, is represented by the following equation (Davidson 1988):

$$I = b_1 e^{b_2 QMD} \tag{2}$$

where

I = Merchantable ingrowth (dbh > 14.2 cm)

QMD = Quadratic mean diameter

b_1, b_2 = Regression coefficients to be estimated

Regression coefficients for equations 1 and 2 are found in Tables 1 and 2, respectively.

The scarcity of data on certain species required grouping species into broad species groups for projection purposes (Table 3). Groups were chosen based on growth and form similarities. Volumes for the species groups are calculated based on the work by Clark et al. (1985).

BIOMASS, NUTRIENT CONTENT, AND ENERGETICS PROJECT

Much of the bottomland hardwood resource in the South is low quality. Small-diameter trees reflect years of past abuse such as livestock grazing, uncontrolled wildfires, and high-grade logging. Approximately 90% of the South's hardwood forests require some form of cultural treatment in order to realize full commercial productivity. This degraded stand condition makes much of the southern hardwood resource particularly attractive for biomass or energy wood production. Whole-tree harvesting is now occurring in stands previously considered to be unmerchantable due to tree size and form. In addition to increasing biomass yield, whole-tree clearcut harvesting will remove large quantities of nutrient elements in the nutrient-rich crown material. Biomass yields from individual trees can be increased 10 to 40% by utilizing crowns, whereas nutrient removal may be increased by greater than 200%, depending upon element, species, and site factors. Concerns have arisen that whole-tree harvesting may lead to decreased forest productivity in successive rotations.

The heterogeneous nature of bottomland hardwood stands necessitates sampling for biomass, energy and nutrient contents on an area basis with plots of sufficient size to adequately represent the species composition of the entire stand. All components comprising the forest site; such as soil, litter layer, understory material, saplings, and overstory;

Table 1. Regression coefficients estimated for the merchantable ingrowth function.

Variable	Metric Units	English Units
Intercept	185.01292	74.87473
QMD	-75.89442	-7.05106

Table 2. Estimated coefficients of the diameter distribution model by species group (English unit coefficients in parentheses).

VARIABLE	GUMS	HARD HARDWOODS	SOFT HARDWOODS	OTHER HARDWOODS
	Probability of staying in the same class			
Intercept	0.67448	0.72671	0.73774	0.77785
	(0.67448)	(0.72671)	(0.73774)	(0.77785)
D	0.00893	-0.00894	0.00069	-0.00028
	(0.02268)	(-0.02270)	(0.00175)	(-0.00071)
D2	-0.00022	–	-0.00011	–
	(-0.00144)	–	(-0.00073)	–
BA	–	0.00181	–	–
	–	(0.00042)	–	–
QMD	0.15066	–	2.00062	–
	(0.01400)	–	(0.18587)	–
	Probability of moving up 1 class			
Intercept	0.00881	0.04201	0.02794	0.00947
	(0.00881)	(0.04201)	(0.02794)	(0.00947)
D	0.00774	0.01077	0.00647	0.00389
	(0.01966)	(0.02736)	(0.01643)	(0.00989)
D2	-0.00005	–	0.00001	–
	(-0.00032)	–	(0.00006)	–
BA	–	-0.00113	–	–
	–	(-0.00026)	–	–
QMD	-1.37901	–	-1.49872	–
	(-0.12812)	–	(-0.13924)	–
	Probability of moving up 2 classes			
Intercept	0.01239	0.08519	0.04952	0.02330
	(0.01239)	(0.08519)	(0.04952)	(0.02330)
D	-0.00121	0.00164	-0.00032	-0.00067
	(-0.00306)	(0.00415)	(-0.00082)	(-0.00170)
D2	0.00003	–	0.00004	–
	(0.00022)	–	(0.00028)	–
BA	–	-0.00117	-	–
	–	(-0.00027)	–	–
QMD	1.12492	–	-0.69278	–
	(0.10451)	–	(-0.06436)	–

Table 3. Species groups used in the model.

Group	Species
Hard Hardwoods	*Quercus* spp. (willow oak, white oaks, red oaks), *Fraxinus* spp. (ash), *Carya* spp. (hickory)
Soft Hardwoods	*Liriodendron tulipifera* L. (Yellow poplar), *Populus* spp. (Cottonwood), *Platanus occidenalis* L. (Sycamore), *Ulmus* spp. (Elm), *Celtis mississippiensis* L. (Hackberry), *Betula* spp.(birch), *Taxodium distichum* L. Richard (bald cypress), *Tilia americana* L. (American basswood), *Acer rubrum* L. (red maple), *Magnolia virginiana* L. (sweetbay)
Gums	*Liquidambar styraciflua* L. (Sweetgum), *Nyssa sylvatica* var. *biflora* L. (blackgum), *Nyssa aquatica* L. (Tupelo gum)
Other Hardwoods	*Carpinus caroliniana* Walter (Ironwood), *Ilex* spp. (Holly), *Morus* spp. (Mulberry), any hardwoods not in above groups.

should be examined, as all contribute a functional role in nutrient cycling. A wide, regional distribution of plots covering major site types and a range of age classes is also desirable to ensure adequate representation of the resource. A comprehensive biomass, nutrient content, and energetics data base for southern bottomland hardwood forests; including all major and minor aboveground biomass, forest floor, and soil; is summarized in Gower (1983), Messina (1983), and Messina et al. (1983). Table 4 and Figure 3 provide the age class and site type distribution of the study plots. An example of the data generated by this extensive survey is presented in Table 5 for river bottom sites.

The data accumulated on southern Coastal Plain hardwood forests indicate that they are very productive, diverse communities. Biomass production and, hence, energy and nutrient accumulation exceed most other naturally regenerated, temperate forest types. The high productivity found on bottomland hardwood forests can be attributed to the hydrology and climate, which complement each other to provide relatively rich growing conditions. Frequent flooding and the associated sedimentation, coupled with the physiographic position of these forests between the terrestrial and aquatic ecosystems, make them receptors and transformers for nutrient subsidies from both sides.

Table 4. Site and age class matrix of study sites used for the biomass, nutrient, and energy content project, where the number preceding each location refers to position in Figure 3.

AGE	BOTTOMLAND	WET FLAT	SWAMP
10	1 Sumter AL	10 Duval FL	18 Columbus NC
	2 Bertie NC	11 St Johns FL	19 Florence SC
20	3 Warren MS	12 Taylor FL	20 Nassau FL
	4 Sumter AL	13 Dorchester SC	21 Glynn GA
40	5 Dallas AL	14 Taylor FL	22 Hertford NC
	6 Southampton VA		
	7 Marion SC	15 Washington AL	23 Craven NC
60	8 Wayne MS	16 Craven NC	24 Taylor FL
	9 Escambia FL	17 Jasper SC	25 George MS

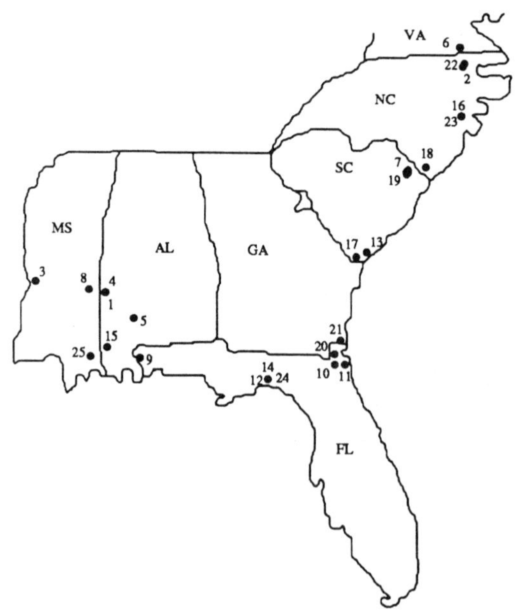

Figure 3. Location of study used for the biomass, nutrient, and energy content project.

Table 5. Dry weight, energy and nutrient contents of site pools in the river bottomland hardwood type.

	Pulpwood-Sawtimber							
Age	Foliage	Branches	Bole[1]	Saplings	Understory-Ground Veg	Forest Floor	Above-Soil Total[2]	Soil[3]
Dry Weight (Mg/ha)								
10	-	-	-	42.6	3.2	10.6	56.4	-
20	1.8	27.4	84.9	27.1	0.9	2.1	144.2	-
40	5.2	56.0	144.2	9.0	2.9	5.0	222.3	
60	6.2	50.4	164.0	15.9	1.5	7.5	245.5	-
Energy Content (KcalX10^6/ha)								
10	14.0	-	-	183.3	14.6	38.2	250.1	-
20	10.5	133.2	390.5	136.0	4.0	21.4	697.4	-
40	21.3	270.8	673.0	54.3	13.4	18.2	1051.0	-
60	25.1	179.0	774.4	86.2	3.0	16.9	1086.2	-
				(kg/ha)				
Nitrogen								
10	-	-	-	141	17	115	273	-
20	25	54	79	42	11	18	229	-
40	101	158	160	20	19	56	514	-
60	93	162	186	36	10	59	546	-
Phosphorus								
10	-	-	-	23	3	12	38	15
20	6	8	12	6	1	2	35	84
40	9	26	29	3	3	5	75	20
60	7	31	22	4	2	6	72	21
Potassium								
10	-	-	-	110	9	53	172	63
20	15	39	96	35	4	5	194	313
40	44	115	168	13	19	22	381	119
60	45	116	232	27	6	17	443	170
Calcium								
10	-	-	-	140	23	23	186	1989
20	19	97	170	103	7	16	412	2895
40	78	317	261	38	6	37	737	2230
60	66	321	734	71	14	91	1297	1636
Magnesium								
10	-	-	-	21	3	21	49	288
20	8	14	33	13	1	5	74	332
40	16	47	35	5	5	8	116	339
60	22	61	28	14	2	13	140	231

[1] To 10-cm diameter outside bark. Values include stemwood and bark.

[2] Includes standing dead material.

[3] Available P and exchangeable K, Ca, and Mg in top 30 cm.

CONCLUSIONS

The inherent productivity of the Coastal Plain bottomland hardwood sites enables them to sustain intensive forest production. However, stand nutrient distribution and accumulation patterns indicate the potential for removing large quantities of nutrients when complete-tree utilization is coupled with short rotations. Managers are considering the fragility of these sites and their inherent nutrient status in light of the data currently available; however, they are also requesting the study of system fluxes, such as nutrient inputs, outputs, and internal cycling. This additional work needs to be done as a logical follow-up to the present studies and should provide a rather complete analysis of bottomland hardwood productivity patterns and management implications in relation to ecosystem function. The data accumulated by the Hardwood Research Cooperative represent some of the most detailed growth and yield, biomass, nutrient, and energy information available on Bottomland, Wet Flat, and Swamp forest site types. They are the result of detailed measurements of standing and cut trees over a wide geographic area. There is considerable variation in the measured parameters among site types and age classes within site types. However, we feel these data contain realistic estimates for fully stocked stands for the site types and age classes sampled.

REFERENCES CITED

Aust, W. M., S. F. Mader, and R. Lea. 1989. Abiotic changes of a tupelo-cypress swamp subjected to helicopter and rubber-tired skidder timber harvest methods. Pages 545-551 *in* Proceedings, Fifth Biennial Silvicultural Research Conference. U.S. Department of Agriculture, Forest Service, Southern Forest Experiment Station, New Orleans, Louisiana.

Clark, A., D. R. Phillips, and D. J. Frederick. 1985. Weight, volume, and physical properties of major hardwood species in the Gulf and Atlantic Coastal Plains. U.S. Department of Agriculture, Forest Service Research Paper SE-250.

Davidson, C. B. 1988. Prediction of total stand ingrowth in southeastern mixed species bottomland hardwood forests. M.S. Thesis. North Carolina State University, Raleigh, North Carolina.

Gardner, W. E., P. Marsh, R. C. Kellison, and D. J. Frederick. 1982. Yields of natural hardwood stands in the southeastern United States. North Carolina State University, Hardwood Research Cooperative Series No. 1, Raleigh, North Carolina.

Gower, S. T. 1983. Distribution of energy in different-aged southeastern bottomland forests. M.S. Thesis. North Carolina State University, Raleigh, North Carolina.

Mader, S. F., W. M. Aust, and R. Lea. 1989. Changes in net primary productivity and cellulose decomposition rates in a water tupelo-bald cypress swamp following timber harvest. Pages 539-543 *in* Proceedings, Fifth Biennial Silvicultural Research Conference. U.S. Department of Agriculture, Forest Service. Southern Forest Experiment Station, New Orleans, Louisiana.

Messina, M. G. 1983. Nutrient content and distribution in natural southern Coastal Plain hardwoods. M.S. Thesis. North Carolina State University, Raleigh, North Carolina.

__________, S. T. Gower, D. J. Frederick, A. Clark III, and D. R. Phillips. 1983. Biomass, nutrient, and energy content of southeastern hardwood forests. North Carolina State University, Raleigh, North Carolina.

Perison, D. 1989. Abiotic response of a black river bottomland to harvest. Pages 97-99 *in* Twenty-sixth Annual Report. North Carolina State University Hardwood Research Cooperative, Raleigh, North Carolina.

Roeder, K. R., and W. E. Gardner. 1984. Growth estimation of mixed southern hardwood stands. North Carolina State University Hardwood Research Cooperative Series No. 3, Raleigh, North Carolina.

Smith, H. D., W. L. Hafley, D. J. Holley, and R. C. Kellison. 1975. Yields of mixed hardwood stands occurring naturally on a variety of sites in the southeastern United States. North Carolina State University, School of Forest Resources, Technical Report No. 55, Raleigh, North Carolina.

Tansey, J. B., and R. C. Kellison. 1985. The supply and availability of hardwoods in the southeastern coastal plain. North Carolina State University Hardwood Research Cooperative Series No. 6, Raleigh, North Carolina.

4. POTENTIAL NITRATE LEACHING LOSSES AND NITROGEN MINERALIZATION IN AN ATLANTIC COASTAL PLAIN WATERSHED FOLLOWING DISTURBANCE: PRELIMINARY RESULTS

David A. Kovacic
Department of Landscape Architecture, University of Illinois,
Urbana, IL 61801

Thomas G. Ciravolo
Kenneth W. McLeod
Jack S. Erwin
University of Georgia, Savannah River Ecology Laboratory, Drawer E,
Aiken, SC 29801

ABSTRACT

Preliminary results of a long-term study are presented that examine the potential for soil nitrate and ammonium losses across four upper coastal plain vegetative habitat types (longleaf pine plantation, upland deciduous forest, bottomland deciduous forest, and cypress-red maple swamp) near Aiken, South Carolina. Ten trenched plots (uniform distance) paired with ten untrenched (control) plots were established on each habitat type. Lysimeter analysis over the first 32 weeks of the study revealed different site responses to disturbance. All trenched sites, except those in the swamp, exhibited significantly greater nitrate losses than the control sites. Results of laboratory soil incubations generally paralleled field lysimeter results. Sites with the highest lysimeter nitrate exhibited the highest amounts of ammonium. Inorganic nitrogen responses to disturbance reflected different microorganism-soil-plant functional relationships within the catena studied. Swamp soils exhibited greater nutrient retention than did soils of the other sites. Lack of nitrate in

---

Ecological Processes and Cumulative Impacts: Illustrated by Bottomland Hardwood Wetland Ecosystems. Edited by James G. Gosselink, Lyndon C. Lee, and Thomas A. Muir. Chelsea, MI 48118. Printed in USA.

swamp soil lysimeter solutions indicated a high capacity to absorb inorganic N. However, aerobic incubations indicated this capacity was closely linked to anaerobic conditions. Human alteration of anaerobic conditions could reduce the resistance of the swamp to nutrient losses.

INTRODUCTION

The southeastern Atlantic coastal plain is extensively managed for production of timber and other cash crops. In an attempt to efficiently utilize land area, bottomland sites are continually being drained and converted to crop production. Wetland habitats are vital, functional components of coastal plain watersheds (Brown et al. 1978, Odum 1978). The destruction of these habitats, which are important sinks and filters for nutrients and biocides of anthropogenic origin (Lowrance et al. 1948a, b; Gambrell et al. 1975; Boto and Patrick 1978; Peterjohn and Correll 1984), is highly detrimental to downstream water quality (Kuenzler et al. 1977, Duda 1982).

Nutrient cycling processes have been studied intensively in southeastern Piedmont forested ecosystems (Gaskin et al. 1983); however, southern Atlantic coastal plain forested watersheds have been largely ignored. It is important that we understand the functions of watershed ecosystems and their responses to disturbance to develop sensible land and water management decisions. The U.S. Environmental Protection Agency is currently faced with the task of determining what constitutes a bottomland hardwood forest. Such a decision must be based on a sound understanding of wetland ecosystem functions (e.g., nutrient removal/transformation, production export, sediment stabilization, and toxicant retention). Information concerning functional characteristics of Atlantic coastal plain ecosystems and the response of these ecosystems to disturbance will facilitate critical management decisions.

The purpose of this ongoing study is to determine the potential for solution losses of nitrate following disturbance along a moisture gradient ranging from upland pine forest to cypress-tupelo swamp ecosystems. Changes in mobile nutrient losses are a useful measure of ecosystem response to disturbance (Vitousek et al. 1981). Such changes can be a sensitive measure of the functional ecosystem state (O'Neill et al. 1977,

Vitousek et al. 1981). The nutrient loss response following a disturbance is a systems-level process that directly affects ecosystem recovery and downstream ecosystem functions (Likens and Bormann 1974, Waide and Swank 1977). Such responses can effectively indicate ecosystem resistance to perturbation. The term resistance is defined as the ability to prevent displacement from a bounded steady state following perturbation (Webster et al. 1975).

In this study, we investigate the resistance of adjacent upland and wetland ecosystems to losses of inorganic N. We have focused on inorganic N because: 1) nitrogen is often limiting to ecosystems, 2) nitrogen is often the most readily lost nutrient following disturbance (Vitousek et al. 1979), and 3) nitrate loss can cause increased cation losses (Likens et al. 1969). Soil lysimeter solutions were analyzed for nitrate and ammonium on trenched and untrenched plots established in four coastal plain community types: longleaf pine, upland deciduous, bottomland deciduous, and cypress-tupelo swamp. Trenching served as a uniform disturbance that interrupted tree root uptake without significantly changing microclimate. Trenching does not duplicate the perturbation effects caused by land clearing, timber harvest, or some natural disturbances. The elimination of root uptake without overstory removal does allow the identification of nitrate movement below the rooting zone, as well as processes that could be important in retaining nitrogen following a large-scale disturbance (Vitousek et al. 1982). Laboratory incubations were also performed on soils from these sites to help determine the processes that contribute to nitrogen losses following the trenching disturbance. We hypothesized the following:

1) Nitrate leaching losses below the root zone would increase in the trenched plots with the degree of response differing among habitats. The degree of response would be characterized by the microorganism-soil-plant functional relationships and would be a relative measurement of ecosystem resistance following a uniform perturbation (Vitousek et al. 1981).
2) The pine habitat would cycle less N through the soil than would deciduous habitats (pine litter contains less N because more N is retranslocated prior to leaf fall); therefore, nitrate accumulations should be lower in the pine habitat than in deciduous habitats following trenching.

3) Total mineralization potentials of inorganic N determined under aerobic conditions in the laboratory would be highest in the bottomland and swamp soils that contain large amounts of organically bound N.

SITE DESCRIPTION

This study was conducted on the U.S. Department of Energy's Savannah River Site (SRS), a National Environmental Research Park located in southwest South Carolina. This area supports several major Atlantic coastal plain plant community types. It offers an excellent opportunity for research on the Atlantic coastal plain (Figure 1).

The four sites were located on Meyer's Branch, a small, third-order (Horton 1945) blackwater stream (Figure 1). Soils on these sites form a catena ranging from dry and well-drained to wet and continually saturated. The National Wetlands Technical Council (NWTC) (Larson et al. 1981) developed a system of six zones that serve as a framework in relating bottomland hardwood community types. In Figure 2, the Meyer's Branch sites are related to an idealized NWTC bottomland zonal sequence (Wharton et al. 1982). The sites studied included a plantation currently supporting longleaf pine (*Pinus palustris*)[1] (NWTC Zone VI); an upland deciduous site dominated by southern red oak (*Quercus falcata*) and nuttall (*Carya tomentosa*) (Zone VI); a bottomland deciduous site dominated by white oak (*Quercus alba*), loblolly pine (*Pinus taeda*), tulip tree (*Liriodendron tulipifera*), red maple (*Acer rubrum*), dogwood (*Cornus florida*), and holly (*Ilex opaca*) (Zones III and IV); and a swamp site dominated by bald cypress (*Taxodium distichum*), water tupelo (*Nyssa aquatica*), and red maple (Zone II).

Soils on the longleaf pine and upland deciduous sites were Grossarenic Paleudults of the Blanton Series[2] (Table 1). Those on the bottomland deciduous sites consisted of Arenic Ochraquults of the Williman Series (these soils are periodically flooded) (Table 1). Soils on the swamp sites were Cumulic Humaquepts of the Pickney Series (Table 1). Swamp site soils are inundated or saturated year round.

---

[1]All taxonomic nomenclature follows Little (1979).

[2]All soil taxonomy follows the U.S. Soil Conservation Service (1975).

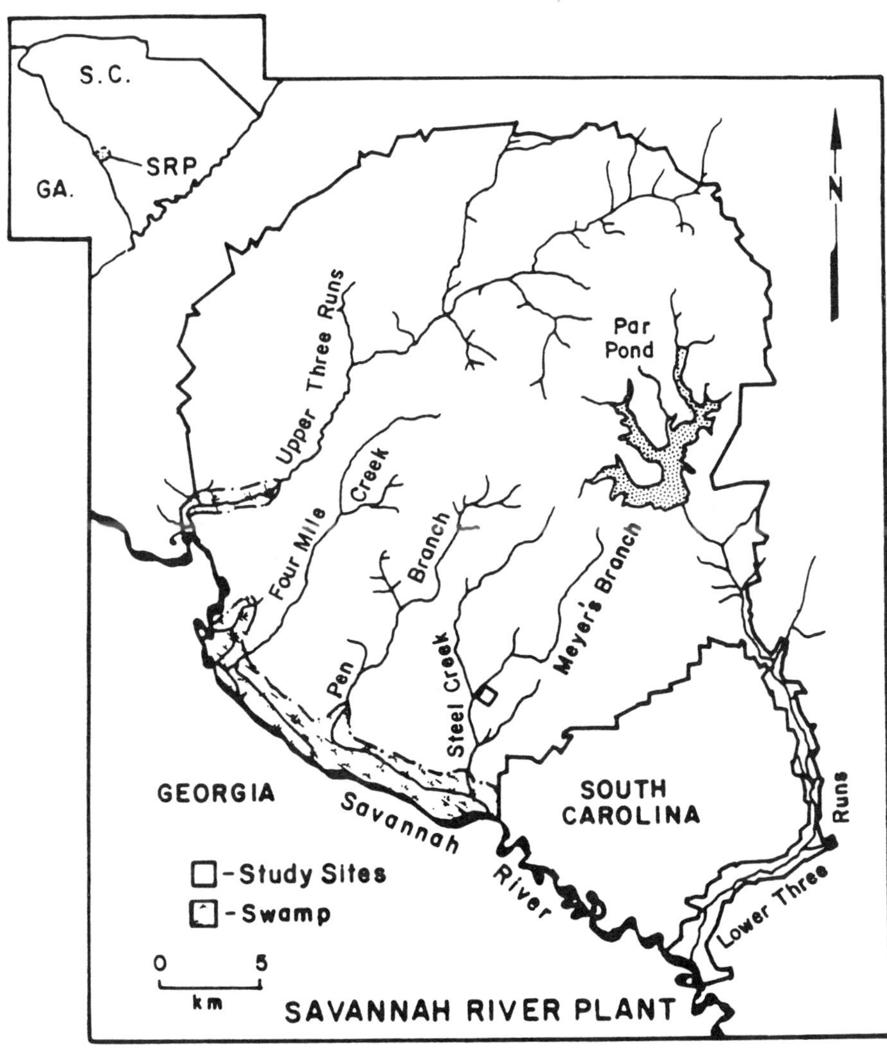

Figure 1. Location of the Savannah River Site and study sites along Meyer's Branch. The four vegetative community types are located in the block designated "study sites."

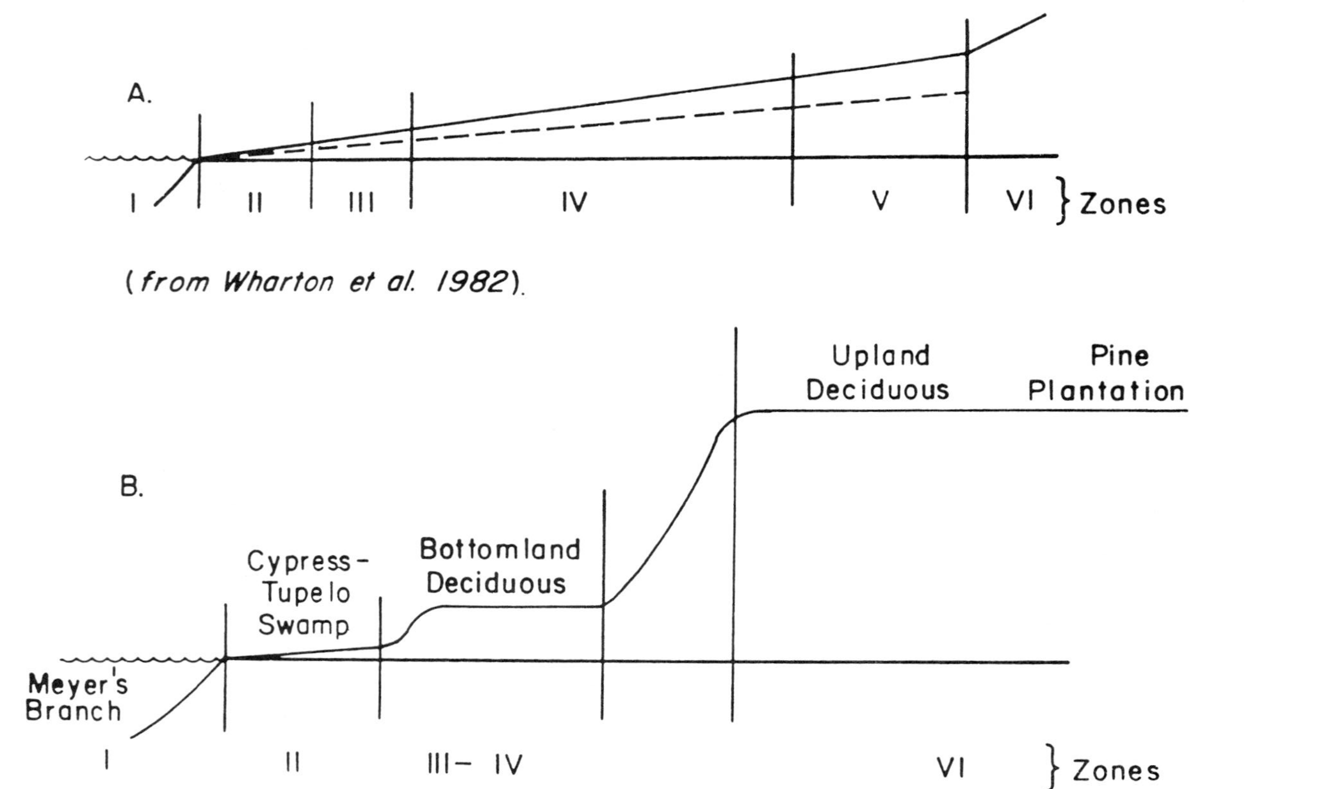

Figure 2. a) Idealized sequence of National Wetlands Technical Council (NWTC) zones from a waterbody to an upland, and b) sequence of NWTC zones along Meyer's Branch.

Table 1. Soils information for four sites along a moisture gradient adjacent to Meyer's Branch.

Vegetative Community	Series	Soil Subgroup	Horizon	Depth (cm)	%Sand	%Silt	%Clay	Texture
Longleaf pine and upland deciduous	Blanton	Grossarenic Paleudult	A	0-18	91.1	6.1	2.8	Coarse sand
			E	18-122	89.2	7.9	2.9	Sand
			Bt1	122-155	88.0	5.8	6.2	Loamy coarse sand
			Bt2	155-183	58.7	12.3	28.9	Sandy clay loam
			C	183-203	71.5	6.0	22.6	Sandy clay loam
Bottomland deciduous	Williman	Arenic Ochraquult	A	0-13	87.8	9.7	2.5	Sand
			E	13-61	88.3	9.8	1.9	Sand
			Btg1	61-89	62.3	7.9	29.8	Sandy clay loam
			Btg2	89-140	68.6	3.5	27.9	Sandy clay loam
			BCg	140-165	80.7	1.6	17.7	Sandy loam
Cypress swamp	Pickney	Cumulic Humaquept	Al	0-20	68.0	21.0	11.0	Black loamy fine sand
			Cg1	20-46	90.0	8.0	2.0	Dark gray loamy sand
			Cg2	46-127	91.0	5.1	3.9	Dark gray sand

METHODS

Ten trenched plots paired with ten untrenched control plots were established on each of the four plant community types. Paired plots were established on similar microsites usually no more than 15 m apart. A machine that cut a 10-cm wide trench was used to trench around a 1.2 by 2.5 m treeless plot to a depth of 1.0 m. Trenching severed any roots that entered the plot and eliminated root uptake of water and nutrients. The trench was lined with two layers of 6-ml polyethylene to prevent solution movement into or out of the plot (Figure 3). A 62-cm wide piece of aluminum flashing was placed outside the polyethylene to prevent root penetration, and the trench was carefully back-filled. It was impossible to trench swamp plots without destroying the integrity of these soils; therefore, an alternate method was developed to stop tree root uptake and soil solution movement. Epoxy-coated, 18-gauge steel cylinders 0.8 m in diameter and 1.0 m long were driven into the swamp soil (Figure 4). Lysimeters were installed to a depth of 62 cm on all longleaf pine and upland deciduous trenched and control plots and to a depth of 46 cm on all bottomland deciduous and cypress swamp trenched and control plots.

Depths of lysimeter placement were determined by the depth of the Bt horizon for the upland and bottomland deciduous sites. It was assumed that placement approximately 15 cm above the top of the Bt horizon would be beneath the major root zones in a zone of relatively rapid leaching. Lysimeters were placed at 46 cm in the swamp soils to allow sampling within a soil horizon that was high in organic matter yet above coarser sands.

Approximately three months elapsed between trenching and the initiation of lysimeter sampling. Lysimeter sampling began in mid March and was continued on a weekly basis. Lysimeters were completely flushed, a vacuum of 30 k Pa was applied to each lysimeter, and samples were taken the following week. Lysimeter solutions were drawn through dedicated tygon access tubes. A positive pressure was applied using a hand pump, and samples were then drawn from the access tube. Any solution in the access tube was flushed before a sample of fresh lysimeter solution was collected in a 25-ml, acid-washed, nalgene vial. Care was taken not to bubble air into the sample, and vials were completely filled to

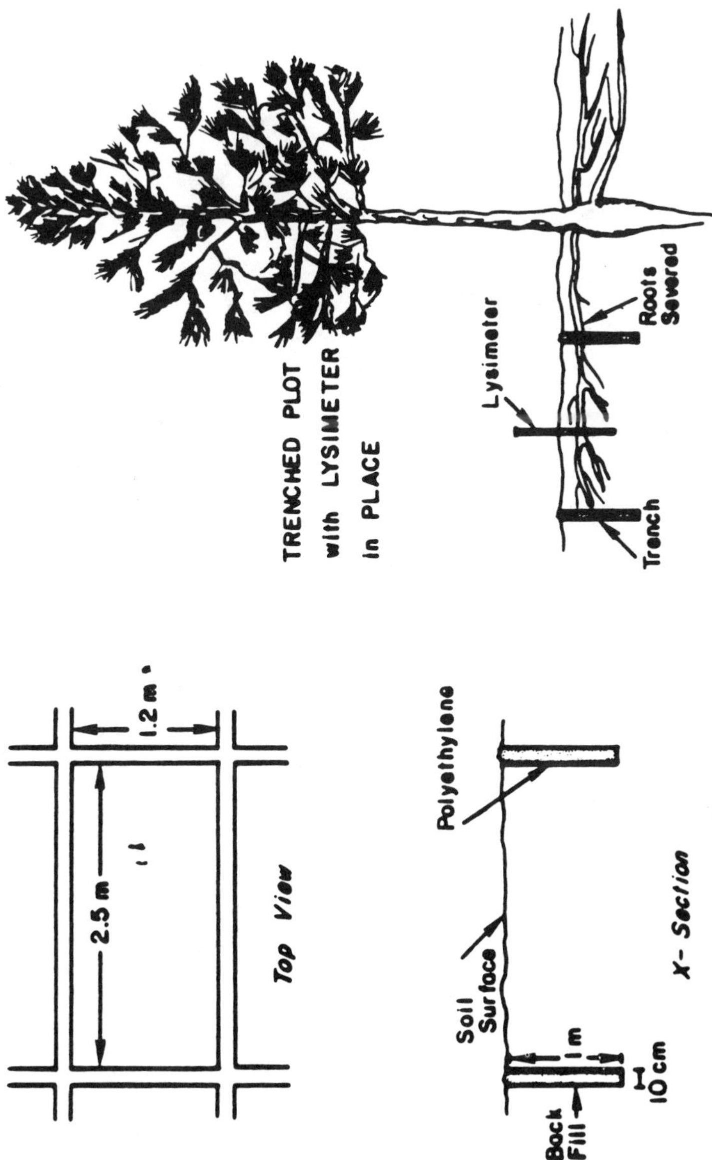

Figure 3. Top view and cross section of a trenched plot with an illustration of a trenched plot and lysimeter in place.

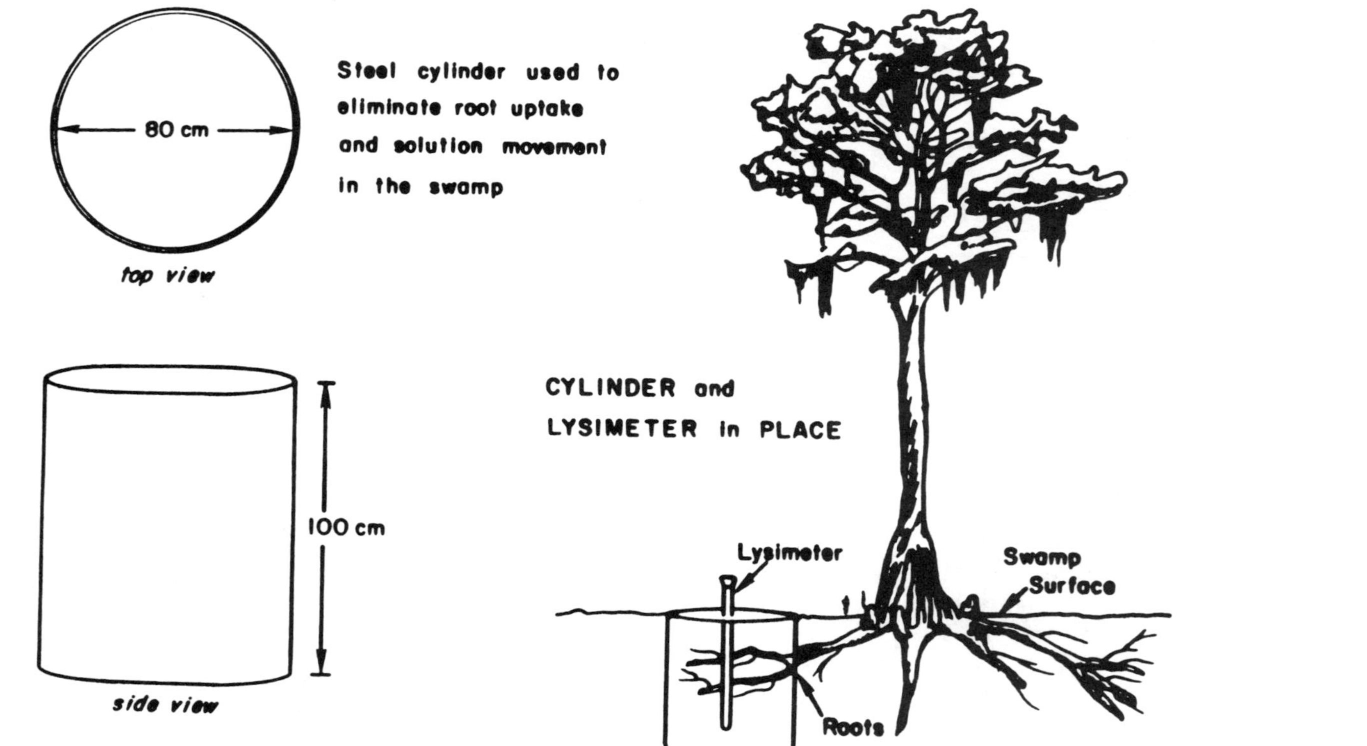

Figure 4. Top and side view of a steel cylinder used to eliminate root uptake with an illustration of a cylinder and lysimeter inplace.

eliminate headspace to avoid oxygenation of the samples. Samples were kept at $4^{o}C$ and analyzed within 24 hours of collection.

Lysimeter solutions were analyzed on a weekly basis for NO_2 + NO_3-N and NH_4-N using a Lachat Quickchem automatic flow injection analyzer (Lachat Chemicals, Inc.; Mequon, Wisconsin, USA). The sodium salicylate method was used for analysis of NH_4-N (as modified by C. Wells, U.S. Department of Agriculture, Forest Service, Southeastern Forest Experiment Station, Forest Science Laboratory, Research Triangle Park, North Carolina, personal communication); NO_2 + NO_3-N was analyzed by the cadmium reduction method (Technicon Corporation 1973).

Laboratory incubations were performed on soils collected from areas adjacent to each of the paired plots across all four sites. A 10-g sample of fresh soil at 60% field capacity was placed in a 200-ml polyethylene container with a plastic lid placed loosely on top. Swamp soils that were continually saturated were kept at their original field moisture level. Moisture was adjusted gravimetrically on a weekly basis. All soils including the saturated swamp soils were incubated aerobically in a dark growth chamber at 20° C.

Ten samples per site were taken at each sampling period 0, 1, 2, 4, and 8 weeks following incubation (a total of 250 soil samples). These were extracted with a KCL (146 g L^{-1}) solution [procedure modified from Bremner (1965)]. Extracts were analyzed for NO_2 + NO_3-N and NH_4-N using the same methods described above.

RESULTS AND DISCUSSION

Preliminary results of lysimeter solution analyses over a 32-week period (from March 15, 1984, to November 15, 1984) revealed different site responses to disturbance (Figure 5). On all sites except the cypress-tupelo swamp, trenched plot lysimeter solutions were higher in nitrate. Pine site responses were minimal (5-10 mg L^{-1} nitrate) by week 32 of sampling. Upland and bottomland deciduous sites showed the greatest increase in nitrate after trenching. The upland deciduous site exhibited an increase in nitrate by the fifth week of collection. Concentrations of upland deciduous site lysimeter solutions peaked (18 mg L^{-1}) between

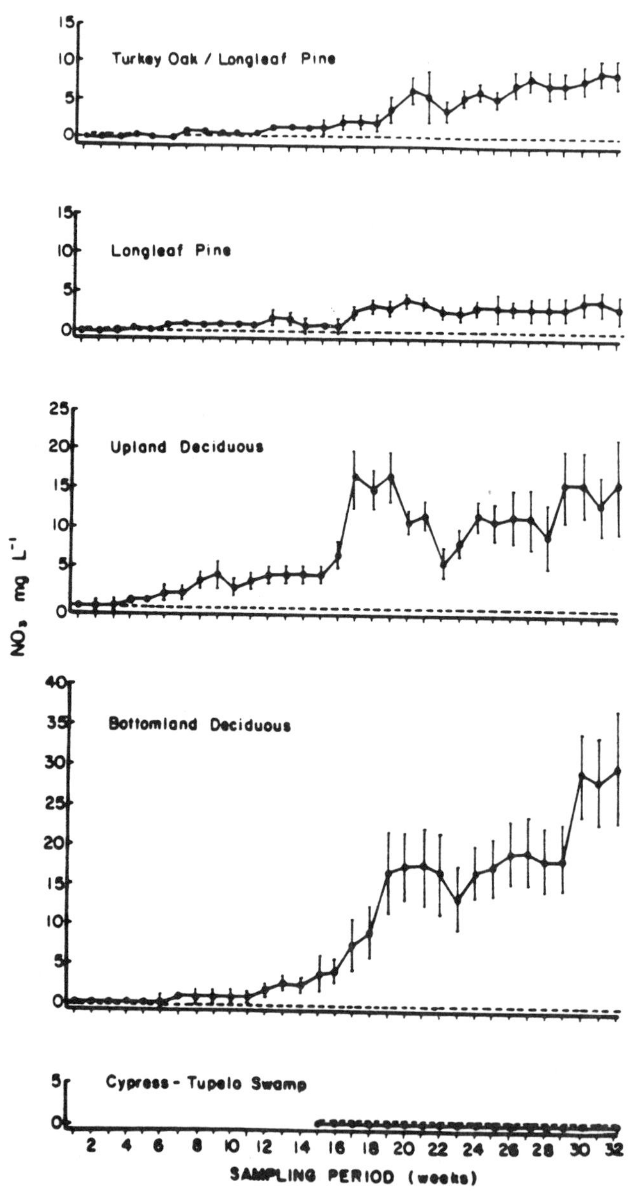

Figure 5. Nitrate concentrations of leachates collected in lysimeters and analyzed on a weekly basis (32 weeks) on four sites along a moisture gradient. Solid lines represent trenched plots, dashed lines represent untrenched control plots. Values reported are means (+/- SE).

weeks 17-19. The bottomland deciduous site responded more slowly, but exhibited the highest lysimeter solution concentration (29 mg L^{-1}, week 32) following trenching. Swamp lysimeter solutions of trenched and untrenched plots contained virtually no nitrate. This was a result of anaerobic conditions which prevent nitrification (Alexander 1977). Untrenched control lysimeter solutions across all habitats contained virtually no nitrate. Results indicate that dramatic potential nitrate losses can occur following trenching. Measured potential losses are conservative in that the trenching did not change the microhabitat conditions as severely as would clear-cutting or drainage; also, trenched plots contained more moisture which would allow greater rates of mineralization.

Preliminary study results revealed that the pine ecosystem exhibited the greatest resistance to nitrate losses below the root zone following disturbance. The upland deciduous ecosystem exhibited moderate resistance to nitrate losses, while the bottomland hardwood ecosystem exhibited the lowest resistance to nitrate losses following trenching. Cypress-tupelo trenched plots lost no nitrate.

Laboratory incubations paralleled our field results in all cases except the cypress-tupelo soils, which yielded 110 µg of nitrate g^{-1} (Figure 6). This was in direct contrast with the low nitrate levels in swamp trenched plot lysimeter solutions. Such a response was due to the effects of oxygenation (during incubation) on soils which are normally anaerobic. The nitrogen response indicated a great potential for nitrate losses from the soils following drainage and clearing (the common agricultural practice).

The inorganic N responses observed following trenching reflect different microorganism-soil-plant functional relationships within sites across the Meyer's Branch catena. Generally, net primary productivity of southeastern swamp and bottomland deciduous sites is greater than that of upland deciduous and upland pine sites (Wharton et al. 1982). The greater productivity of wetland versus upland sites reflects greater nutrient availability on those sites. Silt, clay, loam, and organic substrates of the swamp and bottomland deciduous sites have a greater cation exchange capacity and therefore exhibit greater nutrient retention than do the sandy substrates of the upland sites. Disturbance halts the normal functions of the microorganism-soil-plant unit that has adapted to utilize all available nitrate (as evidenced by the untrenched control plot leachate nitrate concentrations in Figure 5) and thus to maximize production. This unit response is an adaptation to the soil characteristics and to the moisture regime.

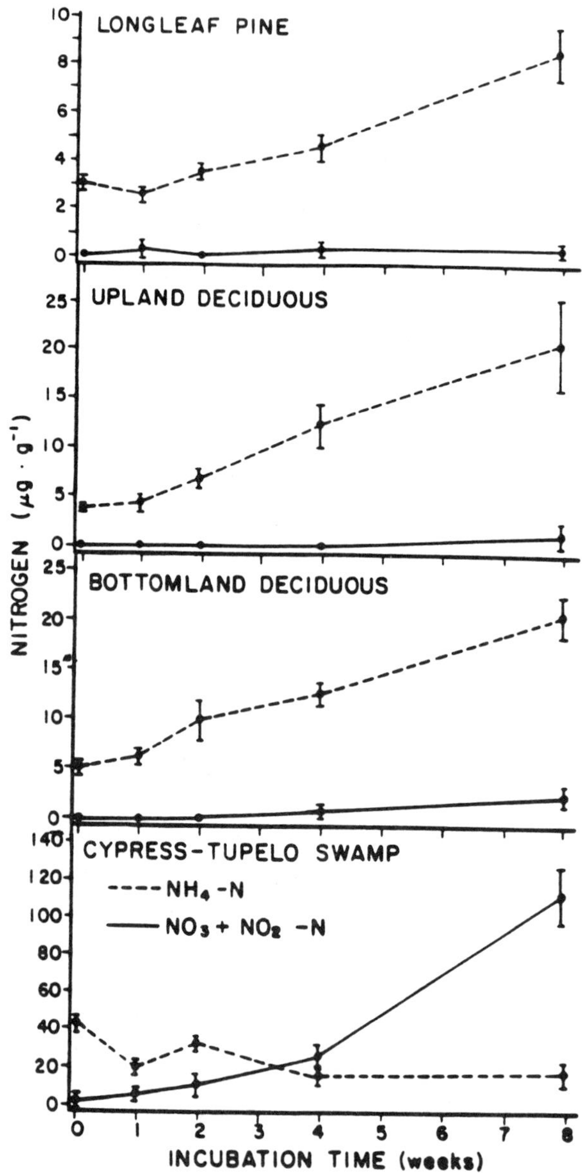

Figure 6. Net $NO_2 + NO_3$-N and net NH_4 accumulation in soil samples collected in four sites along a moisture gradient (June 1984) and incubated for 1, 2, 4, and 8 weeks. Values reported are means (+/- SE).

The absence of nitrate in swamp trenched plot leachates could be misinterpreted as an indication of high resistance to potential nitrate losses following disturbance; however, virtually no nitrate occurs in these anaerobic soils. Anaerobic conditions in swamp sediments slow decomposition and maintain a large nutrient pool of organic N with a long turnover time, as well as an inorganic pool of NH_4-N. A large, standing biomass also maintains a large nutrient pool. Maintenance of large nutrient pools with long turnover times is a requisite condition for high ecosystem resistance (Webster et al. 1975). Ammonium, which is less readily leached than nitrate, is the dominant form of available N in these soils. Initial swamp soil ammonium concentrations prior to incubation were 40 $\mu g\ g^{-1}$ (Figure 6). These soils did not exhibit nitrate losses following trenching (a relatively minor disturbance) because anaerobic conditions were not affected.

Swamp soils are the most fragile of all temperate ecosystem soils. Their fluid to semi-fluid consistency renders them highly subject to disturbance by human activity (e.g., compaction by heavy equipment). Even relatively minor disturbances, such as repeated foot travel, can be deleterious to the structural integrity of these soils. However, as long as anaerobic conditions of swamp soils are not dramatically altered, these systems can be resistant to N losses following disturbance.

Species differences among different zones are the dominant factors that determine ecosystem nutrient cycling responses to disturbances. Deciduous species that lose all of their foliage annually recycle a large fraction of the nutrients lost in leaf fall through the microorganism-soil unit which rapidly frees N, making it available for plant uptake (Gosz 1981, Mellilo 1981). Therefore, greater amounts of inorganic N would be expected in lysimeter leachates of deciduous sites following the elimination of root uptake. Results of our trenched plot study support this premise. Bottomland deciduous sites exhibited the greatest potential nitrate losses because of moist, aerobic, surface soil conditions that favor more rapid rates of nitrification. The continually saturated soil of the cypress swamp inhibited nitrification of the accumulated NH_4-N.

The upland pines shed only a portion of their foliage annually. Before older foliage is shed, much of the N is retranslocated. This results in high C:low N litterfall. Such material immobilizes N, resulting in low microbial activity and low soil inorganic N (Gosz 1981).

CONCLUSIONS

Habitats along a moisture gradient responded to disturbance in distinct ways. Preliminary findings support the premise that deciduous ecosystems recycle more nitrogen through the plant-soil component than do coniferous ecosystems. The longleaf pine sites were the most resistant to nitrate leaching losses below the root zone following disturbance. The bottomland hardwood sites were the least resistant to nitrate losses following disturbance. They exhibited the highest levels of lysimeter solution nitrate leaching. Results indicate a high potential for nitrification of accumulated NH_4-N in swamp and bottomland soils if soils are disturbed. This could result in high nitrate losses following swamp ecosystem disturbances that increase soil oxygenation. The lack of nitrate in the swamp trenched plot lysimeter solutions was attributed to anaerobic conditions. Results underscore the importance of anaerobic conditions (maintained by periods of inundation) that: a) prevent the oxidation of ammonium to nitrate and thus conserve nitrogen, or b) may promote denitrification, thereby reducing potential downstream nitrate losses.

Existing plant assemblages in the natural state usually reflect at least several thousand years of climatic influence. The stability of the swamp and deciduous bottomlands reflect this more or less constant regime to which the present species are adapted. Dramatic changes in the moisture regime and removal of tree species may alter the anaerobic/aerobic conditions to the extent that dramatic losses of nitrate ensue. For example, natural disturbances such as flooding, windthrow, and insect infestation would cause little change in the swamp ecosystem. These disturbances would have minor effects on soil anaerobic conditions; however, changes in the structural integrity of these soils, such as those caused by heavy equipment (which could churn and compact the strata) or by drainage, are the most deleterious disturbances that could occur on these sites. Such activity would drastically reduce the resistance of these wetlands to nitrate losses and would result in ecosystem instability.

ACKNOWLEDGMENTS

We thank Casey Sherrod, Amy Shumpert, and Michael Evans for technical assistance; Jean Coleman and Linda Orebaugh for drafting; and Dan Childers, Barbara Kleiss, Michael Smart, and Lyndon Lee for critical review of this manuscript. Dorita Fransham prepared the rough draft. Meredith McCarthy typed the final manuscript. This study was funded by Department of Energy grant DE-AC09-76SROO-819.

REFERENCES CITED

Alexander, M. 1977. Introduction to soil microbiology. John Wiley and Sons, Inc., New York. 467 pp.

Boto, K. G. and W. H. Patrick, Jr. 1978. Role of wetlands in the removal of suspended sediments. Pages 479-489 *in* P. B. Greeson, J. R. Clark and J. E. Clark, eds. Wetland functions and values: the state of our understanding. Proceedings of the National Symposium on Wetlands. November 7-10, 1978. American Water Resources Assoc., Minneapolis, Minnesota.

Bremner, J. M. 1965. Inorganic forms of nitrogen. Pages 1179-1237 *in* C. A. Black, ed. Methods of soil analysis. Part 2. American Society of Agronomy, Inc., Madison, Wisconsin.

Brown, S., M. M. Brinson, and A. E. Lugo. 1978. Structure and function of riparian wetlands. Pages 17-31 *in* R. R. Johnson and J. F. McCormick, eds. Strategies for protection and management of floodplain wetlands and other riparian ecosystems. U.S. Department of Agriculture, Forest Service, Washington, D.C. GTR-WO-12.

Duda, A. M. 1982. Municipal point source and agricultural nonpoint source contributions to coastal eutrophication. Water Resour. Bull. 18:397-407.

Gambrell, R. P., J. W. Gilliam, and S. B. Weed. 1975. Denitrification in subsoils of the North Carolina Coastal Plain as affected by soil drainage. J. Environ. Qual. 4:311-316.

Gaskin, J. W., J. E. Douglass, and W. T. Swank. 1983. Annotated bibliography of publications on watershed management and ecological studies at Coweeta Hydrologic Laboratory, 1034-1084. U.S. Department of Agriculture, Forest Service, Southeastern Forest Experiment Station. Asheville, North Carolina. Gen. Tech. Rep. SE-30. 140 pp.

Gosz, J. R. 1981. Nitrogen cycling in coniferous ecosystems. Pages 405-426 *in* F. E. Clark and T. Rosswall, eds. Terrestrial nitrogen cycles. Ecol. Bull. 33. Stockholm, Sweden.

Horton, R. E. 1945. Erosional development of streams and their drainage basins: hydrophysical approach to quantitative morphology. Bull. Geol. Soc. 56:275-370.

Kuenzler, E. J., P. J. Mulholland, L. A. Ruley, and R. P. Sniffen. 1977. Water quality in North Carolina Coastal Plain streams and effects of channelization. University of North Carolina Water Resour. Res. Inst. Report 127. 160 pp.

Larson, J. S., M. S. Bedinger, C. F. Bryan, S. Brown, R. T. Huffman, E. L. Miller, D. G. Rhodes, and B. A. Touchet. 1981. Transition from wetlands to uplands in southeastern bottomland hardwood forests. Pages 225-273 *in* J. R. Clark and J. Benforado, eds. Wetlands of bottomland hardwood forests. Proceedings of a workshop on bottomland hardwood forest wetlands of the southeastern United States held at Lake Lanier, Georgia, June 1-5, 1980. Developments in Agricultural and Managed-Forest Ecology, vol. 11. Elsevier Scientific Publishing Company, New York.

Likens, G. E., F. H. Bormann, and N. M. Johnson. 1969. Nitrification: importance to nutrient losses from a cutover forested ecosystem. Science 163: 1205-6.

Likens, G. E., and F. H. Bormann. 1974. Linkages between terrestrial and aquatic ecosystems. BioScience 24:447-456.

Little, E. L., Jr. 1979. Checklist of United States trees (native and naturalized). U.S. Department of Agriculture, Forest Service. Agricultural Handbook 541. 375 pp.

Lowrance, R. R., R. L. Todd, and L. E. Asmussen. 1948a. Nutrient cycling in an agricultural watershed: I. phreatic movement. J. Environ. Qual. 13:22-27.

__________. 1948b. Nutrient cycling in an agricultural watershed: II. streamflow and artificial drainage. J. Environ. Qual. 13:27-32.

Mellilo, J. M. 1981. Nitrogen cycling in deciduous forests. Pages 427-442 *in* F. E. Clark and T. Rosswall, eds. Terrestrial nitrogen cycles. Ecol. Bull. 33. Stockholm, Sweden.

Odum, E. P. 1978. Ecological importance of the riparian zone. Pages 2-4 *in* R. R. Johnson and J. F. McCormick, eds. Strategies for protection and management of floodplain wetlands and other riparian ecosystems. U.S. Department of Agriculture, Forest Service, Washington, D.C. GTR-WO-12.

O'Neill, R. V., B. S. Ausmus, D. R. Jackson, R. I. van Hook, P. van Voris, C. Washburn, and A. P. Watson. 1977. Monitoring terrestrial ecosystems by analysis of nutrient export. Water Air Soil Pollut. 8:271-277.

Peterjohn, W. T., and D. L. Correll. 1984. Nutrient dynamics in an agricultural watershed: observations on the role of a riparian forest. Ecology 65:1466-1475.

Technicon Corporation. 1973. Nitrate in water. Technicon Autoanalyzer II, Industrial Method No. 37-71E. Technicon Industrial System, Tarrytown, New York.

U.S. Soil Conservation Service. 1975. Soil taxonomy -- a basic system of soil classification for making and interpreting soil surveys. Soil Survey Staff, U.S. Department of Agriculture, Washington, D.C. Agricultural Handbook 436. 754 pp.

Vitousek, P. M., J. R. Gosz, C. C. Grier, J. M. Mellilo, and W. A. Reiners, and R. L. Todd. 1979. Nitrate losses from disturbed ecosystems. Science 204:469-474.

__________. 1982. A comparative analysis of potential nitrification and nitrate mobility in forest ecosystems. Ecol. Monogr. 52:155-177.

Vitousek, P. M., W. A. Reiners, J. M. Mellilo, C. C. Grier, and J. R. Gosz. 1981. Nitrogen cycling and loss following forest perturbation: the components of response. Pages 115-127 *in* G. W. Barrett and R. Rosenberg, eds. Stress effects on natural ecosystems. John Wiley and Sons, Ltd. New York.

Waide, J. B. and W. T. Swank. 1977. Simulation of potential effects of forest utilization on the nitrogen cycle in different southeastern ecosystems. Pages 767-789 *in* D. L. Correll, ed. Watershed research in eastern North America. Smithsonian Institution, Edgewater, Maryland.

Webster J. R., J. B. Waide, and B. C. Patten. 1975. Nutrient recycling and the stability of ecosystems. Pages 673-686 *in* F. G. Howell, J. B. Gentry, and M. H. Smith, eds. Mineral cycling in southeastern ecosystems. Energy Research and Development Administration (ERDA) Symp. Ser. CONF-740513.

Wharton, C. W., W. M. Kitchens, E. C. Pendleton, and T. W. Sipe. 1982. The ecology of bottomland hardwood swamps of the Southeast: a community profile. U.S. Fish and Wildlife Service, Washington, D.C. FWS/OBS-81/37. 133 pp.

5. IMPORTANCE OF BOTTOMLAND HARDWOOD FOREST ZONES TO FISHES AND FISHERIES: THE ATCHAFALAYA BASIN, A CASE HISTORY

Victor W. Lambou
U.S. Environmental Protection Agency, P.O. Box 93478,
Las Vegas, Nevada 89193-3478

ABSTRACT

This paper reviews the importance of bottomland hardwood forest zones to fishes and fisheries, using the Atchafalaya Basin as a case history. The Atchafalaya Basin, one of the largest remaining floodplain bottomland hardwood forests in the United States, comprises approximately 72 by 193 km lowland floodplain area in Louisiana with elevations ranging from 15 m to sea level. Fifty-four percent of the 95 species of finfish known to occur in the leveed Atchafalaya Basin use overflow wooded areas for spawning and/or rearing of young, while 56% use these areas for feeding. A total harvest of 8797 $kg \cdot km^{-2} \cdot yr^{-1}$ finfish and crawfish has been documented from the overflow areas of the Basin. Production of red swamp crawfish (*Procambarus clarkii*) in one area of the basin was estimated to be 69,717 $kg \cdot km^{-2} \cdot yr^{-1}$. Total standing crops of finfishes estimated from rotenone sampling ranged from 25,000 to 208,000 $kg \cdot km^{-2}$. Finfishes move out of permanent-water areas into flooded, wooded areas when water stages rise and move back into permanent-water areas when flooding waters recede. The production and yield of finfish and crawfish relative to the annual flooding of the forest are discussed.

Ecological Processes and Cumulative Impacts: Illustrated by Bottomland Hardwood Wetland Ecosystems. Edited by James G. Gosselink, Lyndon C. Lee, and Thomas A. Muir. Chelsea, MI 48118. Printed in USA.

INTRODUCTION

Floodplains support extensive fish populations and sport and commercial fisheries. Composition of fish populations and characteristics of the fisheries are dependent upon water regimes, size of the river system, proximity to estuarine and marine waters, physical and chemical characteristics of the water, and the geographic location of the river basin. In any floodplain there are many different habitats and zones defined by a moisture gradient ranging from the constantly inundated channels and lakes to the dry uplands that are very infrequently inundated. Floodplain zones of bottomland hardwood forests are used extensively by fishes and are very important in determining the size and makeup of floodplain fish populations and the fisheries. The purpose of this paper is to review the importance of bottomland hardwood forest zones to fishes and fisheries, using the Atchafalaya Basin as a case history.

DESCRIPTION OF THE AREA

The Atchafalaya Basin

The Atchafalaya Basin (Figure 1) comprises an 8345 km^2 lowland, floodplain area confined between natural levee ridges that delineate the present and former courses of the Mississippi River. Its overall dimensions are approximately 72 by 193 km, with elevations ranging from 15 m to sea level. The Basin contains one of the largest remaining tracts of floodplain bottomland hardwood forest in the United States.

There are six segments, or units, in the Atchafalaya Basin (Figure 1) that have integrity due to a man-made levee system. These are: 1) the 287 km^2 Morganza Floodway, 2) the 611 km^2 West Atchafalaya Floodway, 3) the 340 km^2 Pointee Coupee Sump Area, 4) the 2129 km^2 Atchafalaya Basin Floodway--historically subject to frequent and prolonged natural flooding, 5) the 259 km^2 leveed Atchafalaya River and other segments located mainly between the upper Atchafalaya River levees and Old River, and 6) the 4719 km^2 East and West Basin areas isolated by levees.

One of the major hydrological features of the Atchafalaya Basin is the system of levees and floodways designed to confine and pass a "project flood" through the Mississippi River Basin of 84,960 m^3s (3 million cfs) at the latitude of Old River. One-half of the "project flood" would be diverted through the Atchafalaya Basin.

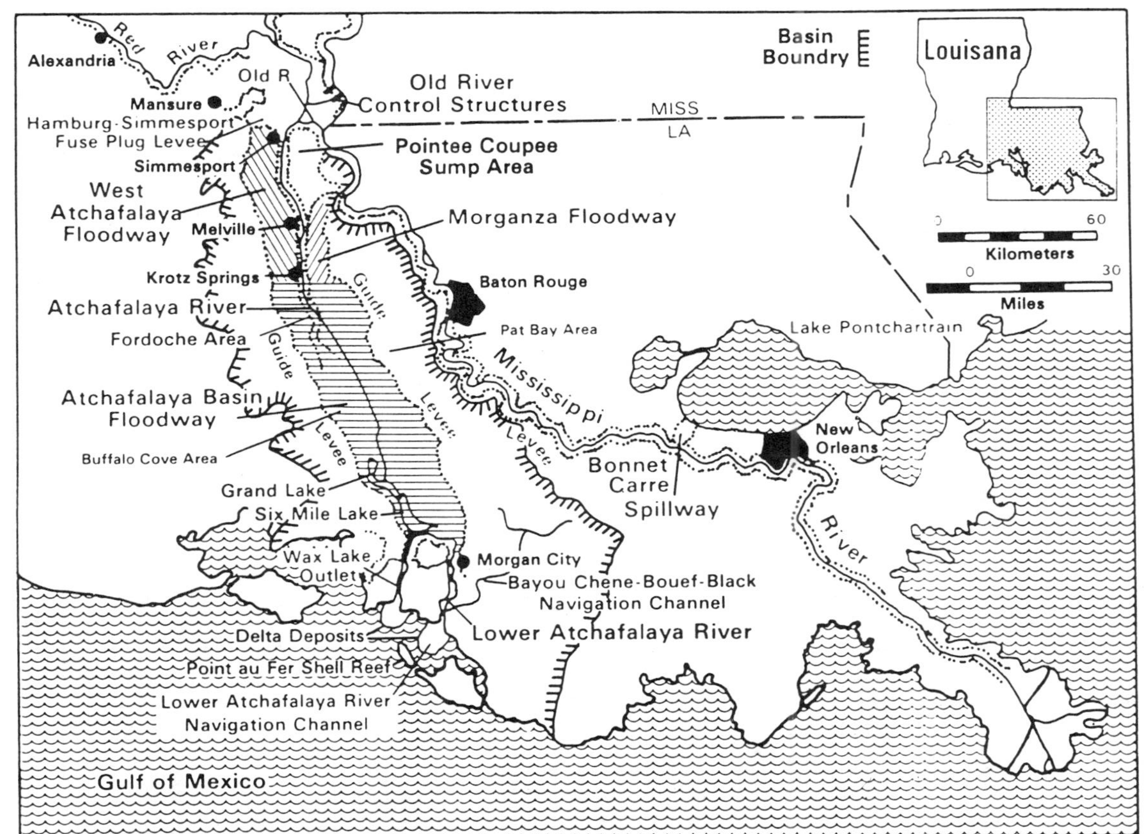

Figure 1. The Atchafalaya Basin, Louisiana.

Water normally enters the Atchafalaya Basin from two major sources. A portion of the Mississippi River's flow enters through the Old River Control Structures and eventually joins with the Red River to form the main stem of the Atchafalaya River. The mainstem flows are then confined by levees until the river enters the Atchafalaya Basin Floodway. There the water spreads out through distributaries and during high water by overbank flows over almost the entire Atchafalaya Basin Floodway. All water, including drainage from the Morganza Floodway, West Atchafalaya Floodway, and Pointee Coupee Sump Area, exits the Basin through the Wax Lake and Lower Atchafalaya River outlets.

The Atchafalaya River receives approximately 30% of the combined flows of the Mississippi and Red Rivers at the latitude of the Old River Control Structures. Atchafalaya River discharges show both seasonal and annual variation. The "average shifted hydrograph" indicates seasonality of the flows (Figure 2). Flows generally begin to increase in the fall or early winter and crest between April and early June. The flows normally drop off sharply to their seasonal lows in late summer and early fall. The cyclic nature of flows also describes the typical annual regime of overflow for the Atchafalaya Basin Floodway. The other segments of the Basin are not subjected to the same type of prolonged overbank flooding because of the levee system.

Of most value to the fisheries is the 2129 km^2 Atchafalaya Basin Floodway, which during normal years has in excess of 1619 km^2 flooded by overflow from the Atchafalaya River. The Atchafalaya Basin Floodway is dissected by an intricate mosaic of large numbers of shallow lakes of various shapes and sizes, bayous, sloughs, distributaries, and canals.

The Fordoche Area

Intensive studies on the relationship of bottomland hardwood forest zones to the aquatic fauna were carried out in the Fordoche Area (Figure 3). This area covers approximately 270 km^2 of the Atchafalaya Basin Floodway south of U.S. Highway 190. It is covered by a bottomland hardwood forest except for permanent waterbodies. Bayou Fordoche is the primary stream flowing through the area. Water levels are controlled primarily by backwater flow from the Atchafalaya River through Bayou La Rose and the West Atchafalaya Basin Protection Levee borrow pit in the southern portion of the area. Local runoff enters the area through the Bayou Courtableau Drainage Structure and from the West Atchafalaya Floodway through natural channels.

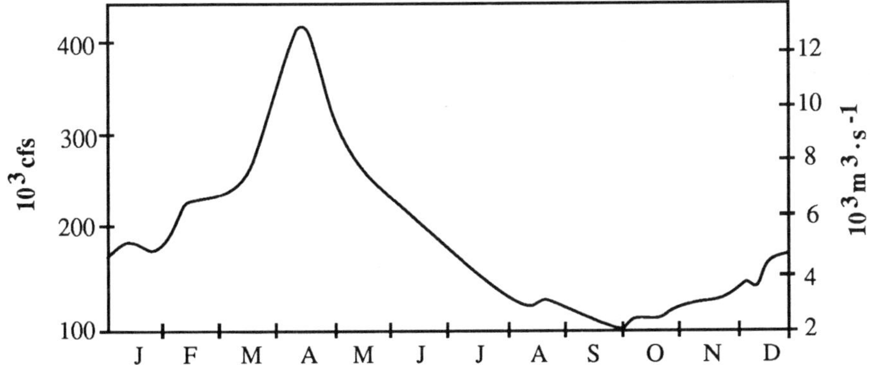

Figure 2. **Average daily shifted hydrograph for the Atchafalaya River at Simmesport, Louisiana, for the period 1949-78, adopted from the U.S. Army Corps of Engineers (USACE) (1982). The hydrograph was computed from daily discharges at the latitude of Old River and adjusted to 30% of the total flow to account for the presence of the Old River Control Structures. Each year's hydrograph was shifted to peak on April 15 (the day the unshifted average hydrograph peaked) before averaging, since daily averages of unshifted hydrographs result in considerably lower peak stages than is representative of actual conditions.**

MATERIALS AND METHODS

General

In this review I have used extensively both published and unpublished data from a variety of sources. Some of the material in this report was summarized from an unpublished manuscript I have prepared on fishes and fisheries in the Atchafalaya Basin (Lambou, MS). The general theory of floodplain fisheries ecology is covered in two excellent publications by Welcomme (1976, 1979).

I define floodplain in the same manner as Leopold et al. (1964) and Welcomme (1979), i.e., to include the river channels, the natural levee regions (where present) that more or less follow the course of present and former river channels and that are normally flooded the shortest time

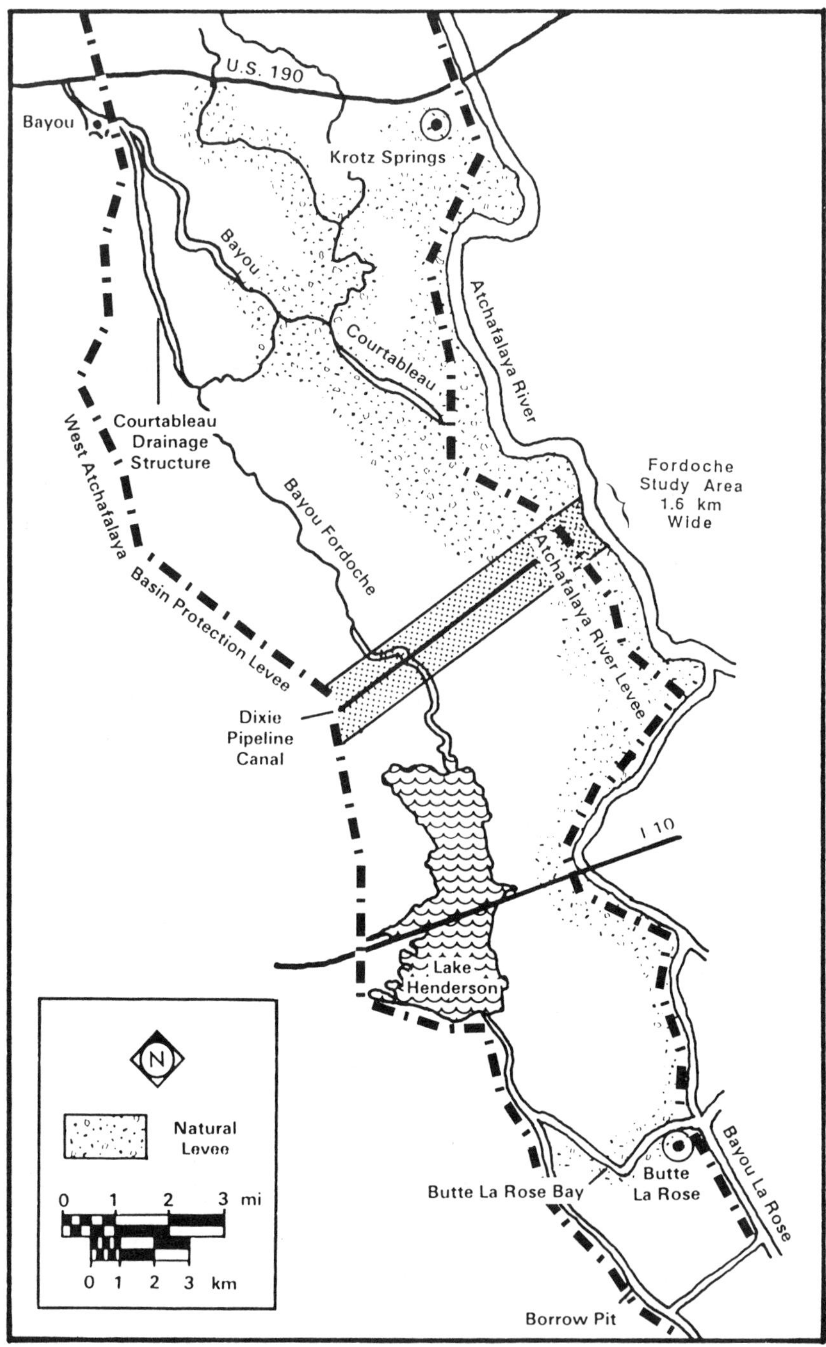

Figure 3. The Fordoche study area in the Atchafalaya Basin Floodway.

annually, and the flats which extend from the natural levees (or river banks where levees are not developed) to the terrace or plateau delimiting the plain. I am also defining bottomland hardwood forest ecosystem in the same general sense, i.e., the floodplain including the forest types described by Larson et al. (1981); but also including the permanent-water areas in the river channels and flats of the floodplain.

Aquatic fauna other than taxonomic true fishes are important in some floodplain fisheries, e.g., crawfishes and crabs in the Atchafalaya Basin. For clarity, true fishes will be referred to in this report as finfishes; fish, fishing, and fishery will be used in a general sense to include crawfishes and crabs as well as finfishes.

Wharton et al. (1981) and Wharton et al. (1982) have reviewed much information on the occurrence, population, and production of aquatic fauna, including crawfishes and finfishes, in bottomland hardwood forest ecosystems, and I will not repeat that information here. Also, considerable information has been collected on the occurrence, population, and production of zooplankton and other invertebrates; nutrient and carbon fluxes; water quality; and hydrology in the Fordoche study area, Atchafalaya Basin Floodway, and other areas of the Atchafalaya Basin. Because of the emphasis on fishes and fisheries in this paper and the lack of space, I will not review that information here.

Some of the measurements presented here were originally made in English units. These have been converted to metric units even though this results in awkwardness in some instances.

Atchafalaya Finfish Occurrence

A listing of finfishes occurring in the leveed Atchafalaya Basin was compiled from data presented by Lambou (1963a), Bryan et al. (1975), Bryan et al. (1976), Horst (1976), and from finfish sampling data presented in this report.

Atchafalaya Basin Fisheries

Fisheries data were obtained from two published surveys of fisheries in the Atchafalaya Basin Floodway. The first, by the Louisiana Department of Wildlife and Fisheries, estimated the commercial harvest during 1962 and the sport harvest during a 9-day period May 11 through 19, 1963 (Lambou 1963a). A more comprehensive survey, conducted as a cooperative study of the Louisiana Department of Wildlife and Fisheries and the USACE, estimated the sport and commercial harvest during the period July 1, 1971, through June 30, 1974 (Soileau et al. 1975). Because the Soileau et al. survey did not record the weight of sport finfishes harvested, a multiplier of 226.75 g was used to convert numbers into weights. This should give a minimum estimate of the weight of sport finfishes harvested. The survey did not record the weight of crabs harvested; therefore, each dozen of crabs harvested was multiplied by 907.26 g, and this should give a very minimum estimate of weight harvested.

Atchafalaya Main Channel Finfish Populations

As part of investigations on pesticide pollution in the Lower Mississippi River System by the former Federal Water Pollution Control Administration, mainstem finfish populations were sampled from Hickman, Kentucky to the Gulf of Mexico during 1966 through 1968. Two sampling stations were located on the Atchafalaya River. The first was a 9.7-km stretch of the River a little north of Simmesport, while the second was a 9.7-km stretch located in the Grand Lake area of the mainstem of the Atchafalaya River about 22.5 km northeast of Morgan City.

Net samples were taken with a standard amount of gear (Table 1) fished for a 5-day period during eight sampling periods per year. Fishing was accomplished by a project biologist and a commercial fisherman who lived and fished in the general area of the sampling station. All gear was fished for 24 hours before it was checked or moved. The nets were moved daily if finfishes were not being caught, since the objective was to catch the maximum number of finfishes. In determining catch per unit of effort, equal weight was given to each of the eight net types given in Table 1, to each sampling period, and to each year of sampling.

Table 1. Sampling gear used during each five-day sampling period to sample finfish populations in the Atchafalaya River near Simmesport and Morgan City, Louisiana.

Number of Nets	Type	Square Mesh Size in Centimeters[a]	Length in Meters
2	Experimental gill net	1.91(0.75)-6.35(2.5)[b]	38.1
2	Trammel net	3.81(1.5)	45.72
2	Trammel net	6.35(2.5)	45.72
2	Trammel net	8.89(3.5)	45.72
2	Hoop net	1.91(0.75)tail	-
2	Hoop net	2.54(1)tail	-
2	Hoop net	5.08(2)tail	-
2	Hoop net	7.62(3)tail	-

[a] Equivalent size in inches is given in parentheses.

[b] Each experimental gill net contained the following mesh sizes in 7.62-m sections: 1.92 cm (0.75 in), 2.54 cm (1 in), 3.81 cm (1.5 in), 5.08 cm (2 in), and 6.35 cm (2.5 in).

Atchafalaya Finfish Standing Crops

Estimates of standing crops of finfishes in waters off the main channel were made in the Atchafalaya Basin Floodway at six sampling sites (Figure 4) by various workers using rotenone poisoning methods (Lambou 1958, Lantz 1970, Sabins 1977, and unpublished data from Louisiana Department of Wildlife and Fisheries). All workers applied sufficient rotenone to obtain the equivalent of approximately 1 $mg \cdot l^{-1}$ of 5% rotenone in approximately .405 ha sampling areas surrounded by a "block-off net" similar to the one described by Lambou (1959a). Finfishes were recovered over a 2-day period.

Fordoche Study

The Fordoche study area located in the Fordoche area (Figure 3) was defined as a 1.6 km wide transect paralleling the Dixie Pipeline Canal beginning at the West Atchafalaya Basin Protection Levee and proceeding easterly up to the 6.10 m NGVD (National Geodetic Vertical Datum)

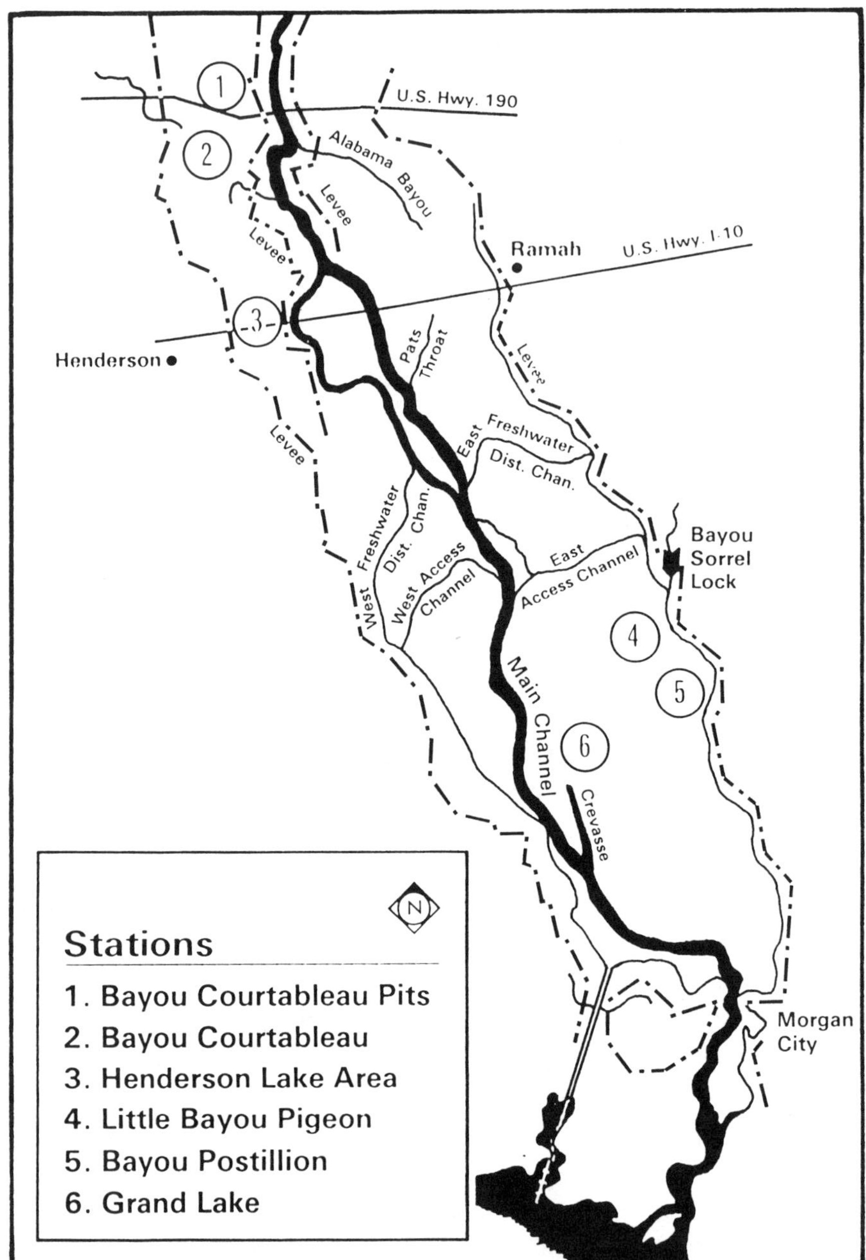

Figure 4. General location of rotenone fish-population sampling stations in the Atchafalaya Basin Floodway.

contour. Six sampling stations were established in permanent-water areas, while 12 sampling stations were located in the overflow wooded area approximately 0.4 km back from the canal (Figure 5). Moveable water-edge sampling stations were established in water depths of less than 0.5 m in the overflow area during each sampling period when some part of the forest was flooded; the exact location of the overflow water-edge sampling stations depended upon the water level at the time of sampling.

Routine sampling was conducted at approximately monthly intervals from November 1979 through August 1980; the number of stations sampled during any time interval depended upon the water stage (Figure 6). An automatic recording water gage was installed in the pipeline canal. The stage hydrograph, the area/elevation curve for the area, and their relationship to the sampling stations are given in Figure 6. The community structure of the woody vegetation at the overflow-water sampling stations was determined during late summer and fall of 1980. All existing woody vegetation was enumerated in 30.48-m by 30.48-m plots at stations O-1 through O-12.

Crawfishes were sampled with funnel traps constructed of 0.32 cm hardware cloth and baited with chicken necks. Each sample consisted of five traps set approximately 10 m apart and fished for 24 hours. Sampling was accomplished during mid-month, normally over a 2-day to 3-day period and in all cases in less than a week's time. There were 17, 14, 19, 19, 17, and 7 samples taken during February, March, April, May, June, and July, respectively. Only red swamp crawfish (*Procambarus clarkii*) were captured in the traps. The carapace length of all captured crawfish was determined to the nearest millimeter, and the weight of each crawfish was estimated from their length-weight relationship. The length-weight relationship

$$\log y = -4.164 + 3.225 \log x \quad (r = 0.994), \tag{1}$$

where y is the weight in g and x is the carapace length in millimeters, was determined from 281 red swamp crawfish.

In May, a 3.048-m by 3.048-m enclosure was created with 0.32-cm, square-mesh seine material in the eastern water-edge area north of the pipeline canal. Through the use of mark and recapture techniques, it was estimated that a red swamp crawfish population density of 16.5 m^{-2}

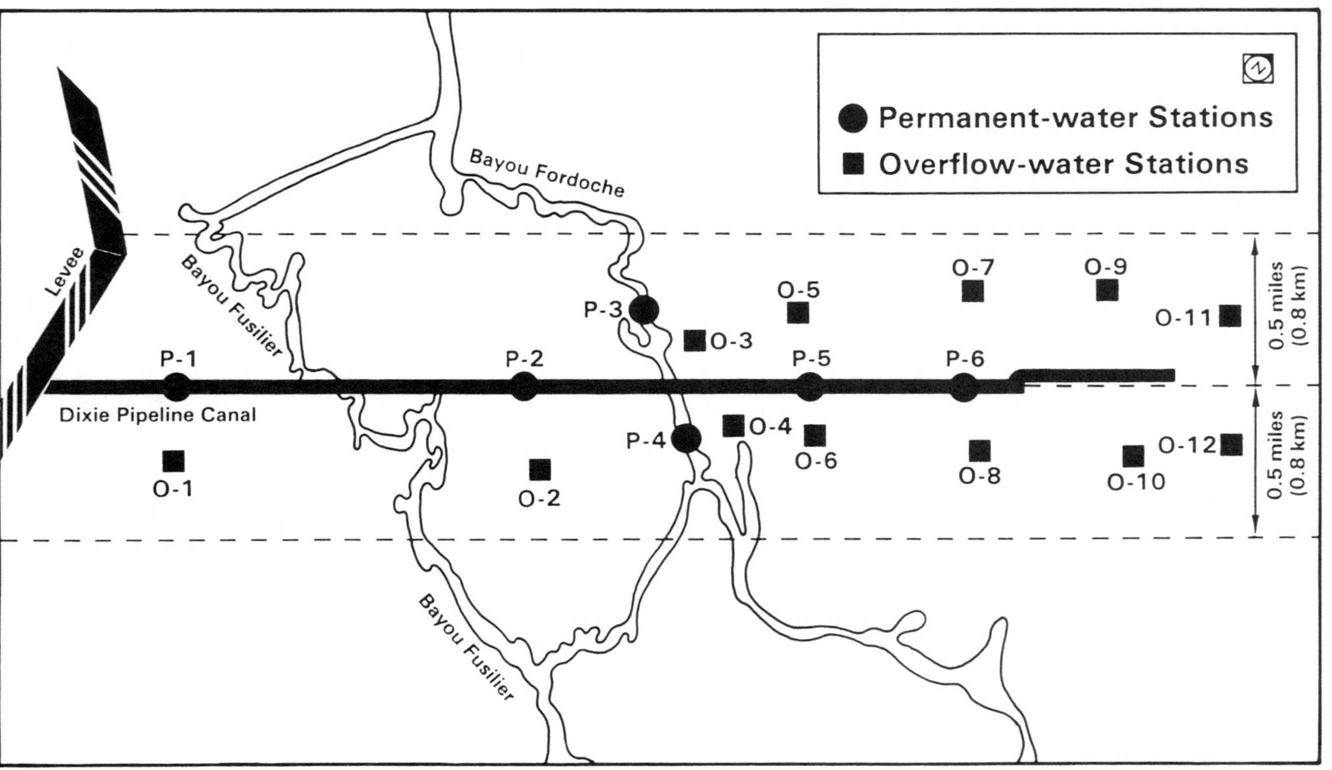

Figure 5. The Fordoche study area showing the location of the permanent-water and overflow-water sampling stations.

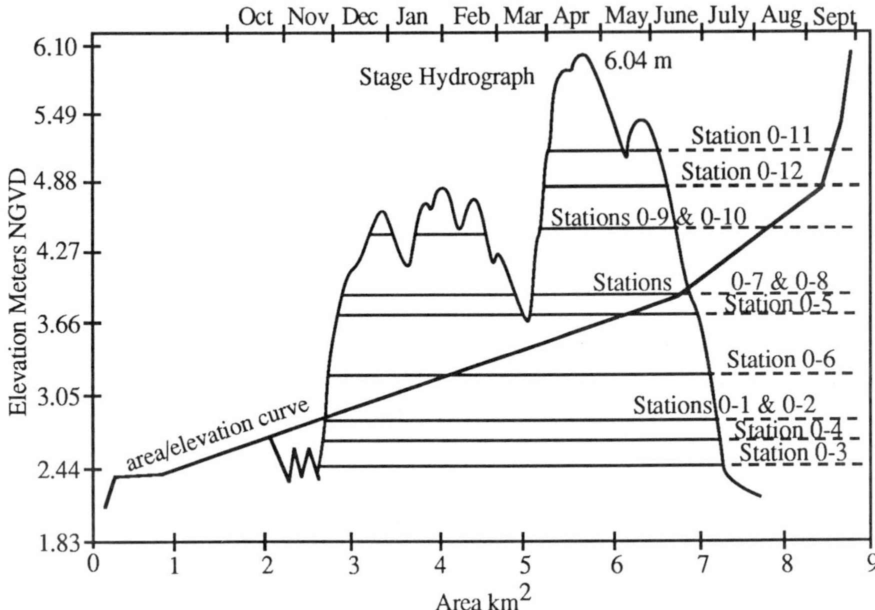

Figure 6. Fordoche study area stage hydrograph during 1977-80 and area/elevation curve and their relationships to overflow-water stations.

occurred in the enclosure. The ratio of the population density ($n \cdot m^{-2}$) in the enclosure to the number of crawfish caught per trap day ($n \cdot TD^{-1}$) in wire traps fished at approximately the same time in the same type of habitat was found to be 0.503. This ratio was used in the formula

$$n \cdot m^{-2} = 0.503 \; n \cdot TD^{-1} \qquad (2)$$

to convert all data on crawfish catch per unit of effort to estimates of population density. Undoubtedly, more enclosure studies at different times of the year would increase the precision of the estimates of population densities. However, Konikoff (1977) reported a correlation coefficient greater than 0.99 between estimates of crawfish population densities from mark and recapture studies and wire trap catch per unit of effort data from the Buffalo Cove area of the Atchafalaya Basin

Floodway. The weight of crawfish in grams per unit area was estimated from

$$g \cdot m^{-2} = n \cdot m^{-2} \, (g \cdot TD^{-1}) \, / \, n \cdot TD^{-1} \quad (3)$$

Estimates of numbers and weights of crawfish in the total study area, as well as in various habitat types, were obtained by weighting the means for the sampling stations occurring in a zone by the area of the zone (Table 2).

Table 2. Area of various zones in the Fordoche study area and the sampling stations located in each zone.

	Area (km^2)	Sampling Station
Permanent-water area	0.221	P-1, P-2, P-3, P-4, P-5, P-6
Overflow area	8.583	- - - -
<3.35 m(<11 ft) NGVD	4.255	- - - -
West of Bayou Fordoche	3.987	0-1, 0-2
East of Bayou Fordoche	0.288	0-3, 0-4
3.35 to 4.27 m (11 to 14 ft) NGVD	2.921	0-5, 0-6, 0-7, 0-8
4.27 to 4.88 m (14 to 16 ft) NGVD	1.089	0-9, 0-10
4.88 to 6.10 m (16 to 20 ft) NGVD	0.318	0-11, 0-12
Total area	8.804	- - - -

Crawfish production was estimated using methods and conventions described by Waters (1977) and Ricker (1958). Cohort growth was estimated from length frequencies of the captured crawfish (Figure 7). Mode lengths plotted against time (Figure 8) showed a reasonable progression in size, i.e., a sigmoid curve with growth relatively slow during the cooler time of year, rapidly increasing as temperatures increased, and slowing down as the crawfish matured. Sheppard (1974) found a somewhat similar growth pattern for crawfish in the Alligator Bayou area of Louisiana, i.e., relatively slow from January to March, rapid from March to June, and slow after June. Mode lengths were converted to weights through the use of formula 1. This gave weights of 0.268, 0.425, 2.210, 4.899, and 5.410 g for February, March, April,

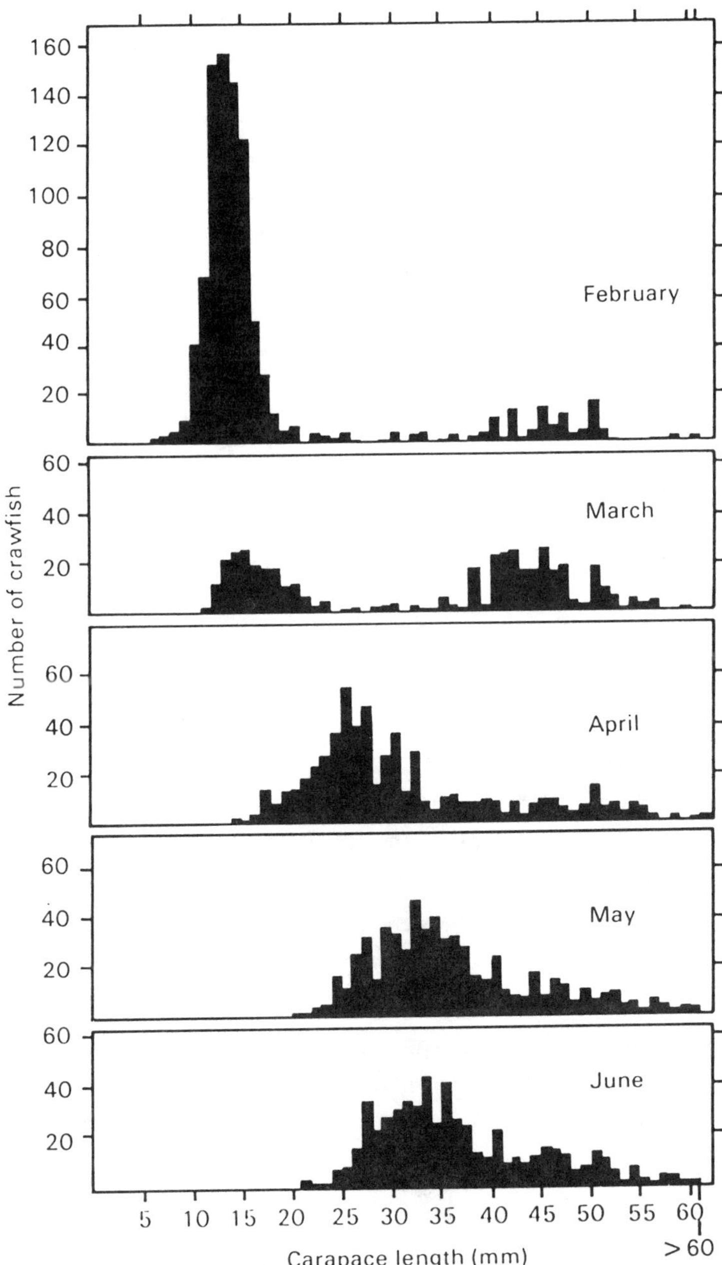

Figure 7. Length frequencies of red swamp crawfish captured in wire traps fished in the Fordoche study area during 1980.

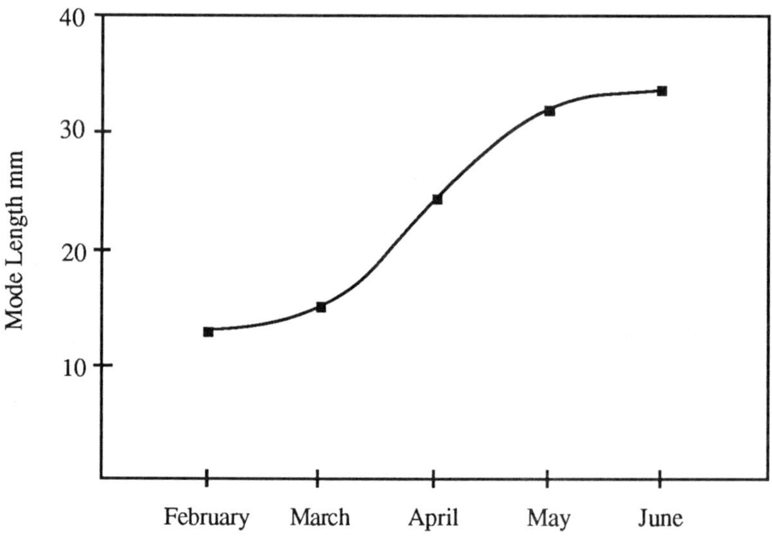

Figure 8. Mode-length growth curve for red swamp crawfish in the Fordoche study area during 1980.

May, and June, respectively. Instantaneous growth rates (G) were calculated from the weights by taking $\log_e$ of the ratio of the weight at the end of an interval to the weight at the beginning. This resulted in estimates of G for four time intervals, i.e., February-March, March-April, April-May, and May-June. Each value of G was considered to be representative of all size crawfish in the corresponding time interval. An estimate of the mean standing crop in each time interval was obtained by averaging the estimated standing crop in the study area at the beginning and end of the time interval. Red swamp crawfish production during each time interval was estimated from

$$P = GB \tag{4}$$

where P is production and B is the mean biomass.

Once the water receded, the number of crawfish burrows in the overflow area was estimated from a survey conducted from September 30, to October 3, 1980. A 45.72-m by 30.48-m grid with sampling points

15.24 m apart, for a total of 12 sampling points, was laid out at each of the overflow sampling stations, i.e., sampling stations O-1 through O-12. All burrows in a 10.1-m by 1-m strip were enumerated at each sampling point giving a total sampling area of 121.2 m^2 at each station. The mean number of burrows in each overflow zone was weighted by the area of the zone in order to estimate the total number of burrows in the overflow area.

The adult finfish population was sampled with a variable-voltage D.C., boat-mounted electroshocker and trammel nets. Each electroshocking sample consisted of all the finfishes recovered from a strip of water that took ten minutes to cover. The trammel nets used were 1.8-m deep and 46-m long with a 2.54-cm, square-mesh inner wall and 10-cm, square-mesh outer walls. The nets were fished for 24 hours. Electroshocking at permanent-water stations was reasonably efficient and gave good reproducibility. However, electroshocking in the overflow areas was very inefficient. The presence of trees, shrubs, stumps, and debris made it extremely difficult to observe and collect the finfishes once they were stunned and hindered movement of the boat through the area. Because of this, electroshocking undoubtedly drastically underestimated the abundance of finfishes in the overflow forested area; however, it probably gave good estimates of the relative abundance of finfishes in different permanent-water areas. Trammel nets were efficient in sampling finfishes in both the permanent-water and overflow-water areas. These nets also had the advantage that they could be used to sample finfishes in the water-edge area which was not possible with the electroshocker. Electroshocking sampling was conducted from February through August 1980, while trammel-net sampling was conducted from March through August, 1980.

Young-of-the-year game finfishes were collected with hand-held dip nets and by seining and electroshocking. Equal amounts of time and effort (two persons for 1 hour) were spent dip netting at each station each month. Bag seines, 6.10 by 1.83 m made of 6.35-mm, square-mesh seine material were used to sample young finfishes in shallow areas. Seining was not performed in bayou or canal areas because of water depth and steepness of the banks. Electroshocking was as previously described. Approximately equal effort was expended in dip netting, electroshocking, seining, or a combination of all three at each station. Dip netting was most effective in capturing young-of-the-year game finfishes for approximately 2 months after the initial spawn was observed.

Young-of-the-year largemouth bass, white crappie, black crappie, and warmouth collected were saved for further analysis and food-habit studies. Young finfishes were allowed to suffocate, as a precaution against regurgitation, before being preserved in 10% formalin. So far resources have allowed only the analysis of the young largemouth bass collected. In the laboratory, young bass were removed from field containers, rinsed in tap water, blotted with absorbent paper towels, weighed on an electrobalance to the nearest 0.1 mg, and both total and standard length to the nearest 0.5 mm determined before transferring each fish to 80% ethanol. When more than 20 young bass were captured at a sampling station, 20 were randomly sub-sampled and analyzed. Gut contents were removed with forceps under a stereo-zoom dissecting microscope, identified, assigned to a size class, and the contents enumerated. The contents were identified to species level whenever possible; however, the advanced state of digestion of some organisms allowed identification to family level only. Identifications were made either on a compound (100 x magnification) or stereo-zoom microscope.

Linear food-selection indices described by Strauss (1979) were calculated for young-of-the-year largemouth bass using the major taxonomic groups present in zooplankton samples and vegetation dip samples from floating water hyacinth-duckweed mats for the selection comparisons. (Results from the sampling of zooplankton and organisms in floating vegetation mats in the Fordoche study area are not presented in this report.) The index

$$L = r_i - p_i \qquad (5)$$

where L is the linear food-selection index and r_i and p_i are the relative abundances or proportion of the *i*th prey item in the gut and habitat, respectively, was calculated for sampling periods when young bass consumed zooplankton (April through June). Values of L range from -1.0 to +1.0, with positive numbers indicating active selection and negative numbers indicating avoidance or inaccessibility of the prey.

RESULTS

Fordoche Study Area Vegetation Zones

The woody vegetation community structure at each of the overflow stations is given in Table 3. Latin names are included in the table, therefore common names only are given in the text. Above the 5.18-m NGVD contour only subjective visual estimates of the forest community are available. The predominant woody vegetation at the 5.49-m and 5.79-m elevation was roughleaf dogwood, green ash, and sugarberry, while American elm (*Ulmus americana*) was common. Small amounts of red maple, black willow, eastern cottonwood, and bald cypress were present. American elm, green ash, and roughleaf dogwood were predominant at the 6.10-m elevation, while sugarberry was common and lesser amounts of bald cypress, American sycamore, and boxelder were present. At the 6.40-m elevation American elm, American sycamore, green ash, and sugarberry were predominant while boxelder was common and lesser amounts of black willow, bald cypress, persimmon (*Diospyros virginiana*), and red maple were present. Above the 6.40-m elevation the forest was a mixture of American sycamore, boxelder, red maple, roughleaf dogwood, blue beech (*Carpinus caroliniana*), and hickory (*Carya sp.*).

The absence of oaks at the higher elevations was probably due to their removal by logging. Some of the area at the 5.18-m elevation had obviously recently been heavily logged resulting in a very open canopy. Undoubtedly, logging has had some effect on the woody-community structure throughout the study area.

The timing and extent of flooding during the growing season has great influence on the structure of the forest community. Therefore, a record of the history of flooding in the study area over a 20-year period was reconstructed from USACE's gaging data at Cleon, Louisiana (Table 4).

Based on the forest community present and the history of flooding, I subdivided the study area into the ecological transition zones described by Larson et al. (1981). I considered all of the area below the 3.35-m elevation to be in their Zone II, i.e., with a soil-moisture regime "intermittently exposed;" the area between the 3.35-m and 4.27-m

Table 3. Woody-vegetation community structure in the Fordoche study area during 1980.

	Station			
	O-1	O-2	O-3	O-4
Station Location[a]	S,W	S,M	N,E	S,E
Station Elevation[b]	9.5	9.4	8.6	9.1
Number of Stems·km^{-2}	1390	2008	1617	2152
Basal Area (m^{2}km^{-2})	4654	5072	4211	5160
Relative Density[c] (Relative Dominance[d])				
Bald cypress, *Taxodium distichum*	5.9(37.4)	2.1(18.4)	2.6(11.0)	2.9(15.4)
Water elm, *Planera aquatica*	37.0(10.1)	29.7(13.0)	11.5(4.4)	31.1(11.3)
Black willow, *Salix nigra*	51.9(52.3)	26.2(60.9)	37.6(79.1)	15.3(69.7)
Swamp privet, *Forestiera acuminata*	2.2(0.2)	12.8(1.2)	36.3(3.3)	26.8(2.0)
Buttonbush, *Cephalanthus occidentalis*	3.0(0.1)	29.2(6.4)	12.1(2.2)	23.4(1.7)
Water locust, *Cleditsia aquatica*	–	–	--	0.5(0.01)
Green ash, *Fraxinus pennsylvanica*	–	–	--	–
Water hickory, *Carya aquatica*	–	–	--	–
Overcup oak, *Quercus lyrata*	–	–	--	–
Sugarberry, *Celtis laevigata*	–	–	--	–
Eastern cottonwood, *Populus deltoides*	–	–	--	–
Nuttall oak, *Quercus nuttallii*	–	–	--	–
Possumhaw, *Ilex decidua*	–	–	--	–
Boxelder, *Acer negundo*	–	–	--	–
Red maple, *Acer rubrum*	–	–	--	–
Roughleaf dogwood, *Cornus drummondii*	–	–	--	–
American sycamore, *Platanus occidentalis*	–		--	

Table 3. (Continued).

	Station			
	O-5	O-6	O-7	O-8
Station Location[a]	N,E	S,E	N,E	S,E
Station Elevation[b]	12.5	10.8	13.0	13.0
Number of Stems·km^{-2}	2193	2296	2018	1472
Basal Area (m^{2}km^{-2})	5460	8500	7452	5751
Relative Density[c] (Relative Dominance[d])				
Bald cypress, *Taxodium distichum*	1.4(6.2)	4.5(32.5)	1.5(21.1)	3.5(11.8)
Water elm, *Planera aquatica*	20.2(13.3)	11.2(5.7)	5.1(2.0)	11.9(4.8)
Black willow, *Salix nigra*	2.8(12.6)	4.5(47.8)	3.6(42.6)	-
Swamp privet, *Forestiera acuminata*	44.1(4.1)	70.9(3.1)	64.8(2.4)	38.5(1.8)
Buttonbush, *Cephalanthus occidentalis*	12.2(2.2)	3.1(0.4)	6.1(0.7)	4.9(0.2)
Water locust, *Cleditsia aquatica*	4.7(2.3)	0.9(1.7)	0.5(0.3)	-
Green ash, *Fraxinus pennsylvanica*	9.9(52.3)	1.8(4.4)	8.7(24.1)	7.0(48.9)
Water hickory, *Carya aquatica*	1.9(3.4)	3.1(4.4)	7.1(5.7)	32.2(28.2)
Overcup oak, *Quercus lyrata*	2.8(4.5)	-	2.6(1.1)	1.4(3.5)
Sugarberry, *Celtis laevigata*	-	-	-	0.7(0.9)
Eastern cottonwood, *Populus deltoides*	-	-	-	-
Nuttall oak, *Quercus nuttallii*	-	-	-	-
Possumhaw, *Ilex decidua*	-	-	-	-
Boxelder, *Acer negundo*	-	-	-	-
Red maple, *Acer rubrum*	-	-	-	-
Roughleaf dogwood, *Cornus drummondii*	-	-	-	-
American sycamore, *Platanus occidentalis*	-	-	-	-

Table 3. (Concluded).

	Station			
	O-9	O-10	O-11	O-12
Station Location[a]	N,E	S,E	N,E	S,E
Station Elevation[b]	15.0	14.9	17.1	16.1
Number of Stems·km^{-2}	1503	1946	1081	1184
Basal Area (m^{2}km^{-2})	6162	5996	2957	4577
Relative Density[c] (Relative Dominance[d])				
Bald cypress, *Taxodium distichum*	-	4.8(16.3)	-	2.6(6.7)
Water elm, *Planera aquatica*	3.4(2.0)	4.2(2.3)	-	-
Black willow, *Salix nigra*	-	1.1(19.6)	-	1.7(13.2)
Swamp privet, *Forestiera acuminata*	52.7(3.3)	65.6(4.2)	-	2.6(0.04)
Buttonbush, *Cephalanthus occidentalis*	1.4(0.02)	2.1(0.1)	-	-
Water locust, *Cleditsia aquatica*	-	-	-	-
Green ash, *Fraxinus pennsylvanica*	20.6(74.9)	15.3(52.7)	21.9(66.0)	28.7(58.1)
Water hickory, *Carya aquatica*	12.3(9.6)	-	-	1.7(0.02)
Overcup oak, *Quercus lyrata*	7.5(5.5)	4.8(4.8)	3.8(1.4)	3.5(3.6)
Sugarberry, *Celtis laevigata*	1.4(2.7)	1.1(0.04)	15.2(4.7)	27.0(7.1)
Eastern cottonwood, *Populus deltoides*	-	-	8.6(14.5)	2.6(3.5)
Nuttall oak, *Quercus nuttallii*	0.7(1.9)	-	-	-
Possumhaw, *Ilex decidua*	-	1.1(0.02)	-	1.7(0.02)
Boxelder, *Acer negundo*	-	-	45.7(12.6)	27.8(7.6)
Red maple, *Acer rubrum*	-	-	1.0(0.2)	-
Roughleaf dogwood, *Cornus drummondii*	-	-	1.0(0.02)	-
American sycamore, *Platanus occidentalis*		-	2.9(0.7)	-

[a] Station codes: S = south of pipeline canal, N = north of pipeline canal, E = east of Bayou Fordoche, M = between Bayou Fourdoche and Bayou Fusilier, W = west of Bayou Fusilier.
[b] Elevation in feet NGVD.
[c] Percentage of total number of stems.
[d] Percentage of total basal area.

Table 4. Number of years various elevations in the Fordoche study area were flooded at selected dates during a 20-year period (1961 through 1980), based on USACE gaging data in the Fordoche Area. The percentage of the time flooded is given in parentheses.

	Elevation NGVD							
	3.35 m (11 ft)	3.66 m (12 ft)	3.96 m (13 ft)	4.27 m (144 ft)	4.57 m (15 ft)	4.88 m (16 ft)	5.18 m (17 ft)	5.49 m (18 ft)
May 1	20(100)	19(95)	19(95)	18(90)	17(85)	14(70)	13(65)	12(60)
May 15	20(100)	20(100)	19(95)	18(90)	16(80)	16(80)	14(70)	9(45)
June 1	20(100)	20(100	17(85)	16(80)	15(75)	13(65)	10(50)	6(30)
June 15	18(90)	15(75)	14(70)	14(70)	12(60)	9(45)	5(25)	3(15)
July 1	14(70)	12(60)	11(55)	8(40)	6(30)	6(30)	4(20)	3(15)
July 15	12(60)	8(40)	7(35)	7(35)	4(20)	4(20)	3(15)	0(0)

elevations to be in their Zone III, i.e., "semipermanently inundated or saturated;" the area between the 4.27-m and 6.10-m elevations to be in their Zone IV, i.e., "seasonally inundated or saturated;" and the area above the 6.10-m elevation to be in their Zone V, i.e., "temporarily inundated or saturated."

Finfish Fauna and Habitats Utilized

The completeness of any listing of the finfish fauna in a floodplain is dependent upon the habitats sampled and the time of year of sampling. The types of sampling methods used will also affect completeness of the listing, e.g., large-mesh gill and trammel nets will not capture most species of minnows and shiners. Since extensive finfish sampling has been conducted using a variety of methods and in most habitats in the leveed Atchafalaya Basin between Morgan City and Simmesport, I feel the listing of finfishes presented in Table 5 is fairly complete. Only common finfish species names are given in the text since Latin names appear in the table.

Table 5. Finfish fauna of the leveed Atchafalaya Basin between Morgan City and Simmesport, Louisiana.[1]

Sturgeons - Acipenseridae
Shovelnose sturgeon,
Scaphirhynchus platorynchus (5)

Paddlefishes - Polyodontidae
Paddlefish,
Polyodon spathula (1,2)

Gars - Lepisosteidae
Spotted gar,
Lepisosteus oculatus (1,2)
Longnose gar,
Lepisosteus osseus (1,2)
Shortnose gar,
Lepisosteus platostomus (1,2)
Alligator gar,
Lepisosteus spatula (1,2)

Bowfins - Amiidae
Bowfin,
Amia calva (1,2)

Tarpons - Elopidae
Ladyfish,
Elops saurus (6)

Freshwater eels - Anguillidae
American eel,
Anguilla rostrata (3)

Snake eels - Ophichthidae
Speckled worm eel,
Myrophis punctatus (4)

Herrings - Clupeidae
Skipjack herring,
Alosa chrysochloris (5)
Gizzard shad,
Dorosoma cepedianum (1,2)
Threadfin shad,
Dorosoma petenense (1,2)
Gulf menhaden,
Brevoortia patronus (6)

Anchovies - Engraulidae
Bay anchovy,
Anchoa mitchilli (6)

Mooneyes - Hiodontidae
Goldeye,
Hiodon alosoides (5)
Mooneye,
Hiodon tergisus (5)

Pikes - Esocidae
Grass pickerel,
Esox americanus vermiculatus (1,2)
Chain pickerel,
Esox niger (1,2)

Carps and minnows - Cyprinidae
Common carp,
Cyprinus carpio (1,2)
Cypress minnow,
Hybognathus hayi (1,2)
Mississippi silvery minnow,
Hybognathus nuchalis (4)
Speckled chub,
Hybopsis aestivalis (4)

Table 5. (Continued).

Carps and minnows (continued)

Silver chub,
Hybopsis storeriana (4)
Golden shiner,
Notemigonus crysoleucas (1,2)
Emerald shiner,
Notropis atherinoides (4)
River shiner,
Notropis blennius (4)
Ghost shiner,
Notropis buchanani (4)
Red shiner,
Notropis lutrensis (4)
Chub shiner,
Notropis potteri (4)
Silverband shiner,
Notropis shumardi (4
Weed shiner,
Notropis texanus (4)
Ribbon shiner,
Notropis fumeus (4)
Taillight shiner,
Notropis maculatus (4)
Blacktail shiner,
Notropis venustus (4)
Mimic shiner,
Notropis volucellus (4)
Pubnose minnow
Notropis emiliae (4)
Bullhead minnow,
Pimephales vigilax (4)
Flathead chub,
Hybopsis gracilis (4)

Suckers - Catostomidae
Blue sucker,
Cycleptus elongatus (4)
River carpsucker,
Carpiodes carpio (1,2)
Spotted sucker,
Minytrema melanops (4)
Smallmouth buffalo,
Ictiobus bubalus (1,2)
Bigmouth buffalo,
Ictiobus cyprinellus (1,2)
Black buffalo,
Ictiobus niger (1,2)

Suckers - Catostomidae (continued)

Chubsucker,
Erimyzon sp. (4)
Quillback
Carpiodes cyprinus (4)

Bullhead catfishes - Ictaluridae
Blue catfish,
Ictalurus furcatus (1,2)
Channel catfish,
Ictalurus punctatus (1,2)
Flathead catfish,
Pylodictis olivaris (1,2)
Yellow bullhead,
Ictalurus natalis (1,2)
Black bullhead,
Ictalurus melas (1,2)
Brown bullhead,
Ictalurus nebulosus (4)
Tadpole madtom,
Noturus gyrinus (1,2)

Pirate perches - Aphredoderidae
Pirate perch,
Aphredoderus sayanus (1,2)

Needlefishes - Belonidae
Atlantic needlefish,
Strongylura marina (2)

Killifishes - Cyprinodontidae
Rainwater killifish,
Lucania parva (1,2)
Golden topminnow,
Fundulus chrysotus (1,2)
Blackspotted topminnow,
Fundulus olivaceus (1,2)
Least killifish,
Heterandria formosa (1,2)

Livebearers - Poeciliidae
Mosquitofish,
Gambusia affinis (1,2)
Sailfin molly,
Poecilia latipinna (1,2)

Table 5. (Concluded).

Silversides - Atherinidae
  Brook silverside,
    Labidesthes sicculus (4)
  Rough silverside,
    Membras martinica (4)

Pipefishes - Syngnathidae
  Gulf pipefish,
    Syngnathus scovelli (4)

Temperate basses - Percichthyidae
  White bass,
    Morone chrysops (1,2)
  Yellow bass,
    Morone mississippiensis (1,2)
  Striped bass
    Morone saxatilis (4)

Sunfishes - Centrarchidae
  Largemouth bass,
    Micropterus salmoides (1,2)
  Spotted bass,
    Micropterus punctulatus (1,2)
  White crappie,
    Pomoxis annularis (1,2)
  Black crappie,
    Pomoxis nigromaculatus (1,2)
  Bluegill,
    Lepomis macrochirus (1,2)
  Redear sunfish,
    Lepomis microlophus (1,2)
  Longear sunfish,
    Lepomis megalotis (1,2)
  Spotted sunfish,
    Lepomis punctatus (1,2)
  Bantam sunfish,
    Lepomis symmetricus (1,2)
  Orangespotted sunfish,
    Lepomis humilis (1,2)
  Green sunfish,
    Lepomis cyanellus (1,2)
  Warmouth,
    Lepomis gulosus (1,2)

Sunfishes - Centrarchidae (continued)
  Flier,
    Centrarchus macropterus (1,2)
  Banded pygmy sunfish
    Elassoma zonatum (4)

Perches - Percidae
  Logperch,
    Percina caprodes (1,2)
  Mud darter,
    Etheostoma asprigene (1,2)
  Swamp darter
    Etheostoma fusiforme (1,2)
  Bluntnose darter,
    Etheostoma chlorosomum (1,2)
  Slough darter,
    Etheostoma gracile (1,2)
  Cypress darter,
    Etheostoma proeliare (1,2)
  Sauger,
    Stizostedion canadense (5)

Drums - Sciaenidae
  Freshwater drum,
    Aplodinotus grunniens (2)

Mullets - Mugilidae
  Striped mullet,
    Mugil cephalus (4)
  White mullet,
    Mugil curema (4)

Gobies - Gobiidae
  Clown goby,
    Microgobius gulosus (4)

Lefteye flounders - Bothidae
  Southern flounder,
    Paralichthys lethostigma (6)

Soles - Soleidae
  Hogchoker,
    Trinectes maculatus (4)

[1] Key: 1, uses overflow wooded areas for spawning and/or rearing of young; 2, uses overflow wooded areas for feeding; 3, spawns in marine waters and adults and subadults use overflow wooded and permanent-water areas for feeding; 4, how extensively overflow wooded areas are used is not known; 5, uses permanent-water areas with current, how extensively overflow wooded areas are used is not known; 6, mainly inhabits marine and/or estuarine waters, a few individuals invade area during low-water periods. See text for source of data listing. Nomenclature from Robins et al. (1980).

Also given in Table 5 are the habitat types utilized by the finfish fauna. This is based on collected data, much of which will be summarized in later sections of this report, and my personal observations in the Atchafalaya Basin over almost a 30-year time span. In listing a habitat type as being utilized, I was conservative and listed it only if its use was extensive and based on appreciable data and/or observations. Many of the species whose use of overflow wooded areas are noted in Table 5 as not being known, in my opinion probably do use overflow areas extensively.

Ninety-five species of finfishes, representing 28 families, are known to occur in the leveed Atchafalaya Basin. The listing of striped bass, fairly recently introduced by humans into the Atchafalaya Basin, is based on Horst's (1976) work. Of the species present, 51 or 54% use overflow wooded areas for spawning and/or rearing of young, while 53 or 56% use these areas for feeding (Table 6). Two species (American eel and striped mullet) which spawn in marine waters use the overflow wooded areas and permanent waterbodies extensively for feeding. It is evident that overflow wooded areas are very important to the finfish fauna of the Atchafalaya Basin.

To the listing should be added three aquatic animals that, even though they are not true fish, are very important to the sport and commercial fisheries and ecology of the area. These are the red swamp crawfish, white river crawfish (*Procambarus acutus acutus*), and blue crab (*Callinectes sapidus*). Both the red swamp and white crawfish use overflow wooded areas for spawning, rearing of young, and feeding. The blue crab spawns in saline waters, and adults and subadults use overflow wooded and permanent-water areas for feeding.

Fisheries

Sport fishing for finfishes in the Atchafalaya Basin is mainly by traditional means for largemouth bass, spotted bass, black crappie, white crappie, white bass, yellow bass, bluegill, redear sunfish, warmouth, channel catfish, and blue catfish. Crawfishes are harvested with baited drop nets or wire traps constructed of 1.905-cm poultry wire. The red swamp crawfish makes up the bulk of the sport harvest; however, at times white river crawfish are common in the catch. Blue crabs are caught mainly in baited traps.

Table 6. Number of species of finfishes using habitat types for a particular purpose in the leveed Atchafalaya Basin between Morgan City and Simmesport, Louisiana.

Key	Habitat Type and Use	Number of Species	Percentage of Total Number of Species
1.	Uses overflow wooded areas for spawning and/or rearing of young.	51	53.7
2.	Uses overflow wooded areas for feeding.	53	55.8
3.	Spawns in marine waters and subadults use overflow wooded and permanent-water areas for feeding.	2	2.1
4.	How extensively overflow wooded areas are used is not known.	31	32.6
5.	Uses permanent-water areas with current, how extensive overflow wooded areas are used is not known.	5	5.3
6.	Mainly inhabits marine and/or estuarine waters, a few individuals invade area during low-water periods.	4	4.2

Lambou (1963a) estimated that 17,723 man-days of hook-and-line sport fishing in the Atchafalaya Basin Floodway yielded a harvest of 200,168 finfishes at the rate of 11.29 per trip during a 9-day period in 1963. According to Soileau et al. (1975), the Atchafalaya Basin Floodway afforded annually 10,274,000 man-hours of recreation (mainly fishing) and a harvest of 2,800,000 finfishes, 700,000 kg crawfishes, and 216,000 crabs during 1971-1974. Because of differences in methodology, it is not possible to make direct comparisons between the two estimates of sport fishing; however, based on my personal observations, I believe sport fishing has increased greatly since 1963, partly because of improved access to the fishing grounds. Lambou (1963a) pointed out that in 1963 access to the Atchafalaya Basin Floodway was a major problem and, even though it has significantly improved, it remains a problem today. There were $1.361 \cdot 10^6$ $kg \cdot yr^{-1}$, or 841 $kg \cdot km^{-2} \cdot yr^{-1}$, of sport fishes harvested from the 1619 km^2 portion of the Atchafalaya Basin subject to overflow during 1971-74 (Table 7). Based on the data presented by Soileau et al. (1975), I estimate that crawfishes made up approximately 53% of the total harvest by weight, and that finfishes made up 47% by weight.

Commercial finfishing in the Atchafalaya Basin is mainly with gill nets, trammel nets, hoop nets, and trot lines; catfish, buffalo fish, freshwater drum, and bullheads make up the bulk of the catch. The channel catfish is the most important commercial finfish. Crawfishes and blue crabs are harvested commercially almost exclusively with baited, 1.905-cm, poultry-wire funnel traps. As with the sport fisheries, the vast bulk of the crawfish catch consists of the red swamp crawfish; the white river crawfish is the only other species of any importance.

It was estimated that 1714 commercial fishermen harvested $6.804 \cdot 10^6$ kg, or 4202 $kg \cdot km^{-2}$, of fishes from the flooded areas of the Atchafalaya Basin Floodway during 1962 (Table 7). This increased to a harvest of $12.882 \cdot 10^6 \cdot yr^{-1}$, or 7959 $kg \cdot km^{-2} \cdot yr^{-1}$, during 1971-74. Finfishes comprised 58.0% of the total weight harvested during 1962, but dropped to 21.1% in 1971-74. Concurrently crawfishes increased from 33% of the catch in 1962 to 78% in 1971-74. The harvest of finfishes decreased by $1.224 \cdot 10^6$ $kg \cdot yr^{-1}$ between the two surveys, while the crawfish harvest increased by $7.711 \cdot 10^6$ $kg \cdot yr^{-1}$.

Because no estimate was made of the sport-fish harvest during 1962, it is not possible to estimate the total harvest for that year. Based on the data from Soileau et al. (1975), the total harvest from the overflow areas of the Atchafalaya Basin Floodway during 1971-74 was $14.243 \cdot 10^6$ $kg \cdot yr^{-1}$, or 8797 $kg \cdot km^{-2} \cdot yr^{-1}$, of which $3.357 \cdot 10^6 \cdot yr^{-1}$, or 2073 $kg \cdot km^{-2} \cdot yr^{-1}$, was finfishes. This harvest compares very well with the harvest from other unfertilized, natural waters. Odum (1971) summarized yield from natural populations of mixed carnivores in unfertilized, natural water as follows: world marine average, 168.1 $kg \cdot km^{-2} \cdot yr^{-1}$; North Sea, 3026.6 $kg \cdot km^{-2} \cdot yr^{-1}$;

Table 7. Annual harvest by sport and commercial fishing from the Atchafalaya Basin Floodway. Based on stage-flow relationships and surface elevations it was estimated that 1619 km^2 of the Atchafalaya Basin Floodway overflowed during the survey years; this was used to determine areal rates.

	1963 Survey[a]				1971-1974 Survey[b]			
	Total Harvest		Carbon Removed by Fishing[c]		Total Harvest		Carbon Removed by Fishing[c]	
	$kg \cdot 10^6$	$kg \cdot km^{-2}$	$kg \cdot 10^6$	$kg \cdot km^{-2}$	$kg \cdot 10^6$	$kg \cdot km^{-2}$	$kg \cdot 10^6$	$kg \cdot km^{-2}$
Sport Fishing								
Finfish	–[d]	–[d]	–[d]	–[d]	0.635[e]	392	0.088	54.5
Crawfish	–[d]	–[d]	–[d]	–[d]	0.726	448	0.036	22.4
Blue crabs	–[d]	–[d]	–[d]	–[d]	0.018[f]	11	0.001	0.6
Subtotal	–[d]	–[d]	–[d]	–[d]	1.361	841	0.125	77.5
Commercial Fishing								
Finfish	3.946	2438	0.548	338.9	2.722	1681	0.378	233.7
Crawfish	2.268	1401	0.113	70.1	9.979	6164	0.499	308.2
Blue crabs	0.136	84	0.007	4.2	0.136[f]	84	0.007	4.2
Others	0.453	280	0.063	38.9	0.009	6	0.001	0.8
Subtotal	6.804	4202	0.731	452.1	12.882	7957	0.885	546.9
Total	--	--	--	--	14.243	8797	1.011	624.4

a From Lambou (1963), represents harvest during 1962.
b From Soileau et al. (1975), represents harvest during period July 1, 1971 through June 30, 1974.
c Fresh weight of finfishes was multiplied by 0.278 to convert to dry weight and this was multiplied by 0.5 to convert to organic carbon; fresh weight of crawfishes and crabs was multiplied by 0.1 to convert to dry organic matter and this was multiplied by 0.5 to convert to organic carbon.
d No estimates were available.
e The survey did not record the weight of sport fishes, therefore the number of fishes harvested was multiplied by 0.5 lbs (226.75 g). This should give a very minimum estimate of weight harvested.
f The survey did not record the weight of the crabs harvested, therefore the dozen of crabs harvested was multiplied by 2.0 lbs (907.26 g). This should give a very minimum estimate of weight harvested.

Great Lakes, 112.1-784.6 kg·km^{-2}·yr^{-1}; African lakes, 224.2-25,219.1 kg·km^{-2}·yr^{-1}; and U.S. small lakes 224.2-17,933.6 kg·km^{-2}·yr^{-1}. The Food and Agriculture Organization (FAO) (1980) stated that the fish yield from relatively natural lakes is: Arctic-Alpine, 70-800 kg·km^{-2}·yr^{-1}; temperate 30-10,000 kg·km^{-2}·yr^{-1}; tropical 500-20,000 kg·km^{-2}·yr^{-1}; and Mediterranean shallow brackish waters 6000-80,000 kg·km^{-2}·yr^{-1}.

However, it probably is more realistic to compare the harvest from the Atchafalaya Basin Floodway to that from other floodplain fisheries rather than to relatively stable, standing waters. Welcomme (1979) examined 21 tropical floodplain fisheries and found the harvest to depend on the maximum area flooded. On the average, he found the harvest to be 3830 kg·km^{-2}·yr^{-1} for the maximum area flooded. Approximately 92% of the deviations from the expected yield could be explained by differences in the number of fishermen. These data included some lightly harvested systems. Welcomme found that for normally exploited systems a harvest of 4000 to 6000 kg·km^{-2}·yr^{-1} for the maximum area flooded could be expected; this is considerably less than the 8797 kg·km^{-2}·yr^{-1} from the Atchafalaya Basin Floodway. However, it appears that in the fisheries examined by Welcomme, the harvest consisted entirely of finfishes. The total harvest of finfishes from the Atchafalaya Basin Floodway was 2073 kg·km^{-2}·yr^{-1} during 1971-74, which is considerably less than Welcomme's figure for normally exploited populations. The total crawfish harvest during 1971-74 was 6612 kg·km^{-2}·yr^{-1}.

The harvest of sport and commercial fishes removed 1.011·10^{6} kg·yr^{-1}, or 624.4 kg·km^{-2}·yr^{-1} of carbon from the overflow area of the Atchafalaya Basin Floodway during 1971-74 (Table 7). Lambou and Hern (1983) estimated that there was a gross water export of 18,619,166 kg·yr^{-1} and a net water export of 2,932,382 kg·yr^{-1} of carbon from the Buffalo Cove and Fordoche areas of the Atchafalaya Basin Floodway during 1976-77. They considered these exports as typical of the habitat in the Atchafalaya Basin Floodway. Measured export of carbon from the Atchafalaya Basin Floodway at Morgan City was not considered typical because of sedimentation in remnant lake and backswamp areas as well as on developing natural levees in the lower portion of the Atchafalaya Basin Floodway. The area of these hydrological subunits is 361 km^{2}, or 16.96% of the total area of the Atchafalaya Basin Floodway; 16.9% of the total carbon removed by fishing amounts to 171,466 kg or 0.92% of the gross and 5.85% of the net export by water from the subunits. Day et al. (1977) estimated that there were 156,000 $kg \cdot yr^{-1}$ of carbon removed by fishing from the 65-km^2 Lac des Allemands in Louisiana, which amounted to 1.96% of total gross export of 8,016,000 $kg \cdot yr^{-1}$ by water from the lake. Even though fishing can remove appreciable quantities of carbon from the carbon cycle, it appears to be a relatively small portion of the total carbon fluxes in the Atchafalaya Basin Floodway.

Crawfish Population Statistics

Red swamp crawfish were abundant in the Fordoche study area from mid-February (the first month of trapping) through mid-June (Table 8). By mid-July the trapable population approached zero (only one red swamp crawfish was captured in 35 trap days of effort). The number and weight of red swamp crawfish peaked in March with a mean number of 3.4 m^{-2} and a biomass of 30.3 $g \cdot m^{-2}$. From a plot of the weighted number of red swamp crawfish per unit area as a catch curve (Figure 9), it appears that the small crawfish were not fully vulnerable to trapping during February because they were not large or active enough. Assuming that there is no recruitment into the trapable population from March through June (which appears to be the case based on an examination of the length frequencies given in Figure 7), the catch curve will represent survival (Ricker 1958). A least square fit of the catch curve for the months of March through June, regressed back to February, gives an estimate of the red swamp crawfish population of 3.47 m^{-2}, or 30,526,712, for February. This is approximately 9% higher than the estimate of 27,837,000, or 3.2 m^{-2}, given in Table 8. The survival rate estimated from the fitted catch curve was 63% from February through June. Another possible source of error in the population estimates could be movement of red swamp crawfish in and out of the study area. I assume that equal movement in and out would negate this consideration; however, in any case, Konikoff (1977) found that crawfishes in the Buffalo Cove Area of the Atchafalaya Basin Floodway moved on the average less than 14 m.

Table 8. Estimates of number and weight of red swamp crawfish in overflow and permanent-water areas of the Fordoche study area during 1980.

	Feb.	Mar.	April	May	June	July
Mean[a]						
$n \cdot m^{-2}$	5.7	3.1	3.2	3.2	3.4	0.02
$g \cdot m^{-2}$	13.1	26.3	19.5	27.3	29.5	0.1
Weighted Total[b]						
$n \cdot 1000$	27,837	30,057	25,005	20,238	19,614	4
kg	82,532	266,785	171,338	209,442	204,503	22
Weighted Mean[c]						
$n \cdot m^{-2}$	3.2	3.4	2.8	2.3	2.2	0.0005
$g \cdot m^{-2}$	9.4	30.3	19.5	23.8	23.2	0.0025
Permanent Water[d]						
$n \cdot m^{-2}$	0.8	2.0	1.6	0.4	3.2	0.02
$g \cdot m^{-2}$	7.4	15.5	11.8	3.4	27.5	0.01
East Edge[e]						
$n \cdot m^{-2}$	42.1	14.6	10.4	15.9	7.6	–
$g \cdot m^{-2}$	46.5	127.5	38.5	96.9	60.2	–
West Edge[f]						
$n \cdot m^{-2}$	0.1	0.2	4.0	0.6	0.2	–
$g \cdot m^{-2}$	0.002	0.5	25.5	9.7	4.2	–
Eastern Overflow[g]						
$n \cdot m^{-2}$	1.2	0.4	3.2	3.3	1.2	–
$g \cdot m^{-2}$	13.3	5.3	24.7	36.2	7.9	--
Western Overflow[h]						
$n \cdot m^{-2}$	0.2	0.2	0.0	0.3	1.6	–
$g \cdot m^{-2}$	2.8	1.6	0.0	2.8	30.5	–
Northern Overflow[i]						
$n \cdot m^{-2}$	1.9	0.4	3.2	2.7	1.8	–
$g \cdot m^{-2}$	23.4	1.8	26.6	32.4	9.0	–
Southern Overflow[j]						
$n \cdot m^{-2}$	0.4	0.3	2.2	2.7	1.1	–
$g \cdot m^{-2}$	3.0	4.3	15.2	27.6	18.6	–
Northeastern Edge[k]						
$n \cdot m^{-2}$	46.9	26.8	13.5	16.4	13.0	–
$g \cdot m^{-2}$	62.4	241.9	52.8	96.3	98.8	–
Southeastern Edge[l]						
$n \cdot m^{-2}$	37.2	2.4	7.2	15.4	2.2	–
$g \cdot m^{-2}$	30.6	13.1	24.1	97.5	21.6	–

a Mean of all sampling stations.
b The sampling data in a habitat zone were weighted by the area of the zone.
c The weighted total was divided by 8.804 km^2, the total area of the study area up to the 6.10-m (20-ft) contour.
d Mean of the sampling stations in permanent-water areas.
e Mean of the sampling stations in the water-edge zone east of Bayou Fordoche.
f Results from the one sampling station located at the western end of the pipeline canal in the water-edge zone.
g Mean of all the sampling stations located in the overflow area east of Bayou Fordoche and excluding the water-edge sampling stations.
h Mean of all the sampling stations located in the overflow area west of Bayou Fordoche and excluding the water-edge sampling station.
i Mean of all the sampling stations located in the overflow area north of the pipeline canal and excluding the water-edge sampling stations.
j Mean of all the sampling stations located in the overflow area south of the pipeline canal and excluding the water-edge sampling stations.
k Mean of the eastern edge sampling stations north of the pipeline canal.
1 Mean of the eastern edge sampling stations south of the pipeline canal.

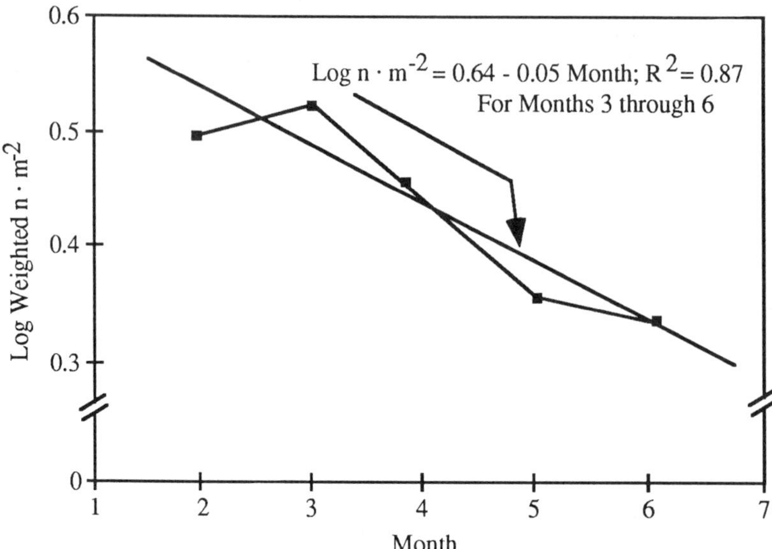

Figure 9. Catch curve for red swamp crawfish from the Fordoche study area during 1980.

Most of the red swamp crawfish were concentrated in the eastern edge zone of the Fordoche study area, evidently moving with the water as the water stages rose or dropped (Table 8). The population was usually higher in the overflow areas east of Bayou Fordoche than west of the Bayou. Also, the population was consistently higher north of the pipeline canal than south of the canal; no differences in vegetation, substrate, and water quality could be detected that might explain the differences.

Konikoff (1977) conducted mark and recapture studies in the Buffalo Cove area of the Atchafalaya Basin Floodway using 1.27-cm, hardware-cloth funnel traps. He estimated that during April, 1976, there were 21.8, 8.77, and 1.58 red swamp crawfish per m^2 in three 800-m^2 grid areas, respectively. Considering that Konikoff was using larger-mesh traps than used in our study and his estimates were not area weighted, his population estimates appear to be in the same general range as our estimates for the Fordoche study area. Konikoff caught a maximum of 40 crawfish per trap day in the Buffalo Cove area. In the Fordoche area, he reported a mean of 0.09 crawfish per trap day for 300 trap days of fishing in the vicinity of Lake Henderson during 1975 and 1976. During May and June of 1975, he reported catches of 3.5 to 6.0 crawfish per trap day in the general vicinity of the Fordoche study area; the highest catch rates occurred next to the West Atchafalaya Basin Protection Levee. Konikoff described crawfish as "not abundant in the flooded woods (1977:5)." It appeared that he did not sample the overflow area east of Bayou Fordoche where the highest population was found.

There were 2,369,368 crawfish burrows, or 0.28 m^{-2}, in the overflow area after the floodwater receded during 1980 (Table 9). Assuming that there was at least one crawfish per burrow, there would be a carryover of at least 2 million crawfish, or a survival of 12% of the population present in the study area during mid-June. However, it is not known how many red swamp crawfish migrated to permanent-water areas. Konikoff (1977) found a mean of 0.02 crawfish burrows per m^2, ranging up to 2.37 m^{-2}, in the Buffalo Cove area. He stated that burrows were not abundant in the Henderson area and were rare in the dewatered woods. If the time water receded from the overflow area of the Fordoche study area is used as a guide to the time of burrowing, crawfish initiated burrowing in mid-June with most taking place from late June through early July. Konikoff first observed burrows in July in the Buffalo Cove area.

The production of red swamp crawfish in the study area was estimated to be 613,791 $kg \cdot yr^{-1}$ or 69,717 $kg \cdot km^{-2} \cdot yr^{-1}$ during 1980 (Table 10). Production was undoubtedly somewhat higher than estimated since production was not measured before February and, as previously discussed, the population in February was probably a little higher than estimated. However, since growth during these periods was relatively slow, total production for the year probably was not appreciably higher. Most of the production (i.e., 84%) occurred from mid-March through mid-May during the period of most rapid growth.

The annual P/B ratio of 3.1 for red swamp crawfish in the study area (Table 10) appears to be reasonable when compared to values for crawfishes reported in the literature. Morrissy (1980) found mean P/B ratios of 2.65, ranging from 2.27 to 3.20, for the Australian crawfish (*Cherax tenuimanus)* in 12 ponds. Under natural conditions, cohort production of the Australian crawfish ranged from 38,400 to 103,300

Table 9. Estimates of number of crawfish burrows in the Fordoche study area during 1980.

	Station												
	0-1	0-2	0-3	0-4	05	0-6	0-7	0-8	0-9	0-10	0-11	0-12	Area[a]
Station Location[b]	S,W	S,M	N,E	S,E	N,E	S,E	N,E	S,E	N,E	S,E	N,E	S,E	–
Date Water Receded	07/07	07/07	07/08	07/07	06/24	07/02	06/21	06/21	06/12	06/12	05/13	06/07	–
Number of Burrows·m^{-2}	0.21	0.23	0.74	0.64	0.64	0.31	0.36	0.26	0.12	0.15	0.00	0.00	–
Weighted Total Number of Burrows[c]	–	–	–	–	–	–	–	–	–	–	–	–	2,369,368
Weighted Number of Burrows·m^{-2}	–	–	–	–	–	–	–	–	–	–	–	–	0.28

a Total overflow area is 8.583 km^2.

b Location codes: S = south of pipeline canal, N = north of pipeline canal, E = east of Bayou Fordoche, M = between Bayou Fourdoche and Bayou Fusilier, and W = west of Bayou Fusilier.

c Number of burrows in each overflow zone was weighted by the area in the zone.

Table 10. Estimates of red swamp crawfish production in the Fordoche study area during 1980.[a]

Time Interval	G	B(kg)	P(kg)	P(kg·km^{-2})	P/B
Feb. - March	0.461	174,659	80,518	9146	0.5
March - April	1.649	219,062	361,233	41,031	1.7
April - May	0.796	190,390	151,550	17,214	0.8
May - June	0.099	206,973	20,490	2327	0.1
Annual	--	197,771	613,791	69,717	3.1

[a] G = instantaneous growth rate, B = mean standing crop, and P = production.

kg·km^{-2}·yr^{-1} while under intensive culture production ranged from 349,000 to 920,000 kg·km^{-2}·yr^{-1}. Momot and Gowing (1983) estimated cohort P/B ratios of 4.5 to 5.7 for the crawfish *Orconectes virilis* in two Michigan lakes. Cohort production ranged from 7832 to 3695 kg·km^{-2}·yr^{-1} in one of the Michigan lakes, while it ranged from 10,850 to 6064 kg·km^{-2}·yr^{-1} in the other lake.

If the crawfish production in the Fordoche study area is considered to be typical of the Atchafalaya Basin Floodway as a whole and of production in years other than when the study was conducted, the crawfish harvest of 6612 kg·km^{-2}·yr^{-1} during 1971-74 (see previous section on fisheries) would represent an exploitation rate of 9% of the annual production.

If the total crawfish production in the study area is converted to organic carbon (fresh weight multiplied by 0.1 to convert to dry organic matter and then multiplied by 0.5 to convert to organic carbon), it would represent a flux of 30,690 kg·yr^{-1}, or 3486 kg·km^{-2}·yr^{-1}, of organic carbon. Lambou and Hern (1983) estimated an areal net water export of 8200 kg·km^{-2}·yr^{-1} from the Fordoche area during 1976-77.

Finfish Population Statistics

Considering the size and importance of the leveed Atchafalaya Basin, there is relatively little quantitative data on finfish populations. Available data include net sampling data from two mainstem stations, net and electroshocking sampling data from the Fordoche study area, and rotenone sampling data from six sampling sites in the Atchafalaya Basin Floodway.

Nets captured a total of 18,050 and 12,633 finfishes, representing 50 species, in the mainstem of the Atchafalaya River at the Simmesport and Morgan City sampling stations (Table 11). At the Simmesport station, nine of the most abundant finfishes in descending order of importance, based on the number caught per net day, were gizzard shad, shortnose gar, threadfin shad, skipjack herring, river carpsucker, blue catfish, white crappie, white bass, and freshwater drum. At the Morgan City station, the nine most abundant finfishes were gizzard shad, striped mullet, spotted gar, smallmouth buffalo, shortnose gar, channel catfish, white bass, freshwater drum, and yellow bass. Based on the weight caught per net day, the nine most dominant finfishes in descending order of importance at Simmesport were shortnose gar, gizzard shad, longnose gar, smallmouth buffalo, river carpsucker, skipjack herring, common carp, blue catfish, and white bass; at Morgan City the most dominant were striped mullet, shortnose gar, smallmouth buffalo, gizzard shad, longnose gar, shortnose gar, common carp, freshwater drum, and blue catfish.

There were differences in the finfish populations at the two stations (Table 11); some are probably due to the closer proximity of the Morgan City station to estuarine waters. For example, the striped mullet that spawns in marine waters was much more abundant at the Morgan City station than at the Simmesport station. The greater abundance of some species (e.g., skipjack herring, goldeye, sauger, blue catfish, and river carpsucker) may be due to the presence of more riverine conditions at the Simmesport station. The river at the Morgan City is more sluggish and lake-like and is subject to tidal influences. And, of course, the abundance of smaller species such as threadfin shad is probably underestimated by net samples at both stations because of the selectivity of nets for larger finfishes. These data show that a varied, abundant finfish population occurs in the mainstem areas of the Atchafalaya River.

Table 11. Total number and catch per net day for some of the more abundant finfishes caught in nets fished in the Atchafalaya River near Simmesport and Morgan City, Louisiana during 1966, 1967 and 1968.[a,b,c]

Species of Fish	Simmesport Station			Morgan City Station		
	n	$n \cdot day^{-1}$	$kg \cdot day^{-1}$	n	$n \cdot day^{-1}$	$kg \cdot day^{-1}$
White bass	536	0.577	0.215	219	0.268	0.047
Yellow bass	195	0.200	0.022	158	0.215	0.016
Blue catfish	877	0.680	0.255	180	0.175	0.074
Channel catfish	285	0.308	0.069	343	0.342	0.072
Black crappie	201	0.163	0.035	122	0.076	0.015
White crappie	634	0.598	0.121	91	0.077	0.013
Longnose gar	355	0.299	0.519	141	0.121	0.420
Shortnose gar	1967	1.602	1.335	402	0.407	0.360
Flathead catfish	109	0.058	0.067	16	0.009	0.010
Spotted gar	89	0.091	0.069	1670	1.585	1.237
Sauger	134	0.142	0.042	5	0.006	0.001
Smallmouth buffalo	691	0.433	0.409	1342	0.993	0.930
Carp	327	0.194	0.278	272	0.182	0.299
River carpsucker	1494	1.057	0.409	40	0.038	0.012
Freshwater drum	783	0.541	0.136	317	0.243	0.081
Goldeye	514	0.468	0.114	8	0.007	0.002
Skipjack herring	1089	1.064	0.346	69	0.080	0.032
Striped mullet	309	0.251	0.158	3379	2.911	1.089
Gizzard shad	5899	4.755	0.855	3360	3.514	0.479
Threadfin shad	868	1.411	0.047	61	0.094	0.003
Total	18,050	15.412	6.091	12,633	11.712	5.522

[a] n = number of fish.

[b] See text for explanation of how catch per unit of effort was calculated.

[c] In addition to the more abundant fishes caught in the nets given in the table, the following numbers of less abundant fishes were caught at Simmesport (S) and Morgan City (MC): largemouth bass: S-39, MC-31; bowfin: S-58, MC-59; southern flounder: S-12, MC-11; alligator gar: S-38, MC-29; American eel: S-3, MC-13; Chain Pickerel: S-3, MC-0; Atlantic needlefish: S-2, MC-1; ladyfish: S-0, MC-1; black buffalo: S-47, MC-32; bigmouth buffalo: S-276, MC-67; bluegill: S-37, MC-50; brown bullhead: S-17, MC-5; black bullhead: S-18, MC-5; yellow bullhead: S-8, MC-19; quillback carpsucker: S-0, MC-1, mooneye: S-5, MC-1; white mullet: S-6, MC-0; paddlefish: S-81, MC-10; golden shiner: S-0, MC-5; shovelnose sturgeon: S-7, MC-0; longear sunfish: S-3, MC-1; redear sunfish: S-5, MC-23; warmouth: S-14, MC-23; spotted sucker: S-$, MC-2; hogchoker: S-1. MC-19; blue sucker: S-3, MC-0; chubsucker: S-2, MC-2; flathead chub: S-0, MC-1; menhaden: S-4, MC-27, flier: S-1, MC-0.

Total standing crops of finfishes in the Atchafalaya Basin Floodway, estimated from rotenone sampling, ranged from approximately 25,000 to 208,000 kg·km^{-2} (Table 12). If data from the Henderson Lake Area, where sampling was the most intensive, are considered representative of the area as a whole, the waters of the type sampled support standing crops of finfishes of approximately 70,000 kg·km^{-2}. This compares favorably with estimates of standing crops of finfishes based on rotenone sampling in other types of water in Louisiana, e.g., standing crop estimates of 8193 kg·km^{-2} for impoundments; 22,585 kg·km^{-2} for Mississippi River Oxbow lakes not subject to overflow; and 49,373 kg·km^{-2} for alluvial floodplain lakes not subject to overflow (Lambou 1959c; Lambou 1960; and Lambou and Geagan 1961). It also compares favorably with estimates of standing crops in other overflow waters. Holcik and Bastl (1976) found, based on mark and recapture and repeated catching methods, a mean of 32,670 kg·km^{-2} for backwaters of the Zofin arm of the Danube. Balon (1967), as reported by Welcomme (1979), found a mean of 23,040 kg·km^{-2} for floodplain lakes of the Danube. Lambou (1959b) found, based on rotenone sampling, 44,509 kg·km^{-2} for Louisiana backwater lakes (this includes samples from Bayou Courtableau and Bayou Courtableau Pits included in Table 12). Welcomme (1979) summarized, from the literature, finfish standing crops in pools and lagoons (presumably subject to overflow) in the floodplain of some tropical rivers. These ranged from 11,000 to 183,500 kg·km^{-2} for six river systems for which there were eight or more samples.

Some caution should be used when comparing rotenone sampling data from different studies and study areas. For one thing, finfishes are usually distributed according to some type of contagious distribution (e.g., see Lambou 1962, 1963b), and the coefficient of variation for finfish rotenone-sampling data is usually quite large. Because of this, large numbers of samples are required to obtain estimates within a fairly small confidence limit (Lambou and Stern 1958; Hayne et al. 1968; Shelton et al. 1982). However, of equal importance in overflow waters is what areal basis should be used to calculate the estimates of standing crop from rotenone sampling data? During the 8 years of sampling in the Lake Henderson area, flooding ranged from 13% to 50% of the total area at the time of sampling. I could find no apparent relationship between the size

Table 12. Estimates of standing crops of finfishes in the Atchafalaya Basin Floodway based on rotenone sampling.

Location[a]	Hen	Hen	Hen	Hen
Source[b]	Lantz	Lantz	Lantz	Lantz
Date	9/11-15/67	8/26-29/68	8/25-29/69	8/10-14/70
Number of Samples	4	4	4	4
Standing Crop (kg·km^{-2})				
Paddlefish	813	-	-	-
Spotted gar	1096	658	131	8641
Longnose gar	413	815	-	-
Shortnose gar	344	-	-	-
Alligator gar	-	-	-	-
Bowfin	557	196	168	1317
Skipjack herring	70	36	-	-
Gizzard shad	23,705	21,097	31,008	28,575
Threadfin shad	2947	105	613	53
Carp	103	322	-	-
River carpsucker	-	-	-	98
Chubsucker	-	-	-	-
Smallmouth buffalo	464	-	608	6647
Bigmouth buffalo	-	1885	344	-
Black buffalo	-	-	-	-
Buffalo fishes	-	-	-	-
Blue catfish	129	34	560	145
Channel catfish	3011	1202	39	1788
Flathead catfish	-	67	-	-
Yellow bullhead	70	135	34	686
Black bullhead	-	308	-	260
Brown bullhead	-	-	-	-
White bass	151	-	28	89
Yellow bass	31	-	-	101
Largemouth bass	2387	1592	2669	2646
Spotted bass	-	-	-	-
White crappie	28	53	-	436
Black crappie	801	873	4121	1149
Bluegill	702	725	1678	2275
Redear	37	11	53	-
Longear sunfish	-	-	-	-
Spotted sunfish	-	-	-	-
Orange-spotted sunfish	-	-	-	-
Green sunfish	-	-	-	-
Warmouth	602	1211	689	185
Flier	-	67	202	-
Freshwater drum	1292	1033	252	1263
Stripped mullet	224	112	568	191
Other fishes	-	-	-	-
Total	39,976	32,536	43,766	56,542

Table 12. (Continued).

Location[a]	Hen	Hen	Hen	Hen
Source[b]	LWF	LWF	LWF	LWF
Date	9/20-23/76	6/27-30/77	7/16-29/79	8/17-20/81
Number of Samples	1	3	3	3
Standing Crop (kg·km^{-2})				
Paddlefish	-	-	-	131
Spotted gar	4909	8279	556	3990
Longnose gar	-	-	-	-
Shortnose gar	-	-	-	8985
Alligator gar	-	4738	-	4671
Bowfin	-	4229	-	-
Skipjack herring	-	-	-	-
Gizzard shad	2275	15,849	16,095	18,084
Threadfin shad	1704	370	714	6073
Carp	14,425	4580	-	788
River carpsucker	-	95	-	-
Chubsucker	-	-	-	-
Smallmouth buffalo	178,551	19,256	-	11,407
Bigmouth buffalo	796	8525	535	9034
Black buffalo	-	1995	28	48
Buffalo fishes	-	-	-	-
Blue catfish	191	333	-	296
Channel catfish	572	1147	485	2031
Flathead catfish	-	-	-	-
Yellow bullhead	-	269	-	-
Black bullhead	-	82	33	35
Brown bullhead	-	-	-	12
White bass	11	120	-	-
Yellow bass	-	48	57	-
Largemouth bass	269	669	457	942
Spotted bass	-	-	-	-
White crappie	303	504	25	1148
Black crappie	975	3161	471	337
Bluegill	1233	1035	799	637
Redear	-	71	-	6
Longear sunfish	-	3	-	-
Spotted sunfish	-	11	-	-
Orange-spotted sunfish	-	-	3	-
Green sunfish	-	8	-	-
Warmouth	168	138	265	73
Flier	-	-	5	-
Freshwater drum	1666	5188	4044	4977
Stripped mullet	168	119	21	-
Other fishes	-	3	2	15
Total	208,216	80,823	24,595	73,717

Table 12. (Concluded).

Location[a]	Hen	ABF	BCP	BC
Source[b]	Mean	Sabins	Lambou	Lambou
Date	67-81	75-77	8/54	8/54
Number of Samples	8 yrs	12	1	2
Standing Crop ($kg \cdot km^{-2}$)				
Paddlefish	118	-	-	-
Spotted gar	3533	11,265	942	22,742
Longnose gar	154	-	-	-
Shortnose gar	1166	-	-	1569
Alligator gar	1176	-	1435	359
Bowfin	808	13,887	3340	762
Skipjack herring	13	-	-	-
Gizzard shad	19,596	13,002	10,132	8070
Threadfin shad	1572	-	179	-
Carp	2527	11,926	-	-
River carpsucker	24	112	-	-
Chubsucker	-	-	-	-
Smallmouth buffalo	27,117	4439		-
Bigmouth buffalo	2640	1323	-	-
Black buffalo	259	-		-
Buffalo fishes	-	-	-	3519
Blue catfish	211	404	-	583
Channel catfish	1284	4416	-	123
Flathead catfish	8	-	-	-
Yellow bullhead	149	2600	-	-
Black bullhead	90	-	-	-
Brown bullhead	2	-	-	-
White bass	50	-	-	112
Yellow bass	30	67	-	-
Largemouth bass	1454	3688	1995	762
Spotted bass	-	79	-	-
White crappie	312	90	-	-
Black crappie	1486	3957	1704	224
Bluegill	1136	2791	5268	504
Redear	22	773	224	11
Longear sunfish	0.4	67	-	-
Spotted sunfish	1	11	22	45
Orange-spotted sunfish	0.4	-	-	-
Green sunfish	1	-	22	-
Warmouth	416	1547	269	78
Flier	34	-	-	-
Freshwater drum	2464	4618	-	1894
Stripped mullet	175	4674	-	-
Other fishes	3	157	-	-
Total	70,021	85,891	25,533	41,371

a Location codes: Hen = Henderson Lake Area; ABF = Samples from the Little Bayou Pigeon, Bayou Postillion, and Grand Lake areas of the Atchafalaya Basin Floodway, BCP = Bayou Courtableau Pitts; and BC = Bayou Courtableau.

b Source codes: Lantz = Lantz (1970); LWF = Unpublished data from the Louisiana Department of Wildlife and Fisheries; Mean = the mean of the 8 years of sampling data from the Henderson Lake area where each year was given equal weight; Sabins = Sabins (1977); and Lambou = Lambou (1958).

of the standing crop and the percentage of the area flooded. Of course, the relationship may be much more complex and undoubtedly is affected by the history of flooding during and previous to the year of sampling, as well as other factors. Undoubtedly, finfishes are concentrated as floodwaters recede; however, high mortality rates occurring as the finfishes are concentrated may compensate to a great extent for apparent increases due to the concentration of finfishes.

Based on standing-crop data presented in Table 12, the major components of the finfish community in the type of habitat sampled in the Atchafalaya Basin Floodway are spotted gar, shortnose gar, alligator gar, bowfin, gizzard shad, threadfin shad, common carp, smallmouth buffalo, channel catfish, black bullhead, largemouth bass, white crappie, bluegill, warmouth, and striped mullet.

The number and weight of finfishes caught per unit of effort by electroshocking and trammel nets in permanent water areas of the Fordoche study area are given in Tables 13 and 14. The more common finfishes in the permanent-water areas were bowfin, warmouth, bluegill, spotted sunfish, largemouth bass, black crappie, gizzard shad, golden shiner, spotted gar, striped mullet, mosquitofish, black bullhead, and yellow bullhead. Based on trammel-net samples (Tables 15 and 16), the more common finfishes in the overflow-water areas were bowfin, flier, warmouth, bluegill, black crappie, spotted gar, black bullhead, and yellow bullhead.

The total catch per unit of effort by electroshocking in permanent-water areas of the Fordoche study area was relatively low during February and March when water stages were of medium height, became extremely low during April and May when water stages were at their maximum for the year, and increased greatly as water stages dropped during June, July, and August (Table 13 and Figure 10). Trammel nets fished in permanent-water areas showed a similar pattern (Table 14 and Figure 10). The total catch per unit of effort for trammel nets fished in the overflow-water areas was much higher during April, May, and June than the catch rate in permanent-water areas (Tables 14, 15, and 16 and Figure 10). During March the catch rate for trammel nets fished in permanent-water areas was higher than the catch rate in overflow-water areas (Tables 14 and 15 and Figure 10). Water stages were dropping relatively fast during March (Figure 6). These patterns seem to indicate that finfishes move out of permanent-water areas into flooded wooded areas when water stages rise and move back into permanent-water areas when floodwaters recede. Many of the finfish species used the eastern water-edge zone extensively; the catch rate was much higher in this zone than in the other overflow-water areas (Table 16 and Figure 10).

The pattern of moving in and out of overflow-water areas was very apparent in the catch rates for flier, warmouth, and spotted gar. However, it was not as pronounced for all species (e.g., gizzard shad). Gizzard shad, as compared to many other species of finfishes common to the area, are relatively intolerant of low dissolved-oxygen levels. The relatively low numbers of gizzard shad in some of the trammel-net samples in the overflow areas may be due to the fairly low oxygen levels known to occur in such habitats during high-water periods in the Atchafalaya Basin Floodway (Hern et al. 1980) and borne out by water-quality measurements taken in the Fordoche study area.

Finfishes are very mobile and some are known to move relatively long distances. Because of the relatively small size of the Fordoche study area, there could have been movement in and out of the study area by some species of finfishes that was not completely balanced during any one sampling period. Also, because of the much larger size of the overflow-water area in the Fordoche study area as compared to the size of the permanent-water area (Table 2), the same catch rate in the overflow-water area as in the permanent-water area would indicate a much larger total population of finfishes, assuming that the catch rates are proportional to population size. Equivalent catch rates for different species of finfishes do not necessarily represent the same relative abundance, since some species are known to be more difficult to capture in nets than others (e.g., largemouth bass).

During 1980, young-of-the-year largemouth bass were first observed in the Fordoche study area during April, while young-of-the-year crappie and warmouth were first observed during May and June, respectively. A total of 413 young-of-the-year largemouth bass were collected from the Fordoche study area of which 232 were used for detailed analysis (Figure 11). Of the total number collected, 5.8% came from bayou areas, 12.3% from canal areas, 22.3% from overflow-water areas exclusive of water-edge areas, and 59.6% from water-edge areas.

Table 13. Number and weight of finfishes caught per sample by electroshocking in permanent-water areas of the Fordoche study area during 1980.[a]

Date	February	March	April
Number of Samples	18	8	6
$n \cdot s^{-1} (kg \cdot s^{-1})$			
Bowfin	0.056 (0.145)	---	---
Pirate perch	---	0.167 (0.006)	---
Brook silverside	---	---	---
Smallmouth buffalo	---	---	---
Bigmouth buffalo	---	---	---
Flier	0.056 (0.002)	---	---
Banded pygmy sunfish	---	---	---
Warmouth	---	1.167 (0.063)	---
Orange-spotted sunfish	---	---	---
Bluegill	0.222 (0.015)	1.833 (0.060)	0.167 (0.000)
Longear sunfish	---	0.083 (0.001)	---
Redear sunfish	---	---	---
Spotted sunfish	---	0.250 (0.003)	---
Bantam sunfish	0.056 (0.001)	0.830 (0.002)	---
Largemouth bass	---	0.417 (0.043)	---
White crappie	0.056 (0.000)	---	---
Black crappie	0.222 (0.011)	0.500 (0.010)	---
Gizzard shad	8.722 (0.817)	4.500 (0.314)	0.167 (0.012)
Threadfin shad	0.167 (0.002)	---	---
Common carp	---	---	---
Golden shiner	---	---	---
Grass pickerel	---	---	---
Black bullhead	---	---	---
Yellow bullhead	---	0.083 (0.011)	---
Spotted gar	0.056 (0.006)	0.083 (0.017)	---
Striped mullet	1.833 (0.453)	---	---
Yellow bass	---	---	---
Mosquitofish	---	---	---
Freshwater drum	---	---	---
TOTAL	11.444 (1.452)	9.167 (0.170)	0.333 (0.012)

Table 13. (Concluded).

Date	May	June	July
Number of Samples	6	6	8
$n \cdot s^{-1} (kg \cdot s^{-1})$			
Bowfin	---	0.583 (1.641)	1.375 (2.675)
Pirate Perch	---	0.050 (0.000)	1.000 (0.004)
Brook silverside	---	0.083 (0.000)	0.375 (0.000)
Smallmouth buffalo	---	---	---
Bigmouth buffalo	---	---	0.250 (0.652)
Flier	---	1.617 (0.052)	3.125 (0.074)
Banded pygmy sunfish	---	---	0.125 (0.000)
Warmouth	0.167 (0.003)	2.550 (0.016)	28.000 (0.474)
Orange-spotted sunfish	---	---	0.125 (0.000)
Bluegill	---	2.217 (0.018)	27.500 (0.433)
Longear sunfish	---	---	0.125 (0.001)
Redear sunfish	---	---	0.250 (0.013)
Spotted sunfish	---	0.117 (0.002)	1.500 (0.023)
Bantam sunfish	---	---	0.625 (0.000)
Largemouth bass	0.167 (0.000)	0.133 (0.014)	6.750 (0.372)
White crappie	---	0.883 (0.004)	---
Black crappie	---	1.783 (0.076)	4.500 (0.359)
Gizzard shad	---	0.167 (0.021)	1.625 (0.145)
Threadfin shad	---	---	---
Common carp	---	---	0.125 (0.131)
Golden shiner	---	2.833 (0.009)	1.250 (0.004)
Grass pickerel	---	---	0.125 (0.000)
Black bullhead	---	0.217 (0.029)	0.125 (0.053)
Yellow bullhead	---	---	---
Spotted gar	---	0.500 (0.402)	1.250 (0.749)
Striped mullet	---	---	0.125 (0.074)
Yellow bass	---	---	---
Mosquitofish	---	5.500 (0.000)	0.875 (0.000)
Freshwater drum	---	---	0.125 (0.014)
TOTAL	0.500 (0.003)	19.417 (2.285)	81.250 (6.250)

[a] n = number of finfish; s = number of samples.

Table 14. Number and weight per net day of finfishes caught by trammel nets in permanent-water areas of the Fordoche study area during 1980.[a]

Date	March	April	May
Number of Samples	6	6	6
$n \cdot s^{-1} (kg \cdot s^{-1})$			
Bowfin	2.167 (4.491)	0.500 (0.416)	1.833 (2.807)
Smallmouth buffalo	0.117 (0.014)	---	0.167 (0.142)
Bigmouth buffalo	---	---	---
Flier	0.333 (0.024)	---	0.167 (0.009)
Green sunfish	0.167 (0.009)	---	---
Warmouth	4.333 (0.765)	0.333 (0.047)	2.000 (0.139)
Bluegill	2.667 (0.260)	---	---
Spotted sunfish	---	---	---
Largemouth bass	---	---	---
White crappie	0.167 (0.094)	---	---
Black crappie	2.333 (0.302)	---	---
Gizzard shad	4.333 (0.385)	---	0.167 (0.009)
Common carp	---	---	0.167 (0.085)
Blue catfish	2.000 (0.217)	---	---
Black bullhead	0.167 (0.014)	---	---
Yellow bullhead	0.833 (0.111)	---	---
Channel catfish	2.500 (0.265)	---	---
Spotted gar	0.667 (0.390)	---	1.000 (0.515)
Longnose gar	---	---	---
Alligator gar	---	---	---
Striped mullet	0.167 (0.104)	---	---
Yellow bass	0.833 (0.097)	---	---
Freshwater drum	1.333 (0.083)	---	---
Spotted sucker	---	---	---
TOTAL	25.167(7.626)	0.833 (0.463)	5.500 (3.707)

Table 14. (Concluded).

Date Number of Samples	June 6	July 7	August 6
$n \cdot s^{-1} (kg \cdot s^{-1})$			
Bowfin	0.333 (0.605)	6.714 (11.481)	6.167 (12.426)
Smallmouth buffalo	---	---	---
Bigmouth buffalo	---	0.143 (0.085)	0.167 (0.047)
Flier	0.500 (0.016)	2.286 (0.067)	0.833 (0.034)
Green sunfish	---	---	---
Warmouth	5.167 (0.467)	9.286 (0.981)	2.833 (0.359)
Bluegill	0.167 (0.008)	0.714 (0.020)	0.167 (0.007)
Spotted sunfish	---	0.143 (0.004)	---
Largemouth bass	---	---	0.167 (0.189)
White crappie	---	---	0.167 (0.005)
Black crappie	---	2.286 (0.274)	1.333 (0.229)
Gizzard shad	---	6.571 (0.483)	32.667 (2.024)
Common carp	---	0.143 (0.108)	---
Blue catfish	---	---	---
Black bullhead	0.167 (0.014)	2.857 (0.911)	0.500 (0.222)
Yellow bullhead	0.167 (0.094)	3.429 (0.769)	4.333 (1.172)
Channel catfish	---	---	---
Spotted gar	4.333 (2.561)	19.143 (7.504)	7.833 (3.780)
Longnose gar	0.167 (0.080)	---	---
Alligator gar	---	0.143 (0.059)	---
Striped mullet	---	---	0.333 (0.146)
Yellow bass	---	---	---
Freshwater drum	---	---	---
Spotted sucker	---	0.143 (0.024)	---
TOTAL	11.000 (3.845)	54.000 (22.772)	57.500 (20.640)

[a] n = number of finfish; s = number of samples.

Table 15. Number and weight per net day of finfishes caught by trammel nets in overflow-water areas (excluding water-edge areas) of the Fordoche study area during 1980.[a]

Date	March	April	May	June
Number of Samples	4	10	10	10
$n \cdot day^{-1} (kg \cdot day^{-1})$				
Bowfin	1.750 (1.106)	2.300 (4.075)	3.200 (7.623)	2.333 (3.566)
Flier	---	0.100 (0.001)	1.200 (0.041)	3.167 (0.093)
Warmouth	2.750 (0.422)	1.500 (0.238)	7.900 (0.931)	5.667 (0.432)
Bluegill	0.250 (0.007)	0.600 (0.033)	0.800 (0.051)	0.500 (0.038)
Redear sunfish	---	0.200 (0.011)	---	---
Spotted sunfish	---	0.100 (0.003)	0.100 (0.003)	---
White crappie	---	---	0.700 (0.043)	0.167 (0.008)
Black crappie	0.250 (0.021)	---	---	---
Gizzard shad	1.500 (0.155)	0.200 (0.014)	---	---
Threadfin shad	---	---	0.100 (0.255)	---
Common carp	0.250 (0.383)	---	---	---
Black bullhead	0.500 (0.177)	0.200 (0.051)	0.900 (0.430)	0.167 (0.047)
Yellow bullhead	0.250 (0.046)	0.400 (0.125)	0.500 (0.153)	0.500 (0.151)
Spotted gar	3.750 (1.470)	2.700 (1.254)	8.500 (4.825)	9.667 (4.848)
Striped mullet	---	0.100 (0.045)	0.100 (0.031)	---
TOTAL finfish	11.250 (3.787)	8.400 (5.852)	24.000 (14.388)	22.167 (9.183)

[a] n = number of finfish.

Table 16. Number and weight per net day of finfishes caught by trammel nets in the eastern edge over flow-water area of the Fordoche study area during 1980.[a]

Date	May	June
Number of Samples	2	2
$n \cdot day^{-1}$ ($kg \cdot day^{-1}$)		
Bowfin	2.500 (5.641)	1.500 (1.984)
Bigmouth buffalo	0.500 (0.269)	---
Flier	6.500 (0.208)	23.500 (0.852)
Warmouth	29.000 (2.517)	10.500 (0.978)
Bluegill	3.000 (0.167)	2.000 (0.084)
Redear sunfish	1.000 (0.062)	---
Spotted sunfish	1.500 (0.084)	---
Largemouth bass	0.500 (0.050)	---
Black crappie	1.000 (0.156)	---
Gizzard shad	1.500 (0.156)	---
Common carp	0.500 (1.233)	---
Grass pickerel	0.500 (0.085)	---
Black bullhead	3.500 (0.836)	0.500 (0.411)
Yellow bullhead	0.500 (0.085)	0.500 (0.057)
Spotted gar	6.000 (2.722)	12.000 (6.534)
Striped mullet	1.000 (0.510)	---
TOTAL	59.000 (14.782)	51.000 (10.921)

[a] n = number of finfish.

Undoubtedly, the flooded forest, especially the water-edge area, is used extensively by largemouth bass for both spawning and rearing of young. Generally, young bass collected from the east water-edge area were larger than those from other habitats (Figure 11). This is probably largely explained by more abundant food supplies occurring in the east water-edge area, e.g., zooplankton were more abundant in this area than in any other habitat in the Fordoche study area (data not included with this report).

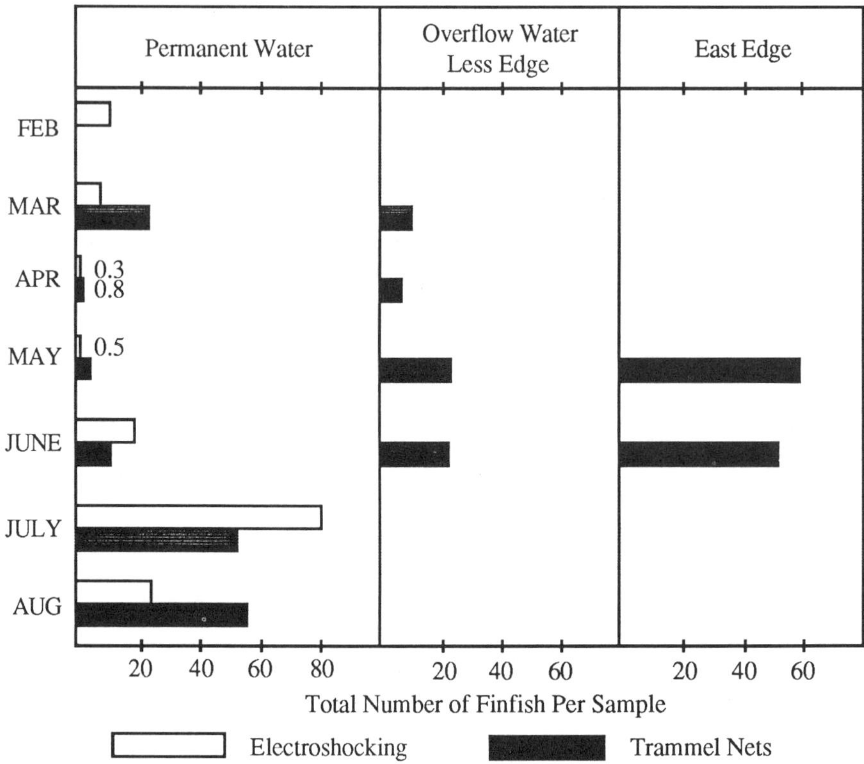

Figure 10. Total number of finfish caught per sample by electroshocking and trammel nets in various zones of the Fordoche study area during 1980.

Food Habits

Even though little has been done to document the food habits of red swamp crawfish in the Atchafalaya Basin Floodway, it is a known vegetation-detritus feeder. Konikoff (1977) examined the stomach contents of 23 large red swamp crawfish from the Buffalo Cove area. Seven were empty and 16 contained plant material and/or plant-like detritus. Of these 16, four contained only detritus, two contained root-like material resembling water hyacinth roots, four contained unidentified plant material, five contained *Polygonum spp*. seeds, and five contained insect parts.

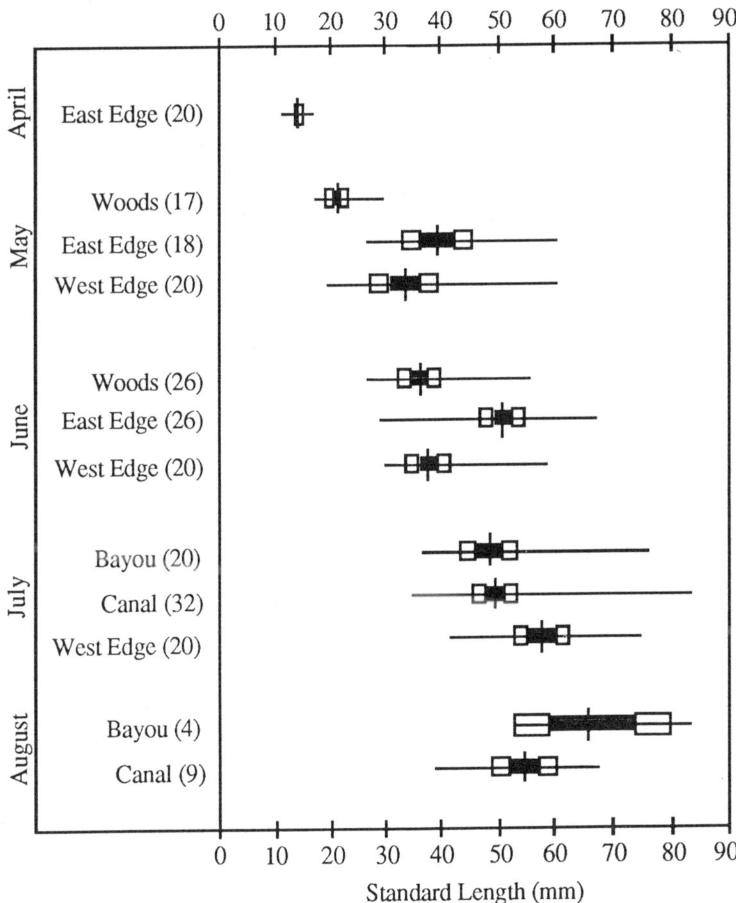

Figure 11. **Mean standard lengths (vertical lines) with standard errors (shaded bars), 95% confidence intervals (white bars), and ranges (horizontal lines) of young-of-the-year largemouth bass collected from the Fordoche study area during 1980. The numbers in parentheses indicate the sample size.**

Levine (1976) studied the food habits of the bullhead minnow and juvenile channel catfish in the Atchafalaya Basin Floodway. He found, from examining a large number of specimens (exact number examined is not stated), that bullhead minnows consumed primarily microcrustaceans (i.e., Cladocera); however, he stated that the numerical preponderance of Cladocera stems from one sample and as such does not represent a true cross-section of food habits. Insects occurred as a small but significant part of the diet; 90% of the guts contained large amounts of "amorphous organic material" and strands of filamentous algae were common. A total of 166 juvenile channel catfish were found to have fed largely upon small crustaceans, with Amphipoda most heavily utilized.

Lambou (1961) examined the stomach contents of 67 largemouth bass collected by artificial bait from The Flats area of the Atchafalaya Basin Floodway during 1958 and 120 collected from the same areas during 1959. Red swamp crawfish were found in 86% of the 28 stomachs containing food collected during 1958; finfishes were found in only 11% of the stomachs. During 1959 red swamp crawfish were found in 59% of the 35 stomachs containing food while finfishes were found in 46% of the stomachs containing food.

Lambou concluded that the difference between the 2 years in the importance of crawfish in the diet of largemouth bass in The Flats was due to the drastic reduction in the abundance of red swamp crawfish in the Atchafalaya Basin Floodway during 1959. That year crawfish from the Atchafalaya Basin Floodway were almost nonexistent in the commercial catch (Lambou 1961); it was described by Viosca as "the recent crawfish catastrophe in the Atchafalaya Floodway" (Anonymous 1959). Lambou (1961) concluded that during 1959, largemouth bass in the Atchafalaya Basin Floodway had to turn to food other than red swamp crawfish, and forage was less abundant. The bass fed less often on forage of smaller size, with the result that the population was in poorer condition. The stomachs of six spotted bass collected by artificial bait from The Flats during 1958-59 were also examined (Lambou 1961). Red swamp crawfish was the only food item in the three stomachs containing food.

Even though no formal detailed laboratory examination of the stomachs of adult finfishes collected from the Fordoche study area was made, gross field examination of the adult finfishes captured showed that most species were feeding very heavily on crawfish during the spring and summer of 1980.

A variety of organisms was consumed by young-of-the-year largemouth bass in the Fordoche study area during 1980 (Table 17). The

Table 17. Percentage of total number of each major taxon present in pooled sample of young-of-the-year largemouth bass stomachs collected from the Fordoche study area during 1980.

	April	May	June	July	August
Rotifera	0.7	<0.1	0.1	–	--
Cladocera	58.0	71.0	36.0	–	--
Copepoda	40.6	24.1	55.8	–	--
Ostracoda	--	0.2	<0.1	–	--
Amphipoda	--	–	<0.1	–	--
Isopoda	0.1	–	–	–	--
Caridea	--	0.7	1.6	52.3	33.3
Unidentified invertebrates	--	<0.1	–	3.1	--
Ephemeroptera	--	<0.1	0.4	0.8	--
Odonata-Anisoptera	--	--	<0.1	–	--
Odonata-Zygoptera	--	–	0.3	–	--
Hemiptera	--	1.5	0.5	3.9	--
Diptera	--	0.4	1.7	4.6	4.2
Osteichthyes	--	1.9	3.3	35.4	62.5
Total Number of Prey Items	969	1337	1990	129	24
Number of Stomachs	20	55	72	72	13
Range of Bass Standard length (mm)	11.2-16.2	16.7-60.2	29.5-69.0	33.0-84.7	40.0-83.5

major taxonomic groups used for food included Cladocera, Copepoda, Caridea (grass shrimp of one species, *Palaemonetes kadiakensis*), and Osteichthyes (finfishes). Cladocera and Copepoda were of major importance on both a numerical and mass basis during April; however, their importance declined as the season progressed (Figure 12). Finfishes and grass shrimp became much more important as the season progressed. The mosquitofish was the primary finfish consumed by young bass.

Percentages calculated from pooled data, as presented in Table 17, can sometimes be misleading because a few fish consuming large numbers of one small food item can heavily influence the calculated percentages. Therefore, mean percentages were calculated to give a picture of the

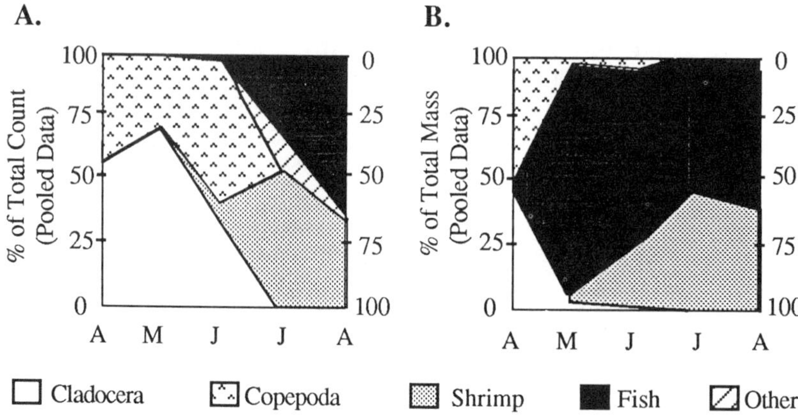

Figure 12. Percentage of total pooled count (A) and total pooled mass (B) for the major food items of young-of-the-year largemouth bass collected from the Fordoche study area during 1980.

distribution of food items in individual bass stomachs collected from the Fordoche study area (Table 18). There were differences in food habits of young bass collected from different habitats. Grass shrimp were much more important as food for bass collected from the west water-edge area than for bass collected from permanent-water areas. Finfishes were much more important in the diet of young bass collected from permanent-water areas than other areas. Linear food selection indices for young-of-the-year largemouth bass showed that during April through June they were selecting for adult cladocerans and copepods for food while avoiding rotifers (Table 19). In areas where water hyacinth mats were present they preferred copepods and selected against cladocerans and amphipods.

Levine (1976) examined the stomach contents of 264 juvenile largemouth bass seined from various areas of the Atchafalaya Basin Floodway during 1973 through 1976. Major foods were crustacea (Mysidacea, Decapoda, Amphipoda, Copepoda, and Cladocera) and insects (Ephemeroptera, Diptera, Hemiptera, and Orthoptera). He stated that the consumption of insects occurred during high water in late spring and early summer, while finfishes were not common dietary items until later in the year.

Table 18. Mean percentage of major food items appearing in stomachs of young-of-the-year largemouth bass collected from the Fordoche study area during 1980.

Month	Area	Number of Fish	Food Item				
			Cladocera	Copepoda	Caridea	Osteichthyes	Other
April	East edge	20	59	40	–	--	1
May	Overflow[a]	17	72	27	–	1	--
	East edge	18	19	17	6	49	9
	West edge	20	12	38	3	25	22
June	Overflow[a]	26	25	43	9	15	8
	East edge	26	1	1	22	63	13
	West edge	20	30	9	23	20	18
July	Bayou	20	–	–	43	44	13
	Canal	32	–	–	37	48	15
	West edge	20	–	–	80	16	4
August	Bayou	4	–	–	11	89	--
	Canal	9	–	–	35	56	6

[a] Overflow-water area exclusive of water-edge area.

DISCUSSION

As shown in Table 6 (and elaborated on more fully in Tables 5, 13, 14, 15, and 16; Figure 10; and the previous text) and pointed out by Wharton et al. (1981) and Wharton et al. (1982), bottomland hardwood forest zones are utilized very extensively by finfishes. Varied and abundant populations of finfishes occur in floodplain areas, and the size and makeup of the populations compare very favorably with that occurring in nonfloodplain habitats (Tables 11 and 12 and previous text).

If the data collected from the Fordoche study area are typical of the Atchafalaya Basin Floodway as a whole and other bottomland hardwood forest floodplain systems, red swamp crawfish and many species of

Table 19. Linear food-selection indices for young-of-the-year largemouth bass collected from the Fordoche study area during time of year (April-June 1980) that zooplankton were important in their diet. Food selection is compared to the possible prey composition of zooplankton and floating vegetation mats (water hyacinth-duckweed mats). A positive value indicates selection while a negative number indicates avoidance or inaccessibility of prey.

Month	Area	Number of Fish	Food Item			
			Cladocera	Copepoda	Rotifera	Amphipoda
Zooplankton						
April	East edge	20	+0.56	+0.36	-0.42	–
May	Overflow[a]	17	+0.79	+0.18	-0.36	–
	East edge	18	+0.60	+0.22	-0.45	–
	West edge	20	+0.09	+0.43	-0.34	–
June	Overflow[a]	26	+0.60	+0.31	-0.69	–
	West edge	20	+0.56	+0.18	-0.81	–
Vegetation Mats[b]						
May	West edge	20	-0.18	+0.49	--	-0.30
June	Overflow[a]	26	-0.06	+0.28	--	-0.07
	West edge	20	-0.18	+0.11	--	-0.08

[a] Overflow-water area exclusive of water-edge area.
[b] Indices were calculated only for areas where vegetation mats occurred.

finfishes (both adults and young-of-the-year) concentrate during high water in the water-edge zone i.e., the flooded forest generally most remote from the permanent-water channels and with depths of less than 0.5 m (Tables 8 and 16, Figure 10 and previous text). The location of the water-edge zone relative to the "ecological transition zones" described by Larson et al. (1981) depends upon the water level which varies within and between years. Therefore, the relative importance of an ecological transition zone to fishes will vary within and between years, depending upon the water level.

In the Fordoche study area, Transition Zone III was most heavily utilized during one part of the year, while Zone IV was most heavily utilized during another part of the year; the intensity of the use of each of these zones was correlated with the water level (Table 8 and Figure 6). However, crawfish burrows in the Fordoche study area were most abundant in Zone II and fairly abundant in Zone III (Table 9 and Figure 6). This shows that using crawfish burrows to delineate bottomland hardwood forest wetlands as suggested by Patrick et al. (1981) may, in some cases, be misleading. Even though red swamp crawfish heavily utilized Zone IV (Table 8), the density of burrows in this zone was either low or nonexistent (Table 9). The abundance of burrows at a particular location in the overflow-water areas appeared to be dependent upon the water level at the time of burrowing which in the Fordoche study area took place mainly during late June and early July.

Finfishes showed patterns of movement very dependent upon changes in water levels. In the Fordoche study area many finfishes appeared to move in and out of the overflow area in response to the rising or falling of the water level.

The yield of finfishes from floodplains to man is, as pointed out by Welcomme (1979), largely dependent upon the maximum area flooded (i.e., 4000 to 6000 $kg \cdot km^{-2} \cdot yr^{-1}$ per maximum area flooded for normally exploited systems). However, as would be expected, the relationship to the area flooded is complex and depends upon present and past flooding history, the time and length of flooding, the amount of permanent water during the low-water season, the kind and types of finfishes present, and the characteristics of the fishery. It is evident to most any observer of the fisheries in the Atchafalaya Basin Floodway and other overflow areas that there are very good, normal, and very bad fishing years that seem to be correlated with past and present annual hydrological regimes. For example, during the fall of 1958, water levels in the Atchafalaya Basin Floodway did not recede to their normal low levels but stayed partially up in the forest. This, coupled with relatively low water levels in 1959, evidently led to a total collapse of the commercial crawfish fishery during 1959 (Lambou 1961, Anonymous 1959).

Holcik and Bastl (1977) found that the finfish yield from the Danube River in Czechoslovakia was correlated with water levels and the flooding of the floodplain. They found positive correlations between yields and water levels in the corresponding and preceding years, while a negative correlation was found between the yields from low-water years in both the corresponding and preceding years. Muncy (1978) found that water levels in the previous 2 years had greater effects on catches than water levels the same year in the Kafue River floodplain fishery in Africa. Welcomme (1975) found that for the Kafue, the central delta of the Niger, and the Shire floodplain systems in Africa the catch in any one year is better explained by a combination of flood history from the preceding years than by any one year.

Of course, observations or correlations between yield and area of the floodplain flooded annually, even though suggestive, do not prove that there is a cause-and-effect relationship. Lambou (1959b) discussed some of the effects of overflow on finfish populations in floodplain systems. Wood (1951) and Wood and Pfitzer (1958) have discussed in some length the effects of fluctuating water levels on finfish populations and fisheries in impoundments. Even though impoundments are a different type of ecosystem, most of the phenomena described applies to floodplain systems. These authors have emphasized that the alternate exposure and flooding of land is very important in maintaining and increasing the productivity of finfish populations. They emphasize that the effects depend more on time, area, and duration of flooding than upon degree of fluctuation as measured in vertical distance.

Welcomme (1979) pointed out that the maximum amount of finfishes in floodplains is present during a flood and the minimum just before the onset of the next flood. He stated that this means there is an overproduction of ichthyomass during floods which persists throughout the period of falling water and which must be lost through total mortality during dry seasons. This implies that the amount of water remaining in the system during the dry season relative to the amount of water present during the wet is one of the major factors determining mortality. He also pointed out that simple, exponential, fish-population dynamic models dealing with mortality and growth, which are adequate for finfish populations in other types of waters, are inconsistent with what is known of the biology of floodplain fishes.

Welcomme and Hagborg (1977) have provided a detailed simulation of the dynamics of floodplain finfish communities. Their simulation indicates that the difference in high-water or flood regime produces great differences in the within-the-year ichthyomass, but that the size of the population carry-over to the following year is largely dependent on the amount of water remaining in the system during the low-water period. They found that the more water remaining during the low-water period the more the difference induced by the high-water regime is transmitted to the following year. The simulation showed the production of finfishes to be 24,100 to 56,400 $kg \cdot km^{-2} \cdot yr^{-1}$ of mean area flooded, depending on the flood regime, with P/B values between 1.35 and 1.77. Production and biomass was maximum during high floods, with P/B ratios increasing as the maximum area flooded decreased.

Lambou and Hern (1983), Hern and Lambou (1979), and Hern et al. (1978) have pointed out that primary production in the Atchafalaya Basin Floodway is primarily above the water surface and on dry land during low-water periods, while energy chains within the water are primarily dependent upon heterotrophic production. One of the keys to the high fish production of the Atchafalaya Basin Floodway is the short, efficient, aquatic, bacteria-detritus-based food chain. Consider the following aquatic heterotrophic-based system very important in the Atchafalaya Basin Floodway:

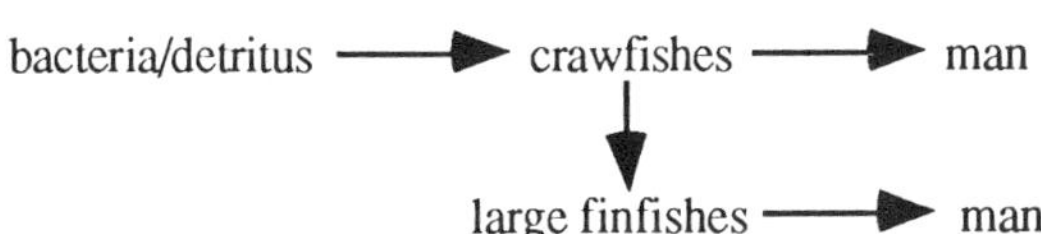

involving two or three transfers of energy, as compared to the following autotrophic-based system typical of many fresh-water lakes:

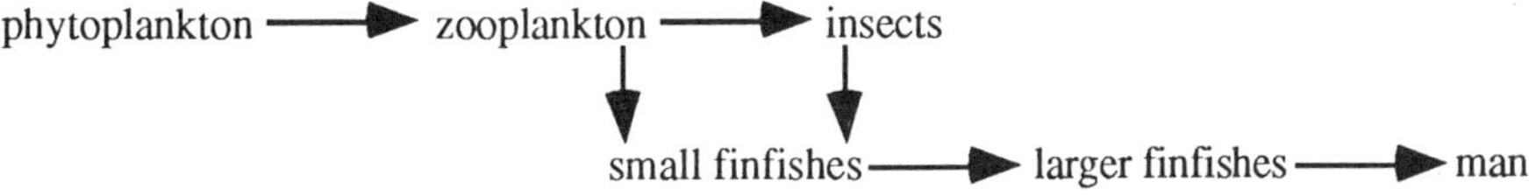

involving four or five transfers of energy. If in these systems 80% to 90% of the energy is lost at each transfer, which is considered typical for food-chain transfers, it can be readily seen that the yield to man of fishes from bottomland hardwood forest ecosystems such as the Atchafalaya Basin Floodway can be much greater than in other types of aquatic ecosystems with a food-chain base of the same size.

Bottomland hardwood forest zones are very important in determining the size and makeup of floodplain fish populations and fisheries. Fish populations (including the finfish populations in permanent-water areas during low-water periods) are dependent upon the forested areas for their food production, feeding, spawning areas, and rearing of young. In turn, the existence of the bottomland hardwood forests depends upon the hydroperiod produced by the water flows in the permanent-water areas and the dissolved and particulate constituents carried by these flows. Thus, by any measure, the two must be considered to be part of the same ecosystem. One difference is that in one case the end product is trees and the products and recreation associated with them, while in the other case the end product is fishes and the products and recreation associated with them. It is not easy to think of fishes as a part of bottomland hardwood forest ecosystems, since in the case of finfishes they are to a great extent utilized (and sampled by biologists) in permanent-water areas. Of course crawfishes, since they are normally caught in the forested areas, are more easily considered as part of bottomland hardwood forest ecosystems. Fishes and the permanent-water areas where they reside for part of the year are as much a part of bottomland hardwood forest ecosystems as are the trees commonly associated with them. Fishes in bottomland-hardwood-forest overflow floodplains are largely a product of the forest.

ACKNOWLEDGMENTS

Although published under one name, a manuscript of this type is the result of the work of many people. I am indebted to my colleagues with the United States Environmental Protection Agency; its predecessor agency, the Federal Water Pollution Control Administration; the Louisiana Wildlife and Fisheries Commission; Coastal Environments Inc., Baton Rouge, Louisiana; the Lafayette Louisiana Office of the U.S. Fish and Wildlife Service; and the Department of Biological Sciences, University of

Nevada, Las Vegas who furnished data, assisted in the collection and analysis of data, and who made many useful suggestions. Many individuals were involved in the collection and analysis of data for the Fordoche Study. These included Johannes L. van Beek, Coastal Environments Inc.; Dugan S. Sabins and Donny Davis, formally with Coastal Environments Inc.; James E. Pollard and Susan M. Melancon, University of Nevada, Las Vegas; Linda S. Blakey, Gary Kork, and Barry Baldigo, formally with the University of Nevada, Las Vegas; and personnel with the Lafayette Louisiana Office of the U.S. Fish and Wildlife Service. van Beek made most of the hydrological measurements for the Fordoche Study and assisted in numerous other ways. Davis made the woody-vegetation measurements, and Sabins conducted the survey of crawfish burrows. Melancon, Kork, Blakey, Pollard, Baldigo, and Sabins did most of the fish and crawfish sampling; personnel from the Lafayette Office of the U.S. Fish and Wildlife Service and Sabins conducted the electroshocking fish sampling. Blakey conducted most of young-of-the-year largemouth bass food-habit study.

The late James Catings of the Federal Water Pollution Control Administration made many of the Atchafalaya Main-Channel fish-population measurements. The Louisiana Wildlife and Fisheries Commission furnished unpublished data on finfish standing crops in the Atchafalaya Basin Floodway based on rotenone poisoning sampling methods. Kenneth E. Lantz and Chuck Killebrew were instrumental in making the data available. Ellis J. Clarain, Jr. and Steven G. Underwood, U.S. Army Corps of Engineers, Waterway Experiment Station, Vicksburg, Mississippi; Dale Hall, U.S. Fish and Wildlife Service, Clear Lake, Texas; Johannes L. van Beek, Coastal Environments Inc., Baton Rouge, Louisiana; Fred Bryan, Louisiana Cooperative Unit, Louisiana State University, Baton Rouge, Louisiana; and Stephen C. Hern and Llewellyn R. Williams, U.S. Environmental Protection Agency, Las Vegas, Nevada reviewed an earlier draft of this manuscript and made many useful suggestions. The information in this document has been funded wholly or in part by the U.S. Environmental Protection Agency. It has been subjected to Agency review and approved for publication.

REFERENCES CITED

Anonymous. 1959. The case of the disappearing crawfish. The Louisiana Conservationist 11(7-8):8-11, 21-22.

Balon, E. K. 1967. Vyvosichtyofauny Dunaja jej sucasy stav a pokus o prosnozu dalsich zmien po vystaube vodnych diel (Evolution of the Danube fish fauna: its recent state and an attempt for the prognosis of further changes after the hydro-electric power building) Biol. Pr. 8(1):5-121.

Bryan, C. F., D. J. Demont, D. S. Sabins, and J. P. Newman, Jr. 1976. Annual report, a limnological survey of the Atchafalaya Basin. Louisiana Cooperative Fisheries Unit. Louisiana State University, Baton Rouge, Louisiana. 285 pp.

__________, F. M. Truesdale, and D. S. Sabins. 1975. Annual report, a limnological survey of the Atchafalaya Basin. Louisiana Cooperative Fisheries Unit, Louisiana State University, Baton Rouge, Louisiana. 203 pp.

Day, J. W., T. J. Butler, and W. H. Conner. 1977. Production and nutrient export studies in a cypress swamp and lake system in Louisiana. Pages 255-269 *in* M. Wiley, ed. Estuarine processes, Vol. 2, Circulation, sediments, and transfer of materials in the estuary. Academic Press, New York, New York.

Food and Agriculture Organization. 1980. Comparative studies on freshwater fisheries. Report of a workshop held at the Instituto Italiano di Idrobiologia. Pallanza, Italy, 4-8 September 1978. United Nations FAO Fish. Tech. Pa., (198). 46 p.

Hayne, D. W., G. E. Hall, and H. M. Nichols. 1968. An evaluation of cove sampling of fish populations in Douglas Reservoir, Tennessee. Pages 244-297 *in* Reservoir fisheries resources symposium. Southern Division, American Fish Society, Washington, D. C.

Hern, S. C., V. W. Lambou. 1979. Productivity responses to changes in hydrological regimes in the Atchafalaya Basin, Louisiana. Pages 93-102 *in* Proceedings of the International Symposium on Environmental Effects of Hydraulic Engineering Works. Tennessee Valley Authority, Knoxville, Tennessee.

Hern, S. C., V. W. Lambou, and J. R. Butch. 1980. Descriptive water quality for the Atchafalaya Basin, Louisiana. U.S. Environmental Agency, Environmental Monitoring Series. EPA-600/4-80-014. 168 pp.

__________, W. D. Taylor, L. R. Williams, V. W. Lambou, M. K. Morris, F. A. Morris, and J. W. Hilgert. 1978. Distribution and importance of phytoplankton in the Atchafalaya Basin. U.S. Environmental Agency. EPA-600/3-78-001. 194 pp.

Holcik, J., and I. Bastl. 1976. Ecological effects of water level fluctuation upon the fish population in the Danube River floodplain in Czechoslovakia. Acta Sci. Nat. Acad. Sci. Bohemoslov. Brno. 10(9):1-46.

__________ and I. Bastl. 1977. Predicting fish yield in the Czechoslovakian section of the Danube River based on hydrological regime. Int. Revueges. Hydrobiol. 62(4):523-532.

Horst, J. W. 1976. Aspects of the biology of the striped bass, *Morone saxatilis* (Walbaum) of the Atchafalaya Basin, Louisiana. M.S. Thesis, Louisiana State University, Baton Rouge, Louisiana.

Konikoff, M. 1977. Study of the life history and ecology of the red swamp crawfish *Procambarus clarkii*, in the Lower Atchafalaya Basin Floodway. Final report of the Department of Biology, University of Southeastern Louisiana to U.S. Fish and Wildlife Service. 81 pp.

Lambou, V. W. MS. Fish and fisheries of the Atchafalaya Basin Louisiana and their relationships to overflow wetlands.

__________. 1958. Fish populations occurring in the backwater lakes of Louisiana. Paper presented at the 88th annual meeting of the Am. Fish. Soc. 39 pp.

__________. 1959a. Block-off net for taking fish population samples. The Proc. Fish-Culturist 21(3):143-144.

__________. 1959b. Fish populations of backwater lakes in Louisiana. Trans. of the Am. Fish Soc. 88(1):7-15.

__________. 1959c. Louisiana impoundments: their fish populations and management. Trans. N. Am. Wild. Nat. Resour. Conf. 24:187-200.

Lambou, V. W. 1960. Fish populations of Mississippi River Oxbow Lakes in Louisiana. Proc. La. Acad. Sciences 23:52-64.

__________. 1961. Utilization of macrocrustaceans for food by freshwater fishes in Louisiana and its effects on the determination of predator-prey relations. The Proc. Fish-Culturist 23:18-25.

__________. 1962. Distribution of fishes in Lake Bistineau, Louisiana. J. Wildl. Manage. 26(2):193-203.

__________. 1963a. The commercial and sport fisheries of the Atchafalaya Basin Floodway. Proc. 17th Ann. Conf. S.E. Assoc. Game and Fish Comm. 17:256-281.

__________. 1963b. Application of distribution pattern of fishes in Lake Bistineau to design of sampling programs. The Proc. Fish-Culturist (April):79-87.

__________, and D. W. Geagon. 1961. Fish populations of alluvial floodplain lakes in Louisiana. La. Acad. of Sciences 24:95-115.

__________, and S. C. Hern. 1983. Transport of organic carbon in the Atchafalaya Basin, Louisiana. Hydrobiologia 98:25-34.

__________, and H. Stern. 1958. An evaluation of some of the factors affecting the validity of rotenone sampling data. Proc. of Southeastern Association of Game and Fish Commissioners 11:91-98.

Lantz, K. E. 1970. An ecological survey of factors affecting fish production in a Louisiana backwater area and river. Fisheries Bull. No. 5, Louisiana Wildlife and Fisheries Commission, Baton Rouge, Louisiana. 60 pp.

Larson, J. S, M. S. Bedinger, F. Bryan, S. Brown, R. J. Huffman, E. L. Miller, D. G. Rhodes, and B. A. Touchet. 1981. Transition from wetlands to uplands in southeastern bottomland hardwoods forest. Pages 225-273 *in* J. R. Clark and J. Benforado, eds. Wetlands of bottomland hardwood forests. Elsevier Scientific Publishing Company, New York.

Leopold, L. B., M. B. Wolman, and J. P. Miller. 1964. Fluvial processes in geomorphology. W. H. Freeman and Co., San Francisco. 522 pp.

Levine, S. J. 1976. Food habits of selected lower basin fishes. Pages 203-212 *in* Bryan et al. Annual report, a limnological survey of the Atchafalaya Basin, Louisiana. Cooperative Fisheries Unit, Louisiana State University, Baton Rouge, Louisiana.

Momot, W. T., and H. Gowing. 1983. Some factors regulating cohort production of the crawfish, *Oreonectes virilis*. Freshwater Biol. 13:1-12.

Morrissy, N. M. 1980. Production of marron in western Australian wheat belt farm dams. Western Australian Marine Research Laboratories, Fisheries Research Bull. No. 24. 79 pp.

Muncy, R. J. 1978. An evaluation of the Zombian Kafue River Floodplain Fishery. Pages 153-164 *in* R. L. Welcomme, ed. Symposium on river and floodplain fisheries in Africa. FAO. CIFA Technical Paper No. 5, CIFA/T5.

Odum, E. P. 1971. Fundamentals of ecology, 3rd Edition. W. B. Saunders Company. 574 pp.

Patrick, W. H., G. Dissmeyer, D. D. Hook, V. W. Lambou, H. M. Leitman, and C. H. Wharton. 1981. Characteristics of wetland ecosystems of southeastern bottomland hardwoods forests. Pages 276-300 *in* J. R. Clark and J. Benforado, eds. Wetlands of bottomland hardwood forests. Elsevier Scientific Publishing Company, New York.

Ricker, W. E. 1958. Handbook of computations for biological statistics of fish populations. Bull. No. 119, Fisheries Research Board of Canada. 330 pp.

Robins, C. R., R. M. Bailey, C. E. Bond, J. R. Brooker, E. A. Lachner, R. N. Leu, and W. B. Scott. 1980. A list of common and scientific names of fishes from the United States and Canada, 4th ed. Am. Fish. Soc. Spec. Publ. No. 12. 174 pp.

Sabins, D. S. 1977. Fish standing crop estimates in the Atchafalaya Basin, Louisiana. U.S. Army Corps of Engineers, New Orleans District, New Orleans, Louisiana. 19 pp.

Shelton, W. C., W. D. Daves, D. R. Bayne, and J. M. Lawerance. 1982. Fisheries and limnological studies on Westpoint Reservoir, Alabama-Georgia. U.S. Army Corps of Engineers, Mobile District, Mobile, Alabama.

Sheppard, M. F. 1974. Growth patterns, sex ratios and relative abundance of crawfish in Alligator Bayou, Louisiana. M.S. Thesis, Louisiana State University, Baton Rouge, Louisiana. 54 pp.

Soileau, L. D., K. C. Smith, R. Hunter, C. E. Knight, D. M. Soileau, W. E. Shell, Jr., and D. W. Hayne. 1975. Atchafalaya Basin usage study. Final Report, July 1, 1971-June 30, 1974. U.S. Army Corps of Engineers. 85 pp.

Strauss, R. E. 1979. Reliability estimates for Ivlev's Electivity Index, the forage ratio, and a proposed linear index of food selection. Trans. Am. Fish. Soc. 108:344-352.

U.S. Army Corps of Engineers. 1982. Atchafalaya Basin Floodway System, feasibility study, Vol. 2, technical appendixes A B, C, and D. U.S. Army Corps of Engineers, New Orleans District, New Orleans, Louisiana.

Waters, T. F. 1977. Secondary production in inland waters. Pages 91-164 *in* A. Macfadyen, ed. Advances in ecological research. Vol. 10. Academic Press.

Welcomme, R. L. 1975. The fisheries ecology of African floodplains. CIFA Tech. Pap. 3. 51 pp.

__________. 1976. Some general and theoretical considerations of the fish yield of African rivers. J. Fish. Biol. 8:351-364.

__________. 1979. Fisheries ecology of floodplain rivers. Longman, Inc., New York. 317 pp.

__________, and D. Hagborg. 1977. Towards a model of a floodplain fish population and its fishery. Environ. Biol. Fish 2(1):7-24.

Wharton, C. H., V. W. Lambou, J. Newsom, P. V. Winger, L. L. Gaddy, and R. Mancke. 1981. The fauna of bottomland hardwoods in Southeastern United States. Pages 87-106 *in* J. R. Clark and V. Benforado, eds. Wetlands of bottomland hardwood forests. Elsevier Scientific Publishing Company, New York.

_____________, W. M. Kitchens, E. C. Pendlenton, and T. W. Snipe. 1982. The ecology of bottomland hardwood swamps of the southeast: a community profile. FWS/OBS-81/37.

Wood, R. 1951. The significance of managed water levels in developing the fisheries of large impoundments. J. Tenn. Acad. Sci. 26(3):214-235.

__________. and D. W. Pfitzer. 1958. Some effects of water-level fluctuations on the fisheries of large impoundments. International Union for Conservation of Nature and Natural Resources. 31, rue Vautier, Brussels, Belgium, seventh technical meeting, Athens, Greece, 11-19 September 1958.

6. COMPOSITION AND REGENERATION OF A DISTURBED RIVER FLOODPLAIN FOREST IN SOUTH CAROLINA

Rebecca R. Sharitz
Department of Botany and Savannah River Ecology Laboratory, University of Georgia, Drawer "E", Aiken, SC 29802

Rebecca L. Schneider
Ecology and Systematics, Cornell University, Ithaca, NY 14853

Lyndon C. Lee
Division of Wetlands Ecology, Savannah River Ecology Labratory, University of Georgia, Drawer "E," Aiken, SC 29802

ABSTRACT

Studies examining plant community structure and regeneration processes related to community structure within a differentially disturbed palustrine wetland in South Carolina are summarized. Based on remote sensing techniques, the vegetation of a 3800 ha floodplain of the Savannah River is identified as chiefly mixed deciduous swamp forest (47%), mixed deciduous bottomland forest (40%), a shrub-scrub community (4%), and two herbaceous associations (5%). Ordination analyses of vegetation data indicate that the community structure is actually more complex, with gradients in species distributions that may be controlled by gradients in hydrology and perturbation. Communities blend into each other, making it difficult to delineate distinct boundaries.

Seed availability, seedling distribution, and seedling survivorship of *Taxodium distichum* and *Nyssa aquatica*, codominants of the mixed deciduous swamp forest, are being examined to determine how regeneration processes may influence species composition. The hydrologic regime strongly affects regeneration by restricting seed distribution, seedling establishment, and seedling growth to specific microsites within the swamp. Alteration of the normal hydrologic regime

---

Ecological Processes and Cumulative Impacts: Illustrated by Bottomland Hardwood Wetland Ecosystems. Edited by James G. Gosselink, Lyndon C. Lee, and Thomas A. Muir.

of the floodplain, through construction of upstream dams on the Savannah River, has changed the natural pattern of annual drawdowns and high discharge events. Such hydrologic changes, especially the desychronization of flooding events, directly influence forested wetland regeneration and may affect community structure.

INTRODUCTION

Bottomland hardwood and swamp forest wetlands occupy the broad floodplains along many of the major rivers of the southeastern United States. According to information developed by the National Wetlands Inventory, Palustrine Forested Wetlands (which include bottomland hardwood forests) cover 20.1 million ha in the United States (W. Wilen, Project Leader, National Wetland Inventory, U.S. Fish and Wildlife Service [FWS], Washington, D.C., personal communication). Sixty-five percent of this area (13.0 million ha) lies in the Southeast. Approximately 16% occurs in South Carolina and Georgia (1.5 million ha and 1.8 million ha, respectively), compared with 12.5% (2.5 million ha) in the lower Mississippi alluvial plain (Figure 1). Although losses of floodplain forests along the river systems of the Atlantic Coastal Plain have been much less severe than losses along the floodplains of the Mississippi River drainage, few areas have escaped impact by man.

A 3800 ha floodplain forest of the Savannah River lies within the boundaries of the U.S. Department of Energy's Savannah River Site (SRS) in South Carolina. Portions of this swamp forest have received cooling water effluents from nuclear production reactors for three decades. Tributary streams have carried reactor cooling waters to the swamp for various periods of time (Four Mile Creek 1955-1985, Pen Branch 1954-1988, Steel Creek 1954-1968; Figure 2). As a result of these reactor effluents, base stream discharge and water temperatures increased. In addition, sediment that has eroded from the upper stream channels has been deposited on the floodplain, forming deltas. Swamp forests in the regions of the deltas initially responded rapidly to these environmental disturbances (Christensen et al. 1984). Extensive tree mortality occurred (Sharitz et al. 1974a), and the wetland forests were replaced by sparse herbaceous and freshwater marsh communities (Sharitz et al. 1974b,

Repaske 1981), which include several thermally tolerant species (Christy and Sharitz 1980, McCaffrey 1982).

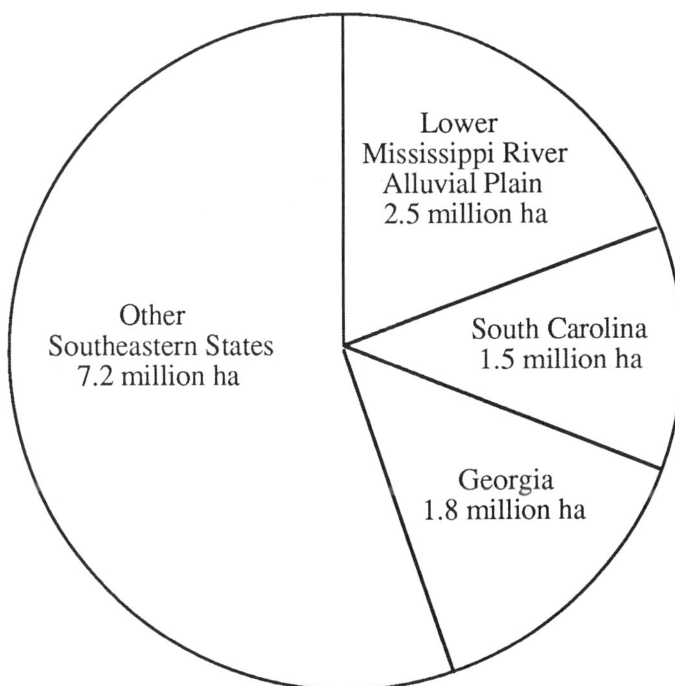

Figure 1. Extent of palustrine forested wetlands in the southeastern United States (Adapted from data supplied by W. Wilen, National Wetlands Inventory).

In addition to the localized effects of cooling water discharges, the SRS floodplain swamp has been affected directly by logging early in the century and indirectly by activities elsewhere in the Savannah River drainage. For example, construction of dams on the river upstream of the SRS has altered the hydrologic regime of the floodplain, especially by modifying the magnitude and timing of major flood events and by maintaining inundated conditions in portions of the forest throughout much of the year (Repaske 1981).

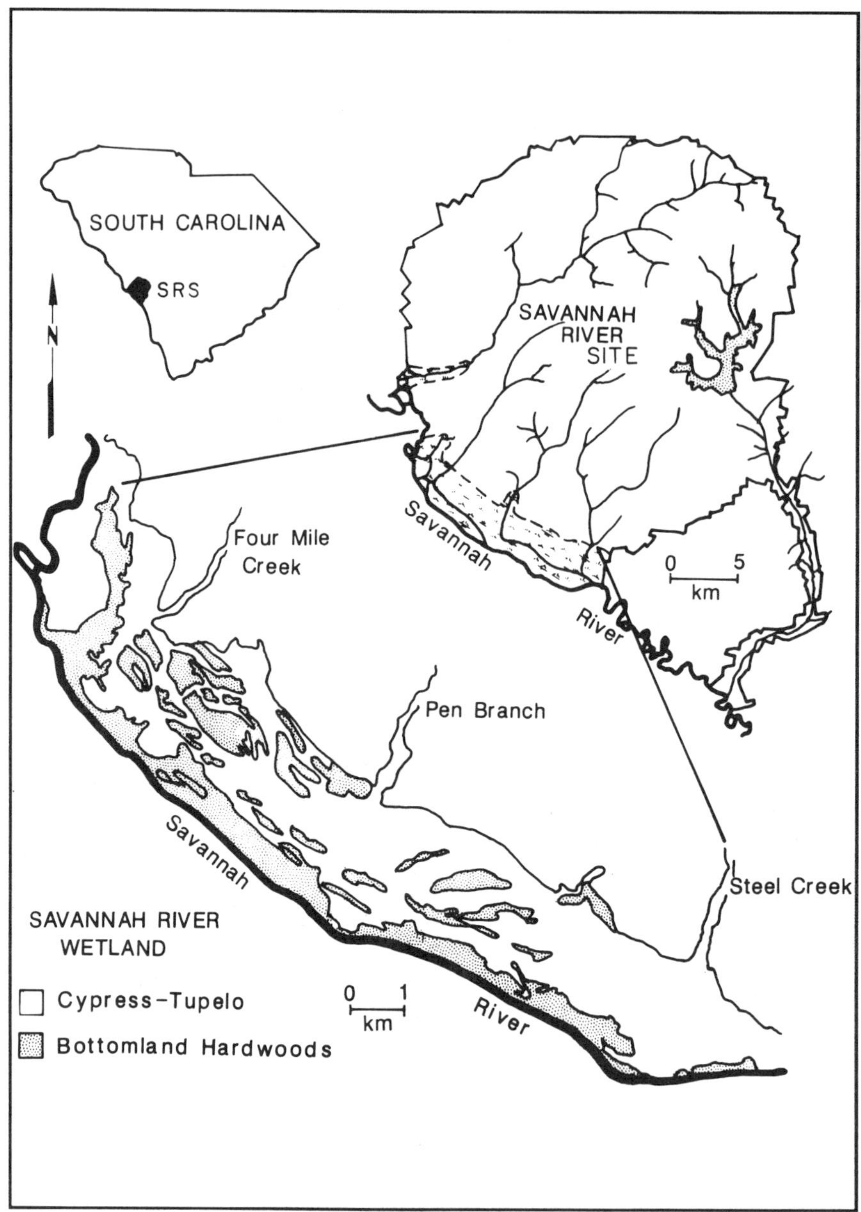

Figure 2. The SRS Savannah River floodplain in South Carolina. Highly disturbed areas occur where major tributary streams enter the floodplain.

The SRS affords an opportunity to study ecosystem processes in degraded wetland forests and to develop recommendations for management. A major research program of the Savannah River Ecology Laboratory is underway to characterize the river floodplain forest and to examine structural and functional responses to disturbance. Studies focus on the wetter end of the soil-moisture/hydrologic gradient, or on bald cypress (*Taxodium distichum*) and water tupelo (*Nyssa aquatica*) forests. At these sites, the soil is inundated on a nearly permanent basis throughout the year. The transition areas from the continuously flooded cypress-tupelo forests to the higher, rarely flooded uplands have also been characterized, and the distributions of these plant communities on the SRS Savannah River floodplain have been mapped. Newly initiated research addresses ecosystem processes throughout the entire bottomland hardwood-swamp forest community gradient.

The initial objective of the research described in this paper is to characterize and map the vegetation of the SRS Savannah River floodplain, including both the relatively natural and the highly disturbed sites. From the results of this community analysis and mapping, studies have been developed to focus on additional aspects of the ecology of this floodplain system, including regeneration processes, physiological responses of major species to stress conditions, and biomass and productivity changes along disturbance gradients. As a second objective, this paper describes studies of regeneration in swamp forests, since this process largely controls species composition within the communities.

COMMUNITY CHARACTERIZATION

Two methods were used to characterize and identify the wetland communities. First, plant community structure of both the relatively natural and the highly disturbed sites was examined using standard sampling and data analysis procedures (Mueller-Dombois and Ellenberg 1974). Throughout the floodplain, 147 quadrats (0.2 ha) were sampled for density and dominance data on woody plants > 2.5 cm dbh (diameter at breast height or above butt swell). Nested sub-plots of smaller size were used for sampling shrubs, woody vines, and herbs. Detrended correspondence analysis (DCA) ordination (DECORANA, Hill 1979a) arrayed the quadrats based upon species importance values calculated from the density and dominance data. A two-way indicator species analysis (TWINSPAN, Hill 1979b) was used to identify major community types.

TWINSPAN results indicated four major woody community types in the SRS Savannah River floodplain (Figure 3). A modified version of the U.S. Fish and Wildlife Service wetlands classification system (Cowardin et al. 1979) was used to name them: 1) two scrub-shrub associations, one dominated by species of willow (*Salix nigra* and *S. caroliniana*) and the other by buttonbush (*Cephalanthus occidentalis*); 2) mixed deciduous bottomland forest associations dominated by oaks (especially *Quercus laurifolia* and *Q. nigra*) and sweet gum (*Liquidambar styraciflua*) in the canopy and ironwood (*Carpinus caroliniana*) in the subcanopy; and 3) mixed deciduous swamp forests dominated by water tupelo and bald cypress. Preliminary DCA analysis indicated that these different communities occur along two gradients that appear to reflect changes in hydrologic characteristics among sites and perturbation history (Figure 3). The swamp and bottomland forest types are interspersed throughout the floodplain, corresponding to the occurrence of frequently flooded sites and less-frequently flooded, slightly elevated ridges and islands, respectively. The scrub-shrub associations appear largely in canopy openings and on disturbed sites. Willow thickets generally occur on slightly elevated and less frequently flooded sites, such as on sand bars in the stream delta areas; whereas, buttonbush is the dominant shrub species in deeper water areas.

In addition to the woody communities, two freshwater marsh community types were identified: 1) nonpersistent emergent marsh dominated by species such as smartweed (*Polygonum* spp.), hydrolea (*Hydrolea quadrivalvis*), and water primrose (*Ludwigia leptocarpa*); and 2) persistent emergent marsh containing species of scirpus (especially knotweed, *Scirpus cyperinus*), cattail (*Typha latifolia*), aneilema (*Murdannia kiesak*), and cut grass (*Leersia oryzoides*). Both herb-dominated communities occur most frequently in the Steel Creek delta, where post-thermal successional recovery of the vegetation following reactor shutdown in 1968 is underway.[1] Additionally, they occur on the

---

[1] Reactor discharges into Steel Creek began again in October 1985, after this paper was completed. An artificial lake constructed upstream on Steel Creek has mitigated temperature increases in the Steel Creek delta.

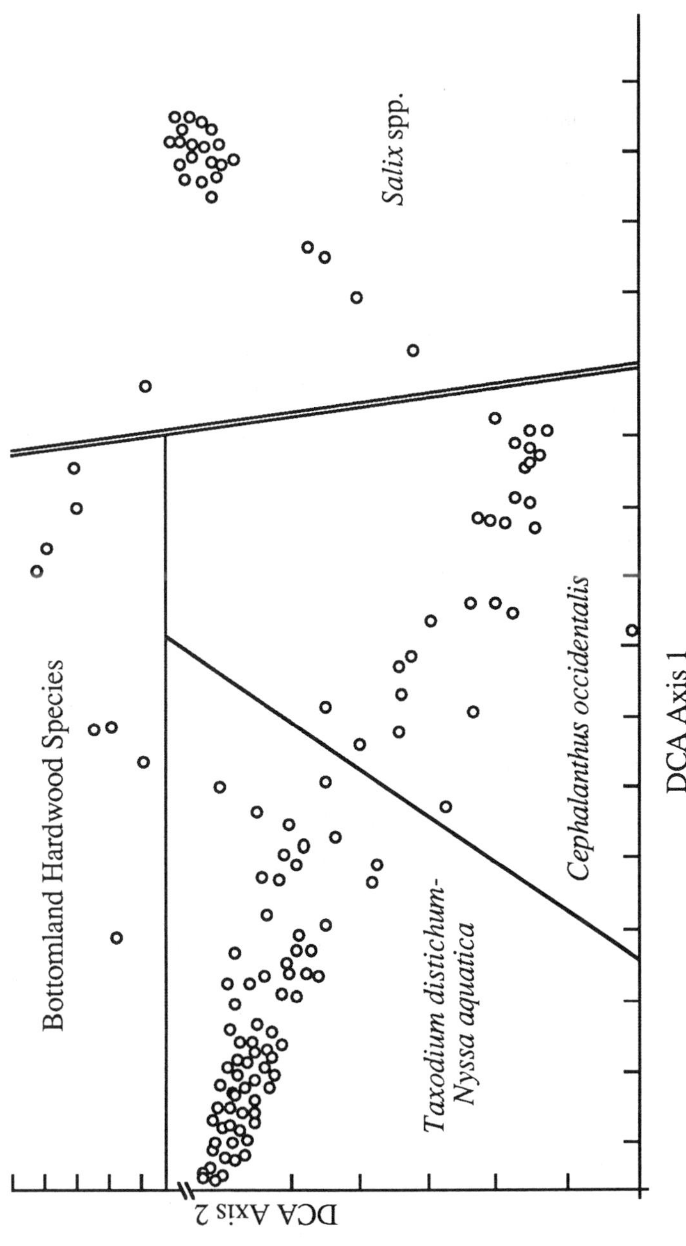

Figure 3. Species assemblages on the SRS Savannah RIver floodplain. TWINSPAN separation is superimposed on a DCA ordination of data from 147 quadrats. See text for description of the analysis.

Four Mile Creek and Pen Branch deltas, in areas where the swamp forest canopy has been totally or partially destroyed but where water temperatures are frequently cool enough to allow extensive growth of herbaceous species.

The second approach to community characterization involved mapping the vegetation using remote sensing data. A map of the entire SRS Savannah River floodplain wetlands (Figure 4) and individual maps of tributary stream delta areas on the floodplain were prepared using low-altitude, high-resolution multispectral scanner (MSS) data (Jensen et al. 1984a, b). The floodplain map was generated from 2438 m above ground level (AGL) data having a 5.6 m x 5.6 m spatial resolution. The deltas of Four Mile Creek, Pen Branch, and Steel Creek were mapped using lower altitude MSS data (1220 m AGL, pixel size of 2.8 m x 2.8 m). Specifications of the scanner system and details of mapping methodology have been previously described (Jensen et al. 1984b).

Six different swamp and bottomland vegetation classes were spectrally discriminated in the SRS Savannah River floodplain (Table 1). The floodplain is covered primarily by mixed deciduous swamp forests (47%) and mixed deciduous bottomland hardwood forests (40%). The remainder of the floodplain contains disturbed shrub/marsh vegetation communities (10%) and occasional needle-leaved evergreen forests (2%) dominated by loblolly pine (*Pinus taeda*). Spectral analysis of the vegetation of the thermal deltas revealed an additional category, algal mats, which occur chiefly in the higher temperature areas.

The vegetation classes discriminated on the basis of spectral reflectance correspond closely to those identified by the TWINSPAN analysis of species importance values, based upon field data. It was necessary, however, to combine willow and buttonbush communities into a scrub-shrub classification for the purpose of mapping the vegetation because these two species could not be distinguished in the multispectral scanner data. An accuracy assessment performed on the Steel Creek delta map indicated that the plant community types were classified and mapped with an overall accuracy of 84% (Jensen et al. 1984b).

In summary, the SRS Savannah River floodplain supports a complex mosaic of vegetation types (Figure 4). Although remote sensing mapping techniques can be used to establish boundaries between these

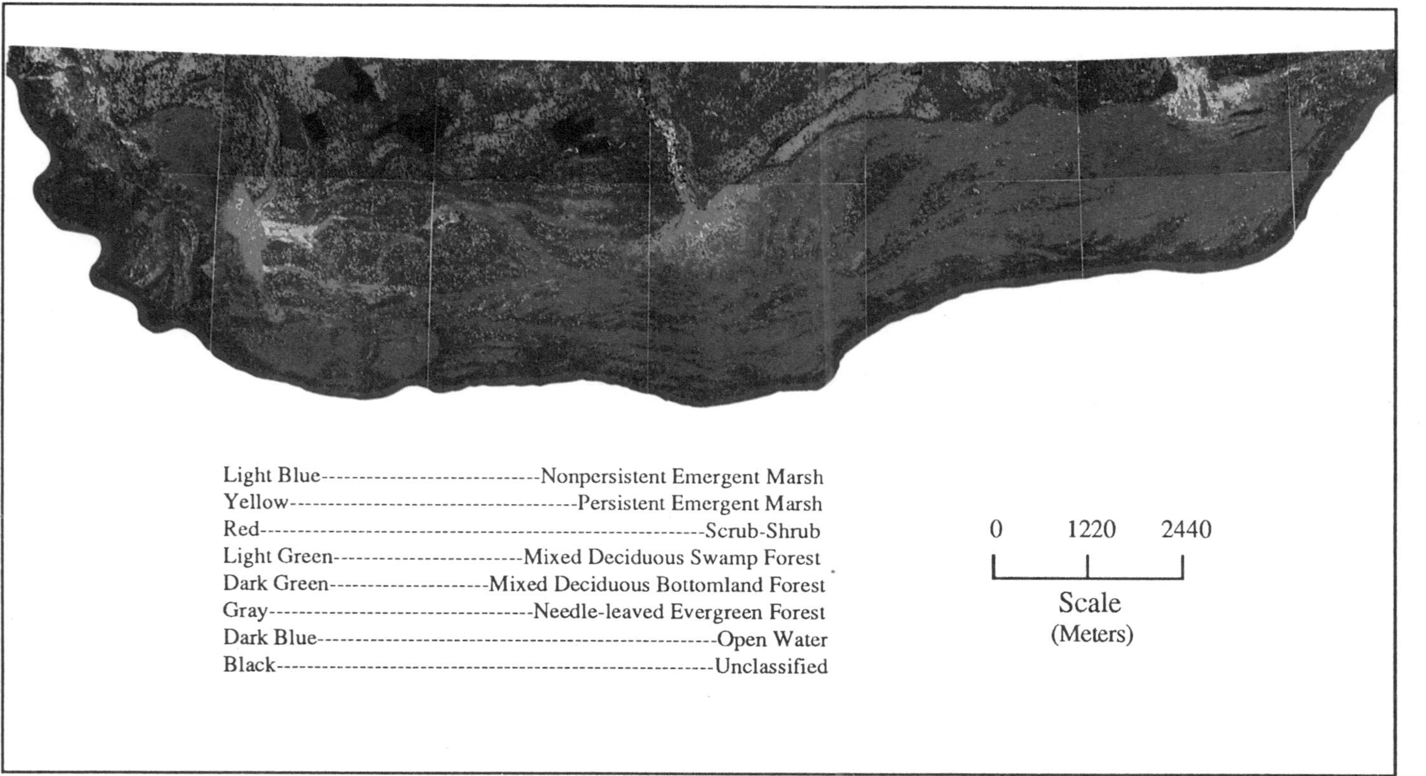

Figure 4. Vegetation map of the SRS Savannah River floodplain (from Jensen et al. 1984a).

community types, there is actually extensive overlap of species composition (Table 2). Relative importance of species changes gradually across environmental gradients (Figure 3) in such a way that ecologically significant boundaries between community types are difficult to distinguish.

Table 1. Areal coverage of major plant community types in the SRS Savannah River floodplain (from Jensen et al. 1984a).

	CLASS AREA (ha)	PERCENT
Persistent Emergent Marsh	54.6	1.4
Nonpersistent Emergent Marsh	152.2	4.0
Scrub-Shrub	154.6	4.1
Mixed Deciduous Swamp Forest	1,793.6	47.2
Mixed Deciduous Bottomland Forest	1,528.1	40.2
Needle-leaved Evergreen Forest	83.4	2.2
Open Water in Swamp	26.7	0.7
Unclassified	8.1	0.2
TOTAL	3,801.3	100.0

Table 2. Mean importance values* of dominant woody species in major SRS Savannah River floodplain communities.

Species	Mixed Deciduous Swamp Forest	Mixed Deciduous Bottomland Forest	Scrub-Shrub
Nyssa aquatica	102.95	7.38	0.29
Taxodium distichum	68.34	11.13	11.30
Fraxinus spp.	9.99	6.65	12.40
Itea virginica	3.46	0.05	4.88
Planera aquatica	2.37	4.21	8.85
Cephalanthus occidentalis	0.65	1.25	75.51
Liquidambar styraciflua	0.20	25.48	-
Quercus laurifolia	0.11	15.98	-
Salix spp.	-	0.17	81.87
Carpinus caroliniana	-	27.75	-
Quercus nigra	-	16.91	-
No. of Plots	34	24	49

$$\text{* Mean Importance Value} = \frac{\text{Relative Density + Relative Dominance}}{N}$$

Along with the community characterization, size class data for the dominant woody species in the floodplain forests were examined. Size frequency distributions of bald cypress and water tupelo from several sites in the mixed deciduous swamp forest indicate a paucity of small individuals of both species. Specifically, less than 16% of 474 bald cypress trees are smaller than 10 cm in dbh (Figure 5). Additional size-age analysis of bald cypress individuals at one of these sites revealed individuals in this < 10 cm size class as old as 60 years. This information implies that very few individuals have been recruited into the population in recent decades. A similar size class distribution also occurs in water tupelo at these sites.

Although such size class data are not as convincing as actual age data, they do suggest that regeneration of the dominant canopy species in the SRS floodplain swamp forest may be somewhat restricted. Limited regeneration of bald cypress and water tupelo is reported in other swamp forests and attributed to their exacting requirements for germination and early seedling growth. For example, Conner et al. (1981) found relatively low densities of small size classes in permanently flooded and naturally flooded sites in Louisiana. On the other hand, comparative data for bald cypress from the Okefenokee Swamp (Schlesinger 1976) revealed a higher occurrence of individuals in the smaller size classes. The apparent reduction in regeneration in the Savannah River floodplain may be related to a variety of factors, including past logging history or environmental disturbances such as alterations in the hydrologic regime. Species regeneration success is a critical component determining community composition; therefore, studies have been developed to examine specific aspects of regeneration in detail.

REGENERATION PROCESSES

Maintenance of a natural swamp forest is largely dependent on the availability and successful establishment of seeds, both those produced in previous years, which are stored in the soil, and seeds currently dispersing into the community. The factors that control seed availability are being studied in a relatively natural mixed deciduous swamp forest (Schneider and Sharitz 1988). Floating seed traps are used to quantify the

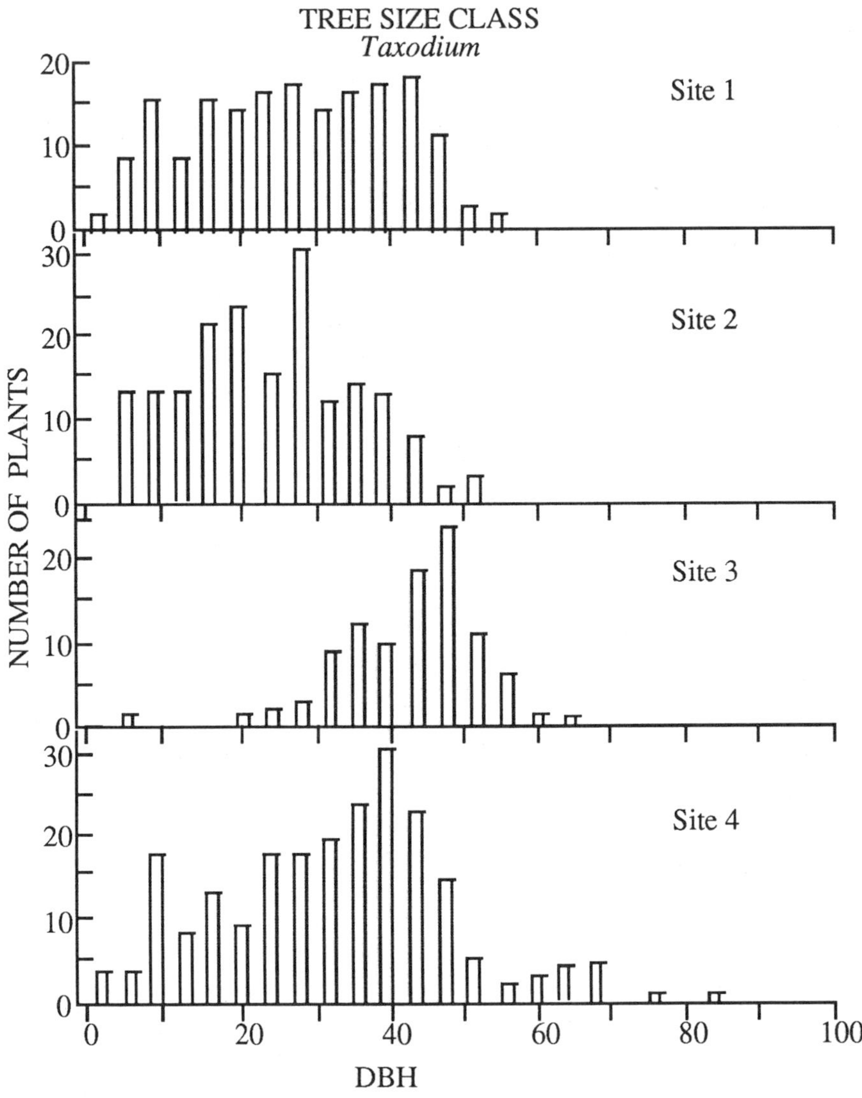

Figure 5. Size frequency distributions of bald cypress (*Taxodium distichum*) at four sites in the SRS Savannah River floodplain.

annual release of seeds by bald cypress and water tupelo. Additional seed traps capture water-dispersed seeds to quantify the potential seed input by water to a given area. The depth, direction, and surface velocity of water flow through the swamp are measured regularly at each trap location. Seed fall for both species typically begins in late September and peaks in November. Water tupelo finishes disseminating seeds in December or January, but bald cypress continues to disperse until early spring (Figure 6). Water typically flows through the swamp during the entire seed dispersal period and transports seeds of both species downstream from their parent trees. The water vector is a secondary mechanism of seed dispersal for several months after actual release from the trees has ceased. Water also changes the patterns of final seed distribution in the community, substantially increasing seed transport to some sites. During major flood events, seeds of several bottomland hardwood species may also be dispersed by water through the floodplain.

Composition of soil seed banks was examined to determine if dispersing seeds are being incorporated into the sediment (Schneider and Sharitz 1986). It was hypothesized that seed banks of the cypress-tupelo community would be more similar to those of the bottomland hardwood community after the winter flooding and secondary seed dispersal by water had occurred. Soil cores were collected in the cypress-tupelo community and on a small bottomland hardwood island directly upstream at three times: in September before seed fall, in December after seed fall but before winter flooding, and in April after the winter inundation. Germination and sieving techniques were used to enumerate seeds in each sample. There was very little similarity in the woody species composition of the seed banks between the two communities (Table 3), although similarity increased slightly after the winter flood. Seed densities in both communities' seed banks increased in December after seed fall but dropped significantly after the winter inundation. It appears that flowing water may transport seeds out of the communities unless they are trapped by knees, logs, or other physical structures.

Given adequate seed availability, the success of seedling establishment represents the next phase in forest regeneration. Several types of microsites, or substrates suitable for seedling establishment, can be distinguished in mixed deciduous swamp forests. Such microsites

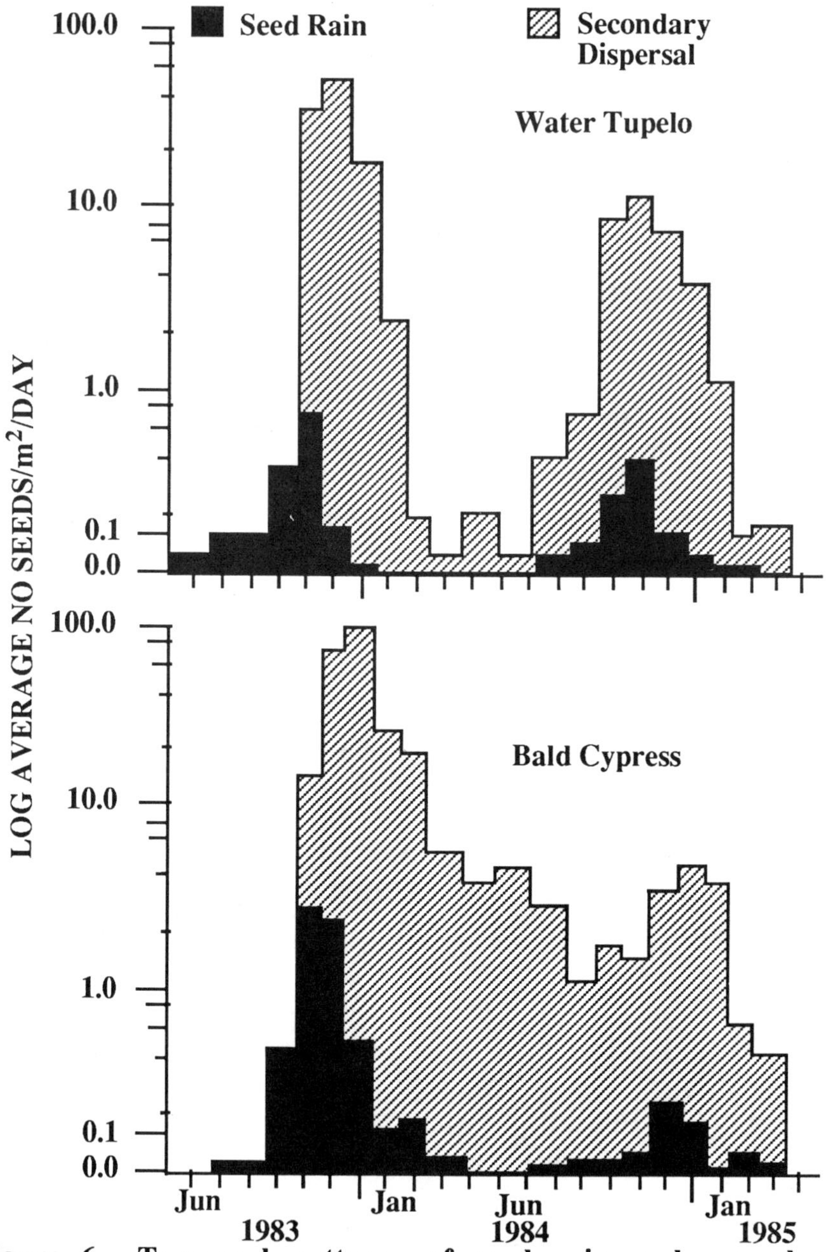

Figure 6. Temporal patterns of seed rain and secondary dispersal by water of bald cypress and water tupelo on the SRS Savannah River floodplain (from Schneider and Sharitz 1988:1058; Copyright © 1988 by the Ecological Society of America, reprinted with permission).

Table 3. Woody species composition of the soil seed banks (based on number of seeds that germinated) from two Savannah River floodplain communities.

Mixed Deciduous Swamp Forest	No. of Seedlings	Mixed Deciduous Bottomland Forest	No. of Seedlings
* *Taxodium distichum*	48	*Liquidambar styraciflua*	58
+ *Nyssa aquatica*	34	*Rubus* sp.	40
* *Cephalanthus occidentalis*	5	+ *Vitis aestivalis*	13
* *Acer rubrum*	3	+ *Berchemia scandens*	13
+ *Fraxinus caroliniana*	1	*Ampelopsis arborea*	13
		+ *Rhus copallina*	8
		* *Acer rubrum*	6
		+ *Vitis rotundifolia*	8
		* *Cephalanthus occidentalis*	5
		Itea virginica	5
		* *Taxodium distichum*	3
		+ *Quercus* spp.	2
		Baccharis halimifolia	1
		Ulmus americana	1

* Species common to both communities.
+ Species found in the soil seed bank and in seed traps collecting water-dispersed seeds.

include the bases of trees and knees (live wood substrates), stumps, branches, twigs, logs (dead wood), and several sediment categories. These different substrates may affect regeneration by differentially trapping water-dispersing seeds and by providing varied conditions for germination and growth. The relative abundance of microsites was quantified along transects at the cypress-tupelo study site and at a disturbed stand that receives some cooling water effluents and increased sediment deposition (Huenneke and Sharitz 1985, Table 4). Woody seedlings were identified and classified by microsite type along additional transects at each location.

The abundance of microsites differed significantly between the two study areas (Table 4). Woody seedlings were distributed non-randomly among microsite types (i.e., not in proportion to the abundance of a given microsite). At the natural site, tree seedlings were concentrated on soil substrates; shrubs and vines occurred more frequently on tree bases,

Table 4. Relative abundance of microsites for seedling establishment in a natural and a disturbed swamp forest (from Huenneke and Sharitz 1985).

Microsite Type	% of Substrate in each Microsite Type: Natural Swamp	Disturbed Swamp
SOIL		
Loose muck	29.6	22.0
Muck over wood	6.5	8.2
Consolidated muck	3.0	36.0
Near tree	2.4	1.4
Near stump	<1.0	<1.0
Near knee	2.4	3.1
Near log	<1.0	<1.0
Near branch	3.8	8.9
	47.7	79.6
LIVE WOOD		
Tree	2.3	1.8
Knee	1.8	1.1
Shrub	≤1.0	≤1.0
	4.1	2.9
DEAD WOOD		
Stump	<1.0	1.2
Log	1.1	2.4
Branch	1.9	9.6
Twigs	10.7	2.2
	13.7	15.4
FINE ROOTS	32.6	<1.0

knees, and logs. Bald cypress and water tupelo seedlings predominated on sediment substrates; however, there were more cypress than tupelo seedlings on knees and stumps (Figure 7). The relative stability and protection provided by a microsite during the winter flood appears to be one determinant of successful seedling regeneration on that substrate.

After germination, differential seedling growth and survivorship determine which individuals become adults. The locations of all bald cypress and water tupelo seedlings were mapped in transects at the disturbed study site. Seedling height, microsite type, and water depth

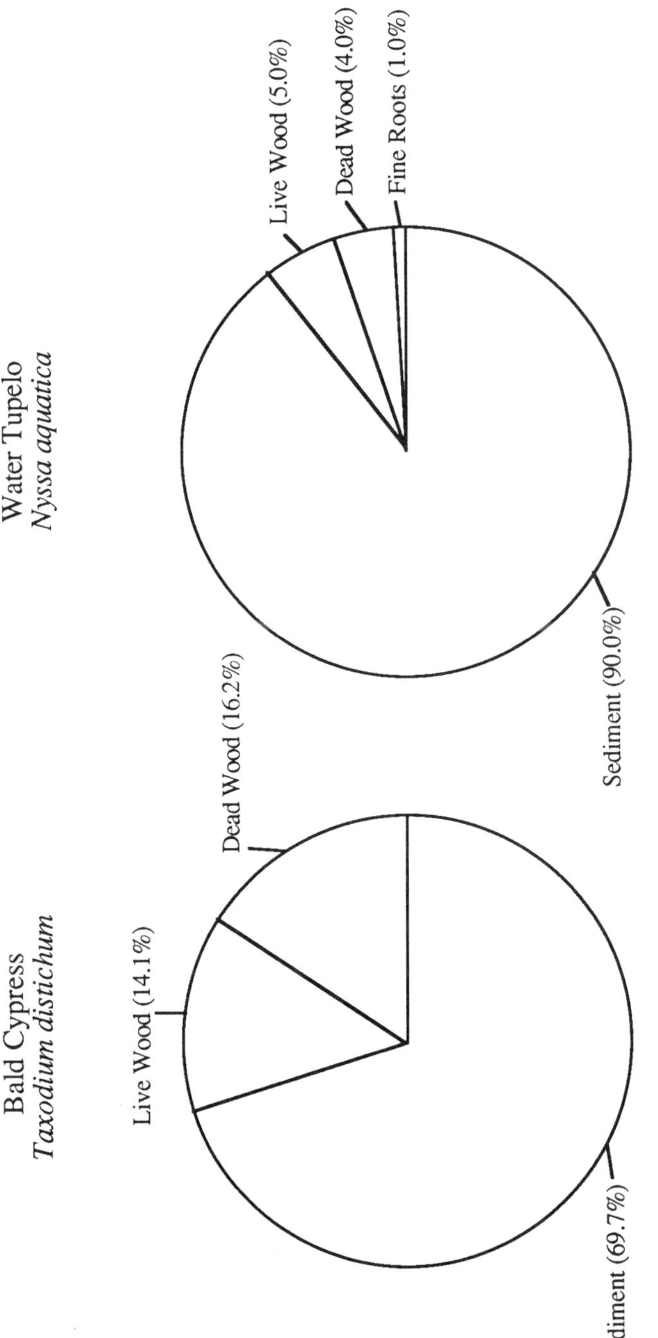

Figure 7. Distribution of bald cypress and water tupelo seedlings on available microsites in a natural mixed deciduous swamp forest. See Table 4 for explanation of categories.

were recorded and the transects were reinventoried three times during the growing season each year. In the first year, bald cypress seedling densities averaged $1.3/m^2$ and water tupelo densities were $0.3/m^2$. Bald cypress seedlings differed in growth among microsites, with the highest growth rates of new seedlings occurring on consolidated muck substrates. Water tupelo seedlings showed fewer differences in growth among microsite types. Differences in growth rates of both species were obscured by 70 - 90% seedling mortality following a late spring flood and an anomalous summer flood (Figure 8) caused by high discharges from reservoirs upstream on the Savannah River. Each flood submerged all seedlings for longer than three days. Thus, variation in large-scale hydrologic regime has greater influence on seedling survivorship than variation among microsites.

DISCUSSION

The vegetation of the 3800 ha SRS Savannah River floodplain is composed of several highly diverse species assemblages. However, these plant species are distributed along environmental gradients that generally correspond to hydrologic regime (especially changes in the frequency and duration of inundation) and to perturbation history (elevated water temperatures and siltation). As a result, there remains considerable overlap in species composition across communities. For example, water tupelo and bald cypress, canopy dominants of the mixed deciduous swamp forest, are present but exhibit lower relative importance in both the mixed deciduous bottomland hardwood and the scrub-shrub communities. In addition, ash (*Fraxinus caroliniana* and *F. pennsylvanica*) and water elm (*Planera aquatica*) appear in the subcanopy of both of the forest types and often occur in the scrub-scrub associations. Such species overlap makes accurate determination of community boundaries a difficult and rather subjective task. Sampling throughout the SRS floodplain has further demonstrated the extent of transition habitats, or ecotones, between major community types. These transitional habitats may be extremely important sites. They support high plant and animal species diversity

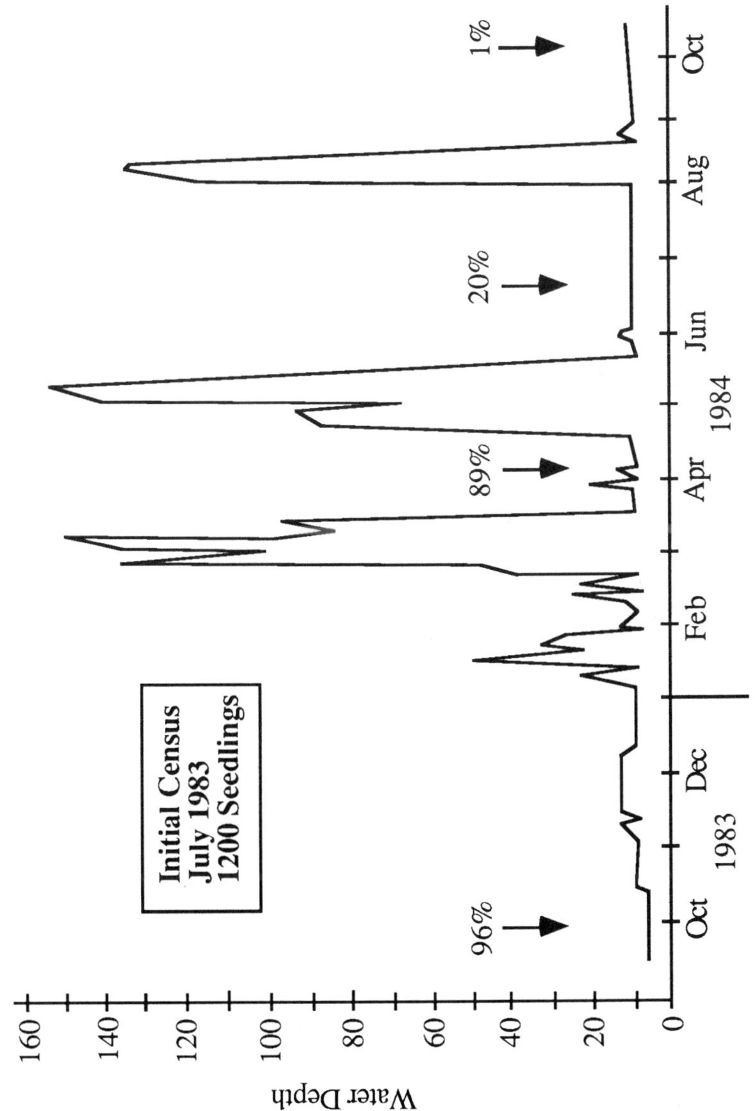

Figure 8. **Relationship between percent seedling survival (shown by numbers and arrows) of bald cypress and water tupelo and flood events in the SRS Savannah River floodplain.**

(Barkman 1978, Brown et al. 1978, Division of Ecological Services 1980) and high net primary productivity (Gosselink and Turner 1978; Brown 1981). In addition, ecotonal areas provide valuable food and cover resources for many wildlife species (Harris et al. 1979, Thomas et al. 1979).

Remote sensing techniques have been extremely successful in mapping these floodplain wetland communities. Multispectral scanner data can be used to discriminate major woody and herbaceous plant communities identified through multivariate analysis of field data. Use of such remote sensing data allows effective, relatively accurate, and rapid mapping of large areas. These features are especially valuable when the areas being mapped are difficult to sample, as are many riverine swamps. However, weaknesses of this mapping technique include limitations in recognition of certain species assemblages and failure to identify transition areas between major communities. For example, willow-dominated and buttonbush-dominated communities could not be distinguished in this multispectral scanner data base, although they are separated across the hydrologic gradient and appear to be major successional shrub species on ecologically different sites.

These problems in defining discrete community types make it difficult to relate floodplain vegetation to the wetland zone concept proposed by Larsen et al. (1981). For example, the mixed deciduous swamp forest dominated by bald cypress and water tupelo, which comprises 47% of the floodplain, classically fits the Zone II definition. This swamp community is dominated by hydrophytic species and inundated on a nearly permanent basis. However, the mixed deciduous bottomland forest, identified independently through both multivariate community analysis and remote sensing techniques, contains elements of several Zones (III-V). Bald cypress and water tupelo also occur in these communities, as do species of oak, sweet gum, and ironwood. On the SRS Savannah River floodplain, it does not appear feasible, either within the data set or in the field, to discriminate plant species subgroups in the mixed deciduous bottomland forest that logically correspond to Zones III-V, as defined for the Mississippi River floodplain. This difficulty may be due to truncation of plant community types into narrow bands across steeper topographic gradients in the Atlantic Coastal Plain, or it may be that such subdivisions in bottomland hardwood forests do not occur commonly in these southeastern floodplains. Even where these zones can be identified, floodplain topography may result in a complex distribution pattern, as suggested by Wharton et al. (1982).

The cypress-tupelo association is the major species complex in the SRS Savannah River floodplain. These species dominate the mixed deciduous swamp forest and also are scattered throughout the mixed deciduous bottomland forest. Examination of size class structure for both species reveals few individuals in the juvenile classes. This small number may be due to a natural but periodic recruitment process or to logging history, or it may indicate that environmental conditions are limiting regeneration of these species. If there are limitations on regeneration success, there will be major changes in community composition as older individuals senesce.

Studies of swamp forest regeneration at the Savannah River Ecology Laboratory indicate that all aspects of the process are strongly dependent on hydrologic conditions. Seeds of the dominant species are adapted to dispersal by water. This vector prolongs the duration of dispersal and increases the distances that seeds are transported. It also changes the final seed distribution patterns, concentrating seeds near knees, logs, and other emergent microsites. Water additionally disperses species across the entire floodplain by transporting seeds among different community types.

Seedling establishment and survivorship also are influenced by hydrologic conditions. Large numbers of seedlings occur on microsites that are stable through the winter flood disturbance and provide protection against floating debris. Differences in microsites are insufficient, however, to protect seedlings from the detrimental effects of large, desynchronized floods that occur during the growing season as a result of upstream reservoir management.

These results demonstrate the importance to southeastern floodplain swamps of wisely managed hydrologic regimes that are synchronous with natural ecosystem processes, including species regeneration. Perturbation of floodplain environments, through the manipulation of water levels, may be responsible for low juvenile recruitment. Floodplain management strategies must account for the close integration of wetland species with their physical environment. The mosaic community structure characteristic of floodplains is strongly dependent upon hydrologic regime as well as on other environmental features. Therefore, management practices that impact a portion of this highly integrated system will necessarily affect physical and biotic processes at several levels of ecosystem organization.

ACKNOWLEDGMENTS

Thanks to K. W. Dyer, D. E. Wickland, L. F. Huenneke, R. L. Sims, D. E. Mahoney, D. M. Potter, W. A. Repaske, A. E. Hodge, C. E. Mitchell, and others who assisted in data collection and analysis, and to J. R. Jensen and E. J. Christensen of the University of South Carolina for analysis of remote sensing data. This research was supported under contract DE-AC09-76SROO-819 between the U. S. Department of Energy and the University of Georgia.

REFERENCES CITED

Barkman, J. J. 1978. Synusial approaches to classification. Pages 111-166 *in* R. H. Whittaker, ed. Classification of plant communities. Dr. W. Junk bv. Publishers, The Hague, Netherlands.

Brown, S., M. M. Brinson, and A. E. Lugo. 1978. Structure and function of riparian wetlands. Pages 17-31 *in* R. R. Johnson and J. F. McCormick, eds. Strategies for protection and management of floodplain wetlands and other riparian ecosystems. Proc. Symp. Callaway Gardens, Georgia. Gn. Technical Report WO-12, U.S. Fish and Wildlife Service, Washington, D.C.

________. 1981. A comparison of the structure, primary productivity, and transpiration of cypress ecosystems in Florida. Ecol. Monogr. 51:403-427.

Christensen, E. J., M. E. Hodgson, J. R. Jensen, H. E. Mackey, Jr., and R. R. Sharitz. 1984. Pen Branch delta expansion. Savannah River Laboratory. E. I. DuPont de Nemours and Company, Aiken, South Carolina. DPST-83-1087.

Christy, E. J., and R. R. Sharitz. 1980. Characteristics of three populations of a swamp annual under different temperature regimes. Ecology 61:454-460.

Conner, W. H., J. G. Gosselink, and R. T. Parrondo. 1981. Comparison of the vegetation of three Louisiana swamp sites with different flooding regimes. Am. J. Bot. 68:320-331.

Cowardin, L. M., V. Carter, F. C. Golet, and E. I. LaRoe. 1979. Classification of wetlands and deepwater habitats of the United States. U.S. Fish and Wildlife Service, Washington, D.C. FWS/OBS-79/31.

Division of Ecological Services. 1980. Habitat as a basis for environmental assessment. 101 ESM (A Portion of the Habitat Evaluation Procedures Documentation). U.S. Fish and Wildlife Service, Washington, D.C.

Gosselink, J. R., and R. E. Turner. 1978. The role of hydrology in freshwater wetland ecosystems. Pages 63-78 *in* R. E. Good, D. F. Whigham, and R. L. Simpson, eds. Freshwater wetlands. Academic Press, New York.

Harris, L. D., D. H. Hirth, and W. R. Marion. 1979. The development of silvicultural systems for wildlife. Pages 65-81 *in* Proceedings of the 28th Annual Forestry Symposium - 1979. Louisiana State University, Baton Rouge, Louisiana.

Hill, M. O. 1979a. DECORANA- a FORTRAN program for detrended correspondence analysis and reciprocal averaging. Available from: Section of Ecology and Systematics, Cornell University, Ithaca, New York.

________. 1979b. TWINSPAN - a FORTRAN program for averaging multivariate data in an ordered two-way table by classification of the individuals and attributes. Available from: Section of Ecology and Systematics, Cornell University, Ithaca, New York.

Huenneke, L. F., and R. R. Sharitz. 1985. Microsite abundance and the distribution of woody seedlings in a South Carolina cypress-tupelo swamp. Am. Midl. Nat. 115:328-335.

Jensen, J. R., E. J. Christensen, and R. R. Sharitz. 1984a. SRS swamp vegetation map. Savannah River Laboratory. E. I. DuPont de Nemours and Company, Aiken, South Carolina. DPST-84-372.

____. 1984b. Wetland mapping in South Carolina using airborne multispectral scanner data. Remote Sens. Environ. 16:1-12.

Larsen, J. S., M. S. Bedinger, C. F. Bryan, S. Brown, R. T. Huffman, E. L. Miller, D. G. Rhodes, and B. A. Touchet. 1981. Transition from wetlands to uplands in southeastern bottomland hardwood forests. Pages 225-273 *in* J. R. Clark and J. Benforado, eds. Wetlands of bottomland hardwood forests. Elsevier Scientific Publishing Company, Amsterdam.

McCaffrey, C. A. 1982. Effects of flooding and sedimentation on germinaton and survival of *Ludwigia leptocarpa (Nutt.) Hara.* MS Thesis, University of Georgia, Athens, Georgia.

Mueller-Dombois, M., and H. Ellenberg. 1974. Aims and methods of vegetation ecology. John Wiley and Sons, New York.

Repaske, W. A. 1981. Effects of heated water effluents on the swamp forest at the Savannah River Site, South Carolina. MS Thesis, University of Georgia, Athens, Georgia.

Schlesinger, W. H. 1976. Biogeochemical limits on two levels of plant community organization in the cypress forest of Okefenokee Swamp. Ph.D. Thesis, Cornell University, Ithaca, New York.

Schneider, R. L., and R. R. Sharitz. 1988. Hydrochory and regeneration of a southeastern riverine swamp forest. Ecology 69:1055-1063.

______________________________. 1986. Seed bank dynamics in a southeastern riverine swamp. Am. J. Bot. 73:1022-1030.

Sharitz, R. R., J. W. Gibbons, and S. C. Gause. 1974a. Impact of production reactor effluents on vegetation in a southeastern swamp forest. Pages 356-362 *in* J. W. Gibbons and R. R. Sharitz, eds. Thermal Ecology. AEC Symposium Series. CONF-730505.

____________, J. E. Irwin, and E. J. Christy. 1974b. Vegetation of swamps receiving reactor effluents. Oikos 25:7-13.

Thomas, J. W., C. Maser, and J. E. Rodick. 1979. Riparian zones. Pages 40-47, chapter 3 *in* Wildlife habitats in managed forests. U.S. Department of Agriculture, Forest Service Agriculture Handbook. No. 553.

Wharton, C. H., W. M. Kitchens, E. C. Pendleton and T. W. Sipe. 1982. The ecology of bottomland hardwood swamps of the Southeast: a community profile. U.S. Fish and Wildlife Service, Washington, D.C . FWS/OBS-81/37.

SECTION II. CUMULATIVE IMPACTS IN BOTTOMLAND HARDWOOD FOREST ECOSYSTEMS AND LANDSCAPES

PROLOGUE TO SECTION II

Perhaps the most difficult kind of human activity to manage in wetlands is the incremental degradation that accompanies many small actions that are individually often fairly minor. Muir, Rhodes and Gosselink discuss regulatory authorities for control of cumulative impacts. Cairns points out the complexity of ecosystem response to multiple stresses, which makes the relationship of stress to response difficult to document; and the fragmented regulatory authority, which makes comprehensive management of the aggregate or cumulative impact on the landscape nearly impossible. Harris and Gosselink document the cumulative landscape impact of bottomland forest clearing on hydrology, water quality, and indigenous biota. Finally, Neal and Jemison discuss a plan that uses a variety of approaches to preserve 100,000 ha of bottomlands in Texas and Oklahoma, mostly from the impact of water control structures.

7. FEDERAL STATUTES AND PROGRAMS RELATING TO CUMULATIVE IMPACTS IN WETLANDS

Thomas A. Muir
U.S. Fish and Wildlife Service,
Washington, D. C. 20240

Charles Rhodes
U.S. Environmental Protection Agency,
Philadelphia, PA 19107

James G. Gosselink
Center for Wetland Resources, Louisiana State University,
Baton Rouge, LA 70803

ABSTRACT

Cumulative impacts are a major source of wetland loss and functional degradation. A number of statutes, and the regulations governing their implementation, require the evaluation of cumulative impacts in the decision process. Despite this requirement cumulative impact management has been ineffective. Various federal programs, both regulatory and non-regulatory, can address cumulative impacts, but effective management requires a focus on large landscape management units and cooperation among federal, state and local agencies, and .

INTRODUCTION

In recent years cumulative impacts have been identified as a major source of wetland loss and functional degradation (Horak et al. 1983a, 1983b, Beanlands et al. 1986, Bedford and Preston 1988, Gosselink and Lee 1989). A key consideration in controlling cumulative impacts is the institutional capacity to respond appropriately, that is, to manage wetland

---

Ecological Processes and Cumulative Impacts: Illustrated by Bottomland Hardwood Wetland Ecosystems. Edited by James G. Gosselink, Lyndon C. Lee, and Thomas A. Muir. Chelsea, MI 48118. Printed in USA.

resources for long-term stability and functional vitality. A number of federal statutes and programs offer opportunities to address cumulative impacts to wetlands through both regulatory and non-regulatory means. The primary regulatory avenues for addressing cumulative impacts have been the Section 404 program of the Clean Water Act and to a lesser degree, the National Environmental Policy Act.

REGULATORY PROGRAMS

Clean Water Act (PL 92-500)

Specific regulations concerning cumulative impacts to waters of the United States, including wetlands, are contained in the Clean Water Act (CWA) Section 404(b) (1) Guidelines (40 CFR Part 230). Section 230.11 (g) and (h) of these Guidelines define cumulative and secondary effects to aquatic ecosystems. Cumulative impacts are defined as the changes in an aquatic ecosystem that are attributable to the collective effect of a number of individual discharges of dredged or fill material. Secondary effects are described as effects on an aquatic ecosystem that result indirectly from the placement of the dredged or fill material. These regulations require the permitting authority (usually the U.S. Army Corps of Engineers [USACE]) to collect information regarding cumulative impacts, document their findings, and consider the results in the decision-making process for individual permit applications, general permits, and in the monitoring and enforcement of existing permits. Section 230.11(g) states specifically that although an individual discharge may, in itself, constitute a minor change, the cumulative effect of numerous such piecemeal changes can result in a major impairment of water resources and interfere with the productivity and water quality of aquatic ecosystems.

Section 320.4 (a) of the Final Rule for Regulatory Programs of the Corps of Engineers (33 CFR Parts 320-330) states that "the decision whether to issue a permit will be based on an evaluation of the probable impacts, including cumulative impacts, of the proposed activity and its intended use on the public interest." Furthermore, Section 320.4 (b) (3) acknowledges that although a particular alteration of a wetland may constitute a minor change, the cumulative effect of numerous changes can result in a major impairment of wetland resources. The regulation requires that a wetland site considered for a Section 404 permit be evaluated with the recognition that it may be part of a complete and interrelated wetland area. District Engineers are required to undertake reviews of particular wetland areas in consultation with other federal agencies, such as the U.S. Fish and Wildlife Service (FWS, as required by the Fish and Wildlife Coordination Act [PL 85-624], as amended by PL 89-72) and the U.S. Environmental Protection Agency (EPA), to assess the cumulative effects of activities in wetlands, where appropriate.

The National Environmental Policy Act (NEPA)

The National Environmental Policy Act (42 U.S.C. 4371 et. seq.) does not specifically address wetland resources but does establish the general basis for addressing cumulative impacts. "Cumulative impact" is defined in the implementing regulations for NEPA (40 CFR Parts 1500-1508) as follows: "the impact on the environment which results from the incremental impact of the action when added to other past, present, and reasonably foreseeable future actions regardless of what agency (federal or non-federal) or person undertakes such other actions. Cumulative impacts can result from minor but collectively significant actions taking place over a period of time" (Section 1508.7). Cumulative impacts are also addressed under the definition of "scope" as it relates to environmental impact statements and assessments (Section 1508.25 (a) and (c)). In determining the scope of an environmental impact statement, agencies should consider three types of actions, "which when viewed with other proposed actions have cumulatively significant impacts and should therefore be discussed in the same impact statement." Impacts include direct, indirect and cumulative. In addition, Section 1502.20 introduces the concept of tiered cumulative impact statements, that is addressing cumulative impacts at a series of increasing geographic scales, such as local, regional and national.

NON-REGULATORY PROGRAMS

Other statutes and programs at the federal level also provide opportunities for addressing cumulative impacts.

Clean Water Act, As Amended By The Water Quality Act of 1987

Advance Identification

Under Section 230.80 of the CWA Section 404 (b) (1) Guidelines, EPA and the permitting authority (generally the USACE), in consultation with the affected state(s), may conduct an advance identification of wetland sites. Under this provision, EPA and the USACE identify and evaluate wetlands within a designated Advance Identification area, to determine if the use of specific wetland sites for dredged or fill material will comply with the Guidelines. Using information gathered through the advance identification study process, wetland sites are mapped and identified as either possible future disposal sites or areas generally unsuitable for disposal site specification. This information is made available to the public as a guide to avoiding dredge and fill activities in the unsuitable areas.

The information developed during the advance identification process is also used by the USACE and EPA to facilitate the Section 404 permit process, but it does not replace it; the identification of any area as a possible future disposal site does not constitute a permit for dredge or fill activities, nor does the identification of an area as generally unsuitable for disposal prohibit applications for permits. Designation of unsuitable wetland sites does, however, put the applicant on notice that a permit for dredge or fill actions will be granted only after the most rigorous review and under the most compelling circumstances.

The advanced identification process is well suited for the identification of cumulative impacts to wetlands. Advance Identification is usually conducted over a large geographic area, often encompassing one or more watersheds. The causes and effects of cumulative impacts are most visible at the watershed level, and the remedies that are most effective in preventing and reversing the degradation of wetlands that results from cumulative impacts are best applied on a watershed or landscape scale (Gosselink and Lee 1989).

Numerous advance identifications have been initiated by EPA and the Corps. In 1990 approximately 60 were completed, underway or proposed. They have varied in size from several to thousands of hectares, and have been the basis for follow up actions by various agencies and organizations to protect valuable wetlands.

The National Estuaries Program

The 1987 amendments to the CWA authorize EPA to convene a management conference for estuaries of national significance and "to develop a comprehensive conservation and management plan...to restore and maintain the chemical, physical, and biological integrity of the estuary..." The law requires priority consideration of certain estuaries listed in the act. The program is managed through EPA's Office of Marine and Estuary Protection under CWA Section 320. This section requires the development of basinwide, comprehensive conservation and management plans for identified estuaries. The Act, along with the Ocean Dumping Ban Act of 1988, identifies a number of estuaries for priority consideration. Management plans are developed for estuaries accepted into the program, and the Office of Marine and Estuary Protection administers a funding program for the purpose of developing management programs and conferences, conducting assessments and implementing control measures. Activities associated with the National Estuaries Program and the Oceans Dumping Ban Act offer opportunities to fund and conduct cumulative impact assessments on an estuary-wide basis.

Near Coastal Waters Program

The Near-Coastal Waters Program is also administered through the Office of Marine and Estuary Protection, and is an initiative developed to focus attention on environmental management of these waters, including coastal wetlands, using existing regulatory authorities. Cumulative impact studies can be conducted in association with the program's demonstration projects.

Non-Point Source Program

The 1987 amendments to the CWA, (subsection 316 (b) (z)), also provide for states to prepare nonpoint source pollution assessments and management plans. State programs must, among other requirements, identify U.S. waters (including wetlands) adversely affected by nonpoint source pollution, and describe processes and legal measures for controlling such pollution. Under Section 303 of the Act, states may designate certain wetlands as outstanding resource waters under their antidegradation policy. Discharges are prohibited into U.S. waters with this designation.

In developing nonpoint management plans, states may coordinate with offices within EPA, such as the Office of Marine and Estuary Protection or the Office of Wetlands Protection. The regulations require that states involve local and private agencies and organizations that have expertise in control of nonpoint source pollution. Management plans must outline best management practices, legal measures for controlling pollution, and a schedule of milestones for program implementation. Such multi-agency approaches that potentially involve federal and state authorities and resources, are an effective means for identifying and controlling cumulative impacts to wetlands.

The Emergency Wetlands Resources Act Of 1986 (PL99-645)

The Emergency Wetlands Resources Act of 1986 was enacted to promote the conservation of wetlands. It directs the U.S. Department of Interior to develop a National Wetlands Priority Conservation Plan. This plan must identify the locations and types of wetlands that should receive priority attention for federal and state wetland acquisition efforts. FWS has prepared a planning framework document, including criteria and guidance intended to meet the requirements under Section 301 of the Act. Criteria to be considered in determining acquisition priorities include (1) functions and values of wetlands, (2) historic wetland losses, including losses of functions and values resulting from cumulative and "piecemeal" impacts, and (3) the threat of future wetland losses. In general, wetlands given priority consideration for acquisition will be those that provide a high degree of public benefits, that are representative of rare or declining wetland types within an ecoregion, and that are subject to identifiable threat of loss or degradation.

Under the Emergency Wetlands Resource Act, states are required to include wetlands in their State Comprehensive Outdoor Recreation Plans.

The Act also requires consistency between these plans and the U.S. Department of Interior's National Wetlands Priority Conservation Plan.

The Emergency Wetlands Resources Act provides an opportunity to identify those wetlands most subject to loss and degradation from cumulative impacts and to acquire them to prevent further losses.

Food Security Act of 1985 (PL 99-198)

The Food Security Act of 1985 denies federal agricultural program benefits to farmers who convert wetlands for crop production. Conversion to croplands is a primary source of the cumulative loss of wetland acreage. The "Swampbuster" provisions of Title XII of the Act make farmers ineligible for price-support payments, farm storage facility loans, crop insurance, disaster payments, and insured or guaranteed loans for any year in which an annual crop is produced on converted wetlands. The Agricultural Stabilization and Conservation Service is charged with identification of those farmers who may be ineligible for agricultural benefits. The Swampbuster provisions of the Act have also encouraged this agency to establish habitat protection programs with the FWS for wetlands held in inventory due to delinquent loans that have been foreclosed.

Coastal Zone Management Act (PL92-583)

The Coastal Zone Management Act authorizes the National Oceanic and Atmospheric Administration (NOAA) of the Department of Commerce to approve and provide grants to States with coastal zone management plans. Twenty-nine States have approved plans. In some States, for example, Louisiana, Coastal Zone Management Plans are required at the county/parish level to assure effective local implementation. The 1980 Amendments to the Coastal Zone Management Act added the concept of Special Area Management Plans (16 U.S.C. Section 1453), and encouraged federal participation in the planning process. Section 1453 states that the purpose of Special Area Management Plans is to "provide for increased specificity in protecting significant natural resources, reasonable coastal-dependent economic growth, improved protection of life and property in hazardous areas, and improved predictability in governmental decision making". The Corps of Engineers has adopted the Special Area Management Plan process in their program, and has used it to conduct studies of extensive areas of wetlands.

National Flood Insurance Act of 1968 (PL 90-448)

The 1973 Amendments to the National Flood Insurance Program require all flood hazard areas to be identified. Communities participating in the program must plan for floods by designating a regulatory "floodway" along rivers and streams to accommodate 100-year floods. Wetlands are often included in these floodways. Recent regulations by the Federal Emergency Management Administration authorize denial of new or renewed insurance coverage to owners of property deemed by state or local authorities to violate state or local laws "intended to discourage or otherwise restrict land development or occupancy in flood-prone areas" (44 CFR Part 73.3, August 25, 1986). To qualify for coverage under the National Flood Insurance Program, localities must develop floodplain regulations or ordinances. Some communities, such as West Bloomfield, Michigan, have adopted regulations that include wetland values and hazard issues. Cumulative impacts of building construction and other activities can and should be considered under the regulations, including the impacts of regulated activities on communities and resources downstream.

Safe Drinking Water Act of 1980 (PL 96-502)

The 1986 Amendments to the Safe Drinking Water Act established the Sole Source Aquifer Demonstration Program for critical aquifer protection areas within designated sole source aquifers (Section 1424 (e)). A critical aquifer protection area is "all or part of a major recharge area of a sole or principal source aquifer which satisfies the criteria established by the administration of Section 1427 (d) of the Safe Drinking Water Act" or, "all or part of an area designated as a sole or principal source aquifer for which an area wide groundwater quality plan was approved under Section 208 of the Clean Water Act."

Generally speaking, the critical aquifer protection areas that may be considered are those in which the sole source aquifer is particularly vulnerable to contamination. In addition, if there is evidence that groundwater in the proposed critical aquifer protection area discharges into a site containing a valuable ecological system, or where significant environmental or social costs and health risks might occur if the site was contaminated, these costs and risks should be considered, and the site protected. Under these circumstances wetlands can be characterized as critical aquifer protection areas.

Protection of these areas can occur through a demonstration program designed by a planning entity with authority over the critical area. Any state, municipal or local government can identify a critical aquifer protection area within its jurisdiction and may apply for the selection of such an area for a demonstration program under Section 1427 of the Safe Drinking Water Act. Cumulative impacts to wetlands and other elements of aquatic systems can be one criterion for identification of critical aquifer protection areas.

Wild and Scenic Rivers Act of 1968 (PL 90-542)

The Wild and Scenic Rivers Act, which was written into law in 1968, and amended in 1986, established a national program to identify rivers and streams that "shall be preserved in free flowing condition" and states that "they and their immediate environments shall be protected..." (16 U.S.C. 1271). Characteristics considered in selection of scenic rivers include scenic, recreational, geologic, fish and wildlife, historic, cultural and similar values.

National scenic rivers are established through two processes: first, through state establishment of a scenic rivers program. The state may then apply to the Secretary of the Interior for designation of a National Scenic River. The second process is a federally-initiated process whereby Congress recommends to the Secretary of Interior that a particular river or stream be given National Scenic River status and the protection this status affords. Presently about half of the states have scenic rivers programs; notable among these states are programs in Maine, Minnesota and Oregon. Studies conducted in both the state-initiated and federally-initiated processes can provide important data regarding cumulative impacts to wetlands, and yield a measure of protection to these aquatic systems.

Coastal Barrier Resources Act of 1982 (PL 97-348)

The Coastal Barrier Resources Act of 1982 established the Coastal Barrier Resources System, which consists of undeveloped coastal barriers on the Atlantic and Gulf coasts. It directs the Secretary of Interior to identify undeveloped barrier islands in order to protect them and all associated aquatic habitats, including the adjacent wetlands, marshes, estuaries and nearshore waters, by limiting federal expenditures that might permit and subsidize development and the subsequent loss of barrier resources. Coastal wetlands can be protected under this program from the cumulative impacts of human development.

The Endangered Species Act of 1982 (PL 93-205)

The purpose of the Endangered Species Act of 1982, as amended in 1988, is to "provide a means whereby the ecosystems upon which endangered species and threatened species depend may be conserved" (16 U.S.C. 1531b). The Act provides for the identification and protection of critical habitat, where appropriate, for endangered and threatened species. While this act has for the most part been used to protect specific sites valuable to threatened and endangered species of animals and plants, it can also be effectively implemented to control cumulative impacts to wetland habitats on which many endangered and threatened species depend for some portion of their life cycle requirements. The CWA Section 404 (b) (1) Guidelines also provide some protection to endangered species: the impacts of the placement of dredged or fill materials in the waters of the United States on threatened and endangered species must be considered when reviewing permits.

Other Statutes, Programs and Organizations

The federal Water Resources Development Act of 1986 (PL 93-205), and Section 10 of the River and Harbors Act of 1899 (33 U.S.C. 622) may also be used to identify and prevent cumulative impacts to wetlands. These, and numerous statutes, programs and efforts underway at the state and local levels, and through private conservation organizations, also should be considered as mechanisms for addressing the cumulative loss of wetland acreage and function.

FAILURE OF CURRENT STATUTES AND PROGRAMS TO PREVENT CUMULATIVE LOSSES

The plethora of regulatory and non-regulatory programs that can potentially act to prevent the cumulative loss of wetland acreage and functions have not been successful in stemming the loss of wetlands nationwide, particularly in the case of bottomland hardwood wetlands. Setting aside political and economic factors that have prevented the full consideration of cumulative impacts, there are several reasons why these programs have been largely ineffective in addressing cumulative impacts. First, agencies usually lack the information necessary to determine the amount and nature of cumulative wetland areal loss and the functions and values of the remaining wetlands within the landscape unit. Second, these agencies individually lack the funds and manpower to undertake a monitoring program sufficient to determine the status of wetland resources. Third, individual agencies lack the regulatory authority to effectively prevent many of the cumulative losses that occur in wetlands. Finally, most regulatory and non-regulatory programs specifically focus on individual wetland sites, not on cumulative effects in the context of some larger landscape unit.

The Section 404 program offers a good example of these difficulties.

1) During the process of permit review generally neither EPA nor the USACE has extensive data on the status of wetlands in the watershed surrounding the permit site, either in terms of losses in acreage or changes in wetlands functional status over time, that can be used to determine if cumulative losses are occurring. Although much useful information is often available in an area, these data have not been analyzed for evaluation of cumulative impacts, and are, therefore, not immediately useful to regulatory personnel. Available data include water quality trends, changes in hydrology, flood damage losses, and data on wildlife and fisheries populations that depend on wetland habitats.
2) Individual agencies, for example, the USACE, EPA, or the FWS, do not have the staff or fiscal resources to gather, analyze, and place in an accessible format, such data for all the geographic areas for which they have responsibility.

3) Regulatory authority under CWA Section 404 is limited to the specific definition of "waters of the United States". This definition may exclude many portions of the landscape critical for wetland functional integrity. In addition, Section 404 specifically exempts from regulation forestry, farming, and ranching, both in wetlands and in uplands. These activities often have significant impacts on wetland functions.
4) Finally, the Section 404 permit process tends to focus on impacts to individual wetland sites. The key criteria used to evaluate whether to approve specific permit proposals to deposit dredged or fill material in wetlands rest on (a) whether or not practical alternatives exist that would result in less adverse impact, and (b) whether the project would cause or contribute to significant degradation of wetlands. For the reasons cited above, this is much more difficult to determine for cumulative effects than for local sites.

Non-regulatory programs face similar problems. Probably the major stumbling block to effective cumulative impact management is the lack of a clearly identified lead agency that can coordinate planning and implementation over large watersheds or other natural landscape units. For example, even federal land management agencies with jurisdiction over federal lands are limited in their ability to provide adequate conservation practices by (1) their narrow mandates, and (2) the small size (in a landscape sense) and often discontinuous boundaries of the property over which they have jurisdiction. As a result data useful in cumulative impact assessments are rarely identified or are inaccessible for decision-making.

POSSIBLE STRATEGIES FOR STEMMING CUMULATIVE LOSSES OF WETLAND ACREAGE AND FUNCTION

Effective control of cumulative impacts in wetlands can be achieved by broadening the scope of information gathered to make individual decisions, and by cooperative efforts by the agencies and organizations concerned with wetlands protection. If data on wetland functions and values which already exist (in the form of US Geological Survey water quality and flow data, wildlife data such as breeding bird surveys, U.S. Forest Service forestry trends, and flood damage statistics such as those collected by the Federal Emergency Management Agency) were to be combined and made accessible to concerned agencies, cumulative changes in wetlands could be more easily and accurately identified.

If federal, state, and local authorities, and private interests combined and coordinated their actions they could be much more effective in stemming cumulative losses. Individually these groups have significant control over small, often fragmented land areas. But collectively, significant portions of most landscapes can be protected. Imagine that the potential jurisdiction of individual agencies and organization are each drawn on a transparent overlay, and placed on a large scale map of a state or natural region. The Section 404 overlay, for example, would show a dendritic pattern, largely following the pattern of streams and rivers. Land management agencies would have discrete patches of lands and waters shown on their individual overlays. An overlay illustrating state authorities might show a mix of areas that are either owned by the state, or fall under a state regulatory program. An overlay for private conservation organizations would show individual patches owned or protected by easement. Taken together, and placed over the base map, these overlays would illustrate that a significant portion of most landscapes can be afforded a measure of protection against the loss of wetland acreage or function.

Thus, by broadening their data acquisition and analysis, and by cooperation and coordination, agencies and organizations concerned with wetlands protection can effectively protect wetlands from piecemeal losses of area and function, without a large increase in fiscal or staff resources.

REFERENCES CITED

Beanlands, G. E., W. J. Erckmann, G. H. Orians, J. O'Riordan, D. Policansky, M. H. Sadar, and B. Sadler, eds. 1986. Cumulative environmental effects: a binational perspective. Canadian Environmental Assessment Research Council/U.S. National Research Council, Ottawa, Ontario, and Washington, D.C.

Bedford, B. L., and E. M. Preston, eds. 1988. Cumulative effects on landscape systems of wetlands. Environ. Mgt. 12 (5) : 561-772.

Gosselink, J. G., and L. C. Lee. 1989. Cumulative impact assessment in bottomland hardwood forests. Wetlands 9: 89-174.

Horak, G. C., E. C. Vlachos, and E. W. Cline. 1983a. Fish and wildlife and cumulative impacts: is there a problem? Eastern Energy and Land Use Team, U.S. Fish Wildl. Serv. Prepared by Dynamac Corp., Fort Collins, CO.

________________________________. 1983b. Methodological guidance for assessing cumulative impacts on fish and wildlife. Eastern Energy and Land Use Team, U.S. Fish Wildl. Serv. Prepared by Dynamac Corp., Fort Collins, CO.

8. GAUGING THE CUMULATIVE EFFECTS OF DEVELOPMENTAL ACTIVITIES ON COMPLEX ECOSYSTEMS

John Cairns, Jr.
University Center for Environmental and Hazardous Materials Studies and Department of Biology, Virginia Polytechnic Institute and State University, Blacksburg, VA 24061

ABSTRACT

Estimates of hazards to complex ecosystems and to resource management have been fragmented so that precisely judging the aggregate impact of a variety of human-induced stresses is presently nearly impossible. Agencies charged with managing these natural resources have different missions, information-gathering responsibilities, and views of the management problem. Scientifically determining the effects of a potentially hazardous chemical tested under laboratory conditions in isolation from other impacts is not likely to produce information that will protect the resource. Although the literature is replete with studies of the effects of individual chemicals on specific organisms, rarely are mixtures of chemicals and other stresses tested on a community or ecosystem, or a surrogate of either. It is unlikely that such additional stressors as acid rain or airborne heavy metals will be considered seriously in the estimate because the developers are not responsible for these impacts. State regulatory agencies may not feel it their responsibility to protect ecosystems beyond the political boundaries of a particular state, and federal regulatory agencies may individually feel that they are only responsible for a specific segment of the resource management responsibility. Ultimately this is neither good science nor good politics since natural resources, such as bottomland hardwood ecosystems, respond to the aggregate and cumulative impact of all the stresses to which they are exposed. Fragmentation of the responsibility for their well-being may be administratively convenient, but it is not the best way to protect the resource.

Ecological Processes and Cumulative Impacts: Illustrated by Bottomland Hardwood Wetland Ecosystems. Edited by James G. Gosselink, Lyndon C. Lee, and Thomas A. Muir.

INTRODUCTION

Managing cumulative impacts of anthropogenic activities on complex ecosystems requires the following:

1) prediction of hazards or risks of various societal activities to these ecosystems before the bottomland hardwoods and other parts of the ecosystem are exposed to them;
2) validation of these predictions in a scientifically justifiable way;
3) assessment of the impact of long-term, continuous exposure to contaminants and other types of anthropogenic stress, together with the aggregate of natural impacts, including infrequent episodic events such as floods, hurricanes, etc;
4) integrated resource management that reduces or eliminates fragmentation of data gathering and fragmentation of institutional or organizational responsibilities for these complex ecosystems;
5) establishment of appropriate geographic boundaries for the study of these complex ecosystems so that subunits, such as bottomland hardwoods, are always examined in the larger context of the entire ecosystem in which they occur; and
6) a data acquisition system for environmental quality control that provides real time information as to whether the complex ecosystems are within predicted boundaries and alerts managers immediately when they are not.

This discussion deals only in part with bottomland hardwood forests, since evaluating cumulative impacts to complex ecosystems mandates that both larger ecological systems and larger institutional systems be examined. Although it is difficult to consider larger systems in a relatively short discussion, some of the factors impeding skillful evaluation of cumulative impacts on complex ecosystems are examined.

Toxicity testing can be traced to Aristotle, who over 2000 years ago placed freshwater organisms in sea water and observed their response. However, the formal discipline known as toxicology did not arise until the early 1800's in response to the development of organic chemistry (Zapp 1980). The potential for both benefits and penalties when these chemical substances were used was quickly recognized. Simultaneously, the possibility of adverse effects of these substances on organisms other than humans was recognized, but it was not until the late 1940's and early 1950's that the effects on non-human organisms became of substantial interest.

For non-human organisms, the field of aquatic toxicology developed much more rapidly than did terrestrial toxicology. One of the first aquatic toxicity tests was formalized by Hart et al. (1945), later adopted somewhat more efficiently in a peer-reviewed journal by Doudoroff et al. (1951). At the same time, the methodology presented by Hart et al. was modified to become one of the first formal, professionally endorsed standard methods for toxicity testing by the American Society for Testing and Materials.

The most common test organisms in the early development of the field were fish. They were vertebrates, edible, had recreational and commercial value, and were selected as test organisms because non-scientists could recognize their names and were able to see them. No one asked whether fish were the best organisms to test, for example, because they had a large biomass, were important in fixing or converting energy, or were most representative of ecosystem response.

Although early toxicity tests were quantitative, inexpensive, and gave highly replicable results when lethality was used as an end point, they had a number of deficiencies:

1) Most laboratory tests were low in environmental realism.
2) Species were tested in isolation from each other, and usually only one life history stage of a single species was involved in a specific test.
3) Validation of the accuracy of predictions of safety and harm was based more on circumstantial evidence (no observed fish kills, etc.) than on a scientifically justifiable testing of a hypothesis.
4) Responses measured were those characteristic of single species (e.g., lethality, reproductive success, respiration, etc.), and the relationship of these response thresholds to end points characteristic of communities or ecosystems (e.g., nutrient and energy cycling and other functional attributes) was rarely considered and poorly documented.

Furthermore, information on anthropogenic stresses to which ecosystems are exposed is fragmented and not gathered systematically. More importantly, as Harper (1982) notes, the concept of species that ecologists routinely use in describing ecological communities may be ecologically inadequate for the task and will probably remain so as long as ecological work continues to be basically descriptive rather than predictive. While individual species' responses may be an index of ecosystem processes (Levin et al. 1984), measurements of multispecies responses or ecological processes (e.g., decomposition, primary productivity, and nitrogen fixation) are clearly more appropriate indicators of ecological effects (Cairns 1980, 1983, 1985; National Research Council 1981).

FRAGMENTATION OF RESPONSIBILITY FOR RESOURCE MANAGEMENT

Resource management at federal and state levels of political organization is equally incomplete. Various governmental agencies are charged with managing a component of a resource but not the entire ecosystem. When part of an ecosystem (such as national parks) is the responsibility of a single agency, that portion is not the same as the entire ecosystem boundary, especially for large mammals such as elk (*Alces*). Therefore, the management task is complicated by the fact that a portion of the ecosystem is not under the control of the agency responsible for managing it. Similarly, ecosystems such as bottomland hardwoods are subject to episodic events (e.g., flooding) that an agency such as the U.S. Army Corps of Engineers (USACE) may be charged with preventing or reducing. In carrying out channelization, dam construction, and other activities that are physical (as opposed to ecological) means of flood control, the agencies are following a congressional mandate that does not adequately consider impacts to all parts of the ecosystem.

Scott et al. (1985) provide a classic example of the type of fragmentation that results when integrated resource management is absent. Long-term deterioration of portions of a cypress-tupelo (*Taxodium distichum-Nyssa aquatica)* wetland described in their paper resulted from an interaction between thermal loading (from nuclear production reactors owned by the U.S. Department of Energy [DOE] and operated by Dupont) and flooding (controlled upstream by the USACE). Evidence suggests that continued deterioration of portions of this wetland could be mitigated by eliminating large, episodic releases of water during the growing season. The USACE regulates water levels primarily for recreation on an upstream reservoir, and DOE and Dupont operate the reactors to produce plutonium and tritium. The narrow focus of agency tasks probably hampers the integration of management decisions to optimize ecosystem condition. Agency administrators are quite capable of broader focus but may be hampered because the United States Congress allocates funding for specific purposes for each administrative unit. This reduces territorial battles between organizations but does not enhance ecosystem management.

ECOSYSTEM RISK ANALYSIS

Both highly industrialized and developing countries are faced with serious problems as a consequence of failing to develop the scientific base necessary to accurately predict risks to ecosystems. They have also failed to develop the regulatory and political systems necessary to implement integrated resource management of large, natural systems. The crisis appears to have developed relatively recently because, in the past, terrestrial systems were damaged in relatively small patches and were surrounded by natural systems that enhanced recovery once the stress was removed or reduced. Similarly, waste discharges into rivers and other waterbodies were sufficiently small and spaced at sufficient intervals to enable the natural assimilative capacity of aquatic ecosystems to restore the system to equilibrium before the next event. There were some notable exceptions to this, such as the Cloaca Maxima (one of the first sewage collection and discharge systems) of Rome during the height of the Roman Empire, and damage to the tidal Thames at London hundreds of years ago that caused sheets soaked in vinegar to be hung in Parliament to offset the stench from the nearby river. In the latter case, political action was not effective for many decades, but the recent remarkable restoration of the Thames is now one of the leading examples of effective, integrated resource management using presently available scientific information in a cost-effective way.

Although the workshops focus was on bottomland hardwood ecosystems, it is my opinion that these forests cannot be managed in isolation from other ecosystems and other regional problems. Ultimately, some groups must face the problem of generating the scientific information for both predicting and validating cumulative impacts of anthropogenic origin on natural ecosystems and separating these from those of natural origin. This information must be coupled with integrated resource management by the governmental agencies charged with maintaining the ecosystems' well being. If this is not done, the fate of the bottomland hardwoods ecosystems will be sealed regardless of how well this important fragment of a larger ecosystem is understood.

FUNDING SCIENTIFIC STUDIES OF SOCIETY'S CUMULATIVE IMPACT ON NATURAL ECOSYSTEMS, INCLUDING BOTTOMLAND HARDWOOD FORESTS

Large systems research is not only at least an order of magnitude more expensive than traditional research but requires long-range, uninterrupted funding at substantial levels if projects are to succeed. Successful systems-level projects, such as the Hubbard Brook (Bormann and Likens 1979) and Coweeta studies (Swank and Crossley 1988), provide useful models. The studies of Robert J. Livingston of Florida State University on the Apalachicola River and Bay show that long-term studies can be carried out by a persistent individual with a variety of funding sources. However, finding a sufficient number of dedicated, creative individuals willing to devote a major portion of their professional careers to such projects is not likely and, additionally, unpredictable events suggest that systems studies should probably be an institutional responsibility.

A good model for institutional systems research is the Ohio River Valley Water Sanitation Commission, which has routinely monitored water quality for significant portions of the Ohio basin over a long period of time. The governments of eight states (Illinois, Indiana, Kentucky, New York, Ohio, Pennsylvania, Virginia, and West Virginia), with approval of the United States Congress met in 1948 and pledged co-operation in pollution abatement through the establishment of an interstate agency (Cleary 1967). Figure 1 illustrates an attempt to integrate resource (e.g., the Ohio River) management by orchestrating interactions among various interest groups. Although much remains to be done, resource health has markedly improved despite a concomitant increase in use. Integrated resource management has also been successful on the River Thames (Gameson and Wheeler 1975). Both these examples are not

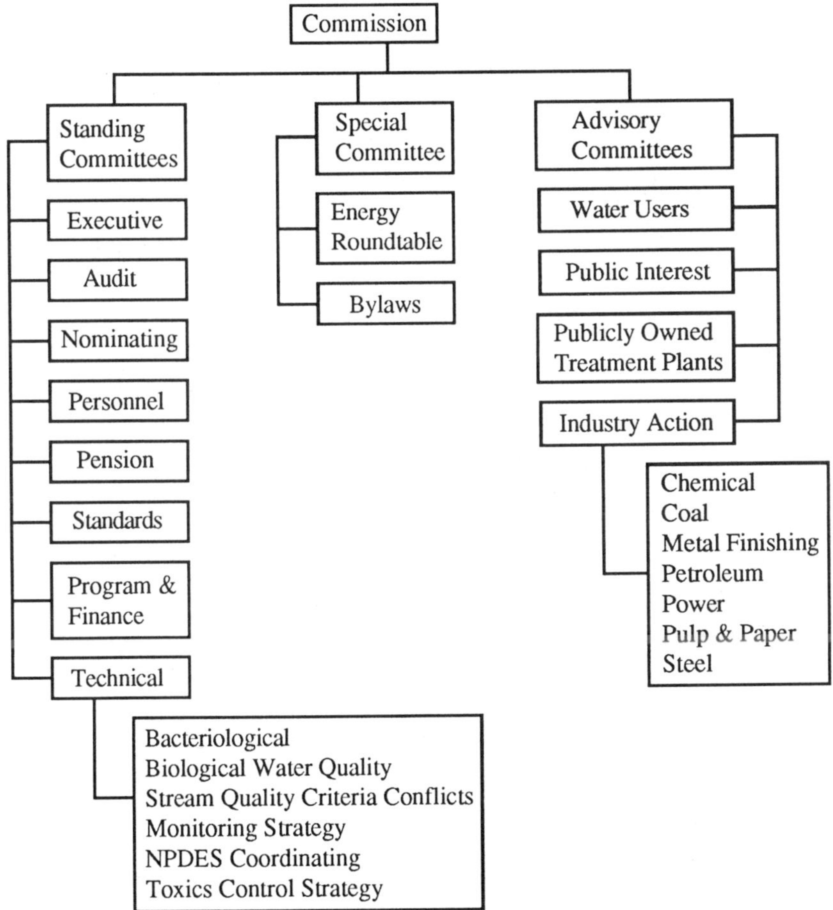

Figure 1. Ohio River Valley Water Sanitation Commission Committee structure.

ultimate models but rather indications that the process does work. The quantity of biological data from these model programs does not match chemical/physical water-quality data, but the institutional responsibility for gathering certain types of data is exemplary.

It is clear that significant new research money is unlikely to become available through federal agencies, except possibly the U.S. Department of Defense. The distribution of existing funds is unlikely to change drastically since the constituency already developed in specific, strongly traditional research areas will fight fiercely to retain the present distribution. Reductions in federal research funding will only exacerbate the competition for the remaining funding among the traditional disciplines. As a consequence, it is unlikely that the large systems studies needed to generate information for effective integrated resource management will be available unless new funding develops. Private foundations might fund a portion of the necessary research, but there is little evidence that they are capable of, or willing to, supply funds on the scale that would be necessary to provide the regional information essential to manage large systems.

The only other major source of funding is the private sector. This at first seems an extremely unlikely source of substantial, long-range funds, but it bears further examination because obtaining the necessary amounts from the other sources seems even more improbable. First of all, if integrated resource management includes industrial use of resources in such a way that their ecological integrity will not be seriously impaired, industry has a major stake in seeing that management practices are based on the best available information. The value of integrated environmental management is so evident that one wonders why we have such difficulty implementing this goal. The more obvious benefits are: a) cost-effectiveness; b) improved, long-term protection of the resource; c) enhanced possibilities for effective multiple use; d) more rapid and effective restoration of damaged resources to usable condition; and e) reduced expenditure of energy on conflicts over use and the possibility of redirection of these energies and funds to resource management.

The major obstacle, of course, is lack of mutual trust. Industry, regulatory agencies, and environmental groups are more accustomed to court cases than collaborative efforts. One of the unfortunate consequences of this lack of mutual trust is the prescriptive way in which laws and regulations are written. They are so detailed and specific that they discourage innovative, creative approaches to solving problems. As a consequence, the regulated (public and industries) and regulators are frequently pitted against each other in ways that prevent even simple solutions to obvious problems. It is increasingly common to find that the cost of the legal battle may markedly exceed the cost of solving the problem. Whatever happens in court, there is usually little or no new scientific information generated that will improve resource management. Additionally, money is diverted from environmental quality control into legal fees.

BARRIERS TO INTEGRATED RESOURCE MANAGEMENT

Common Barriers:

1) "Turf" or organizational territoriality.
2) Insufficient funding.
3) Resistance to change.
4) Inability to compromise.
5) Restrictive laws and regulations.
6) No benefits associated with interdisciplinary activities.
7) Lack of mutual trust.
8) Lack of leadership commitment.
9) Limited statutory authority.
10) Motivational barriers.
11) Lower priority than daily operations.

Less Common Barriers:

1) Misunderstanding of objectives of other agencies (the hidden agenda problem).
2) Lack of consistency in involving private enterprise in resource development.
3) Loss of control due to size and scope of project.
4) Public system does not reward innovation as well as private enterprise.
5) Geographic location of headquarters not appropriate for large system management.
6) Problem of getting subordinates to "think system."
7) Problems associated with new efforts where the outcome appears highly uncertain.
8) Elitist attitude (my discipline is superior to your discipline).
9) Private rights versus public good.

NEED TO RECOGNIZE DIFFERENT LEVELS OF RISK IN INTEGRATED RESOURCE MANAGEMENT

Managing large, natural systems effectively requires a larger data base, validated models describing how these systems work, and an equally large and complex information base on institutional arrangements. Furthermore, while management problems have much in common from one region to another, each ecological system is unique and requires substantial, site-specific modification of the general model or models. Generation of both types of information will require major adjustments in the attitudes of academic institutions, industrial organizations, and governmental agencies.

While the development of mutual respect and a working relationship among these disparate components may now seem visionary, we have all witnessed adjustments that would have been unthinkable before the energy crisis of the early 1970's. Many of these adjustments were made reluctantly and not always effectively, although in the aggregate they have been enormously effective as evidenced by the U.S.'s increased energy efficiency during the late 1970's. The present budget crisis in the federal government, foreign competition for American industry, and declining enrollments and steady state or reduced budgets for academic institutions may provide the impetus for a collaborative relationship that would have been unthinkable during the tumultuous period of the 1960's and 1970's.

On the negative side, court cases are becoming increasingly common, and professional liability insurance costs are skyrocketing. Even in the absence of professional liability lawsuits, it is difficult to get information about the operation of large systems without taking some risks. In an era when governmental units as small as towns can be taken to court for negligence and lose large sums of money, it will not be easy to proceed in unknown areas, both environmental and institutional, without taking some risks. The probability of environmental hazard can be determined somewhat in surrogate systems, but this information is less useful if it is not validated in a natural system. As a consequence, risk may be reduced but cannot be eliminated.

If we are to improve environmental and institutional information gathering on the management of large systems, a highly flexible approach will be needed that will permit rewards for taking carefully calculated risks in order to generate new information. In this regard, one might distinguish between three levels of risk and the information to be generated in each:

1) The outcome is fairly certain. This involves a low level of risk and minimal generation of new information. For example, methodology developed elsewhere and proven successful could be utilized in a new system with the expectation that results would be quite similar but not identical. The uncertainty would revolve around the degree to which effects were likely to occur rather than which types of effects would occur. This is

assuming, of course, that a reasonable match has been made between the kinds of systems being modeled and other considerations. In short, the appropriateness of the model being used is assumed to be high. The rewards could take a variety of forms, such as variance from certain laws and regulations, tax breaks, subsidies from foundations or the federal government, etc. For example, a damaged ecosystem might be restored with methodology proven effective elsewhere in a comparable ecosystem. The outcome is fairly certain, but not entirely so since each ecosystem has some unique characteristics.

2) Intermediate certainty that the outcome will be as predicted. This situation might arise where a model is being used that was developed on a dissimilar system, and the assumption is made that there is a substantial correspondence between the two systems based on reasonable, but purely theoretical grounds. If the understanding of the two systems is sufficiently good to justify these assumptions, the risk might be described as intermediate. Most of the events likely to occur can be predicted but not always accurately in terms of the degree of impact. The likelihood of some unpredicted events occurring is also moderately high. On the other hand, the new information could explain where the model was in error and which assumptions were or were not justifiable. A study could furnish much new, critical information. There would also be considerable cost savings on future management efforts as a consequence of not needing to "reinvent the wheel." On the other hand, there might be significant hidden costs because all events were not accurately predicted. Under these circumstances, the rewards for taking risks in order to generate new information should be as proportional as possible to the increased risks being taken. Society should be flexible and generous toward the risk takers charged with generating new information that will ultimately benefit society. Clearly, this will not happen until the public is more knowledgeable about integrated resource management. For example, suppose a mining company wanted to restore a surface mined area to its predisturbance ecological condition using a scientifically plausible but somewhat innovative

methodology. If properly designed, useful scientific and management information would be generated even if the experiment failed (i.e., next time do not try this approach or do so but with certain modifications). At best, a damaged ecosystem would be restored to pre-damaged condition in a more timely and cost-effective way than was previously possible. The industry has saved money, the regulatory agency has fulfilled its mission, and new information useful to society has been generated with no use of public or research foundation funds. However, in research the outcome is not always the expected one. Even though useful information has been generated, the ecosystem could still be damaged and the regulatory agency might appear inept. Some incentive must be provided to the industry (e.g., tax breaks) and to the regulatory agency (e.g., developing methodology in the mission statement) so that alternatives to prescriptive legislation are more attractive. Otherwise the information and methodology for effective integrated resource management will take longer to generate and will require more public and foundation financial support.

3) The outcome is highly uncertain but much new information will be generated. An example of this might be a hazardous waste storage site involving a new substance and a new method. The waste product was from a process highly desirable to society (e.g., an inexpensive form of energy) and was generated within the geographic boundaries of the system in question, thus directly benefiting the citizens taking the risk. One might add the caveat that a variety of more-or-less traditional tests at single species level, etc., would be carried out before embarking on this project. However, there is high uncertainty about the degree to which system response can be predicted from laboratory tests low in environmental realism. As a consequence, field testing within the larger system must be undertaken, to validate the predictions based on the laboratory tests where the effects can be contained to a much greater degree than in natural systems. Since such studies would benefit the entire country and since most of the risks would be taken within the system itself, it

seems reasonable to subsidize this type of information gathering from outside the system as well as from within the system.

RESOURCE FORM AND SERVICES

It is important to distinguish carefully between resource or ecosystem form (i.e., ecological structure) and services (i.e., ecological function). Environmental toxicology has demonstrated repeatedly that it is possible to impair the ability of organisms to function properly without actually killing them. In short, a variety of response thresholds exist, the ultimate being lethality. At lower concentrations of chemicals or reduced stress, various physiological, behavioral, or ecological processes may be diminished although the structure (e.g., organisms) will continue to survive at least for the near term. If both structure and services are to be preserved in original or present condition, a different management strategy must be adopted and a greatly modified monitoring system put in place to ensure that this is being accomplished. It might be easy to preserve the form of an ecosystem, such as bottomland forests, while losing some or all of its important ecological contributions to the larger ecosystem of which it is a part. Both qualitative and quantitative identification of these services or functions is an essential component of resource management.

Most of the management plans I have seen assume that if the form remains undisturbed the function will also, despite evidence to the contrary. It is my impression that with some notable exceptions the public is mostly interested in services in terms of recreational fisheries, etc. However, in respect to bottomland forests and other increasingly rare and unusual ecosystems, a good case could be made for maintaining both form and services. This aspect deserves attention because alternative species or ecosystems might fulfill some of the key functions of bottomland ecosystems, or it may be that modifications of the larger system (of which this component ecosystem is a part) have so altered it that the maintenance of the original services is impossible. Thus, the form may be difficult to maintain, but alternative species might be found to provide comparable services or ecosystem function. To quote Paul Ehrlich "the essence of tinkering is to make sure that none of the parts are lost" (Ehrlich and Ehrlich 1981). In short, until we are certain that the original services that were lost can be replaced with other species, we should not permit the existing ones to be "lost" or destroyed.

These bottomland hardwood forest workshops provided lists of critical functions that deserve attention. These lists are useful in category 1 below. Other aspects of function that deserve additional attention are given below.

1) A generalized list of important functional attributes of bottomland hardwoods should be available to all those charged with management responsibilities. It should be made clear that site-specific characteristics should influence the use made of this list.
2) Guidelines should be provided for selecting or identifying unique functions that may differ from site to site. These guidelines should furnish adequate details about the selection process so that unique regional or site-specific attributes will be identified.
3) Validation of the selection process is essential. In short, there should be scientifically sound evidence: a) that the "right" (i.e., critical) functions were selected; b) that the quality and quantity of information are suitable for management decisions; and c) that the measurements will provide an early warning of deleterious conditions.
4) It is also essential to determine that functions are "key" or "representative" and are both scientifically sound and cost-effective for continuous monitoring.

EPISODIC EVENTS

Measurement of ecosystem response to infrequent but important events (e.g., floods, forest fires, etc.) may appear inappropriate because the duration of a particular set of administrators in a particular position may not cover more than one episodic event and the appropriate response is difficult to communicate even in the rare instance when an attempt is made. Generally, routine housekeeping events take precedence over preparing for episodic events, despite the fact that the latter may be more important in determining ecosystem survival than any other single forcing factor. As a consequence, effective integrated resource management must include an "institutional memory" to cope with these episodic events, since they may be pivotal in maintaining resource integrity and continual delivery of services.

DOES INTEGRATED RESOURCE MANAGEMENT REDUCE CONFLICT?

Initially, integrated resource management may appear to increase conflicts, since identifying important characteristics and uses of a resource must be done at the outset to develop a management plan. However, this apparent increase in conflict is merely due to raising issues that might not otherwise be immediately apparent. In the long term, conflict should be reduced as a consequence of getting the major issues resolved through conflict resolution or some other appropriate means before the situation becomes polarized and institutional commitments and budgets have been prepared. For example:

1) Integrated resource management should enhance multiple use. This should reduce conflict by redistributing use patterns and making compromise more likely on a particular use for a particular area.
2) Integrated resource management should improve overall resource quality. This should make everyone happy even though some compromise is necessary.

Ecosystems do not have an infinite capacity for assimilating anthropogenic wastes. If an ecosystem is "fully loaded" (i.e., at capacity limits), new uses are precluded if ecosystem damage is to be avoided. Perhaps a free market (i.e., user bidding) approach to assigning non-destructive assimilative capacity privileges would provide funding for integrated resource management and would permit industries and municipalities with the most effective waste treatment systems to outbid those with antiquated systems. This would permit reallocation within a free market system and furnish resources for ecosystem management.

Integrated resource management should be far less costly and more effective than a fragmented approach; therefore, further development activities on many complex ecosystems will only be possible without serious ecosystem damage if the present fragmented approach, both scientific and institutional, is replaced with integrated resource studies and management. From a scientific standpoint, the traditional tools of ecology should not be abandoned, but we should recognize that they are not designed for this purpose. Additional methods based on system level attributes must be put in place. Additionally, evidence on present condition should be gathered in a systematic, orderly way and be accompanied by development of a predictive capability for determining the consequences of various courses of action, including impairment of environmental quality, etc. Finally, none of the scientific information will be used effectively unless the now fragmented institutional responsibilities for resource management are integrated. While this is by no means a simple task, there is evidence that it can be accomplished with skill and persistence.

Despite the cautions about and current limitations of artificial intelligence (Sheil 1987), predictions of the outcomes of various decision scenarios described in this manuscript should lend themselves beautifully to decision-making strategies or expert systems that are now being considered for agricultural systems (Applied Artificial Intelligence Reporter 1987) and natural resource management (Coulson et al. 1987). Information gathered from a number of similar ecosystems could be assembled into a type of blueprint that would help to guide and inform various management decisions, based on past successes and failures with similar systems. This would also allow for the incorporation of information on episodic events and how such events could alter resource decisions. While such a system would serve as only one of many potential resource management tools, it offers a potential solution to some of the problems inherent in integrated resource management.

ACKNOWLEDGMENTS

Funds for the preparation of this manuscript were provided, in part, by the Mobil Foundation. I am indebted to Michael L. Scott for useful comments on a draft of this manuscript. Darla Donald, University Center for Environmental and Hazardous Materials Studies Editorial Assistant, helped prepare the manuscript for publication. Betty Higginbotham prepared the final draft on the word processor.

REFERENCES CITED

Applied Artificial Intelligence Reporter. 1987. Agriculture managers cultivate expert systems. November, p. 4.

Bormann, F. H., and G. E. Likens. 1979. Pattern and process in a forested ecosystem. Springer-Verlag, New York. 253 p.

Cairns, J., Jr. 1980. Guest editorial: Beyond single species toxicity testing. Mar. Environ. Res. 3:157-159.

__________. 1983. Are single species toxicity tests alone adequate for estimating environmental hazard? Hydrobiologia 100:47-57.

__________., ed. 1985. Multispecies toxicity testing. Pergamon Press, New York. 261 p.

Cleary, E. J., Jr. 1967. The ORSANCO story: water quality management in the Ohio Valley under an interstate compact. The Johns Hopkins University Press, Baltimore, Maryland. 335 p.

Coulson, R. N., L. J. Folse, and D. K. Loh. 1987. Artificial intelligence and natural resource management. Science 273:262-267.

Doudoroff, P., B. G. Anderson, G. E. Burdick, P. S. Galtsoff, W. B. Hart, R. Patrick, E. R. Strong, E. W. Surber, and W. M. Van Horn. 1951. Bio-assay for the evaluation of acute toxicity of industrial wastes to fish. Sewage and Industrial Wastes 23:1380-1397.

Ehrlich, P. R., and A. Ehrlich. 1981. Extinction. Ballantine Books, New York.

Gameson, A. L. H. and A. Wheeler. 1975. Restoration and recovery of the Thames Estuary. Pages 72-101 *in* Recovery and restoration of damaged ecosystems. J. Cairns, Jr., K. L. Dickson, and E. E. Herricks, eds. University Press of Virginia, Charlottesville, Virginia.

Harper, J. L. 1982. After description. Pages 11-25 *in* E. I. Newman, ed. The plant community as a working mechanism. Blackwell Scientific Publishers, United Kingdom.

Hart, W. B., P. Doudoroff, and J. Greenbank. 1945. The evaluation of the toxicity of industrial wastes, chemicals and other substances to fresh water fishes. Waste Control Laboratory, Atlantic Refining Co., Philadelphia, Pennsylvania.

Levin, S. A., K. D. Kimball, W. H. McDowell and S. F. Kimball. 1984. New perspectives in ecotoxicology. Environ. Management 8:375-442.

National Research Council. 1981. Testing for effects of chemicals on ecosystems. National Academy Press, Washington, D.C. 103 p.

Scott, M. L., R. R. Sharitz and L. C. Lee. 1985. Disturbance in a cypress-tupelo wetland: an interaction between thermal loading and hydrology. Wetlands 5:53-68.

Sheil, B. 1987. Thinking about artificial intelligence. Harvard Business Review. July-August, p.91-97.

Swank, W. T., and D. A. Crossley, Jr. 1988. Forest hydrology and ecology at Coweeta. Springer-Verlag, New York. 469 p.

Zapp, J. A., Jr. 1980. Historical consideration of interspecies relationships in toxicity assessment. Pages 2-10 *in* Aquatic toxicology. J. G. Eaton, P. R. Parrish, and A. C. Hendricks, eds. American Society for Testing and Materials, Philadelphia, Pennsylvania.

9. CUMULATIVE IMPACTS OF BOTTOMLAND HARDWOOD FOREST CONVERSION ON HYDROLOGY, WATER QUALITY, AND TERRESTRIAL WILDLIFE

Larry D. Harris
Department of Wildlife and Range Sciences, 118 Newins-Ziegler Hall, University of Florida, Gainesville, FL 32611

James G. Gosselink
Marine Sciences Department, Center for Wetland Resources, Louisiana State University, Baton Rouge, LA 70803

ABSTRACT

Bottomland hardwood forested wetlands, through their ecological processes, perform valuable services for human society by moderating severe floods, maintaining water quality, and protecting balanced, indigenous populations of biota. These wetlands are being rapidly converted to other uses (primarily agriculture), especially in the southeastern United States and the Mississippi River Valley where only a small fraction of the original forest remains. Although forest loss at local sites may have only a moderate impact on the natural services of the bottomland hardwood ecosystem, the total, or cumulative, impact of many small conversion actions may be significant. The cumulative influence of many flood-control projects increases stream efficiency. As a result, local bottomland hardwood forests are flooded less frequently. However, during severe floods downstream stages are higher, flooding is more frequent, and streams are increasingly unstable. As watersheds are cleared, especially as the stream-edge floodplain forests are cleared, erosion increases and streamwater quality degrades. The first effects on native biotic diversity due to cumulative bottomland hardwood forest loss are felt by individual species, which suffer abnormalities in reproduction, range restriction of sensitive species, and loss of genetic heterogeneity.

Ecological Processes and Cumulative Impacts: Illustrated by Bottomland Hardwood Wetland Ecosystems. Edited by James G. Gosselink, Lyndon C. Lee, and Thomas A. Muir. Chelsea, Mi 48118. Printed in USA.

As the cumulative stress level increases, the whole ecosystem is modified. Trophic webs become simplified, biotic community composition and structure are modified, and native species are lost due to invasion by exotic species, parasites, and diseases. Finally, the ecosystem collapses and is transformed with resulting extinctions of native species and depletion of migrant bird populations.

ECOLOGICAL FUNCTIONS AND PROCESSES

The Seasonal Pattern of Process

In discussions of natural ecosystems, the terms "functions, processes, and structures" are often interchangeable with the term "values" (goods, services, amenities). We reserve the term value to refer to characteristics of bottomlands that benefit humans in some way; that is, value is an anthropocentric term. Bottomland forest values may change as our perception and technology change even though the functions and structures of the forest remain the same. The relationship between ecosystem functions and human values that derive from them is illustrated in Figure 1. The driving solar and hydraulic energy for the system and the structural sediment create an environment within which bottomlands function to build vegetation and animal communities, process chemical elements, and store surface and groundwater. These functions lead to human values of at least three types: 1) biological values--biomass production, fish and wildlife habitat, and maintenance of biological diversity; 2) water quality protection and erosion control; and 3) water storage values--flood modification, low flow augmentation, and deep aquifer recharge. Functions and values of bottomland hardwood forest ecosystems have been recognized in a number of recent publications (Johnson and McCormick 1978, Greeson et al. 1979, Wharton 1980, Brinson et al. 1981, Clark and Benforado 1981, Conner and Day 1982, Wharton et al. 1982, Office of Technology Assessment (OTA) 1984, Taylor et al. 1984). The following discussion borrows heavily from Gosselink et al. (1981), Harris (1982, 1984), Harris and Vickers (1984), and Taylor et al. (1990).

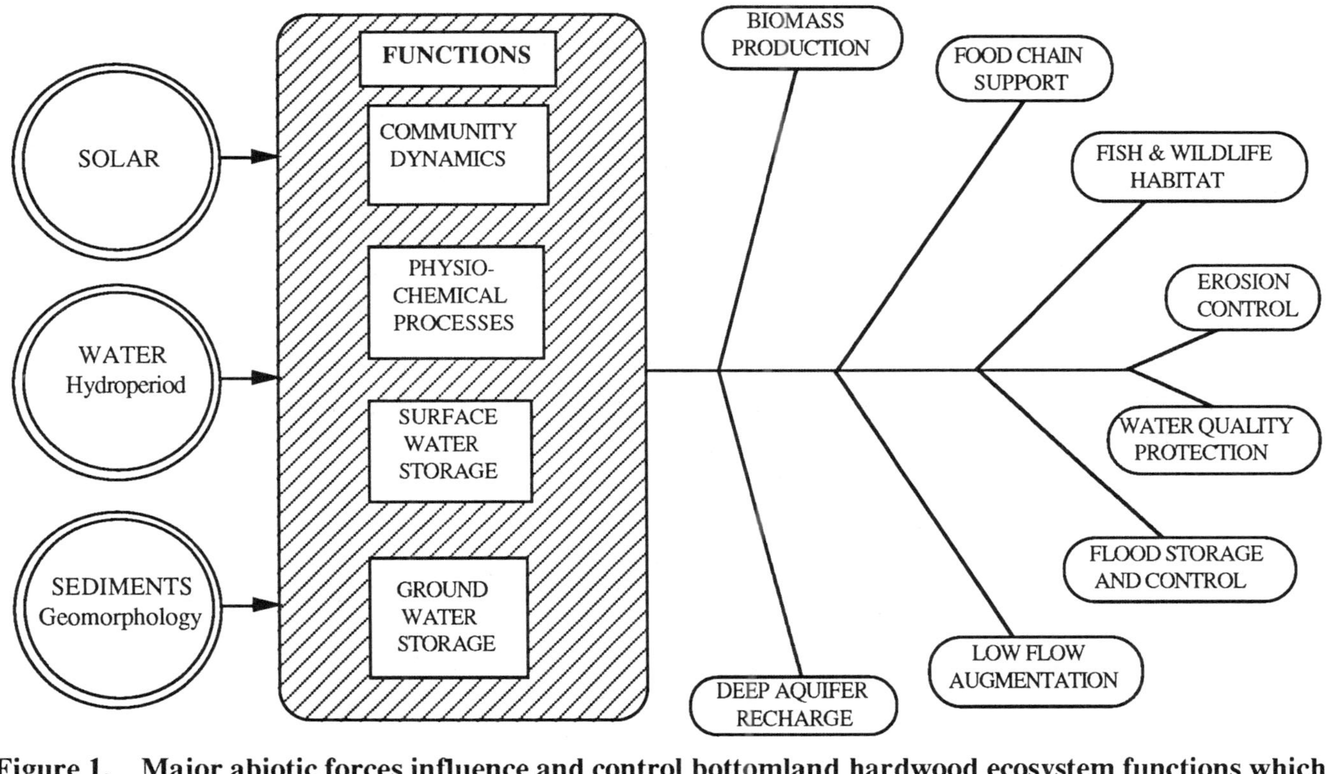

Figure 1. **Major abiotic forces influence and control bottomland hardwood ecosystem functions which in turn produce goods and services that are of value to humans (from Taylor et al. 1990).**

Bottomland hardwood forests are most extensively developed along the broad floodplain of the lower coastal plain in the southern and southeastern United States. They are inundated when rising floodwaters overflow the river banks. Flood depth, duration, and frequency are key variables that determine the kind of plant and animal species found on the floodplain. In general, the smaller the watershed and the closer it is to the headwaters of the river, the more frequent and shorter the duration of flooding. This is because small watersheds respond rapidly to local rainfall regimes. Conversely, the broad, flat floodplain of the Lower Mississippi River is responsive to precipitation patterns over nearly one-half of the contiguous United States. The Mississippi River thus tends to follow a seasonal pattern of late winter and spring flooding. Flooding seldom occurs during the summer and fall, and when it does it is localized in response to upland runoff from local precipitation.

Because bottomlands are predominantly located in the subtropical zone below 35° north latitude, principles of tropical ecology are more applicable than is generally the case in North America. First, the seasonal cycle tends to be defined more by the occurrence of precipitation, flooding, and surface water level than by temperature. Second, there is a high incidence of broad-leaved, evergreen plants, many of which flower and/or bear fruit during winter. A third relevant principle, not specific to the tropics, depends on the structure of rivers and riverine systems. Near headwaters the nature of streams and rivers is dominated by the surrounding landscape. Conversely, the relative influence shifts downstream near the mouth, and it is the nature and behavior of the river that dominates the surrounding landscape. Because of this the floodplain forest is as much a part of the river as the channel.

Finally, the wildlife habitat function of bottomland hardwood forest ecosystems can only be appreciated fully when viewed in the context of the coastal plain landscape. Having evolved under a heavy influence of lightning and wild fire, the preColumbian upland forest was dominated by fire-adapted pines (*Pinus* spp.) with an understory of wire grass (*Aristida stricta*) (Delcourt and Delcourt 1977, Croker 1979). Few woody, angiospermous (flowering) trees or shrubs occurred in these pine-dominated uplands. Only in wet areas along rivers or other refugia protected from fire could the more fire-sensitive hardwoods predominate. Thus, it was a conifer-dominated landscape with primarily linear, riparian bottomland hardwood forests. These riparian forests were distributed dendritically and thus facilitated the dispersal (and dispersion) of animals throughout the coastal plain (Figure 2). Many of the animal-related ecological functions ascribed to bottomland hardwood forest ecosystems derive from this linear interspersion among the pinelands.

During late winter and spring, floodwaters carry large amounts of nutrient-rich sediment and organic materials which are deposited on the floodplain as the overflowing water slows. These deposits form a rich subsidy of allochthonous nutrients and energy that support the high biological productivity of these systems. Primary productivity is high relative to surrounding upland systems. These subsidies support a detritus and detritivore food web of secondary productivity throughout the animal community. As water levels fluctuate so does the interface between aquatic and terrestrial systems. Many animal species move with this shallow, highly productive zone, giving rise to the concept of an oscillating migratory ecotone.

Although the bulk of on-site primary and secondary productivity occurs during summer, bottomland hardwood forests perform many ecological functions later in the year as well. In late summer and fall when surrounding uplands approach dormancy and have lost much of their habitat value, a new productivity phase is beginning in the bottomlands. Acorns and other hard seeds and nuts cover the ground and attract large numbers of mast-consuming resident species such as white-tailed deer (*Odocoileus virginianus*), turkey (*Meleagris gallopavo*), and black bear (*Ursus americanus*), as well as large numbers of overwintering migrant birds from throughout eastern North America.

The biomass-storage function (i.e., production during summer when consumer levels are low and storage until winter when consumer levels are high) continues through the winter until rising water levels initiate yet another phase. As surface water invades the floodplain, forest animals again respond to the migrating ecotone. Terrestrial species, many amphibious furbearers, aquatic finfish, shellfish, and mast-foraging waterfowl follow the water's edge. Organic sediments and detritus that have undergone months of aerobic oxidation are brought into solution; they further enrich the water and provide subsidies to downstream areas far removed from the local site.

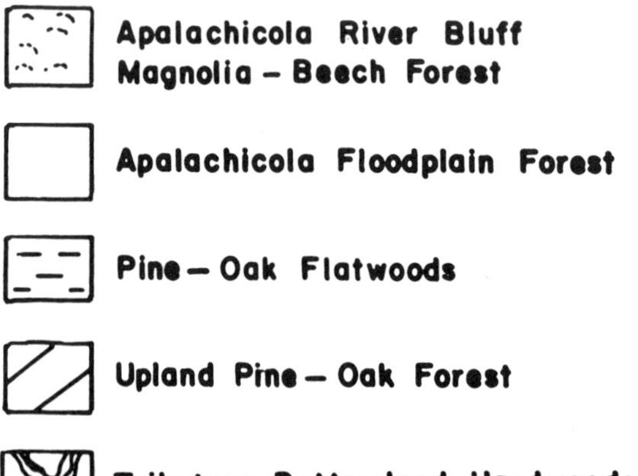

Figure 2. In the pre-Columbian landscape hardwoods were restricted to the floodplains by frequent fire that occurred in the pine-dominated landscape. The dendritic pattern is a common distributional pattern that facilitated animal movement throughout the landscape (after Delcourt and Delcourt 1977).

Human Values from Bottomland Hardwood Forests

The importance of the natural forested floodplain to human society has become increasingly clear as our understanding of these natural processes and the consequences of their loss has developed. The value of bottomland hardwood ecosystems in terms of biota, water quality, and flood storage will be discussed below.

Biotic Values

Societies depend on the renewable resources of the living world for food, shelter, recreation, and maintenance of the genetic diversity that supports life. The most obvious of these resources in bottomland hardwood ecosystems is the forest itself, which produces a timber crop valued from $3 billion to $8 billion per year. Harvesting under proper guidelines need not cause serious habitat deterioration and can complement other resource uses. This timber value results from the high natural productivity of bottomland hardwood forests, a consequence of the regular, natural fertilization by floodwaters, the regular availability of water for life processes, and the increasing scarcity of cypress (*Taxodium distichum*) and hardwood timber.

Bottomland hardwood ecosystems are valuable for their wildlife habitat, for commercial harvest of fish and wildlife species, for hunting and recreation, and for the maintenance of biological diversity. These values result from the structural diversity of woody plant communities, the presence of surface water and abundant soil moisture, the spatial complexity and interspersion of habitat features, and the linear nature of bottomlands that intensifies the "edge" or "ecotone" effect and provides corridors for wildlife movement. The bottomlands of Louisiana yield a larger annual harvest of furs than any state in the United States, sometimes as high as 65% of all United States production. Revenue from this harvest frequently exceeds $10 million per year (O'Neil and Linscombe n.d., Tiner 1984). The annual crawfish harvest in Louisiana is valued at about $11 million, mostly from bottomland swamps. Most of the commercially harvested fish of southeastern river systems spawn and feed in the riparian zone. Many require this shallow flooded zone for successful spawning, and their commercial and sportfishing value is closely keyed to the frequency of spring floods.

The value of bottomland hardwood forests for hunting is high. They support white-tailed deer populations at least double those of equivalent areas of upland forest (Table 1). Because of the diversity of food and cover resources, small and medium-sized mammals abound (Table 2).

Table 1. Estimated white-tailed deer carrying capacity in common southeastern habitats and percent of annual deer harvest deriving from bottomland hardwood forests in Mississippi and Louisiana (from Glasgow and Noble 1971).

Habitat Type	ha/deer
Longleaf-Slash Pine	13
Loblolly-Shortleaf Pine	13
Upland Hardwoods	8
Bottomland Hardwoods	5

State	% Land in bottomland hardwood forest	Harvest from bottomland hardwood forest (% of total)
Miss. (1970-71)	15	45
LA (1964-70)	31	57

Resident bird populations are larger in bottomland and coastal hardwoods than in surrounding upland environments during all seasons, but this is especially apparent during winter months (Figure 3). When overwintering North American migrant birds arrive on the coastal plain they do not disperse throughout upland and bottomland environments equally but discriminate heavily in favor of bottomlands (Figure 3). During peak winter months (November-February) 35 birds/ha occur in bottomland hardwood forests, compared to summer densities in the northern states of only about 6 birds/ha. Thus, 1 ha of bottomland hardwood may serve as overwintering habitat for the following year's breeding stock on 6 ha of northern forest.

Just as bottomland hardwood forests serve as critical overwintering habitat for animals that disperse northward during summer, they also serve as breeding habitat for species that otherwise live in the uplands.

Table 2. Percent trapping success of 10 species of mammals in three habitats in southwestern Georgia and northwestern Florida (McKeever 1959).

Species	Pines	Upland Hardwoods	Bottomland Hardwoods
Didelphis virginiana (Opossum)	1.4	1.5	3.3
Sylvilagus floridanus (Cottontail Rabbit)	0.6	1.1	0.8
Sylvilagus aquaticus (Swamp rabbit)	-	-	0.1
Sciurus niger (Fox squirrel)	0.4	0.2	0.1
Urocyon cinereorgenteus (Gray fox)	0.2	1.1	0.2
Vulpes fulva (Red fox)	0.1	-	-
Procyon lotor (Raccoon)	0.8	2.4	3.2
Mephitis mephitis (Striped skunk)	1.6	1.1	0.6
Spilogale putorius (Spotted skunk)	-	-	0.1
Lynx rufus (Bobcat)	0.9	0.9	0.9
Total Percent Catch	6.0	8.3	9.0

Amphibians that require surface water for reproduction are known to migrate to cypress ponds during the breeding season, and under normal circumstances even larger numbers emigrate back into the flatwoods later in the year (Harris and Vickers 1984). Perhaps these populations of insectivores occurring in and radiating out from swamp forests exert a biological control function. For example, outbreaks of pine sawfly (*Hymenopterous* sp.) rarely if ever occur in pine plantations near cypress ponds.

Beyond their value as habitat for resident and overwintering species, the linear nature of riparian hardwood forests greatly facilitates seasonal and longer term dispersal of individuals in addition to the even more critical biological resource, genetic diversity. Small riparian "stringers"

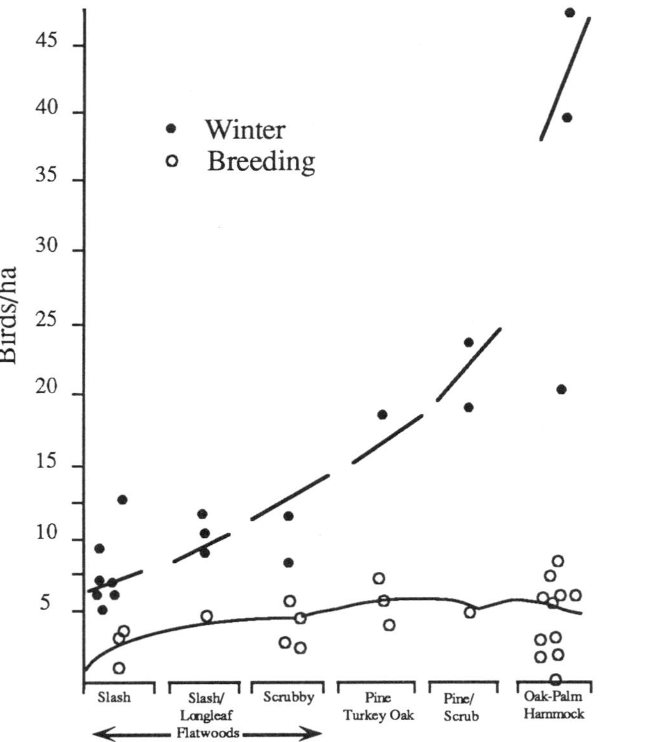

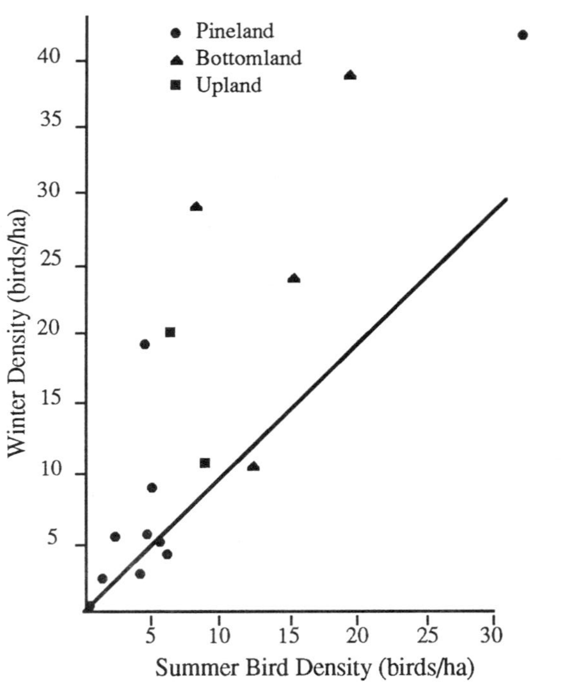

Figure 3. Winter versus summer bird densities in various habitats in northern Florida. Left panel displays the results of the 40 different studies or published reports of censuses. Right panel shows the results of 10 studies in which both winter and summer counts were conducted. Both panels reflect the higher densities of hardwood habitats (data compiled by R. Repenning).

"stringers"are known to serve as essential dispersal corridors that allow gray squirrel (*Sciurus carolinensis*) to occur in pinelands that are otherwise unsuitable habitat (McElfresh et al. 1980). Nixon found similar results for squirrels in Illinois, and the same is true for numerous species of breeding birds (Tassone 1981). Even small blocks of native forest may be able to support bird species more characteristic of much larger block sizes when the small tracts are interconnected by corridors (MacClintock et al. 1977, Robbins 1979).

Water-Quality Protection and Erosion Control

Bottomland hardwood forests change the water that flows across the floodplain. During periods when the forest is not flooded, runoff from adjacent uplands is filtered as it flows through the bottomland. Most particulate inorganic sediment is removed. Upland runoff from agricultural fields usually carries high concentrations of inorganic nutrients, particularly nitrate and phosphate. These nutrients are adsorbed on the forest floor, taken up by the forest vegetation, and released later as organic plant detritus (Lowrance et al. 1984c, Peterjohn and Correll 1984). Although bottomland hardwood forests are believed to form a similar function when backwater flooding occurs during winter and spring, the effectiveness is probably less because biological activity of the forest is greatly reduced during this season. Thus, the riparian system filters and transforms nutrients, removing the more toxic dissolved inorganic ions and releasing them as non-toxic particulate organic material that constitutes a food source for instream organisms. This provides the benefits of water-quality improvement and increased stream productivity.

Flood Storage Values

One of the major values of forested floodplains to society is water storage during severe floods. This reduces flood crests downstream by prolonging the flood periods at lower, safer levels. Economic benefits result from less flood damage, lower insurance costs, and savings in flood-control construction. During periods of flooding, some bottomland areas may serve to recharge aquifers, thus making more groundwater available for human use elsewhere. It has been suggested that the slow release of water from bottomland to shallow subsurface flows augments stream flow during low-flow periods. This has been challenged by those who point out that bottomland trees transpire at high rates, probably using any water that might have been available to augment stream flow (Bell and Johnson 1974, Carter et al. 1979). Evidence suggests that during low river stages some riparian systems continue to provide water to the stream through seepage from groundwater.

In sum, the values of bottomland ecosystems are significant, but most do not accrue to the owner of the real estate. Rather, they go to broader segments of society such as commercial fishermen downstream, city dwellers whose drinking water is of good quality, individuals who do not experience flooding because of the mitigating effects of the upstream bottomlands, or to northern residents who await the return of spring migrant birds. The discrepancy between land ownership and value accrual is a major problem in management of this resource.

Another major difficulty in dealing with the use and misuse of bottomlands is the incremental nature of the changes that are occurring from gradual human destruction of such habitats. These cumulative changes are watershed-wide phenomena that resist analysis as long as the focus is limited to a specific individual site or current human activity.

RATE OF CONVERSION OF BOTTOMLAND FORESTS

National Perspective

Numerous estimates of original wetland acreage in the United States exist, and it is not possible to reconcile these often disparate and conflicting reports. The reason is at least twofold. Clear definition of the term "wetlands," of which lands were included and which excluded, was rarely given, and the categories used in one survey were rarely commensurate with those of another. Thus, it is not now possible to split, lump, and compare estimates within or among various categories.

The second reason for uncertainty involves the method of estimation. For example, in the first Congressional mandated national survey in 1906 "The following letter and blank for statistical information were sent to one or more persons in each of the counties in the several States east of the one hundred and fifty meridian..." (Wright 1907:7).

Similarly, the second survey of lands compiled for the Secretary of Agriculture in 1921-22 was neither comprehensive of all states nor direct in the reporting of statistics (Gray et al. 1923). Personal biases and "guesstimates" by numerous individuals of various persuasions were involved in both of these early works. Before we progress further definition of wetland is necessary.

We subscribe to the definition given by Cowardin et al. (1979) referred to as a multi-parameter definition because it is based on three parameters or site characteristics rather than a single measure--such as presence or absence of water. Wetlands are lands transitional between terrestrial and aquatic where the water table is at or near the surface and where:

1) a non-soil substrate is saturated with water or covered by shallow water at some time each year, and/or
2) the substrate is predominantly undrained hydric soil, and/or
3) the vegetation is predominantly hydrophytic, at least periodically.

This definition has the advantage of being responsive to short-term fluctuations or changes in water (presence or absence), medium-term changes or conditions (presence or absence of hydrophytic vegetation), and long-term changes or conditions (soil classification).

The following statistics for forested wetlands combine those dominated by trees (woody vegetation with a single stem above 2 m in height) with those dominated by shrubs and/or small, stunted trees.

We conclude that approximately 80 million ha of wetlands of all types existed in the conterminous 48 states at the time of European settlement. This is 7.5% less than the estimate of Roe and Ayres (1954). Support for this estimate derives from published land records. For example, by June 30, 1906, 33.25 million ha of bottomlands had already been ceded to only 15 states under provisions of the 1849, 1850, and 1860 swamp and overflowed lands acts (Wright 1907). Fully 50% of this acreage (17.4 million ha) occurs in the states of Florida, Arkansas, and Louisiana (Table 3). This 33 million-ha transfer does not include wetlands in the original 13 states nor the 20 other predominantly western states.

It was again observed in 1922 that "nearly all of the land unfit for cultivation without drainage is in the southern states and one-half of the remainder is in the three lake states. Nearly all of the wetland in the South; except the Florida Everglades and prairies, tidal marshes, and Gulf coastal prairies; is forested and requires both drainage and clearing..." (Gray et al. 1923:425). Based on statements such as this we assume that two-thirds of the wetlands were forested, and thus approximately 53 million ha of forested wetland existed in the 48 conterminous states prior to European settlement.

Table 3. Swamp and overflowed lands ceded to 15 states prior to June 30, 1906, in accordance with the Swamplands Acts of 1849, 1850, and 1860 (data from Wright 1907).

State	Hectares	State	Hectares
Alabama	216,000	Michigan	2,952,000
Arkansas	3,503,000	Minnesota	2,215,000
California	836,000	Mississippi	1,459,000
Florida	9,014,000	Missouri	1,960,000
Illinois	1,611,000	Ohio	48,000
Indiana	558,000	Oregon	213,000
Iowa	1,851,000	Wisconsin	1,944,000
Louisiana	4,857,000	Total	33,237,000

Drainage and loss of wetlands accelerated in the mid nineteenth century, much earlier than in the 1930's as is commonly reported. This judgement is based on historical landuse patterns and the indirect as well as direct effects. For example, the original Swamp and Overflowed Lands Act was passed in 1849 specifically "to aid the state of Louisiana in constructing the necessary levees and drains to reclaim the swamp and overflowed lands..." (Wright 1907:5).

Similarly, throughout the later 1800's Florida contracted with numerous companies to drain swampland on the specific condition that it would be drained (Blake 1980).

Indirect actions that might have significantly reduced the acreage of flooded lands in the 1800's include removal of snags from rivers by the U.S. Army Corps of Engineers (USACE) and the over-exploitation of beaver (*Castor canadensis*) by trapping. Trees that fall into stream channels and/or lodge in logjams cause impeded flow and an increase in area of flooded area during high water stages. Between 1874 and 1912 the USACE removed an average of 31 snags per km of river per year based on nine representative rivers in Alabama, Florida, Georgia, Mississippi, and South Carolina (Sedell et al. 1981). This intensity of snag removal to improve navigability probably led indirectly to a substantial reduction in area of functional bottomland hardwood forests.

Beavers were a major factor in the creation and maintenance of forested wetlands (Atwood et al. 1970, Novak 1987). Heavy trapping of beaver from the sixteenth century onward no doubt reduced their densities to a fraction of pre-Columbian levels and led to near extirpation from many areas by 1900 (Obbard et al. 1987). Beaver dams and resulting impoundments perish quickly without active maintenance, and thus wetland acreage would have decreased.

Despite our knowledge of factors such as these, it is impossible to quantify the impact or trace the magnitude of wetland loss before 1850. Frayer et al. (1983) estimated recent loss rates of 0.42% per year for all wetlands during the period from the mid-1950's to the mid-1970's. During the same period, forested wetlands decreased from 27.0 million ha in the conterminous 48 states to 24.4 million ha, a loss rate of 0.48% per year.

Abernethy and Turner (1987) used U.S. Department of Agriculture (USDA) Forest Service forest survey statistics to evaluate the status of forested wetlands. They concluded that by 1980 only 23.1 million ha of forested wetlands remained in the conterminous United States (Figure 4), about 8% of the total forested area and about 2% of total land area. The majority (57%) occurs in the southeastern United States. But while forested wetlands constitute only 8% of United States forested land, losses of forested wetlands constitute 37% of all forest loss. The destruction of bottomland hardwood forests is perhaps five times higher than the loss of other forest types (Abernethy and Turner 1987). We suspect that even this statistic is conservative because of large areas of

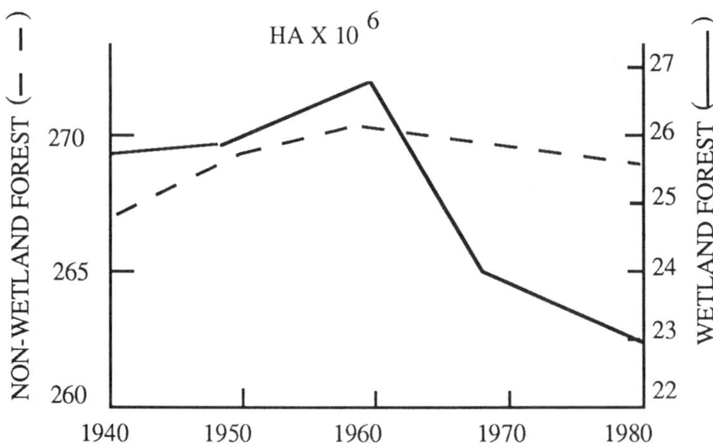

Figure 4. Comparison of total area estimates for wetland versus non-wetland forests in the United States (excluding AZ, AL, CA, HA, NM, PR). From U.S. Forest Service, forest survey data in Turner and Abernethy (1987).

scrub-shrub wetlands (e.g., titi [*Cyrilla* sp.] swamps and pocosins) that are being converted to pine plantations and therefore are not subtracted from the forested wetland area.

Southern Region Trends

If the national losses of wetlands and/or forested lands are dramatic, then the losses in the Mississippi Valley and the southeastern states are cause for alarm (Figures 5 and 6). Less than 5% of Iowa's natural wetlands remain; Ohio, Indiana, and Illinois probably have lost over 50% of all wetlands; and 98% of southern Illinois bottomland hardwood forests have been destroyed (Table 4) (Tiner 1984). Only 20% of the formerly expansive Lower Mississippi Valley hardwoods remain, and the overwhelming majority of these have been cut-over at least once and presently occur in small fragments (Figure 6).

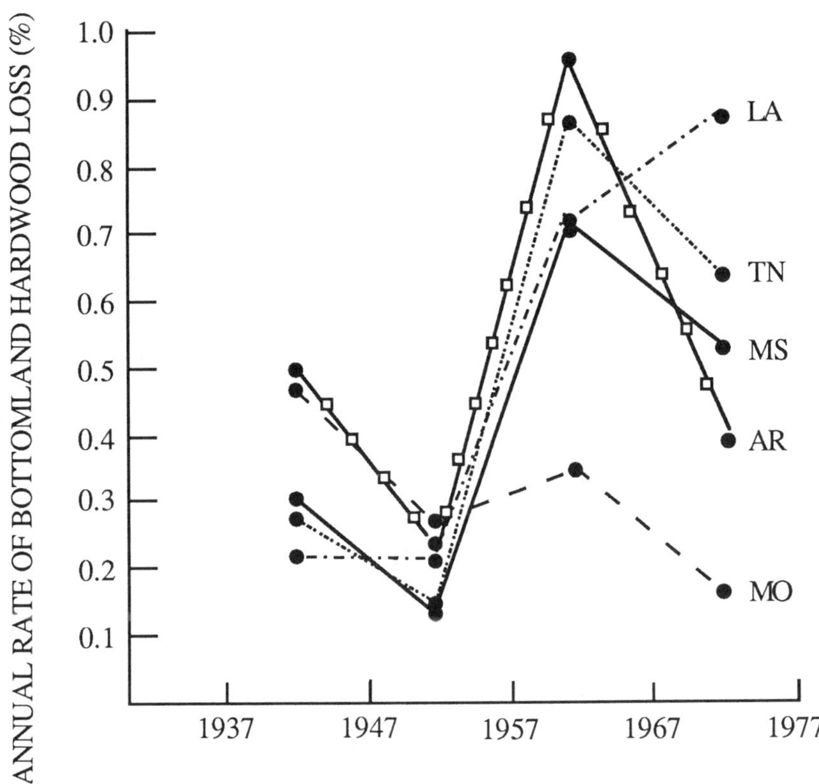

Figure 5. Annual rates of loss of bottomland hardwood acreage in the Lower Mississippi River Valley (as percentage of total land area) (after MacDonald et al. 1979).

The loss of bottomland hardwood area is a critically important measure of landuse and ecological change. Yet it is by no means the only measure. We have briefly mentioned the issue of forest fragmentation. Few if any data exist on this phenomenon, yet it is of equally critical importance for certain ecological functions (see Harris 1984). Similarly, the degradation of internal stand characteristics; such as abundance of very large trees, trees with broken tops, cavity trees, fallen tree boles, and preferred fruit-, nut-, and berry-producing species; has reduced habitat quality for wildlife. We have no large-scale database with which to assess these trends.

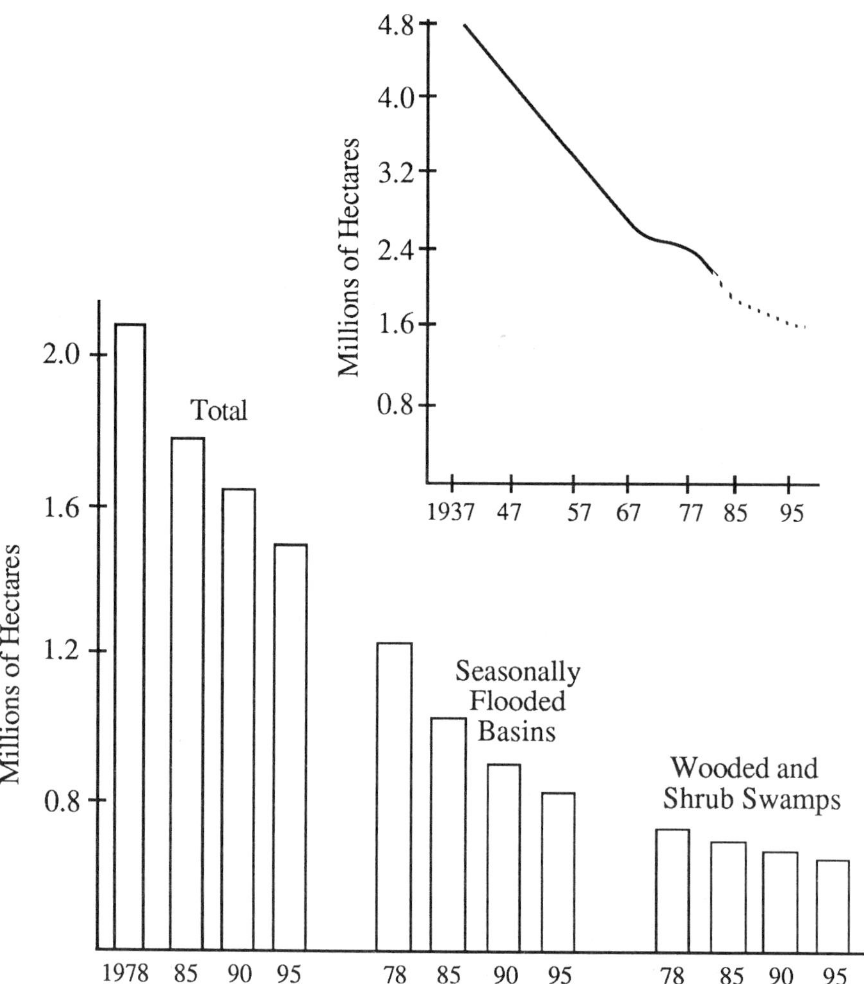

Figure 6. **Upper right panel shows time trend of bottomland forest area in the lower Mississippi River alluvial plain. The bar graph shows the comparative areal loss for seasonally flooded basins versus wooded and shrub swamps (data from MacDonald et al. 1979).**

Table 4. Loss of wetlands in selected states as reported in the published literature.

	% loss	Source
Southern Illinois bottomland hardwoods	98	IL Dept. Cons. (in Tiner) 1984
Mississippi alluvial plain	78	MacDonald et al. 1979
Louisiana's forested wetlands	50	Turner and Craig 1980
North Carolina pocosins	33	Richardson et al. 1981
Ohio and Mississippi River wetlands	37	KT Department of Fish and Wildlife (in Tiner 1984)

CUMULATIVE IMPACTS OF HUMAN ACTIVITIES IN BOTTOMLAND HARDWOOD FORESTS

Nature of the Issue

Although a few states such as California have cumulative impacts legislation, the National Environmental Policy Act (NEPA) of 1970 and regulations issued pursuant thereto provide the broadest single mandate for their control. The Council on Environmental Quality (CEQ) defines a cumulative impact as (40CFR part 1508. 7 & 1508.8):

> the impact on the environment which results from the incremental impact of the action when added to other past, present, and reasonably foreseeable future actions regardless of what agency (federal or non-federal) or person undertakes such other actions. Cumulative impacts can result from individually minor but collectively significant actions taking place over a period of time. Effects include:
>
> a) Direct effects, which are caused by the action and occur at the same time and place.
>
> b) Indirect effects, which are caused by the action and are later in time or farther removed in distance, but are still reasonably foreseeable. Indirect effects may include growth-inducing effects and other effects related to induced changes in the pattern of landuse, population density or growth rate, and

> related effects on air and water and other natural systems, including ecosystems.

To date, the methodologies and actual assessments of cumulative impacts have been marginally effective. For at least 100 years, Cartesian or reductionist approaches in the biological sciences have placed emphasis on simple cause-effect relations between observable variables. This has meant a commitment to evaluation of simple, additive, direct, on-site effects as opposed to the analysis of consequences to complex ecological processes. Little attention has been given to effects that may occur off-site or may occur in time-delayed, interactive, multiplicative, or synergistic ways. In the words of Horak et al. (1983: 11):

> one of the biggest hazards is the expectation that traditional, deterministic procedures can be transferred from current practices to fulfill the requirements of cumulative impact assessment. The demand for cumulative impact assessment requires a complete restructuring of the problem itself; an articulation of the assumptions driving the assessment; new techniques and tools for aggregating diverse impacts; and a search for standards or criteria of significance in order to judge overall, long range impacts.

Horak et al. (1983) developed criteria that tend to maximize 1) emphasis of multiple projects and actions; 2) consideration of off-site impacts and effects; 3) interaction and synergism among actions, impacts, and effects; 4) ability to aggregate effects; 5) consideration of ecological functional aspects; 6) consideration of ecological structural aspects; 7) ability to predict; and 8) adaptability (Figure 7).

Concurrent with developments in cumulative impact analysis, foundations of relevant ecological theory have been advanced. Odum (1969) and Margalef (1975) described trends and/or patterns that ecosystems would manifest after stress or perturbation. More recently, "stress ecology" (Barrett et al. 1976, Odum et al. 1979, Barrett and Rosenberg 1981, Loucks 1985, Odum 1985) has laid considerable groundwork for anticipating cumulative impacts. The notion that ecosystems manifest a linked set of observable responses to disturbance is increasingly accepted by ecologists. Even more specifically, the ecosystem distress syndrome (analogous to Selye's General Adaptation Syndrome [Selye 1956]) is proposed as a generalizable, three-part sequence of reaction: 1) initial effects or "alarm" reactions, 2) feedback

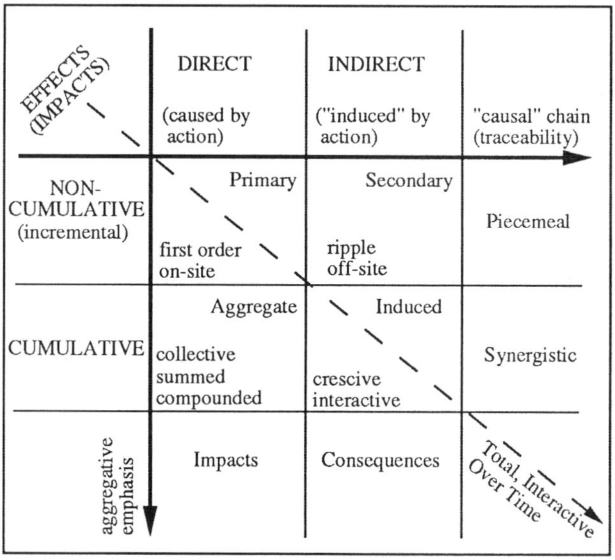

EFFECTS (IMPACTS)	DIRECT (caused by action)	INDIRECT ("induced" by action)	"causal" chain (traceability)
NON-CUMULATIVE (incremental)	Primary first order on-site	Secondary ripple off-site	Piecemeal
CUMULATIVE	Aggregate collective summed compounded	Induced crescive interactive	Synergistic
aggregative emphasis	Impacts	Consequences	Total, Interactive Over Time

Figure 7. Cumulative impact analysis must consider more than on-site, immediate, direct effects; it must move toward the righthand side of the chart and trace through probable chains of consequence. Similarly, it can not be assumed that effects will be simply additive; many of them will be multiplicative and require analysis of interaction effects between different system components (from Horak et al. 1983).

mechanisms or "coping" reactions, and 3) breakdown of the system (Rapport et al. 1985). Under first-level "alarm" reactions one might predict abnormalities in reproduction, abnormal fluctuations in sensitive populations, changes in the distribution of sensitive species, and increased nutrient leaching from the ecosystems. Second-level "coping" reactions may reveal the replacement of stress-sensitive species with functionally similar but more resistant species, development of external loops for processing nutrients, changes in primary production and respiration and thus changes in ratios such as production:biomass and production:respiration. Third-level responses, involving ecosystem transformation and collapse, imply that the system has changed to such an extent that it cannot regain its former state. Such is the case with species extinction and degradation of soil substrates such that recolonization cannot occur.

The following information concerning hydrology, water quality, and native biotic diversity will demonstrate the nature of cumulative impacts of human activity in bottomland hardwood ecosystems.

Effects on Hydrology

Hydrology is a watershed-level phenomenon, and the types of human activities that affect hydrology may occur anywhere on the watershed, not just in the bottomland region. Historically, there has been considerable concern and research regarding the influence of upland clearing on streamflow. Generally, upland clearing increases the "flashiness" of streamflow. Water runs off cropland faster than off forest land, and therefore less water infiltrates the soil (Ursic 1965, U.S. Soil Conservation Service 1972). In contrast, bottomlands act primarily as temporary reservoirs between the uplands and downstream areas during floods, storing water that would otherwise add to the peak flow. As the flood peak passes, this water is released back into the stream, so that the flood peak is lower and longer when there is a well-developed floodplain (Figure 8) (Mitsch and Gosselink 1986). Therefore, bottomlands influence the timing, magnitude of discharge, and stage of the stream.

In this context the influence of forest clearing *per se* on the floodplain is not as important as such activities as leveeing and damming and is less devastating than activities that alter drainage patterns. As long as water can flow freely, floodplains serve as reservoirs regardless of the plant cover. Vegetation does change the friction that water encounters as it floods the bottomland, but this effect is probably less important to flood control than the area available to be flooded. Therefore, the most irreversible and serious hydrologic impacts in bottomlands are flood- and navigation-control structures.

The relationship of bottomlands to hydrology is reciprocal: when control structures or upstream modifications change hydrology, the bottomland hardwood forests downstream are influenced. Both plant and animal species are determined by the flooding regime (Larson et al. 1981). Thus, changes in stage and hydroperiod eventually lead to a change in the character and productivity of the bottomland forest, both on the site of human intervention and downstream on land influenced by the intervention.

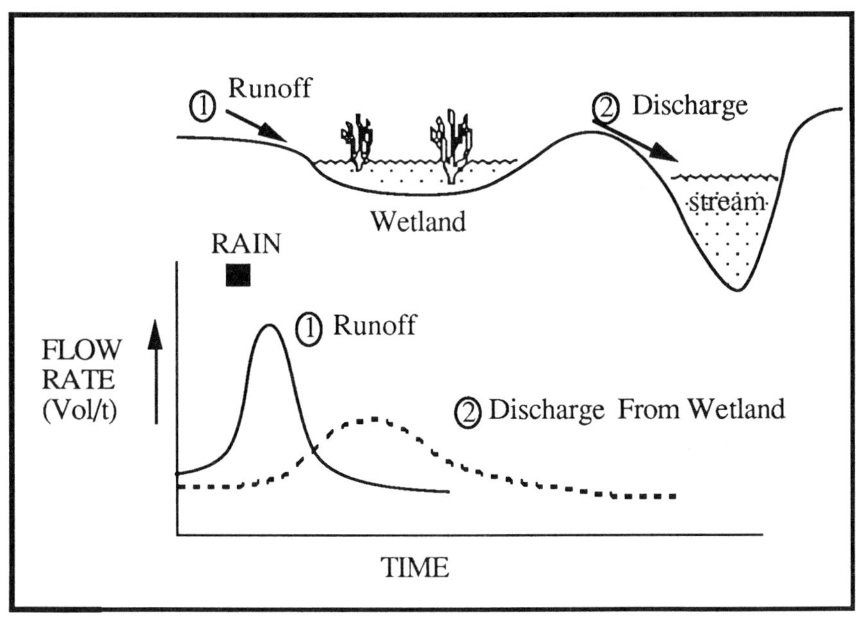

Figure 8. Generalized curves depicting the role of riparian forest wetland in dampening the "flashiness" of storm flow (from Mitsch and Gosselink 1986:402; Copyright © 1986 by Van Nostrand Reinhold, reprinted with permission).

Data from Bayou Bartholomew in northern Louisiana are used to illustrate both the cumulative trends and the problems of data analysis. Bayou Bartholomew drains an area between the Ouachita and Mississippi Rivers that is nearly all bottomland. Historic U.S. Forest Service surveys (MacDonald et al. 1979) reveal a rapid decrease in bottomland forest in this watershed from 1957 to 1985 (Figure 9). During the same period, the annual mean discharge of Bayou Bartholomew has apparently increased. Since variance in rainfall and installation of upstream flood-control structures change discharge patterns, a reliable analysis of the trends depends on construction of annual stage-discharge curves. Both the U.S. Geological Survey and the USACE routinely construct these curves in order to estimate discharge from the primary stage data, but published systematic analyses of how the curves are changing are few.

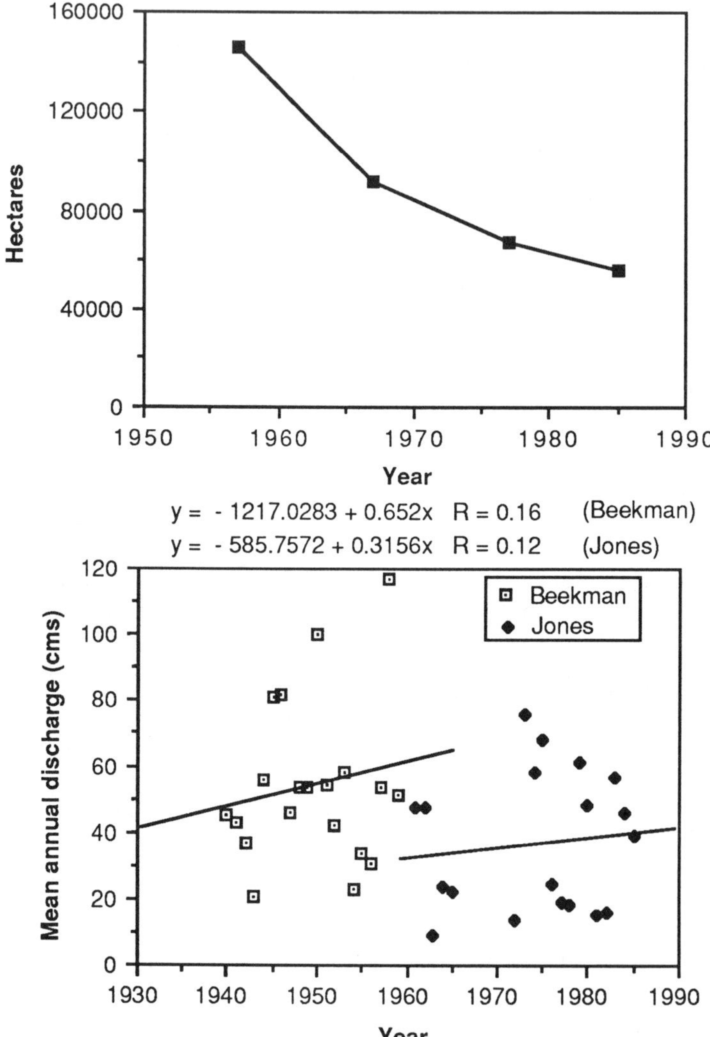

Figure 9. Upper panel depicts the forested area in the Bayou Bartholemew Louisiana bottomlands (data from MacDonald et al. 1979). Lower panel is the mean annual discharge at two locations (Beekman and Jones) on the bayou. The data record at each station is incomplete (cms = cubic meters per second).

A notable exception to this generalization is the analysis performed by Belt (1975) on the Mississippi River at St. Louis. The stage-discharge curve of the river was fairly stable from the 1860's through the 1920's when major flood-control projects on the river began. Belt plotted the changes through time from a "base rating curve" for 1903, 1906, 1926, and 1927 to demonstrate that the stage-discharge relationship has become increasingly unstable since the early 1930's (Figure 10). Several factors contribute to this instability.

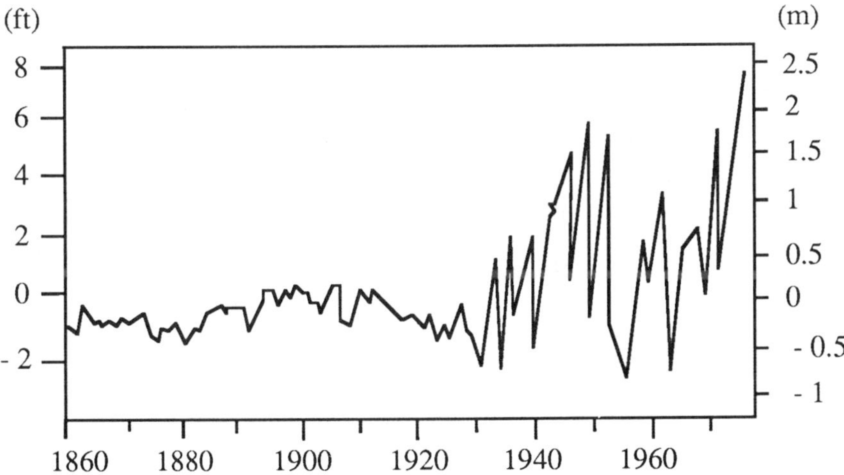

Figure 10. Change in stage from base rating curve of the Mississippi River at St. Louis (after Belt 1975).

Unmodified streams typically develop a predictable relationship between depth, width, and discharge (Leopold et al. 1964), but this relation no longer pertains to the Mississippi River. Currently at peak flood discharges the river stage is much higher than in earlier years and the stage discharge relationship is unpredictable. In the 1973 flood peak the stage at peak discharge was 3.3 m above the baseline. Belt (1975) attributes these changes in the stage-discharge relationship primarily to confinement of the river within levees (that is, to a loss of floodplain). "Under natural conditions, the Mississippi eroded its bottom and banks during flood peaks, making room for some of the floodwaters. The rest spilled out over the natural reservoir into the floodplain. Since 1837, the channel has lost about a third of its volume so now, during a flood on the man-modified Mississippi, the stages are higher for a given discharge" (1975:684). Estimates for the Mississippi River based on the change in forested floodplain suggest that its water detention time (floodplain storage capacity/river discharge) has decreased from about 60 days historically to about 12 days at present (Gosselink et al. 1981).

The Mississippi River has also been straightened by the construction of meander cutoff channels that have considerably shortened the river and made it steeper and hydraulically more efficient. The consequence of both trends is higher river stages and flashier floods.

The economic consequences of these changes in the river are enormous. Increased river stages for a given flood mean either more frequent and severe economic losses from flooding or more construction to protect against flooding, such as the building of higher levees. Ironically, the same modifications that have led to higher flood stages have also led to lower stages at times of minimum discharge (Belt 1975). Because of the increased hydraulic efficiency of the straighter river, the river bottom tends to silt up during low flow periods. Associated navigation problems in some reaches of the river are causing additional economic cost.

We have focused on the consequences of decrease of floodplain storage along the Mississippi River because of the thoroughness of Belt's (1975) analysis. Most other major rivers and streams with significant floodplains have also experienced major flood and navigation works. Wherever this has occurred, stage-discharge records are available so that cumulative impacts of bottomland modification on downstream hydrology can be analyzed.

The projected economic consequences of floodplain loss on the Charles River, MA were considered so great that the USACE recommended purchase of the floodplain as a means to control flooding in Boston rather than construction of structural devices (USACE 1972). Ogawa and Male (1983) simulated peak flows under different conditions for this river system (Figure 11) and concluded that flood peaks depended on the amount of floodplain loss and that downstream wetlands were more effective in reducing downstream flooding than were upstream wetlands. The larger the magnitude of a flood event, the greater the effect of wetland loss on peak flow modification (Ogawa and Male 1983).

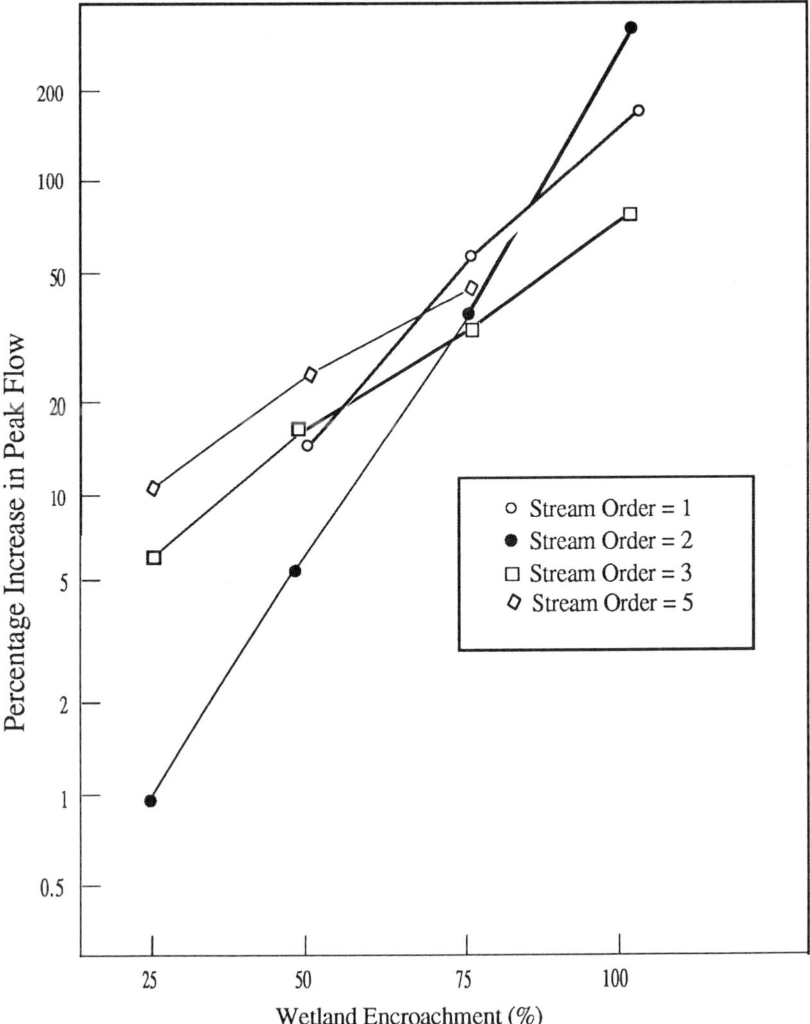

Figure 11. Variation in relationship between percentage increase in peak flow and wetland encroachment (loss) due to stream order for the 130-year basin-averaged recurrence interval flood (from Ogawa and Male 1983).

Effects on Downstream Water Quality

Cumulative impacts are watershed-level phenomena, but we have little experience in quantifying them. Impacts on water quality downstream from bottomland hardwood forests exemplify the difficulty by illustrating the numerous different processes involved. In the following paragraphs we discuss the major processes controlling water quality, and present a simple model that summarizes how they interact.

Land clearing, especially for row-crop production, accelerates erosion and runoff from both uplands and bottomlands. Because most watersheds contain upland areas, it is difficult to separate the contribution of upland agriculture to water quality from that of bottomlands. Moreover, most studies on the relation of landuse to water quality have been carried out in upland systems. Here the effect of forest removal is clear. Because the data are derived from nationwide sources, Omernik's (1977) study serves as a good example. Runoff from 928 nonpoint-source watersheds revealed that stream levels of orthophosphorus, total phosphorus, inorganic nitrogen, and organic nitrogen all increased sharply when watersheds were cleared of vegetation, and their concentrations were directly proportional to both the percentage of the watershed cleared and to the degree of disturbance (Figures 12 and 13). Stream phosphorus and nitrogen concentrations were 10 times higher when the watershed was farmed than when it was forested.

Omernik's sites include some streams draining bottomlands. For example, all of the Mississippi sites are in the Yazoo Basin which consists predominantly of Mississippi River bottomlands. The analysis does not separate bottomlands from uplands, so it is not possible to determine whether they respond to clearing in the same way as uplands. Other studies provide reason to conclude that they do. For example, Murphree et al. (1976) showed that erosional losses were high for cleared Sharkey clay bottomlands of the Mississippi Delta on slopes of only 0.2%. Whereas erosional losses were only 0.2 metric tons/ha/yr from forested watersheds (Table 5), and only slightly higher from revegetated fields and pastures, bottomland fields lost 29 metric tons/ha/yr, 150 times the rate for forested areas but lower than the loss from an upland cultivated field (48.8 metric tons/ha/yr).

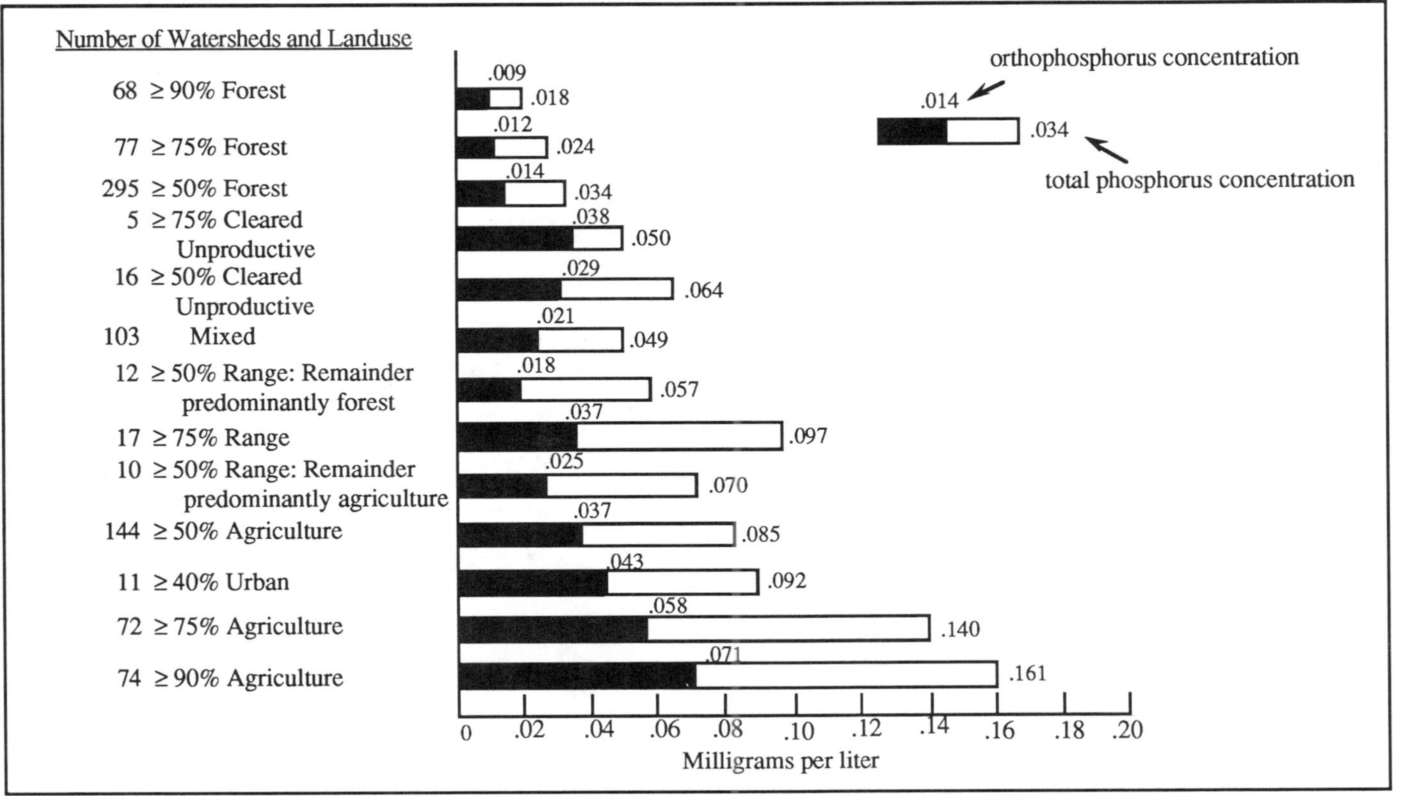

Figure 12. Relationship between general landuse and total phosphorus and orthophosphorus concentrations in streams (from Omernik 1977).

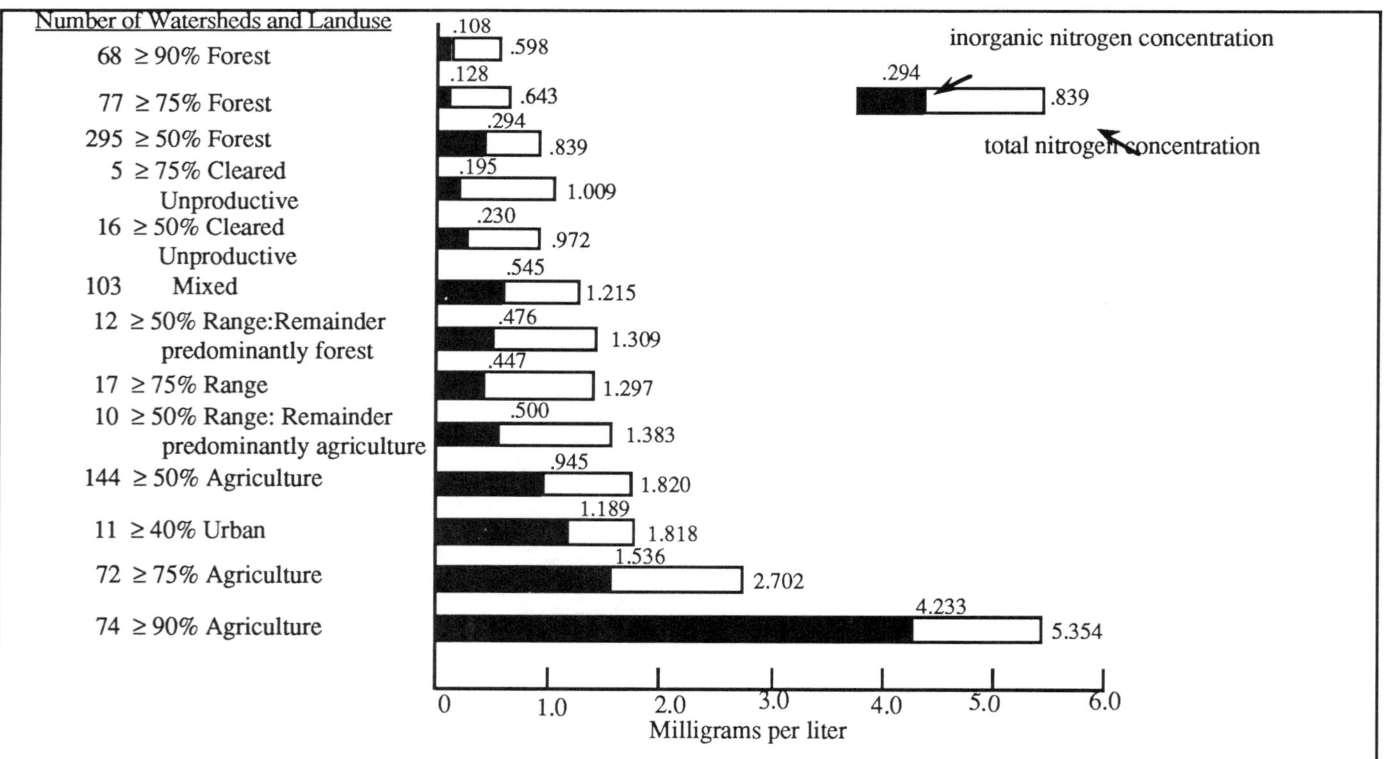

Figure 13. Relationship between general landuse and total nitrogen and inorganic nitrogen concentrations in streams (from Omernik 1977).

Table 5. Annual sediment yields from watersheds under different land uses (data from Ursic 1965 and Murphree et al. 1976).

Land Use			Sediment yield (MT/ha/yr)
Upland	-	Pine-hardwoods	0.045
	-	Pine plantation	0.045
	-	Hardwoods	0.22
	-	Abandoned Field	0.29
	-	Pasture	3.61
	-	Cultivated	48.80
Bottomland	-	Cultivated	29.00

Bottomland hardwood forests trap and transform materials. For example, in a series of studies from the Southeast Watershed Research Laboratory and the University of Georgia, the role of the riparian zone in stream water-quality maintenance was quantified (Asmussen et al. 1979; Lowrance et al. 1983, 1984a, 1984b, 1984c). The study site was a small watershed on the Atlantic and Gulf Coastal Plain. About 45% of the watershed was occupied by row crops; the riparian zone made up about 30% of the watershed, and all waterborne nutrients from the agricultural fields had to pass through this zone before reaching the stream. Water quality in the stream was good despite the prevalence of cropped fields. In fact, Asmussen et al. (1979) showed that streamflow outputs of nitrate from the watershed were less than inputs in precipitation. Lowrance et al. (1984c) found that N, P, Ca, K, and Cl were all stored within the riparian zone (or denitrified in the case the case of nitrogen) so that little of the nutrients leaching from the fields reached the stream (Table 6). In this system nearly all the water and nutrients moving off the cropland were in subsurface flow. Lowrance et al. (1983) estimated that complete replacement of the riparian zone by the mix of landuse on the rest of the watershed would result in a twenty-fold increase in the nitrate-N load to the stream and increase instream concentrations from 0.2 mg nitrate-N/l to over 4 mg/l. Of the common nutrients studied, the stream loading of all would increase except organic N, inorganic P, and total P.

Richardson (1985) reported that the phosphorus retention capacity of freshwater wetlands is directly related to the soil content of extractable aluminum, and that although these wetlands may retain phosphorus extremely efficiently for several years, their efficiency decreases rapidly when loading rates are high. In contrast, much of the nitrogen reduction in the riparian zone is attributed to high rates of denitrification so that there is little accumulation (Lowrance et al. 1984c).

A significant amount of the nutrients trapped in the riparian zone is stored in aboveground plant parts (Table 6). This is a relatively short-term storage and does not contribute to long-term storage in a mature system. But while the incoming nutrients are predominantly inorganic, they leave as organic compounds. Thus, riparian zones also transform nutrients. Generally cropland runoff is high in inorganic ions such as nitrate and phosphate. In contrast, output from the riparian zone is high in organic N and P. In the Georgia study, whereas the N input from the cropland was 82% inorganic, the N output to the stream was 80% organic (Lowrance et al. 1983).

Table 6. Nitrogen inputs and outputs (kg/ha/yr) and above-ground storage (kg/ha) in the riparian zone of a small watershed (from Lowrance et al. 1984c).

	Inputs			Outputs			Balance
	Precipi-tation	Subsur-face	N fixa-tion	Stream-flow	Denitri-fication	Above-ground storage	Input - (output + storage)
N	12.2	29.0	10.6	13.0	31.5	51.8	-44.5
P	3.5	2.1	-	3.9	-	3.8	-2.1
CA	5.2	47.4	-	31.8	-	40.3	-18.5
MG	1.4	18.1	-	15.0	-	6.1	-1.6
K	3.9	19.5	-	22.2	-	18.6	-17.4
CL	21.4	83.5	-	97.0	-	-	+7.9

The transformation of nutrients is further confirmed by Mulholland and Kuenzler (1981), who showed that organic carbon export from forested wetland watersheds was higher than from upland watersheds, and by Elder (1985), who showed that as water flows through the bottomland forests of the Apalachicola River it does not change much in total N and P content, but inorganic forms (especially dissolved inorganic nitrogen) diminish while dissolved and particulate organic forms are augmented (Table 7).

Table 7. Yields of nitrogen and phosphorus species, June 3, 1979 to June 2, 1980 (from Elder 1985).

Species	Nutrient input to Apalachicola subbasin kg/ha/yr	Net yield (output) from Apalachicola subbasin kg/ha/yr
Dissolved Inorganic N (DIN)	1.8	-1.1
Dissolved Organic N (DON)	1.5	3.5
Particulate Organic N (PON)	0.9	2.3
Dissolved Phosphorus (DP)	0.1	0.3
Inorganic Phosphorus (SRP)	0.05	-0.3
Particulate Phosphorus (PP)	0.2	0.4

Widespread clearing on lowlying floodplains seldom occurs without extensive channelizing and/or tiling to increase the drainage rate. Channelized water bypasses the riparian zone, routing runoff nutrients directly into receiving streams. Because the energy of the flowing water is concentrated it tends to increase erosion. Thus, there is a synergism between canal density and wetland loss. This has been shown in such diverse wetland systems as the floodplain of the Chowan River, a tidal river draining about 12,700 km^2 of southern Virginia and eastern North Carolina (Figure 14) (Duda 1982) and the Barataria Basin, a large coastal watershed of the Mississippi River (Figure 15) (Gael and Hopkinson 1979).

Figure 14 shows higher sediment transport by streams with extensive channelization in their watersheds compared with four

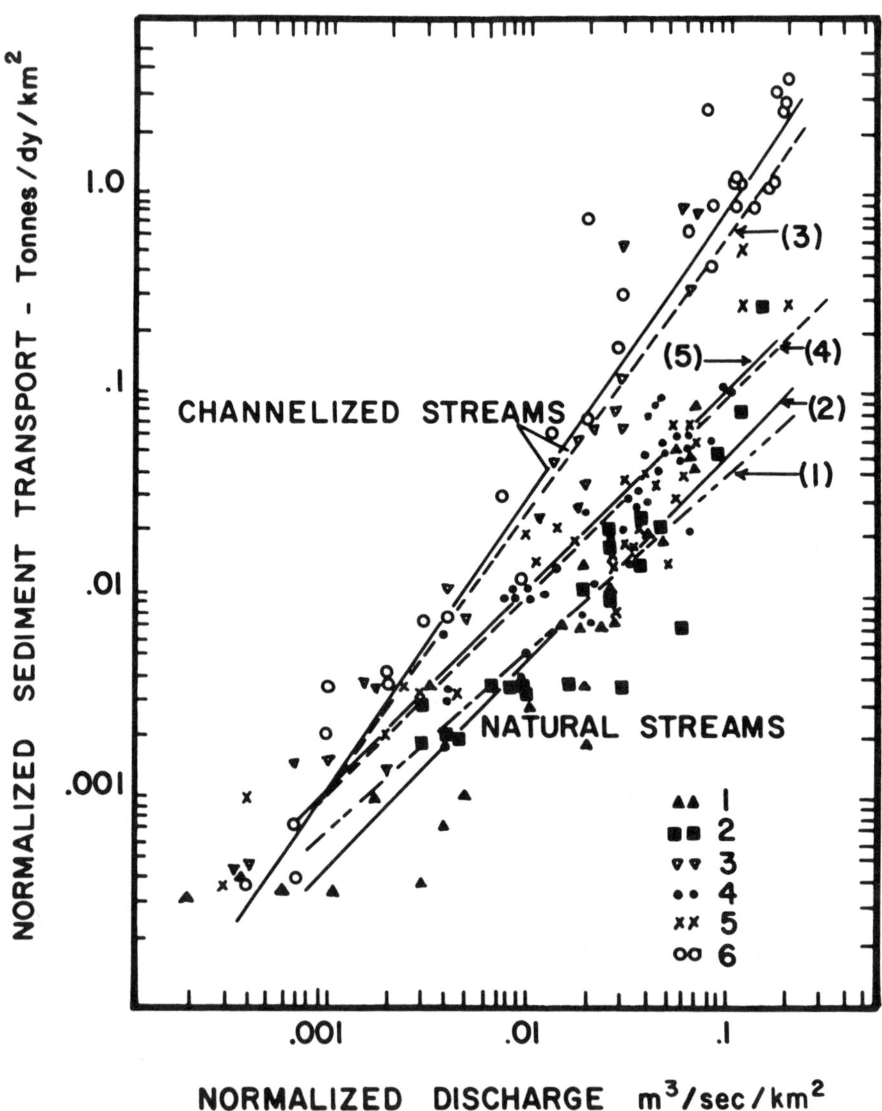

Figure 14. Variation of normalized sediment transport with normalized flow for coastal plain streams with and without extensive drainage improvements (from Duda 1982, page 403; Copyright © 1982 by American Water Resources Association, reprinted with permission).

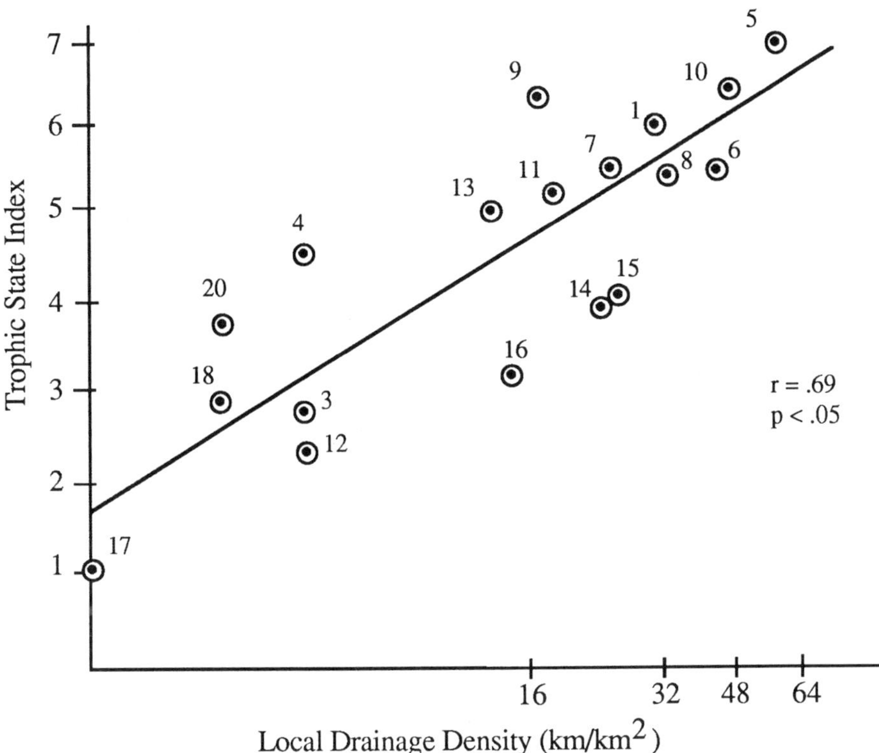

Figure 15. Local drainage density versus trophic state index (TSI) for 20 drainage subbasins in the Barataria Basin, Louisiana. TSI is modified from Brezonik and Shannon (1971). Low numbers indicate oligotrophic conditions; high numbers indicate eutrophy (from Gael and Hopkinson 1979).

unmodified streams. Figure 15 shows the deterioration of water quality that accompanied canalization in the Louisiana coastal zone. This relationship was also reported by Livingston (1975). Extensive agricultural drainage in the Apalachicola Basin has reduced salinity and increased sedimentation in Apalachicola Bay to the point that shrimp resources are no longer plentiful.

The interactions of upland runoff, riparian storage and transformation of nutrients, and accelerated drainage are complex. In a forested watershed, nutrient runoff from the upland is small and most of it is trapped in the riparian zone. As upland is cleared, sediment and nutrient delivery to the riparian zone increases rapidly. The increased load to the riparian zone is at first efficiently trapped and transformed, but trapping efficiency decreases as the load increases. As the riparian zone itself is cleared, its efficiency as a sediment and nutrient filter diminishes, and instead of trapping materials it becomes a source to the stream. Finally, channeling shunts water away from the riparian zone, dumping unfiltered water directly into the receiving stream.

Unfortunately, historic stream water-quality data are generally fragmentary, making it difficult to trace the cumulative interaction of all these factors. Omernik's (1977) study mimics a time series by taking an instantaneous snapshot of a great number of watersheds in various stages of clearing. One of the major unknowns in water quality data is the effect of dams and reservoirs on the historic record. Dams dramatically change sediment delivery patterns (Meade and Parker 1985; Figure 16), and this must be taken into consideration if one is to reconstruct the historic effects of bottomland clearing.

Native Biotic Diversity

While Alfred Russell Wallace was defining the faunal regions of the world, the American Joel Asaph Allen (1871, 1892) was delineating faunal regions of southeastern North America. Allen labelled the assemblage of bird and mammal species found in the southern and southeastern coastal plain the "Louisiana fauna" and distinguished it from the "Carolinian fauna" to the north and the "Sonoran fauna" to the west. Other workers, such as Dice (1943), applied different names to roughly the same region (viz., Austroriparian) (Figure 17). Darlington (1957:421-422) describes faunal regions and fauna by the following analogy.

> Suppose a carpet were woven...with very different threads not following a common pattern but variously scattered or concentrated in different places in the fabric. Such a carpet would have no regular pattern, but it might have a pattern of irregular, differently colored blotches formed by predominance of differently colored threads in different places. The pattern might be real and very obvious even though the different blotches might not be sharply limited and might run together at the edges. Plants and

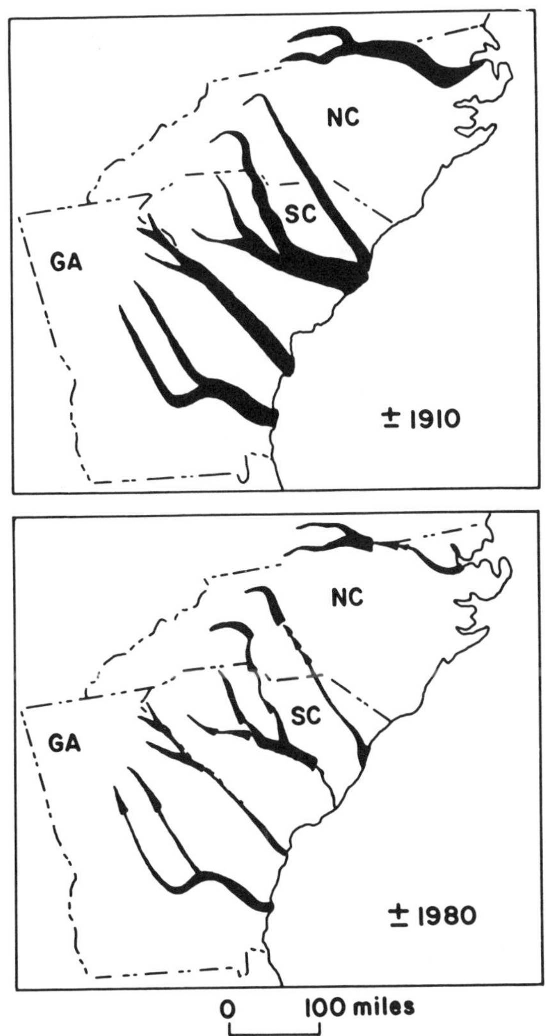

Figure 16. Average suspended sediment discharges of major rivers in Georgia and the Carolinas during two periods (about 1910 and about 1980) that indicate the decrease in sediment loads caused by several reservoirs constructed during the intervening years (from Meade and Parker 1985).

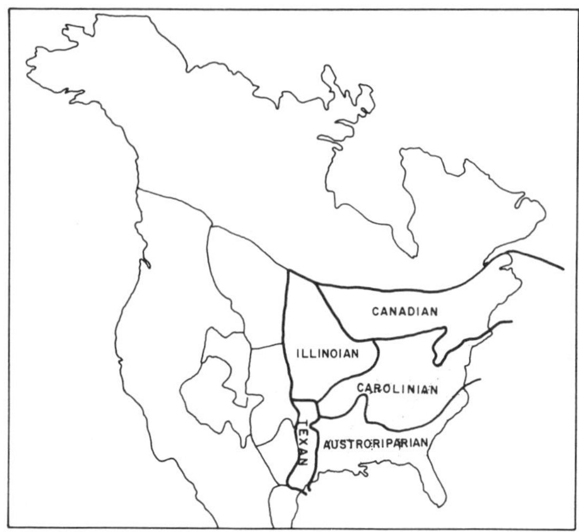

Figure 17. Faunal regions of the southeastern United States (after Dice 1943).

> animals do form such a carpet over the world. It is thick in some places and thin in others, and different sorts of plants and animals are variously scattered or concentrated in different places in it. The regional faunas, which determine the faunal regions, tend to be great concentrations of animals in favorable places...and main faunas are often separated by unfavorable areas where animals are few...and where transitions occur. The pattern formed in this way by the distribution of animals is real and very obvious even though the regional faunas...run together at the edge.

Our purpose in introducing this description of faunal regions is to establish the basis for native biotic diversity and to allow distinction between native biotic diversity and species diversity. Native biotic diversity refers to the plants and animals that are characteristic of a region and help distinguish that region from others. Knowing what species and what relative abundance of species would occur in a region without human influence provides the basis for standards and parameters such as Balanced Indigenous Population (BIP) and Balanced Biological Community (BBC) as specified in the U.S. Clean Water Act of 1977 (PL

95-217). Moreover, because native plant and animal species coevolved under specific sets of selective pressures, preservation of organisms in the absence of similar pressures does not meet the spirit of preserving native genetic resources. Conservation of biota will be successful in the long run only in a climate of continuing evolution, and this means that species must be conserved at a stage when they still possess their intrinsic genetic variability rather than waiting until they are endangered. The faunal assemblage represents one hierarchical level of genetic diversity, and protection of its integrity would seem at least as critical as saving the last individual of a species. Within faunal and floral regions, many different community types occur within which organisms interact most frequently and consistently. Thus, it is within these communities that coevolutionary processes are most important. Maintenance of ecological processes at the landscape level is necessary in order to conserve this degree of ecological diversity. About 100 species of non-native vertebrates have now established wild, breeding populations in the southeastern region. From this it should be obvious how a specific human intrusion might greatly increase the species diversity of the region while critically eroding native biotic diversity.

There are different levels at which species are considered. An endemic species that occurs only in the region and is the only representative of a Genus or Family constitutes a higher level of genetic divergence and probably warrants greater concern than a species that occurs widely throughout North America and has numerous sibling species. Similarly, a keystone species may be judged to be more critical than one that has no unique ecological role. A species that represents little more than a Pleistocene relict marooned in an obscure location would not justify the same attention as a more recent evolutionary line.

The most fundamental level of biotic diversity occurs at the chromosomal level within individual species. This diversity may be manifest as distinct races (or demes), or it may appear as a gradient referred to as clinal variation. Quantitative measures of genetic diversity at this level are available (e.g., heterozygosity) and can be used to monitor change of genetic diversity over time.

An ideal approach to the maintenance of biotic diversity would encompass all levels mentioned (Figure 18). Moreover, it seems ideal to

NATURAL BIOTIC DIVERSITY

	α "Within"	β "Between"	γ "Among"
Genetic	Polymorphism heterozygosity	Clines	Ecotypes
Species	"Interior" Species	"Exterior" Edge Species	Wide-ranging "Shepherds"
Ecological	Structural	Juxtaposition "Distinctiveness"	Landscape

Figure 18. Maintenance of biotic diversity must be approached at several levels. Alpha diversity refers to the component that occurs within specific communities, beta diversity occurs between communities, and gamma diversity occurs at the landscape or regional level. Similarly, the hierarchical levels of diversity range from the chromosomal to the faunal and community levels.

approach the problem from the highest level and work downward since the levels are somewhat hierarchical. Maintenance of native biotic diversity at the highest levels (region and landscape) first will go a long way to maintaining diversity at lower levels. The converse is not the case.

Examples follow that show the spectrum of impacts on native biotic diversity resulting from human activity in and around bottomland ecosystems. Generally, the first responses of animals to the increased stress of cumulative impacts can be characterized as "alarm" reactions (Rapport et al. 1985).

1. Abnormalities in reproduction are one of the most common and early responses to stress. Quantitative indices of reproductive success (or failure) are available and can be used as first indications of system stress. Such an analysis has been

done for ospreys (*Pandion haliaetus*) along the Atlantic coast. Although it can not be shown with certainty that low reproduction in the 1960's was due to the pesticide DDT, the evidence is great. Osprey productivity during 1969-1972 was about half the lower probable bound of normal productivity (Verner et al. 1985).

It is believed that agricultural pesticides from converted bottomlands washed into the aquatic food chain and curtailed reproduction in eastern brown pelicans (*Pelecanus occidentalis*) in the 1950's and 1960's. The last year of known nesting by native pelicans in Louisiana was 1966, and the species was locally extirpated shortly thereafter (Schreiber 1978). Pelicans have since been reintroduced into Louisiana and are again successfully reproducing now that the pesticide is no longer in use.

Adults of long-lived species may continue to inhabit submarginal habitat long after it has ceased to be suitable as breeding habitat. In the southern bald eagle (*Haliaeetus leucocephalus*), for example, established territory holders continue to breed, but no new recruitment occurs under stressful circumstances.

2. Range restriction of sensitive species follows quickly after reproduction declines. This is consistent with biogeographic principles which hold that organisms are most secure in areas where optimal combinations of ecological factors exist. Two striking examples of this phenomenon are the southern bald eagle and the swallow-tailed kite (*Elanoides forficatus*). The southern bald eagle was formerly distributed across the continent from northern California and New Mexico to Virginia. By 1980, the bird was restricted to a small percentage of its former range, and perhaps only 150 pairs existed outside of Florida. Swallow-tailed kites are even more characteristic of bottomland hardwood forests than the southern bald eagles. Their breeding range formerly extended up the Mississippi and Ohio River Valleys to Ohio and Michigan. Their current breeding range,

dangerously fragmented in the southeastern coastal plain (Figure 19), is only a shadow of the original one.

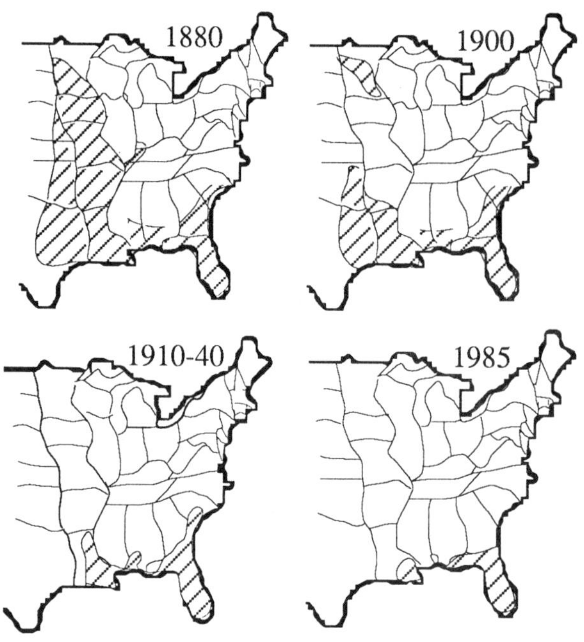

Figure 19. Progressive restriction and fragmentation of the breeding range of swallow-tailed kites in North America. This species is a prime example of cumulative impacts of bottomland forest depletion and alteration (from Cely 1979).

3. Loss of genetic heterogeneity occurs as populations become small and/or isolated and the consequences of genetic drift and inbreeding become manifest. For example, loss of libido and fertility are among the early manifestations of inbreeding and often precede endangerment. Recent studies reveal an extremely low sperm fertility rate (only 5%) in two adult male Florida panthers (*Felis concolor coryii*; Roelke 1986).

The now critically endangered red wolf (*Canis rufus gregoryi*) formerly ranged throughout the southeastern bottomlands, especially coastal areas. By the mid 1960's the population was reduced to only a fraction of its former range. Under heavily depleted and stressed conditions, the remaining wolves of coastal Louisiana fell subject to interbreeding with the opportunistic coyote (*Canis latrans*). Thus, a somewhat different problem, that of genetic "swamping" by an alien gene stock, came to threaten the few remaining wolves that still existed in the wild. All remaining wolves were captured and put in a captive-breeding program in the hope of reestablishing a wild population after environmental conditions are improved in the American Southeast. The coyote is not a forest species, rather, it is a species of more open grasslands, prairies, and meadows. As the eastern forests became progressively more fragmented, the ecological barriers that formerly separated coyotes from wolves were broken (Gipson 1974). The coyote colonized these now open landscapes and dominated the stressed wolf populations.

After "alarm" reactions, responses to continual or cumulative stresses can be characterized as "coping" reaction (Rapport et al. 1985).

1. Shortened food chains and simplified trophic structures are commonly observed under stressed conditions with accompanying reduction in numbers of top carnivores. This results from at least two different processes. Associated with the generalized pyramid of numbers is the inverse pyramid of habitats. This principle holds that organisms of lower trophic levels are more site specific than those of higher trophic levels. Thus, while organisms in the higher trophic levels are less abundant, their home ranges are large and they must integrate resources from several community types.

   Animals of these higher trophic levels also seem to acquire their energy and nutrients from a diversity of energy-flow channels. A diverse food energy source is the seasonally migratory ecotone which results from fluctuating water levels in the bottomlands. Under normal conditions, an aerobic, primary-production-based food chain occurs side by side with a detritivore-based food chain. Thus, at the lower trophic levels animals are not only of different species but of different ecological guilds. At approximately the tertiary level, vertebrates such as wading birds and furbearers begin to rely on both energy pathways and thus integrate variance within and between pathways. Twenty-six of North America's 30 furbearer species (order Carnivora) are either true carnivores or omnivores. Most of these species have amphibious life habits and exist at the interface of the aquatic and terrestrial subsystems. They no doubt utilize both aquatic and terrestrial energy sources and are adversely affected by stresses on either energy pathway. For example, when northern Florida cypress ponds were stressed by additions of sewage, the food chains and trophic webs became simplified at the expense of furbearer carnivores (Figure 20).

The effects of continuing habitat loss and fragmentation are further exemplified by the high proportion of threatened or endangered species that depend on forested wetlands. Although wetlands compose only 3.5% of the land area of the United States, 35% of all rare, threatened, and endangered wildlife species occur there or depend on these ecosystems for survival (Kusler 1977). In the lower coastal plain of the southeast (Southern Georgia and Florida), the number of amphibious and reptile species is about equal to the number of mammal and breeding bird species (i.e., 153 species in Florida). Amphibians are obligatorily linked to water during their egg and larval stages, and many reptiles are functionally tied to wetlands. Eighteen percent of all Florida amphibians and 35% of all Florida reptile species are considered to be in a category of concern (e.g., rare, endangered, or status undetermined; McDiarmid 1978). These percentages are much higher than those for birds, mammals, and fish.

2. Loss of faunal identity and "integrity" is a common situation in human-dominated environments, as a result of forest fragmentation, that is, the occurrence of numerous small, residual tracts of forest more or less surrounded by vegetation or landuse of a greatly different type. As early

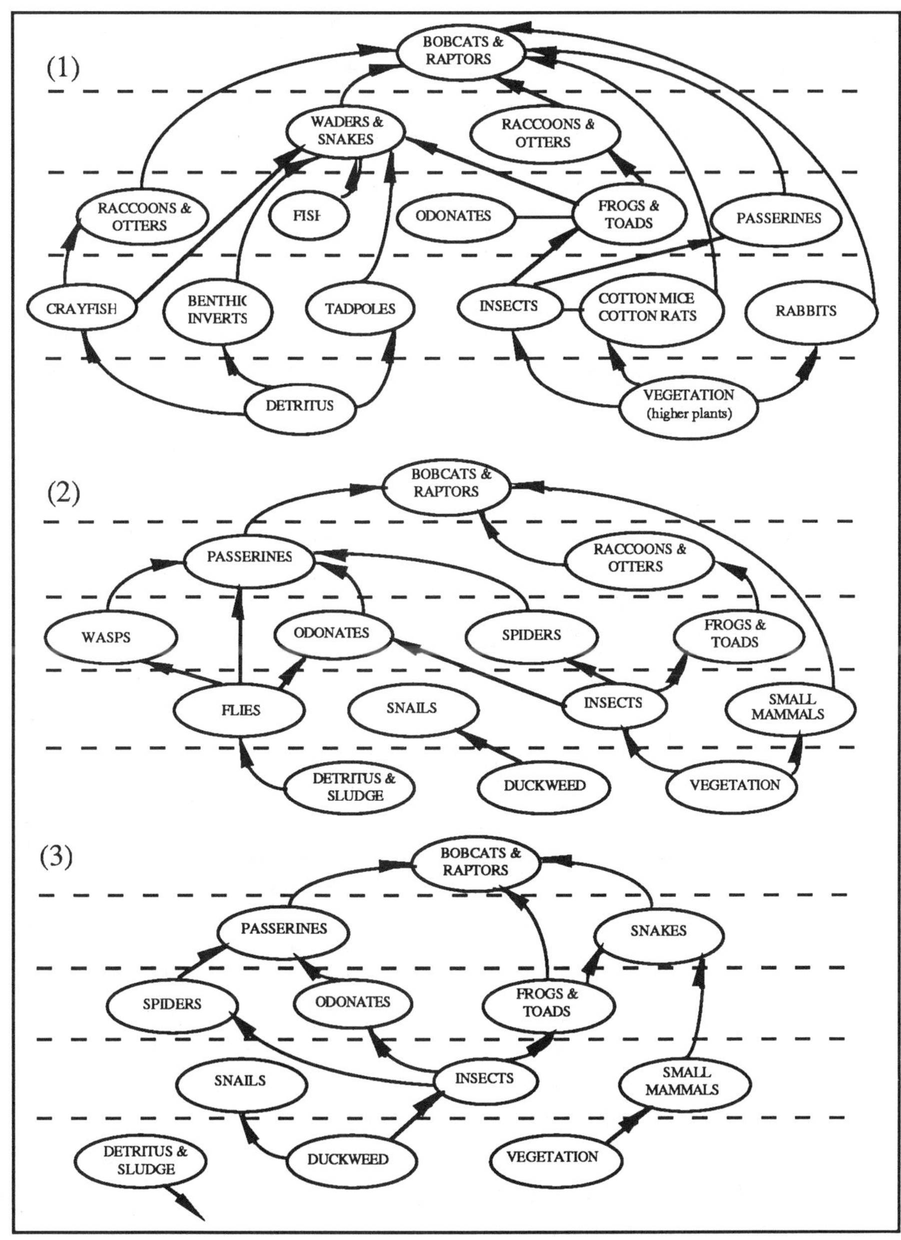

Figure 20. Food webs of cypress ponds. (1) Natural pond draws equally upon primary production and detritus pathways. (2) Simplified food web of nutrient enriched pond.and (3) sewage-effluent stressed pond (from Harris and Vickers 1984).

as 1855, de Condolle observed that "the breakup of a large landmass into smaller units would necessarily lead to the extinction or local extinction of one or more species and the differential preservation of others" (Browne 1983:159).

One reflection of the intrinsic richness of habitat is the species-area curve. The numbers of bird species that breed in forest tracts of various sizes reveal that: 1) bottomland hardwood forests support a richer avifauna than cypress-dominated swamps, 2) both bottomland hardwood forests and cypress swamps support a richer avifauna than upland hardwoods; 3) within the state of Illinois, and perhaps generally, bottomland hardwood forests support a richer avifauna than more northern forests; 4) for tracts less than a few hundred hectares in size, decreasing the tract size by 75% would reduce the number of breeding species by about 50%; and 5) for large tracts of several thousand hectares, reducing the tract size by 90% would reduce the number of breeding species by perhaps 33% (Figure 21).

Species that require large tracts of closed-canopy forest are referred to as "interior" or "area-sensitive" species; in other words, they seem to be negatively affected by clearings, roads, powerlines, or other edge environments. Most interior species tend to be habitat specialists. Because they occur "within" a tract of forest they contribute to the "Alpha" component of diversity. Perhaps 20% of the vertebrate species native to an area would be of this type. It is this alpha component of diversity that distinguishes one community type from another. The most common consequence of forest fragmentation is the loss of these characteristic species (e.g., Bachman's warbler [*Aimorphilia aestivalix*], Swainson's warbler [*Limnothlypis swainsonii*], and Kentucky warbler [*Oporornis formosus*]).

A second group of species that is heavily impacted by fragmentation, but for different reasons, is the group of large, generally carnivorous, and generally wide-ranging species that utilize extensive landscapes in their day-to-day movements. The home-range size for a single individual of these bird, mammal, and sometimes reptile species may be as large as 65,000 ha. The

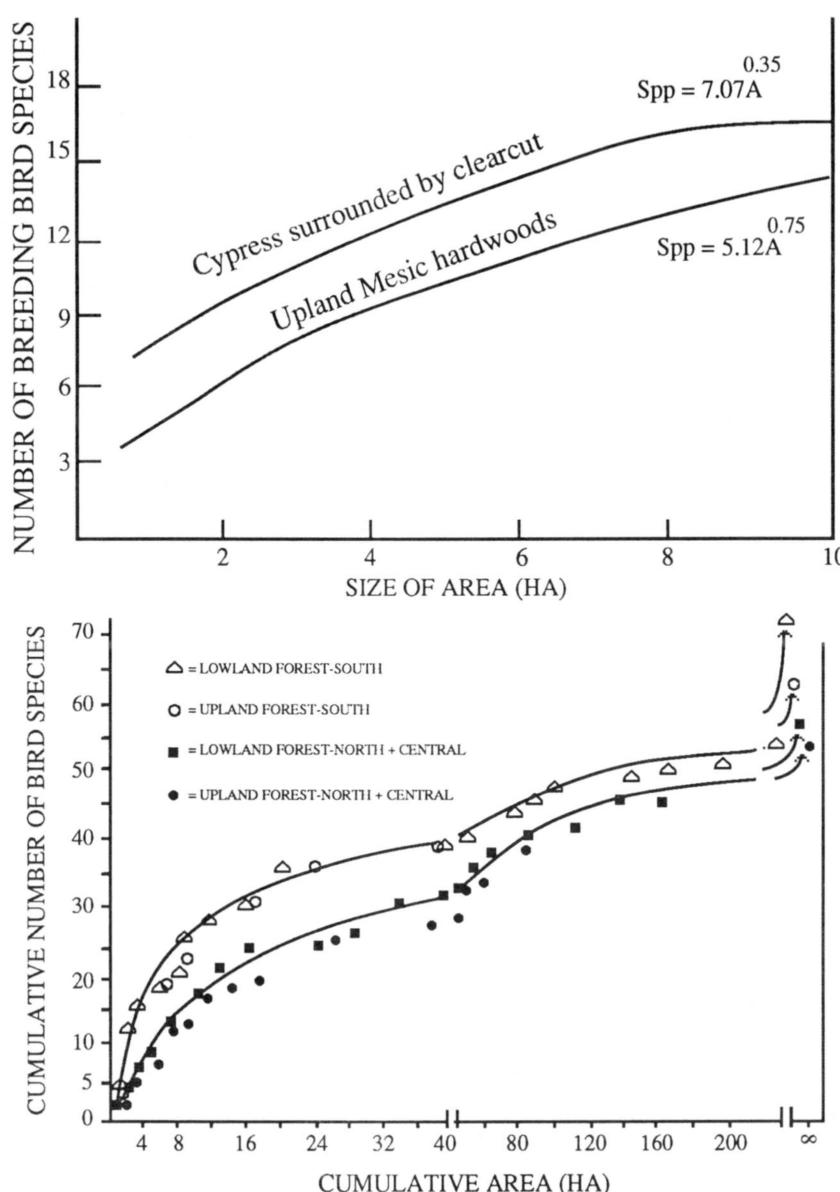

Figure 21. **Number of breeding bird species in forest tracts of different size. Upper panel reflects the relation for relatively small tracts in northern Florida; lower panel reflects the relation for larger tracts in Illinois (lower panel from Graber and Graber 1976).**

area necessary to maintain a minimum viable population of these wide ranging species is many times the home range size. The principal cause of death of these species is negative interactions with humans, e.g., road kills, etc. Without a system of travel corridors that allows these creatures unmolested passage from one major refugium to another, they will probably not occur in our future landscapes (Harris 1985).

Losing species from both ends of the spectrum (i.e., within-habitat "specialists," and among-habitat top carnivores) leaves an array of generalists that tend to be early-successional, edge, common, and frequently alien species. This heavily impacted and truncated faunal array does not meet the spirit of BIP or BBC as already introduced in law via the Clean Water Act.

3. Mortality and species loss due to parasitism is a predicted consequence of community stress (Odum 1985) and is easily demonstrated in converted bottomlands. This example also demonstrates how indirect effects can be just as consequential as direct effects.

Early explorers described a flocking bird, the "buffalo bird," in close association with the American bison (*Bison bison*) in grassland and prairie habitat. "The bird originally attended and followed the herds of buffalo, and possibly the antelopes and other large game" (Friedmann 1929:150). It so happens that this bird is North America's only obligate nest parasite--meaning that it has lost the genetic information necessary to build nests. Of necessity it puts its eggs in other birds' nests for incubation and care of the young. This usually means failure for the eggs and the chicks of the host species whose nest is parasitized. The buffalo bird of the plains in pioneer days is the brown-headed cowbird (*Molothrus ater*) of the farm pastures and forest edges of today. The species has expanded its range throughout the United States, but does particularly well in the deep South where rice fields provide abundant food resources. It now parasitizes large proportions of the nests of native passerine birds and is believed to be a major factor in the decline of certain species (Brittingham and Temple 1983).

Recall that the cowbird was formerly a bird of the open prairie and even now it does not invade the interior of large forest stands. Despite the fact that it remains a bird of clearings and edges, patch cutting has allowed it to become one of the more common birds throughout the White River National Wildlife Refuge, the largest remaining tract of bottomlands in the United States. Thus, fragmentation and conversion of bottomlands to cereal grains creates ideal habitat for a parasite and indirectly leads to the demise of native faunal diversity.

4. Loss of avian species because of cutting forests to stream edges has been a problem in many areas even though progressive foresters have now accepted the multiple merits of retaining streamside buffer zones and trees (leave strips). Many forestry agencies and companies now follow policies that allow only low-intensity cutting near streams, if any at all. The basis for determining buffer strip width has been almost totally dominated by surface runoff considerations. But it is also clear that width largely determines the value of the buffer strip for wildlife. For example, in Virginia the Acadian flycatcher (*Empidonax virescens*), American redstart (*Setophaga ruticilla*), hooded warbler (*Wilsonia citrina*), Louisiana waterthrush (*Seiurus motacilla*), northern parula (*Parula americana*), and red-eyed vireo (*Vireo olivaceus*) have strong affinities for leave strips. Parula warblers occur only in the widest leave strips; blue jays (*Cyanocitta cristata*), yellow-throated vireos (*Vireo flavifrons*), and Louisiana waterthrush occur only infrequently when strips are below 60 m in width; and pileated woodpeckers (*Dryocopus pileatus*), hairy woodpeckers (*Picoides villosus*), and Acadian flycatchers rarely occur below 50-m strip width (Tassone 1981). Heavily converted landscapes that do not maintain leave strips of at least 50 m will not support this array of breeding bird species.

The final ecological response to continuing stresses associated with cumulative impacts is system collapse and transformation (Rapport et al. 1985). When combined stresses exceed certain limits, ecosystems may temporarily or permanently lose capacity to perform certain biological functions. With severe impacts the ecological function of predation may be impaired and/or changed and top carnivores may be lost. Ecosystem collapse and transformation may be repairable but on time scales of thousands of years. Logging in steep terrain can lead to soil slippage and erosion so that thousands of years are required for the site to rebuild a comparable vegetative cover with attendant animals. In other cases, such as loading with toxic substances and/or channelizing a bottomland site, perhaps only centuries would be required for ecosystem regeneration, given that the stressors were removed and rehabilitative measures executed. Other practices, such as clearcutting a site, might only require 100-150 years to regain essential system integrity. Clearly, actions that only temporarily impact the system, such as hunting or trapping of animals, select-cutting, or small patch clearcutting of trees, may fall within the domain of renewable resource management. Actions and/or indirect consequences of human actions (e.g., soil slippage) so severe as to exterminate a species or permanently change the site are outside the domain of renewable resource management. Policies that allow actions that greatly alter the options and opportunities of future generations of humans seem arrogant at best and foolhardy at worst. E. O. Wilson (1988) has observed that the single action for which future generations are most likely not to forgive us is the extinction of species.

Species extinction has occurred because of our mismanagement of bottomland hardwood forests, and the processes and consequences continue today. North America's largest woodpecker, the ivory-billed woodpecker (*Campephilus principalis*), and the Carolina parakeet (*Conuropsis carolinensis carolinensis*) (North America's only parrot) are the most commonly cited examples of early extinctions in bottomland hardwood forests. The ivory-billed woodpecker played an ecological role by shredding, stripping, and reducing to small particle size large standing dead trees and branches, and probably served as an important cavity builder for middle-sized mammals too large to use small woodpecker cavities. Its extermination by over-zealous logging of mature and climax forest habitats has no doubt impaired certain ecological functions (e.g., large-dimension dead wood decomposition and cavity formation) that continue to impact our lives today (e.g., reduced wood duck [*Aix sponsa*] abundance). The Carolina parakeet preferred the seeds of cypress as food and served as an effective seed dispersal agent that could move seed against the current. Now, left only to the agent of flowing water, cypress dispersal and colonization is presumably unidirectional with the flow of water. Extinctions continue today. Bachman's warbler appears to be the most recent, or the next, victim of extinction.

Depletion of North American migrant bird populations continues to occur and may be tied to depletion and management of southern bottomland hardwood forests. Populations of mallards (*Anas platyrhynchos*) and pintails (*A. acuta*) are lower now (1986) than in any year since comprehensive censusing was begun in the 1950's (Figure 22). Although the reduction in number of prairie potholes in the breeding range is commonly believed to be the causative agent, there is increasing attention to problems in the overwintering grounds. Trauger and Stoudt (1978:199) stated:

> The poor relationship between waterfowl populations and wetlands habitat on the study areas is apparently due to the presence of more ponds than needed to support current duck numbers in the Canadian parklands during the post-drought era... Rather, production appears to be more directly related to the relative size of the breeding population than to numbers of ponds... Factors limiting continental waterfowl populations, including human exploitation, away from the breeding grounds are implicated.

Heitmeyer and Fredrickson (1981:54) went further in search of explanations for the alarming decrease in North American mallards, concluding that "winter wetland indices explained more of the variation in mallard age ratios [index of productivity] than breeding ground wetland indices." Unfortunately, no studies conclusively tie mallard numbers to overwintering bottomland hardwood area or quality. The reductions in densities of mallards and other waterfowl are no doubt the result of confounded, indirect, and cumulative impacts from several human land-use and resource management decisions.

About 150 species of North American breeding birds overwinter in the southeastern coastal plain or farther south in the neotropics. The

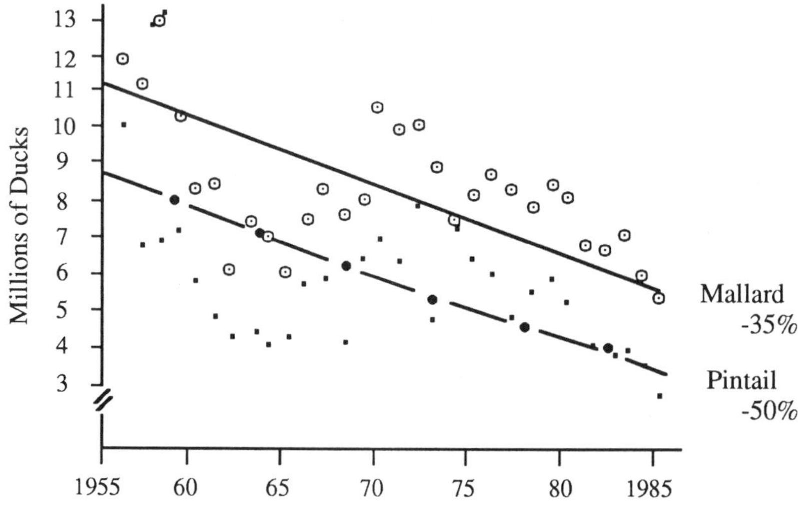

Figure 22. Mallards, pintails, and several other species of North American waterfowl have declined dramatically. Mallards overwinter in bottomland hardwood ecosystems of southern states, and of all variables analyzed in one study, water conditions in these overwintering areas explained the greatest amount of variance in the following year's duck production (from USFWS and CWS 1985).

abundance of many of these species is dramatically declining (Figure 23). As with waterfowl, it cannot be "proven" that this reduction is specifically tied to wintering habitat conditions. Yet, it is true that "when only the neotropical migrants are considered, the percentage change in the breeding population is much more drastic than for all species combined" (Robbins 1980:32). If the explanation were totally in the breeding habitat, then it should impact non-migrants as well as migrants. The effect is clearly more severe on migrants. Similarly, if the problem were somewhere in the tropics (e.g., a factor such as moist forest liquidation), then it should only impact those species that overwinter in the tropics. This is not the case since many declining species overwinter on the southeastern coastal plain. Thus, there is circumstantial evidence that general declines in migratory bird populations may result from conditions in southern United States bottomland hardwood forests and coastal marshes.

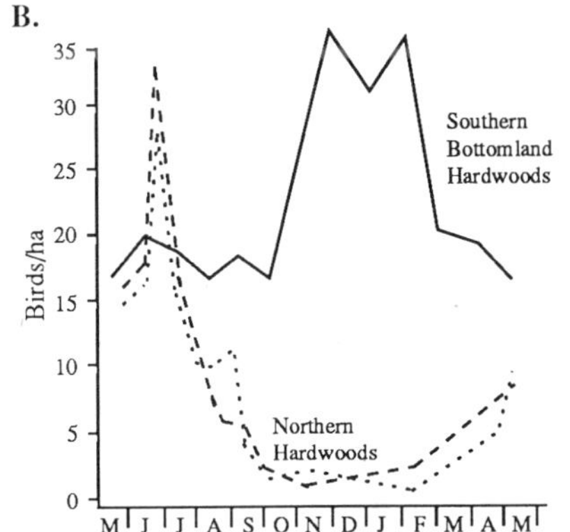

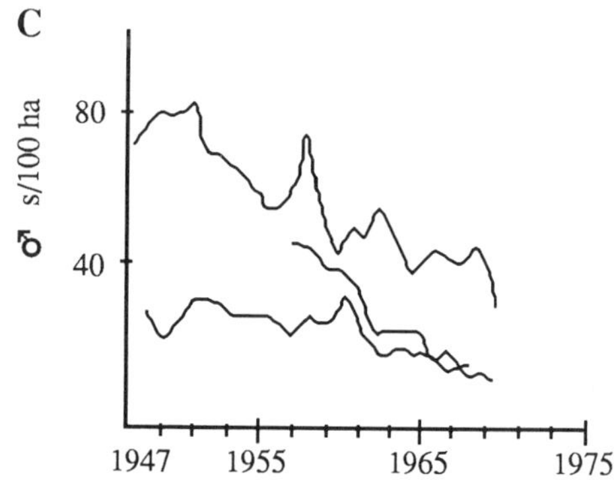

Figure 23. A. About 150 species of North American birds migrate to or through southern wetland ecosystems. Sprunt (1967) estimates that 90% of all eastern North American bird species utilize southern wetlands, many of them as overwintering habitat. Southern bottomland hardwood forests support their highest densities of birds during winter, out of phase with northern forests such as Hubbard Brook (Keast 1980). B. U.S. song bird density in northern hardwood forests peaks in early summer. In southern bottomland forests is highest during the winter (northern data from Holmes and Sturges 1973; southern data from Kennedy 1977). C. The decline (as measured by territorial males) of several species of migrant birds in three study areas on Maryland breeding grounds (Briggs and Criswell 1979).

SUMMARY

Bottomland hardwood ecosystems are probably the most productive of North American inland ecosystems. This is most apparent at the secondary productivity level (that of animal communities). The unusual productivity results from the occurrence of bottomland hardwood forests at low latitudes with high solar input, less dramatic fluctuations in day length and temperature than farther north, and major river systems. This leads to 1) an abundance of woody plants supporting fruit, seed, and arthropod production during winter months; and 2) annual flooding that facilitates both aquatic and terrestrial energy pathways through the receipt of allochthonous silt and organically-bound energy that facilitates both detritivore and primary producer energy flow pathways.

The rate of decline of bottomland hardwood area is dramatic. Forested wetlands constitute only 8% of 1980 United States forested areas, but account for 37% of forest loss. About 57% of forested wetland acreage occurs in 12 southeastern states. Some 50% of the 1940 area was lost by 1985. Only one-fifth of the natural forested wetlands of the Lower Mississippi River Valley remain. States such as Florida, with rapid human population growth, are losing forests at the rate of 60,000 ha/year. Attendant to this loss is the fragmentation of remaining areas into small, disjointed, and sometimes isolated patches. Selective logging and other past forestry practices have degraded internal stand quality for numerous plant and animal species.

The combined impacts of converting bottomland hardwood forests to other uses, such as agriculture and construction of dams, levees, channels, and canals, have seriously impaired the capacity of these ecosystems to perform their historical ecological functions. The human values that society formerly derived from these systems (e.g., flood control, water-quality maintenance, endangered species habitat) have been seriously compromised by these cumulative effects.

Cumulative effects can be easily described after the fact. They can also be predicted, but their nature and magnitude generally can not be specified in detail. Because cumulative impacts by definition occur both on and off site and represent compounded and multiplied effects of incremental actions, they cannot be proven to result from, or to be directly linked to, individual actions. Cumulative impact analysis on large-scale ecosystems is not amenable to the cause-effect, single factor, experimental science that currently guides policy and decision processes. As a result, increased emphasis must be placed on concepts such as ecosystem integrity as implied in the legislative phrases BIP and BBC.

Continuing to allow the conversion of southern bottomland hardwood forests will affect not only the ecological and economic well-being in the American South, but also in the entire United States and in other countries (about 150 species of animals migrate internationally through the southern United States). The consequences not only affect our immediate future, but also curtail the options and opportunities available to future generations.

ACKNOWLEDGMENTS

This manuscript was prepared with support from the Environmental Research Laboratory, U. S. Environmental Protection Agency, Corvallis, OR.

REFERENCES CITED

Abernethy, Y. and R. Turner. 1987. U. S. forested wetlands; status and changes 1940-1980. BioScience 37:721-727.

Allen, J. A. 1871. On the mammals and winter birds of East Florida, with an examination of certain assumed specific characters in birds, and a sketch of the bird-faunae of eastern North America. Bull. Mus. Comp. Zool. II(3):161-450.

__________. 1892. The geographic distribution of North American mammals. Bull. Am. Mus. Nat. Hist. 4:199-243.

Asmussen, L. S. E., J. M. Sheridan, and C. V. Booram, Jr. 1979. Nutrient movement in streamflow from agricultural watersheds in the Georgia Coastal Plain. Trans. Am. Soc. Agric. Eng. 22:809-915, 821.

Atwood, S. S., et al. (committee) 1970. Landuse and wildlife resources. National Academy of Sciences, Washington, D.C. 262 pp.

Barrett, G. W., G. M. VanDyne, and E. P. Odum. 1976. Stress ecology. BioScience 26:192-194.

_____________, and R. Rosenberg (eds). 1981. Stress effects on natural ecosystems. Wiley and Sons, New York.

Bell, D. T., and F. L. Johnson. 1974. Groundwater level in the floodplain and adjacent uplands of the Sangamon River. Trans. Ill. St. Acad. Sci. 67:376-383.

Belt, C. B., Jr. 1975. The 1973 flood and man's construction of the Mississippi River. Science 189:681-684.

Brezonik, P. L., and E. E. Shannon. 1971. Trophic state of lakes in North Central Florida. Publication No. 13. Florida Water Resources Research Center. 102 pp.

Briggs, S., and J. Criswell. 1979. Gradual silencing of spring in Washington. Atl. Nat. 32:19-26.

Brinson, M. M., B. L. Swift, R. C. Plantico, and J. S. Barclay. 1981. Riparian ecosystems: their ecology and status. U.S. Fish and Wildlife Service, Washington, D.C. FWS/OBS-81/17. 155 p..

Brittingham, M., and S. Temple. 1983. Have cowbirds caused forest songbirds to decline? BioScience 33:31-35.

Browne, J. 1983. The secular ark. Yale University Press, New Haven. 273 pp.

Carter, V., M. S. Bedinger, R. P. Novitski, and W. O. Wilen. 1979. Water resources and wetlands. Pages 344-376 *in* P. E. Green, J. Clark, and J. Clark, eds. Wetland functions and values: the state of our understanding. American Water Resources Association, Minneapolis, Minnesota. 674 pp.

Cely, J. 1979. Status of the swallow-tailed kite and factors affecting its distribution. Pages 144-150 *in* D. Forsythe and W. Ezell Jr., eds. Proceedings of the 1st South Carolina Endangered Species Symposium, South Carolina Wildlife and Marine Resources Department, Columbia.

Clark, J. R. and J. Benforado, eds. 1981. Wetlands of bottomland hardwood forests. Elsevier Scientific Publishing Co., Amsterdam. 401 pp.

Conner, W., and J. W. Day, Jr. 1982. The ecology of forested wetlands in the southeastern United States. Pages 69-87 *in* B. Gopal, R. Turner, R. Wetzel, and D. Whigham, eds. Wetlands: ecology and management. International Science Publishers, Jaipur, India. 314 pp.

Cowardin, L., V. Carter, F. Golet, and E. LaRose. 1979. Classification of wetlands and deepwater habitats of the United States. U.S. Fish and Wildlife Service, Washington, D.C. Pub. No. FWS/OBS-79/31.

Croker, T. 1979. The longleaf pine story. J. Forest Hist. 23:32-43.

Darlington, P. 1957. Zoogeography: the geographical distribution of animals. John Wiley and Sons, New York. 675 pp.

Delcourt, H., and P. Delcourt. 1977. Presettlement magnolia-beech climax of the Gulf Coastal Plain: quantitative evidence from the Apalachicola River bluffs, north-central Florida. Ecology. 58:1085-1093.

Dice, L. 1943. The biotic provinces of North America. University of Michigan Press, Ann Arbor.

Duda, A. M. 1982. Municipal point source and agricultural nonpoint source contributions to coastal eutrophication. Water Resour. Bull. 18:397-407.

Elder, J. F. 1985. Nitrogen and phosphorus speciation and flux in a large Florida river wetland system. Water Resour. Res. 21:724-732.

Frayer, W., T. Monahan, D. Bowden, and F. Graybill. 1983. Status and trends of wetlands and deepwater habitats in the conterminous United States, 1950's to 1970's. Published for U.S. Department of the Interior, Fish and Wildlife Service by Department of Forest and Wood Sciences, Colorado State University, Fort Collins, Colorado. 31 pp.

Friedmann, H. 1929. The cowbirds, a study in the biology of social parasitism. Chas. Thomas Publishing Co., Baltimore. 421 pp.

Gael, B., and C. Hopkinson. 1979. Drainage density, land-use and eutrophication in Barataria Basin, Louisiana. Pages 147-163 *in* J. Day, Jr., D. Culley, Jr., R. Turner, and A. Mumphrey, Jr., eds. Environmental conditions in the Louisiana coastal zone. Louisiana State University Division of Continuing Education, Baton Rouge.

Gipson, P. 1974. Food habits of coyotes in Arkansas. J. Wildl. Manage. 38:848-853.

Glasgow, L., and R. Noble. 1971. The importance of bottomland hardwoods to wildlife, Pages 30-43 *in* Proceedings of the Symposium on Southeast Hardwoods, Dothan, Alabama, U.S. Department of Agriculture, Forest Service.

Gosselink, J., W. Conner, J. Day, Jr., and R. Turner. 1981. Classification of wetland resources: land, timber and ecology. Pages 28-48 *in* B. Jackson and J. Chambers, eds. Timber harvesting in wetlands. Louisiana State University Division of Continuing Education, Baton Rouge. 166 pp.

Graber, J., and R. Graber. 1976. Environmental evaluations using birds and their habitats. Biol. Notes 97, Ill. Nat. Hist. Survey. Urbana. 39 pp.

Gray, L., O. Baker, F. Marschner, B. Weitz, W. Chapline, W. Shepard and R. Zon. 1923. The utilization of our lands for crops, pasture and forests. Pages 415-506 *in* Agriculture Yearbook. U.S. Department of Agriculture, Washington, D.C.

Greeson, P., J. Clark, and J. Clark, eds. 1979. Wetland functions and values: the state of our understanding. American Water Resources Association, Minneapolis, Minnesota. 674 pp.

Harris, L. 1982. Bottomland hardwoods, valuable, vanishing, vulnerable. Florida Cooperative Extension Service, Gainesville. Special Publication 28. 18 pp.

__________. 1984. The Fragmented forest; application of island biogeography principles to preservation of biotic diversity. University of Chicago Press. Chicago 216 pp.

__________, and C. Vickers. 1984. Some faunal community characteristics of cypress ponds and the changes induced by perturbations. Pages 171-185 *in* K. C. Ewel and H. T. Odum eds. Cypress swamps. University Presses of Florida, Gainesville.

__________. 1985. Conservation corridors, a highway system for wildlife. ENFO, Nov. 1985. Florida Conservation Foundation, Winter Park, Florida 10 pp.

Heitmeyer, M., and L. Fredrickson. 1981. Do wetland conditions in the Mississippi delta hardwoods influence mallard recruitment? Transactions of the North American Wildlife Natural Resources Conference 46:44-57.

Holmes, R., and F. Sturges. 1973. Annual energy expenditure by the avifauna of a northern hardwood ecosystem. Oikos 24:24-29.

Horak, G., E. Vlachos, and E. Cline. 1983. Methodological guidance for assessing cumulative impacts on Fish and Wildlife. Unpublished report to U.S. Department of the Interior, Fish and Wildlife Service by Dynamic Corp., Fort Collins, Colorado. 102 pp.

Johnson, R. R., and J. F. McCormick, eds. 1978. Strategies for protection and management of floodplain wetlands and other riparian systems. U.S. Department of Agriculture, Forest Service, Washington, D. C. GTR-WO-12,

Keast, A. 1980. Synthesis: ecological basis and evolution of the nearctic - neotropical bird migration system. Pages 559-576 *in* A. Keast and E. Morton, eds. Migrant birds in the neotropics: ecology, behavior, distribution and conservation. Smithsonian Institution Press, Washington, D. C.

Kennedy, R. 1977. Ecological analysis and population estimates of the birds of the Atchafalaya River basin in Louisiana. Ph.D. dissertation, Louisiana State University, Baton Rouge.

Kusler, J. A. 1977. Wetland protection: a guidebook for local governments. Environmental Law Institute, Washington, D. C.

Larson, J., M. Bedinger, C. Bryan, S. Brown, R. Huffman, E. Miller, D. Rhodes, and B. Touchet. 1981. Transition from wetlands to uplands in southeastern bottomland hardwood forests. Pages 225-273 *in* J. Clark and J. Benforado, eds. Wetlands of bottomland hardwood forests. Elsevier Scientific Publishing Co., Amsterdam. 401 pp.

Leopold, L., M. Wolman, and J. Miller. 1964. Fluvial processes in geomorphology. W. H. Freeman and Co., San Francisco.

Livingston, R. 1975. Resource management and estuarine function with application to the Apalachicola drainage system. Pages 3-17 *in* Proceedings of Estuarine Pollution Control and Assessment. U.S. Environmental Protection Agency. Washington, D. C.

Loucks, O. 1985. Looking for surprise in managing stressed ecosystems. BioScience 35:428-432.

Lowrance, R., R. Todd, and L. Asmussen. 1983. Waterborne nutrient budgets for the riparian zone of an agricultural watershed. Agric. Ecosyst. and Environ. 10:371-384.

__________________________________.1984a. Nutrient cycling in an agricultural watershed: I. Phreatic movement. J. Environ. Qual. 13:22-27.

__________________________________.1984b. Nutrient cycling in an agricultural watershed: II. Streamflow and artificial drainage. J. Environ. Qual. 13:27-32.

Lowrance, R., R. Todd, J. Fail, Jr., O. Hendrickson, Jr., R. Leonard, and L. Asmussen. 1984c. Riparian forests as nutrient filters in agriculture watersheds. BioScience 34:374-377.

MacClintock, L., R. Whitcomb, and B. Whitcomb. 1977. Evidence for the value of corridors and minimization of isolation in preservation of biotic diversity. Am. Birds 31:6-16.

MacDonald, P., W. Frayer, and J. Clauser. 1979. Documentation, chronology, and future predictions of bottomland hardwood habitat losses in the lower Mississippi Alluvial Plain. U.S. Fish and Wildlife Service, Vols. 1 and 2. Jackson, Mississippi. 133 pp. and 259 pp.

Margalef, R. 1975. Human impact on transportation and diversity in ecosystems. How far is extrapolation valid? Pages 237-241 *in* Proceedings of the 1st International Congress of Ecology. Structure, functioning and management of ecosystems. The Hague, Wageningen.

McDiarmid, R. 1978. Amphibians and reptiles, Vol. 3, Rare and endangered biota of Florida. University Presses of Florida, Gainesville, 74 pp.

McElfresh, R., J. Inglis, and B. A. Brown. 1980. Gray squirrel usage of hardwood ravines within pine plantations. Louisiana State University Forestry Symposium 29:79-89.

McKeever, S. 1959. Relative abundance of twelve southeastern mammals in six vegetative types. Am. Midl. Nat. 62:222-226.

Meade, R., and R. Parker. 1985. Sediment in rivers of the United States. Pages 49-60 *in* U.S. Geological Survey, National Water Summary 1984. Washington, D. C. Water-Supply Paper 2275.

Mitsch, W. J., and J. G. Gosselink. 1986. Wetlands. Van Nostrand Reinhold Co., New York. 539 pp.

Mulholland, P., and E. Kuenzler. 1979. Organic carbon export from upland and forested wetland watersheds. Limnol. Oceanogr. 24:960-966.

Murphree, C., C. Mutchler, and L. McDowell. 1976. Sediment yields from a Mississippi delta watershed. Proceedings of the 3rd Federal Inter-Agency Sedimentation Conference. Water Resources Council. pp. 1-99 to 1-109.

Novak, M. 1987. Beaver. Pages 283-312 *in* M. Novak, J. A. Baker, M. E. Obbard, and B. Malloch, eds. Wild furbearer management and conservation in North America. Ontario Trappers Association, North Bay, Ontario.

Obbard, M. E., J. G. Jones, R. Newman, A. J. Satterthwaite, and G. Linscombe. 1987. Furbearer harvests in North America. Pages 1007-1034 *in* M. Novak, J. A. Baker, M. E. Obbard, and B. Malloch, eds. Wild furbearer management and conservation in North America. Ontario Trappers Association, North Bay, Ontario.

Odum, E. 1969. The strategy of ecosystem development. Science 164:262-270.

Odum, E. P., J. T. Finn, and E. H. Franz. 1979. Perturbation theory and the subsidy-stress gradient. BioScience 29:349-352.

Odum. E. P. 1985. Trends expected in stressed ecosystems. BioScience 35:419-422.

Office of Technology Assessment. 1984. Wetlands: their use and regulation. U.S. Congress, Office of Technology Assessment, Washington, D.C. OTA-0-206. 209 pp.

Ogawa, H., and J. Male. 1983. The flood mitigation potential of inland wetlands. University of Massachusetts Water Resources Research Center Publication No. 138. Amherst. 164 pp.

Omernik, J. M. 1977. Nonpoint source-stream nutrient level relationships: a nationwide study. Corvallis Environmental Research Laboratory, Office of Research and Development, U.S. Environmental Protection Agency, Corvallis, Oregon. EPA-600/3-77-105.

O'Neil, T. and G. Linscombe. n.d. The fur animals, the alligator, and the fur industry in Louisiana. Wildl. Educ. Bull. 106. Louisiana Wildlife and Fisheries Commission, New Orleans. 66 pp.

Peterjohn, W. T., and D. L. Correll. 1984. Nutrient dynamics in an agricultural watershed: observations on the role of a riparian forest. Ecology 65(5):1466-1475.

Rapport, D., H. Regier, and T. Hutchinson. 1985. Ecosystem behavior under stress. Am. Nat. 125:617-640.

Richardson, C. J. 1985. Mechanisms controlling phosphorus retention capacity in freshwater wetlands. Science 228:1424-1427.

______________, R. Evans, and D. Carr. 1981. Pocosins: an ecosystem in transition. Pages 3-9 *in* C. J. Richardson, ed. Pocosin wetlands. Hutchinson Ross Publishing Co., Stroudsburg, Pennsylvania. 364 pp.

Robbins, C. 1979. Effect of forest fragmentation of bird populations, *in* R. DeGray and K. Evans, eds. Proceedings of the workshop on management of north central and northeast forests for nongame birds. U.S. Department of Agriculture, Forest Service, Washington, D.C. GTR NC51.

__________. 1980. Effect of forest fragmentation on breeding bird populations in the Piedmont of the mid-Atlantic region. Atl. Nat. 33:31-36.

Roe, H., and Q. Ayres. 1954. Engineering for Agricultural drainage. McGraw-Hill. New York. 501 pp.

Roelke, M. 1986. Medical management, biomedical findings, and research techniques. Pages 7-14 *in* Survival of the Florida panther: a discussion of issues and accomplishments. Conference Proceedings, Florida Defenders of the Environment, Tallahassee, Florida.

Schreiber, R. 1978. Eastern brown pelican. Pages 23-25 *in* H. W. Kale II, ed. Vol. 2, Birds, rare and endangered biota of Florida. University Presses of Florida, Gainesville.

Sedell, J., F. Everest, and F. Swanson. 1981. Fish habitat and streamside management: past and present. Proc. Soc. Am. For. Ann. Meet. 41-52.

Selye, H. 1956. The stress of life. McGraw-Hill Book Co., Inc., New York. 325 pp.

Shaw, S., and C. Fredine. 1956. Wetlands of the United States, their extent and their value to waterfowl and other wildlife. U.S. Department of the Interior, Fish and Wildlife Service, Circ. 39. 67 pp.

Sprunt, A. 1967. Values of the South Atlantic and Gulf Coast marshes and estuaries to birds other than waterfowl. Pages 64-72 *in* Proceedings of the Marsh and Estuary Management Symposium, Louisiana State University, Baton Rouge, Louisiana.

Tassone, J. 1981. Utility of hardwood leave strips for breeding birds in Virginia's central Piedmont. MS thesis. Virginia Polytechnic Institute and State University, Blacksburg. 83 pp.

Taylor, J., M. Cardamone, and W. Mitsch. 1990. Bottomland hardwood forests their functions and values. Pages 13-86 *in* J. G. Gosselink, L. C. Lee, and T. A. Muir, eds. Ecological processes and cumulative impacts: illustrated by bottomland hardwood wetland ecosystems. Lewis Publishers, Inc., Chelsea, Missouri.

Tiner, R., Jr. 1984. Wetlands of the United States and recent trends. U.S. Fish and Wildlife Service, National Wetlands Inventory Washington, D. C. 59 pp.

Trauger, D., and J. Stoudt. 1978. Trends in waterfowl populations and habitats on study areas in Canadian parklands. Transactions of the North American Wildlife and Natural Resources Conference 43:187-205.

Turner, R. E., and N. J. Craig. 1980. Recent areal changes in Louisiana's forested wetland habitat. Louisiana Acad. Sci. 43:61-68.

Ursic, S. J. 1965. Sediment yields from small watersheds under various landuses and forest covers. Pages 48-58 *in* Federal Inter-Agency Sedimentation Conference Proceedings, U.S. Department of Agriculture Miscellaneous Publication 970.

U.S. Army Corps of Engineers. 1972. Charles River Watershed, Massachusetts. New England Division, U.S. Army Corps of Engineers, Waltham, Maryland.

U.S. Fish and Wildlife Service (USFWS) and Canadian Wildlife Service (CWS). 1985. Status of waterfowl and fall flight forecasts. Washington, D. C. 32 pp.

U.S. Geological Survey. Annual. Water Year Records. Washington, D.C.

U.S. Soil Conservation Service. 1972. Hydrology. Section 4, National Engineering Handbook, Washington, D. C.

Verner, J., R. Pastorok, J. O'Connor, W. Severenghaus, N. Glass, and B. Swindel. 1985. Ecological community structure analyses in the formulation, implementation, and enforcement of law and policy. Amer. Stat. 39:393-402.

Wharton, C. 1980. Values and functions of bottomland hardwoods. Transactions of the North American Wildlife and Natural Resources Conference 45:341-353.

__________, W. Kitchens, E. Pendleton, and T. Sipe. 1982. The ecology of bottomland hardwood swamps of the Southeast: a community profile. U.S. Fish and Wildlife Service, Washington, D. C. FWS/OBS-81/37.

Wright, J. 1907. Swamp and overflowed lands in the United States, ownership and reclamation. U.S. Department of Agriculture, Office of Experimental Stations, Wshington, D.C. Circular 76. 23 pp.

10. THE TEXAS/OKLAHOMA BOTTOMLAND HARDWOOD FOREST PROTECTION PROGRAM

James. A. Neal
U.S. Fish and Wildlife Service,
Nacogdoches, TX 75962

Ernest. S. Jemison
Tishomingo National Wildlife Refuge,
U.S. Fish and Wildlife Service,
Tishomingo, OK 73460

ABSTRACT

The bottomland hardwood ecosystem of Texas and Oklahoma supports a diverse flora and fauna: 24 plant communities (some restricted to these two states), 110 plant species of special concern, high densities of various wildlife species including a number of waterfowl species, and 125 animal species of special concern.

Conversion of floodplain forests to other land uses places these forests among the most severely altered ecosystems in the United States. These biotic resources are threatened by water-control structures, agriculture, destructive forestry practices, mining and petroleum extraction, development, pollution, and minor floodplain modifications. As compared to most of the southeastern U.S. bottomlands, which are principally threatened by agriculture, Texas and Oklahoma bottomlands are threatened mostly by water-control structures (primarily reservoirs). To counter these impacts, the U.S. Fish and Wildlife Service is proposing to preserve 100,000 ha of bottomlands in Texas and Oklahoma through a variety of means.

Ecological Processes and Cumulative Impacts: Illustrated by Bottomland Hardwood Wetland Ecosystems. Edited by James G. Gosselink, Lyndon C. Lee, and Thomas A. Muir. Chelsea, MI 48118. Printed in USA.

INTRODUCTION

The U.S. Fish and Wildlife Service (FWS) has long been concerned with the preservation of bottomland hardwood forests. Over 30 national wildlife refuges (NWR) have been established from eastern Texas and southeastern Oklahoma to North Carolina to preserve overflow bottoms. Eleven refuges, which preserve over 91,000 ha of bottomlands, have been established within the Lower Mississippi River Delta.

Preservation activities have been aided and often supplemented by The Nature Conservancy (TNC) through its Rivers of the Deep South Program, made possible by a $15 million grant from the Richard King Mellon Foundation. It is anticipated that approximately 190,000 ha of ecologically valuable river habitat in Georgia, Florida, Alabama, Louisiana, and Mississippi will be preserved through this program and with other funds from federal and state governments (Blair 1981).

Preservation of bottomland hardwoods in Texas and Oklahoma has been accomplished largely by the U.S. Forest Service (USFS), the National Park Service, the Texas Parks and Wildlife Department (TPWD), the Oklahoma Department of Wildlife Conservation (ODWC), and very recently the FWS (Little River NWR in Oklahoma and the Little Sandy conservation easement in Texas).

In 1977, an effort was initiated to rank important waterfowl habitats in the United States. Habitats were divided into 33 categories, and each category was rated on its importance to the nation's waterfowl resources. The Lower Mississippi River Delta, the only bottomland hardwood habitat listed, was rated seventh (FWS 1978).

Recent efforts have been made to establish an updated habitat preservation strategy for waterfowl. As a result of the efforts of an interdisciplinary team of waterfowl and wetland biologists, 44 recommendations were made concerning habitat strategies for nine waterfowl species identified as National Species of Special Emphasis (FWS 1984 and 1985 supplement). Twenty recommendations dealt specifically with acquiring land interests. From these efforts, 11 habitat areas of general concern were proposed as a replacement for the 33-habitat category system. Bottomland hardwood forests of Arkansas, Louisiana, Mississippi, Missouri, Iowa, Illinois, Kentucky, Tennessee, Oklahoma, and Texas were proposed as Category No. 3 (Figure 1). This

PRIORITY	NAME
1.	CENTRAL VALLEY
2.	PRAIRIE POTHOLES AND PARKLANDS
3.	BOTTOMLAND HARDWOODS
4.	ATLANTIC COASTAL PLAIN
5.	GULF COAST
6.	ALASKA AREA
7.	INTERMOUNTAIN WEST
8.	PLAYA LAKES
9.	KLAMATH BASIN
10.	UPPER PACIFIC COAST
11.	SAN FRANCISCO BAY

Figure 1. Recommended waterfowl habitat acquisition areas, January 1985. Area delineations are general.

preservation category includes a significant portion of habitat of the breeding and wintering wood duck (*Aix sponsa*) population, and winters 30% of the North American mallard (*Anas platyrhynchos*) population. This report recommends that 120,000 ha of easements be obtained to protect these habitats and their waterfowl resources (FWS 1984 and 1985 supplement). Nearly identical recommendations are proposed in the North American Waterfowl Plan (U.S. Department of the Interior and Environment Canada 1986).

BIOLOGICAL RESOURCES TO BE PROTECTED

Twenty-four bottomland community types were identified in the Texas/Oklahoma bottomland ecosystem, most of which seem to occur throughout the southeast and in the same topographic/physiographic positions (FWS 1985). Thirty-four community types have been described in the southeast (Neal and Haskins 1986). The list includes several community types that are not often identified as "bottomland" habitats but are included here because of their hydric and ecological connections to the broad river floodplains. The flatland hardwood community type (swamp chestnut oak [*Quercus michauxii*] - willow oak [*Q. phellos*] - laurel oak [*Q. laurifolia*]) is apparently unique to Texas (Marks and Harcombe 1981), and a "western" bottomland community, dominated by cedar elm (*Ulmus crassifolia*) - sugarberry (*Celtis laevigata*) - willow oak, is rarely found in other areas of the southeast. In contrast, several floodplain community types in the southeast are absent in Texas and Oklahoma. Bottomland habitats of Texas and Oklahoma support 110 plant species of special concern (Brabander et al. 1985, FWS 1985).

Faunistically, there are even fewer differences between Texas/Oklahoma bottomlands and other southeastern bottomlands. The waterfowl resources, which are the FWS justification for protecting the bottoms, are essentially the same -- the important species are the wood duck (breeding and wintering) and the mallard (wintering only). With the exception of the black bear (*Ursus americanus*), which for all practical purposes has been extirpated from Texas and Oklahoma, most of the major animal species are the same. The bottomlands of the two states support 125 animal species of special concern for at least a portion of their life cycle (Brabander et al. 1985, FWS 1985).

A number of floral and faunal species might qualify as functional indicators for bottomlands. The following species are heavily dependent on floodplain habitats of the southeast for all or a critical portion of their life cycle: 1) water hickory or bitter pecan (*Carya aquatica*), 2) overcup oak (*Quercus lyrata*), 3) water-elm or planer tree (*Planera aquatica*), 4) western mayhaw (*Crataegus opaca*), 5) green ash (*Fraxinus pennsylvanica*), 6) giant cane (*Arundinaria gigantea*), 7) inland sea oats (*Chasmanthium latifolium*), 8) green-backed heron (*Butorides striatus*), 9) wood duck (*Aix sponsa*), 10) American swallow-tailed kite (*Elanoides forficatus*), 11) prothonotary warbler (*Protonotaria citrea*), 12) Swainson's warbler (*Limnothlypis swainsonii*), 13) gray squirrel (*Sciurus carolinensis*), 14) river otter (*Lutra canadensis*), 15) swamp rabbit (*Sylvilagus aquaticus*), 16) Mississippi mud turtle (*Kinosternon subrubrum*), 17) mud snake (*Farancia abacura*), 18) canebrake rattlesnake (*Crotalus horridus*), and 19) mole salamander (*Ambystoma talpoideum*). Because animal species are dynamic, it is preferable to utilize plant communities as system indicators.

PROBLEMS/THREATS AND RESULTING LOSSES

Threats

Threats that result in the modification of bottomland hardwoods take seven general forms: 1) water-control structures, 2) agriculture, 3) destructive forest practices, 4) mining and petroleum extraction, 5) development, 6) pollution, and 7) other threats (FWS 1985).

Water-Control Structures

The principal threats to bottomland hardwoods in east Texas and Oklahoma, in contrast to the Mississippi Delta and a number of other major southeastern river systems, are from water-control structures, primarily reservoirs. Other water-control structural modifications include channelization, diversion, levees, dredging, snagging and clearing, channel construction, and drainage.

A total of 42 major reservoirs, totaling 269,572 ha, have been constructed in east Texas, primarily for flood control, water supply, and recreation. The largest reservoirs are Livingston (33,400 ha), Sam Rayburn (46,300 ha), and Toledo Bend (73,500 ha). A total of seven additional reservoir sites and two saltwater barriers are authorized federal projects within the study area; 25 additional sites have been selected for potential projects by other entities (Texas Department of Water Resources 1984). Three sites are under construction.

On the basis of projected future needs and possible sources of water supply, the Texas Department of Water Resources has initiated a surface-water allocation procedure that has resulted in the scheduling of water development projects over the years 1984-2030. Within the study area, 22 reservoirs are projected to be needed by the year 2030.

At least 21 of 62 key bottomland tracts in Texas (see below in Preservation Program to Counter Threats/Losses) would be partially or totally inundated as a result of these projects. Almost all of the 62 areas would experience indirect impacts.

Reservoir construction has two major direct impacts on natural bottomland ecosystems: 1) destruction of the bottomland ecosystem as a result of inundation, i.e., conversion of a riverine system to an open water lacustrine system; and 2) modification and instability of plant communities along the periphery of the lake as a result of fluctuating water levels. The indirect downstream effects are often more destructive, albeit not immediate. These downstream effects include: 1) reduced flooding below the dam, which modifies the bottomland hydroperiod and results in subtle vegetational changes toward more xeric species; 2) reduction of silt and associated nutrient inputs to downstream bottomlands; 3) excessive bed and bank scour resulting from irregular releases and loss of bank-stabilizing vegetation, with accompanying modifications to the aquatic flora and fauna; 4) disruption of normal feeding and spawning cycles of fish that use floodplains; 5) elimination of high flows into bottomlands, which prevents the input of bottomland nutrients into the aquatic system; 6) reduced instream flows to coastal estuaries, causing major changes in the salinity gradient; 7) encouragement of downstream conversion of bottomland forests to agricultural lands or monoculture pine plantations as a result of reduced flooding; and 8) potential negative effects to plant communities as a result of flooding of forests during the growing season.

Agriculture

Most of the clearing of floodplain habitats for agricultural land (crops and pasture) is a result of water-control projects. There are increasing demands to place more lands into agricultural production to feed the growing populations of the nation and foreign countries. Agriculture in the southern states has experienced tremendous growth since the mid-60's, about four times the rate of increase in the rest of the country (Healy 1982). Most of this increased production has been in soybeans.

Texas and Oklahoma have followed the trend of other southern states toward increased agricultural production but to a lesser extent. In 1982, 31,000 ha of the Pineywoods region of east Texas were in soybean production (almost all of this in floodplains); down slightly from 33,600 ha the year before (Texas Crop and Livestock Reporting Service 1983).

Forestry

There are also increasing demands for forest products in the South. In 1976, 45% of the nation's softwood timber came from the South; by 2030, the percentage is projected to be 51% (Healy 1982). Most of the forests of the Pacific Northwest have been cut over, and the southern climate is much more favorable for intensive silviculture.

Most of the increased demand is for pine timber, very little of which grows in natural bottomlands. After the construction of flood-control projects, downstream bottomland hardwoods are often subjected to clearing and conversion to pine plantations. This results in impacts to the ecosystem similar to those listed for agriculture. Pine plantations support decreased wildlife populations and no waterfowl.

Besides conversion of bottomlands to pine plantations, forest practices have some negative impacts on hardwood systems. Present timber management is oriented toward a short rotation period leading to a decrease in old-growth, bottomland forests. Higrading of a timber stand (i.e., cutting only the best trees) and clear cutting of large acreages are detrimental to wildlife.

Mining and Petroleum Extraction

Coal production in east Texas (almost all of it lignite) has risen from approximately 2 million metric tons in 1971 to nearly 30 million metric tons in 1981. Estimates are that production will be up to approximately 40 million metric tons, with 11,250 ha disturbed in 1985 (Crawford 1982).

Production is projected to increase to over 81 million metric tons in 1990 (Kaiser et al. 1980).

Impacts of lignite mining and production on bottomland communities involve several major effects (Cloud 1978, Espey, Huston, and Associates 1983). The most detrimental impact is an indirect one; the mines serve power plants that require cooling-water reservoirs. The impacts of these reservoirs on bottomland ecosystems are essentially the same as other reservoirs.

The direct impacts, resulting from the actual mining of the coal, are less severe but significant. Mining regulations prohibit most floodplain mining, therefore floodplain ecosystems are not usually involved. However, some mining of smaller riparian zones is possible, and some streams may be diverted to permit mining of these areas. Here the forest community will be destroyed. This will be permanent since present reclamation procedures favor revegetation with cultivated pasture grasses of little potential benefit to most floodplain fauna.

Other minerals mined in eastern Texas are sand and gravel, kaolin and other clays, and iron. Generally, the extraction of these minerals produces localized impacts that are outside the floodplain. However, sand and gravel operations can result in almost total destruction of the bottomland community in small areas.

Oil and gas exploration and production began in eastern Texas in the early 1930's and have continued to date. Production of oil and gas in Texas increased steadily to a high of 1.3 billion barrels in 1972 and has since declined. The 1980 production estimate was 931 million barrels. This decline is expected to continue, but the annual percentage of decline should decrease as unconventional oil and gas sources are tapped (Fisher 1981).

The major impacts of oil and gas exploration on bottomland ecosystems are: 1) surface disruption from the drilling and production process, 2) impact of service roads to the well site and for seismic activities, and 3) impact of brine discharges on the aquatic environment. The actual surface area disturbed by drilling is small, but where large numbers of wells are drilled in an area the cumulative impacts can be significant. Brine discharges are largely controlled but in the past have had significant impacts on the aquatic environment and have caused the irretrievable loss of some small wetlands areas.

Development

Development of recreational homesites, urbanization, and industrialization are causing significant losses of bottomland hardwoods in certain areas. These activities cause direct losses of bottomlands and have been especially severe in the southern portions of east Texas near Houston. Some of the key bottomland areas have been irretrievably affected by residential development. Losses from these developments are likely to increase.

In the southern United States, the demand for "a place in the country" has led to urban sprawl that is more extensive than in the northeast or west (Healy 1982). This has led to another related impact, land ownership fragmentation. The increasing demand for "ranchettes" and small farms has led to the subdivision of many large tracts of land formerly under a single owner. Land fragmentation makes land preservation efforts much more difficult.

Pollution

Pollution of bottomland hardwoods and other wetlands takes many forms, but the results are essentially the same -- the alteration of the aquatic biota at all trophic levels. Pollution of bottomland hardwood ecosystems comes from six sources: 1) increased sediments from erosion; 2) brine wastes from oil wells and natural sources; 3) pesticides, herbicides, and fertilizers from agricultural operations; 4) municipal wastes from sewage treatment plants and wastes from septic tanks; 5) industrial wastes from refineries and factories; and 6) mine waste runoff. The major sources of water pollution in east Texas bottomlands are municipal wastes and sediments from erosion. Quality of most stream segments is believed to be improving, at least in eastern Texas.

Other Threats

Other threats are largely a result of roads, transmission lines, and pipelines. Because of the large number of streams and rivers in the eastern parts of Texas and Oklahoma, and the importance of this area for energy production, it is inevitable that numerous pipelines and transmission lines traverse rivers. Impacts are local and minor, except where large numbers of separate crossings lead to pronounced cumulative impacts. These intrusions into the bottomlands make management of wildlife resources more difficult.

Rate of Loss

General

All of the above threats have led to losses of bottomland habitats. Bottomland hardwoods have been increasingly threatened and modified, and only a small portion of the bottomland system remains today.

Conversion of floodplain forests to other land uses places them among the most severely altered systems in the United States. Of the 26.7 million ha of the continental United States originally classified as riparian forest, only 34% (8.9 million ha) remained in 1971. From the 1950's to the 1970's, there was a net loss of 2.4 million ha of palustrine forested wetlands (Frayer et al. 1983).

Clearing of bottomlands has been especially severe in the Lower Mississippi River Delta of Arkansas, Louisiana, and Mississippi. In the early 1930's, 4.8 million ha of the Lower Mississippi River Delta were classified as forest. By 1974, 1.9 million ha had been converted to other uses. From the mid 60's to mid 70's, the region's forests declined at a rate of 105,000 ha annually (Sternitzke 1976).

Losses in East Texas and Southeast Oklahoma

Losses of bottomland hardwoods in eastern Texas and Oklahoma are not well documented. Data comparable to that of the Lower Mississippi River Delta region are not available.

The best data available on trends is provided by USFS status and trends reports. Data from 1975 (Murphy 1976), including both commercial and non-commercial acreage, show that bottomland forests covered 882,686 ha in Texas; 1981 data (Thomas 1984) for Oklahoma includes 161,956 ha in bottomlands. Commercial bottomlands in Texas decreased by 17.97% between 1935 and 1975, with a 10% decrease occurring from 1965 to 1975. In Oklahoma, there was a 9% decrease from 1936-1981. Data from the TPWD (Frye 1986) suggest that there has been a 63% loss of the original bottomland component. Based on estimates of original bottomland acreage derived from soil surveys, and on present estimates of bottomlands from National Wetland Inventory maps, only about 15% of Oklahoma's original bottomland forest remains today. It is notable that the oak-gum-cypress bottomlands (a more mature type) have declined, while the elm-ash-cottonwood type (an early succession type) has increased. Of the 728,000 ha of bottomlands in east Texas (excluding the Post Oak Region), only 18,650 ha are at least "medium stocked" with desirable trees (less than 3% of the total bottomland area). Approximately 292,000 ha are so poorly stocked as to need regeneration (Murphy 1976). Further human development and resulting habitat loss in the area is nearly certain.

PRESERVATION PROGRAM TO COUNTER THREATS/LOSSES

To counteract the continuing development and land alteration, the FWS has proposed a program to protect bottomland habitats in eastern Texas and Oklahoma (Figure 2) and has identified the minimum level of protection necessary to preserve as much as possible of the remaining high-quality bottomland habitat (FWS 1985, 1986; Brabander et al. 1985). The FWS has established a goal of 100,000 ha in Texas and Oklahoma to be protected by a variety of means (FWS 1986). This represents less than 10% of the estimated bottomlands that exist in the two states and is, in effect, a minimum goal.

Alternative methods considered in developing the proposed program include (a) leasing the project lands; (b) easements; (c) fee acquisition by FWS, state, or private conservation groups; (d) preservation by regulatory agencies or local zoning requirements; (e) increased management of existing sites; and (f) increased use of wildlife extension efforts. The recommended proposal utilizes all of the alternative solutions to protect the bottomland resources of eastern Texas and Oklahoma.

The magnitude of this program dictates that no one type of preservation tool be exclusively considered. Rather, a method approach is required, involving all possible means, varying from fee acquisition to increased wildlife extension efforts designed to show landowners the values and means of preserving waterfowl. The type of preservation recommended for specific sites will depend on several factors, e.g.,

Figure 2. Project area bottomland hardwoods of Texas and Oklahoma.

wishes of the landowner, type of preservation proposed to accomplish program objective, identified habitat management needs, and potential land uses. The proposed program will utilize both perpetual conservation easements and fee preservation. In all cases the least interest acquired in property will be perpetual development and grazing rights. Other additional rights, such as hunting or other special uses, will be negotiated on a case-by-case basis. Cost of these less-than-fee rights will vary, depending on the present highest and best use of the area and future development potential. Less-than-fee preservation is often the most cost-effective method of preserving land.

The specific sites identified in this concept plan are vital for maintaining populations of the mallard and wood duck. A total of 75 sites (62 in Texas) of widely varying quality have been identified (Figures 3 and 4). In evaluating the 75 sites, a number of criteria were utilized to establish whether or not a particular area would qualify for preservation

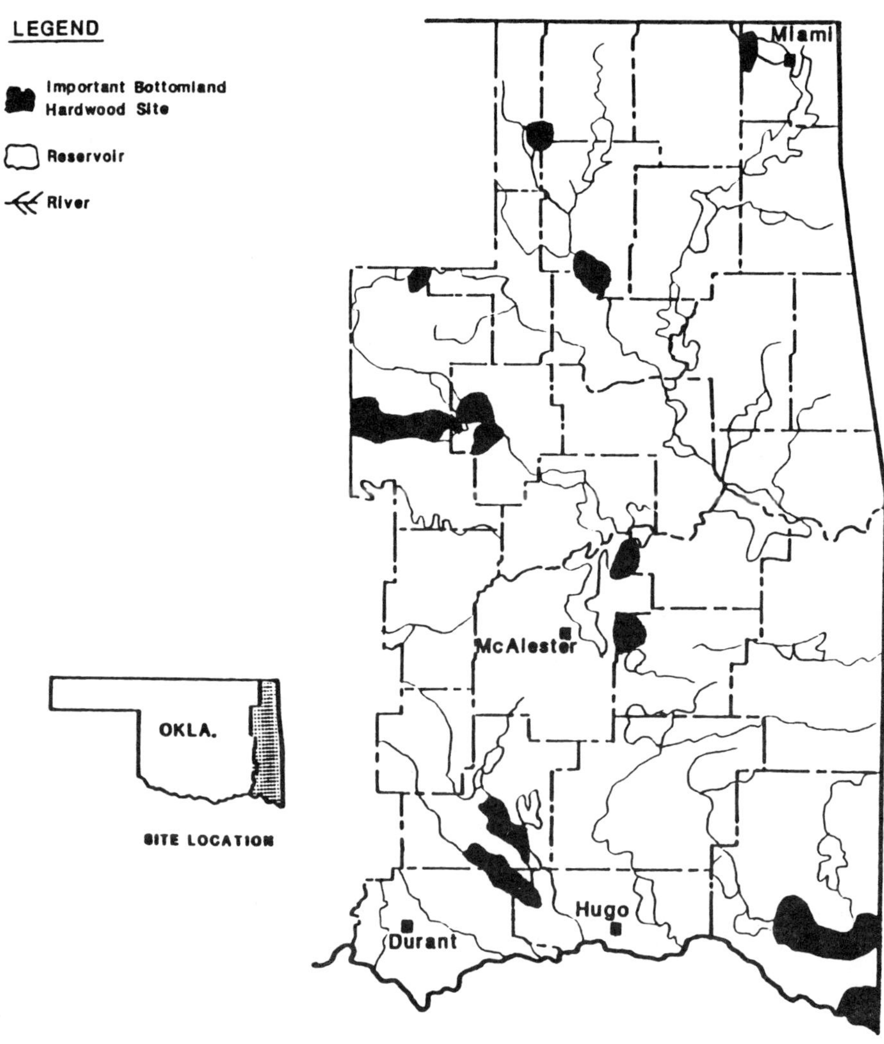

Figure 3. Areas identified as important bottomland hardwood sites, Oklahoma.

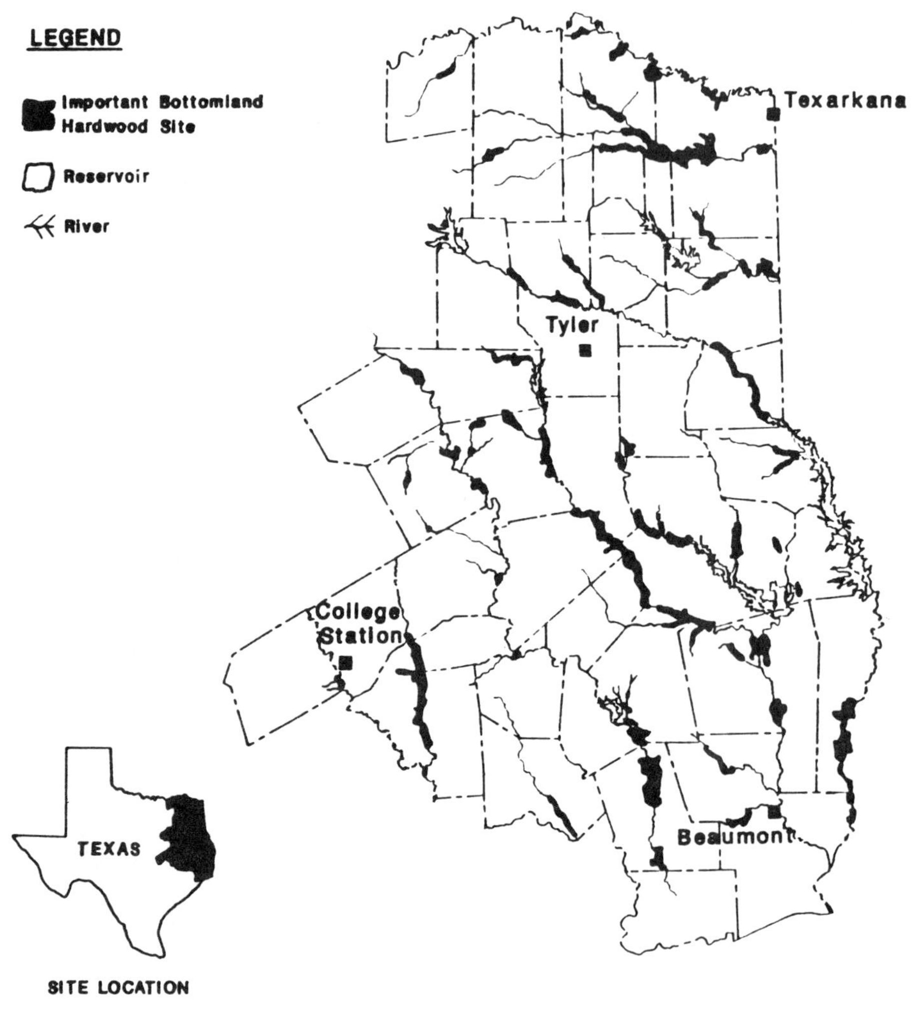

Figure 4. Areas identified as important bottomland hardwood sites, Texas.

and the priority grouping for each site. To qualify for consideration, it was determined that the sites should contain natural habitat of benefit to waterfowl (notably the mallard and wood duck) that requires little or no additional restoration.

The following criteria were utilized to evaluate each site: 1) hydrological regime (i.e., presence of beaver ponds or other permanent water sources, frequency and duration of flooding, etc.); 2) habitat diversity and quality (i.e., number of habitat types and number of woody, subcanopy, and herbaceous species); 3) waterfowl utilization and production; 4) degree and imminence of threats; and 5) presence of federal endangered or threatened species and state species of special concern. Data were also compiled on use of the sites by other wildlife species, degree of disturbance, proximity to other units benefiting waterfowl, and opportunities for management. Utilizing the above criteria, the sites were placed in six priority categories. Not all of the sites are being recommended for preservation, and other areas may be listed as additional information becomes available.

The FWS will attempt to protect sites that are well distributed throughout the study area, along all or most major river systems, and in a diversity of wetland habitat types. The agency will request funding from the Migratory Bird Conservation Commission, from monies received from the sale of Duck Stamps, and from Land and Water Conservation Funds, to preserve these high-quality, bottomland habitats. The active involvement of TPWD, ODWC, TNC, and others is considered vital to achieve the program objectives.

A variety of means will be used to encourage the public to participate in this preservation program. Educational films, hearings, and informational meetings and presentations will be utilized. Copies of the concept plans and land protection plan have been provided upon request to any interested party.

CONCLUSION

Bottomland hardwoods support a diversity of wildlife species far greater than adjacent upland systems. Over 230 plant and animal species of special concern in Texas and Oklahoma occupy these floodplain forests. In addition, floodplain wetlands are critical to certain game species, such as waterfowl (primarily mallards and wood ducks), wild turkey (*Meleagris gallopavo*), gray squirrels, and white-tailed deer (*Odocoileus virginianus*).

The bottomland hardwood system is among the most threatened wetland systems in the United States. Bottomland hardwood forests are threatened by various factors, most notably water-control structures and conversion to agricultural systems (principally for growing soybeans). Only 34% of the presettlement riparian forests were still extant in 1971. Palustrine forested wetlands experienced a net loss of 2.4 million ha from the 1950's to 1970 (Frayer et al. 1983). Losses in Texas and Oklahoma have paralleled this nationwide decline. In Texas, there was an 18% loss from 1935-1975, and a 9% loss in Oklahoma from 1936-1981. It is estimated that only about 15% of Oklahoma's presettlement bottomlands and about 37% of Texas presettlement bottomlands remain today (Brabander et al. 1985, Frye 1986).

In order to counter this trend, the FWS is proposing that 100,000 ha in eastern Texas and southeastern Oklahoma be protected. A variety of protection tools will be required to protect these bottomland forests. To be successful, this protection program will require the participation of state and federal conservation agencies and private conservation groups.

To address such issues as thresholds for loss, cumulative impacts, and indicator functions in a comprehensive manner, additional information is our first need! National workshops such as the three sponsored by the U.S. Environmental Protection Agency provide a great deal of the needed information. In addition, regional or state meetings could be very useful. In this regard, TPWD, FWS, and USFS sponsored a symposium to address the status and ecology of bottomland hardwoods in Texas (McMahan and Frye 1986). Meetings such as these could be the beginning of the formation of state or regional study groups. The highly successful land preservation group in Tennessee composed of federal, state, local, and private land management and conservation groups might serve as a model for other regional or state study groups. Meetings of state science academies might also organize special symposia dealing with bottomland hardwood systems. Additional support for specific research studies, such as those of Gosselink and Lee (1989) on cumulative input assessment, is greatly needed.

REFERENCES CITED

Blair, W. D. 1981. The Conservancy's Richard King Mellon Grant: great expectations. The Nature Conservancy News 31(2):4-5.

Brabander, J. J., R. E. Masters, and R. M. Short. 1985. Bottomland hardwoods of eastern Oklahoma. U.S. Fish and Wildlife Service, Division of Ecological Services, Tulsa, Oklahoma.

Cloud, T. J. 1978. Texas lignite: environmental planning opportunities. U.S. Fish and Wildlife Service, Fort Worth, Texas. FWS/OBS-78/26.

Crawford, R. A. 1982. Extraction of coal and uranium resources. Pages 47-54 *in* J. T. Baccus, ed. Texas wildlife resources and land use. Texas Chapter of the Wildlife Society, Austin, Texas.

Espey, Huston and Associates, Inc. 1983. Impacts of lignite development in Texas. Texas Energy and Natural Resources Advisory Council, Austin, Texas.

Fisher, W. L. 1981. Texas energy: trends and outlook to year 2000. Pages 371-375 *in* Mike Kingston, ed. Texas Almanac. A. H. Belo Corporation, Dallas, Texas.

Frayer, W. E., T. J. Monahan, D. C. Bowden, and F. A. Graybill. 1983. Status and trends of wetlands and deep water habitats in the conterminous United States, 1950's to 1970's. Department of Forest and Wood Sciences, Colorado State University, Fort Collins, Colorado.

Frye, R. G. 1986. Bottomland hardwoods - current supply, status, habitat quality and future impacts from reservoirs. Pages 24-28 *in* C. A. McMahan and R. G. Frye, editors. Texas Parks and Wildlife Department, Austin, Texas. PWD-RP-7100-133-3/87.

Gosselink, J. G., and L. C. Lee. 1989. Cumulative impact assessment in bottomland hardwood forests. Wetlands 9:83-174.

Healy, R. G. 1982. Land in the South: Is there enough to satisfy demands? Conservation Foundation Newsletter. September.

Kaiser, W. R., W. B. Ayers, Jr., and L. W. La Brie. 1980. Lignite resources in Texas. Bureau of Economic Geology, Austin, Texas.

Marks, P. L., and P. A. Harcombe. 1981. Forest vegetation of the Big Thicket, southeast Texas. Ecol. Monogr. 51:287-305.

McMahan, C. A., and R. G. Frye, eds. 1986. Bottomland hardwoods in Texas. Texas Parks and Wildlife Department, Austin, Texas. PWD-RP-7100-133-3/87.

Murphy, P. A. 1976. East Texas forests: status and trends. U.S. Department of Agriculture, Southern Forest Experiment Station. New Orleans, Louisiana.

Neal, J. A., and J. W. Haskins. 1986. Bottomland hardwoods: ecology, management, and preservation. *In* D. L. Kulhavy and R. N. Conner, eds. Wilderness and natural areas in the eastern United States: a management challenge. Center for Applied Studies, School of Forestry, Stephen F. Austin State University, Nacogdoches, Texas.

Sternitzke, H. S. 1976. Impacts of changing land use on Delta hardwood forests. Journal of Forestry 74:25-27.

Texas Crop and Livestock Reporting Service. 1983. 1982 Texas county statistics. U.S. Department of Agriculture and Texas Department of Agriculture, Austin, Texas.

Texas Department of Water Resources. 1984. Water for Texas: a comprehensive plan for the future. Vols. 1 and 2. Austin, Texas.

Thomas, C. E. 1984. Oklahoma midcycle survey shows changes in forest resources. U.S. Department of Agriculture, Southern Forest Experiment Station, New Orleans, Louisiana.

U.S. Department of the Interior and Environment Canada. 1986. North American waterfowl management plan. Washington, D.C.

U. S. Fish and Wildlife Service. 1978. Bottomland hardwood preservation program: habitat category 7, Lower Mississippi River Delta. Atlanta, Georgia.

______1984 and 1985 (supplement). Summary of waterfowl habitat recommendations. Washington, D.C.

______ 1985. Texas bottomland hardwood preservation program: category 3. Division of Realty, Albuquerque, New Mexico.

______ 1986. Land protection plan: bottomland hardwood protection program: category 3, Texas and Oklahoma. Division of Realty, Albuquerque, New Mexico.

SECTION III. REPORTS OF WORKGROUPS IN THE EPA-SPONSORED BOTTOMLAND HARDWOOD FOREST WORKSHOPS

PROLOGUE TO SECTION III

In this section the deliberations of the participants in workgroups in all three workshops are summarized in chapters on bottomland hardwood forest hydrology, soils, water quality, vegetation, fisheries, wildlife, culture/recreation/economics, and ecosystem processes and cumulative impacts. A summary chapter concludes the book. The chapters vary widely in their emphases, depending on the composition of the workgroups. All contain a wealth of information reflecting the broad expertise and experience of the participants. The interplay of academic scientists with agency personnel brought theory into a close association with practice. Perhaps this is the major strength of this collection of chapters.

11. BOTTOMLAND HARDWOOD FOREST ECOSYSTEM HYDROLOGY AND THE INFLUENCE OF HUMAN ACTIVITIES: THE REPORT OF THE HYDROLOGY WORKGROUP

James G. Gosselink
Center for Wetland Resources, Louisiana State University,
Baton Rouge, LA 70803

B. Arville Touchet

U.S. Department of Agriculture, Soil Conservation,
Alexandria, LA 71301

Johannes Van Beek
Coastal Environments, Inc.,
Baton Rouge, LA 70802

David Hamilton
National Ecology Research Center,
U.S. Fish and Wildlife Service,
Fort Collins, CO 80525

with Panel
Thomas Cavinder, William Conner, John Elder, Beverly Ethridge,
David Hamilton, Kelly Hendricks, Terry Huffman, Phillip Jones,
John Kittleson, Edwin Miller, Richard Novitski,
Sam Patterson, Susan Ray

ABSTRACT

Hydrology is the primary physical driving force that determines structure and ecological processes in wetlands. Key hydrologic elements of bottomland hardwood forest ecosystems are storage volume; frequency, duration, depth and timing of surface floods; flow velocity; soil saturation; and infiltration rate. These elements control hydrologic services that benefit society: flood stage reduction, flow velocity reduction, base flow augmentation, and ground water recharge. Human

Ecological Processes and Cumulative Impacts: Illustrated by Bottomland Hardwood Wetland Ecosystems. Edited by James G. Gosselink, Lyndon C. Lee, and Thomas A. Muir. Chelsea, MI 48118. Printed in USA.

activities in bottomlands may change key hydrologic elements, thus modifying human services. Common human activities and their influence on hydrology and on hydrologic services are summarized, both at the local site level and at the drainage basin level. Finally, a four step assessment model is presented, that produces a qualitative estimation of the direction and intensity of the expected change in hydrologic services resulting from site isolation, channel modification, clearing or draining.

INTRODUCTION: THE VALUE OF BOTTOMLAND HARDWOOD FOREST-STREAM ECOSYSTEMS AS A PROVIDER OF HYDROLOGIC SERVICES

Hydrology is the primary physical driving force that determines structure and ecological processes in wetlands. It is also true, however, that streamside wetlands modify the hydrologic characteristics of adjacent streams. Some of these characteristics (e.g., flood storage capacity) result in services to society (e.g., reduction in the costs of downstream flood control and water treatment) and are therefore valuable processes that should be conserved for economic as well as ecological reasons. Human activities that modify hydrology directly affect the delivery and quality of such services.

Figure 1 summarizes diagrammatically ways in which man's activities affect the ecology of bottomland hardwood ecosystems through hydrologic alteration. Channel flows from upstream and overland flows from adjacent uplands are primary water sources to the bottomland hardwood floodplain-stream ecosystem. When these flows or the ecosystem itself are modified by human activities, the services delivered also change, not only the directly influenced hydrologic ones but also other services that result indirectly from ecosystem processes (shown by the lower box in Figure 1).

Bottomland hardwood forest-stream ecosystems are connected through their hydrology to much larger regional landscape systems. This property, termed "openness," means that human activities at one site may result in ecological changes elsewhere. They may modify the bottomland hardwood floodplain directly (**a** in Figure 1); the stream itself, (**b** in Figure 1); or both indirectly through changes in upland land use (**c** in

Figure 1). An understanding of these spatial relationships is important to the following discussion of processes in bottomland hardwood ecosystems.

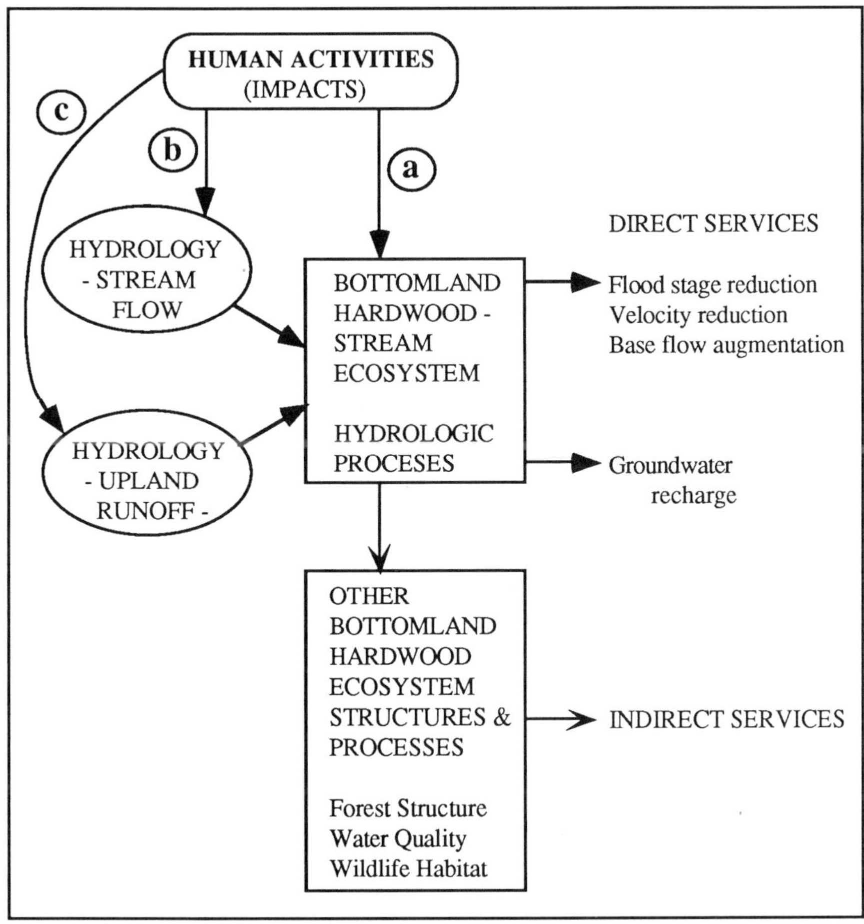

Figure 1. Conceptual diagram of interrelations of hydrology in bottomland hardwood forest/stream ecosystems to human use (amenities and services), and to human activities. See text for explanation.

In this chapter we discuss streamside bottomland hardwood ecosystem hydrology, the specific hydrologic services that floodplain ecosystems perform for society, and how human activities affect both the bottomland hardwood ecosystem and its hydrologic services. Finally, we present a model for impact assessment. We do not discuss further the relationship of hydrology to the water quality and biotic ecosystem services which are discussed in Chapters 12 through 18.

HYDROLOGY OF BOTTOMLAND HARDWOOD FOREST-STREAM ECOSYSTEMS

The basic hydrologic components and properties of streamside forested floodplains are illustrated in Figure 2. The key elements are: storage volume; frequency, duration, depth and timing of surface floods; flow velocity; degree and duration of soil saturation; and infiltration rate. Each of these parameters assumes more or less importance at some time in the cycle in which floodwater flows out of stream banks across the floodplain and returns to the stream as the flood recedes.

A process closely related to hydrology is sediment flux, deposition, and erosion. As rising water overtops stream banks it spreads out and slows down, depositing entrained sediments in a broad blanket across the floodplain with coarse sediments dropping first, close to the stream, and finer sediments settling in slower backwater areas (Klimas 1988). As water moves back into the stream after a flood, it usually erodes in distinct channels. Thus, deposition tends to be in a broad sheet, while erosion is highly localized. Usually bottomland hardwood systems are net sinks for sediments, with deposition exceeding erosion (Mitsch 1977, Novitski 1979).

For both upland runoff and overbank flooding bottomland hardwood ecosystems act as a transformer, reducing flow velocity and peak stage heights during high water. Under some circumstances, during low flow, the floodplain forest releases water slowly thereby augmenting minimum flows in the adjacent stream. The degree of transformation depends on structural and ecological characteristics that are highly site specific.

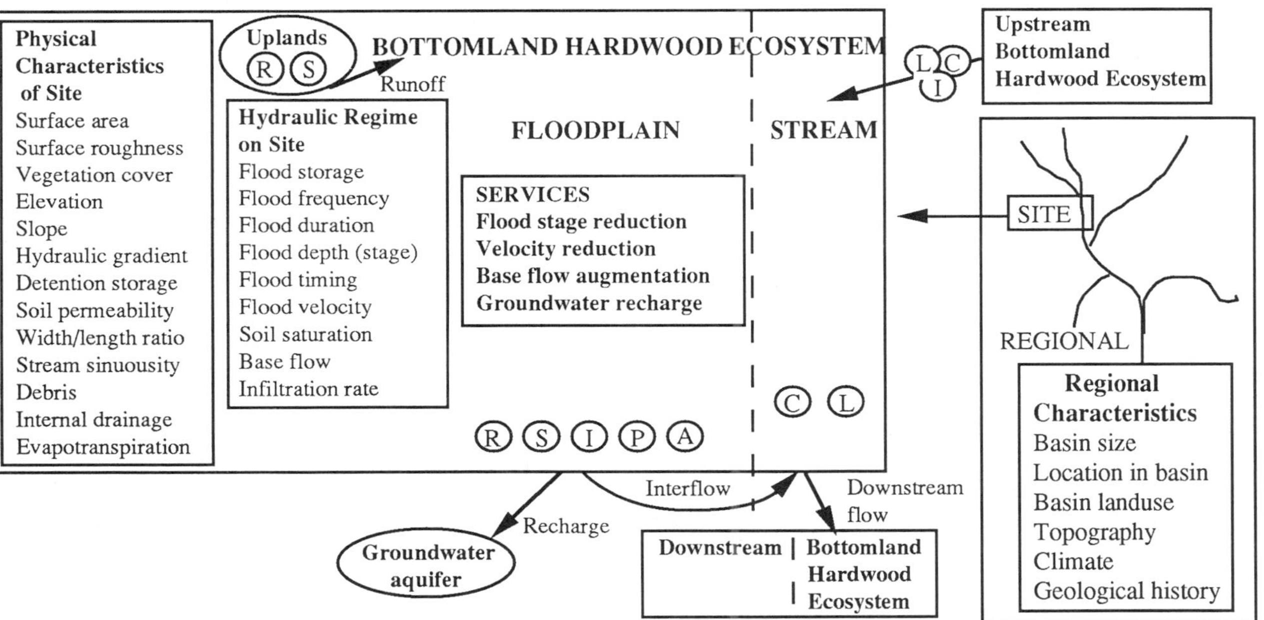

Figure 2. **Conceptual model of a bottomland hardwood forest ecosystem, illustration some of the complexity. Physical characteristics determine the hydrologic regime on site, which in turn affects the physical and biotic processes of the ecosystem. Letters in circles refer to human activities: R = conversion of site to rice, S = conversion of site to spybeans, I = construction of impoundments, C = channelization, L = levee construction, P = conversion of site to pine plantation, A = conversion of site to aquaculture.**

Variability

The influence of site-specific variables can be organized into four categories. These include, first, the broad geometric or topographic limits that determine water storage capacity; second, the microtopographic relief that determines hydrologic roughness and thereby influences flow velocity and flood duration; third, the type of vegetation cover, which also affects flood duration through influences on both evaporation and transpiration; and fourth, the influence of floodplains on water routing by their soil properties, particularly permeability, which mediates exchanges with groundwater aquifers.

Site-specific variability influences the magnitude of forested floodplain effects on stream hydrology, but the net results in most cases, relative to a channelized system, are lowered flood crests, slower and prolonged flows, delayed peaks downstream, larger base flows, and a reduction in the frequency and duration of flood stage exceedance. Most of the effect of the bottomland hardwood floodplain is felt in the adjacent stream or downstream, but minor changes in upstream flows also occur when stream gradients are modified by floodplain storage.

Superimposed on the site-specific relationship between the floodplain forest and the adjacent stream are watershed level influences. The most important of these are size and position.

The volume of upland flow across the site is determined by the width of the watershed (i.e., the area that drains across the site), the topography, and the vegetation cover, which determines the proportion of precipitation that infiltrates the soil. Position in the watershed determines the frequency, stage, volume, and duration of flows from upstream. Generally, low-order streams drain smaller watersheds, are flashier (i.e., respond more quickly to local precipitation), and have steeper slopes than higher order streams. Low-order streams can be said to determine the characteristics of their floodplains, while floodplains along high-order streams have more influence on their adjacent streams (Harris and Gosselink, Chapter 9).

Finally, at a regional scale, climate and geologic factors must be considered. The recent weather history determines the saturation of the soil and magnitude of runoff. Geomorphic characteristics of the region determine the types of soils and substrates; their permeability; their erosion potential; the existence and depth of aquifers; and the sediment load reaching the stream.

Unifying Concepts

It has been difficult to find unifying concepts around which to generalize and compare bottomland hardwood wetland systems that exhibit the range of variability discussed above. One theme that has gained widespread support is that of a zonal classification based on gradient. As a floodplain slopes from the stream upward to the adjacent uplands, the slope gives rise to a gradient of flooding depth, duration, and frequency, which is roughly correlated with changes in the dominant tree species. The gradient can be divided into more or less discrete zones.

The zonal concept of bottomland hardwood ecosystems was first described in some detail in a 1980 bottomland hardwood forest workshop (Clark and Benforado 1981). Shown in Table 1 are the average percentages of the growing season during which saturated soil conditions occur in the various zones from permanently-flooded (Zone 1) aquatic habitat through the bottomland hardwood floodplain (Zones 2-5), to upland (Zone 6), where soils are seldom saturated. The relative flooding percentages have been found to be accurate and meaningful in comparisons of bottomland hardwood ecosystems within regions of similar geomorphology and hydrology, as, for example, the Mississippi River alluvial floodplain (Leitman et al. 1983, Leitman 1984). This has not been the case, however, when comparisons have been attempted between regions of very different geomorphology, as with the Mississippi River floodplain and the South Atlantic Coastal Plain, for example. A second drawback to widespread application of the zonal concept is the requirement for detailed topographic, soils, and evapotranspiration data that are not available at most sites.

The Flood Tolerance Index (FTI) (Theriot 1988) has been proposed as an alternative indicator of the hydrologic regime which requires less direct information on flooding characteristics. This index uses the known flood tolerance of dominant tree species to construct bottomland hardwood zones. It should be noted, however, that the FTI is influenced by processes other than flooding and therefore provides only a rough inference of the hydrologic regime. In addition, the FTI does not provide

Table 1. Concluded.

Soil moisture regime	Forest types[1]
V. Temporarily inundated or saturated - soils are inundated or saturated by surface water or groundwater periodically for short periods during growing season but not totaling more than 1 month for the entire growing season of the prevalent vegetation. Such conditions typically occur with a frequency ranging from 11 to 50 years per 100 years. The total duration for such seasonal events typically ranges from 2 to 12.5 percent of the growing season	V. Loblolly Pine (81 - wet site variants) Loblolly Pine - Hardwood (82 - wet site variants) Mesquite (68 - wet sitevariants) Southern Redcedar (73 - wet site variants) Swamp Chestnut Oak - Cherrybark Oak (91 - wet site variants)
	Terrestrial or upland forest types
VI. Intermittently inundated or saturated - soils are rarely inundated or saturated by surface water or groundwater during the growing season of the prevalent vegetation, except during exceptionally high floods or extreme wet periods. Such conditions typically occur with a frequency ranging from 1 to 10 years per 100 years. The total duration of such seasonal events is typically less than 2 percent of . the growing season	VI. Cherrybark Oak (91 - dry site variants) Live Oak (89) Loblolly Pine (81 - dry site variants) Loblolly Pine - Hardwood (82 - dry site Loblolly Pine - Shortleaf Pine (80 - dry Longleaf Pine (70 - dry site variants) Longleaf Pine - Scrub Oak (71) Longleaf Pine - Slash Pine (83 - dry site variants) Mesquite (68 - dry site variants) Sand Pine (69) Shortleaf Pine (75) Shortleaf Pine - Oak (76) Slash Pine (84 - dry site variants) Southern Redcedar (73 - dry site variants) Southern Scrub Oak (72) Swamp Chestnut Oak - Cherrybark Oak (91 - dry site variants) Sweetgum - Yellow Poplar (87 - dry site variants) Virginia Pine (79) Virginia Pine - Oak (78)

1 T. T. Huffman and S. W. Forsythe 1981.

2 Forest number, society of American Foresters (1980). Placement of each forest type is based on maximun tolerance to soil moisture. Forest types may therefore be found associated with less frequently inundated sites; however, these associations are not shown. No attempt is made to order forest types within each soil moisture regime due to the high within-group variability.

3 Perennial plant species having an estimated areal coverage per hectare greater than or equal to 30 percent.

Table 1. Concluded.

Soil moisture regime	Forest types[1]
V. Temporarily inundated or saturated - soils are inundated or saturated by surface water or groundwater periodically for short periods during growing season but not totaling more than 1 month for the entire growing season of the prevalent vegetation. Such conditions typically occur with a frequency ranging from 11 to 50 years per 100 years. The total duration for such seasonal events typically ranges from 2 to 12.5 percent of the growing season	V. Loblolly Pine (81 - wet site variants) Loblolly Pine - Hardwood (82 - wet site variants) Mesquite (68 - wet sitevariants) Southern Redcedar (73 - wet site variants) Swamp Chestnut Oak - Cherrybark Oak (91 - wet site variants)
VI. Intermittently inundated or saturated - soils are rarely inundated or saturated by surface water or groundwater during the growing season of the prevalent vegetation, except during exceptionally high floods or extreme wet periods. Such conditions typically occur with a frequency ranging from 1 to 10 years per 100 years. The total duration of such seasonal events is typically less than 2 percent of . the growing season	Terrestrial or upland forest types VI. Cherrybark Oak (91 - dry site variants) Live Oak (89) Loblolly Pine (81 - dry site variants) Loblolly Pine - Hardwood (82 - dry site Loblolly Pine - Shortleaf Pine (80 - dry Longleaf Pine (70 - dry site variants) Longleaf Pine - Scrub Oak (71) Longleaf Pine - Slash Pine (83 - dry site variants) Mesquite (68 - dry site variants) Sand Pine (69) Shortleaf Pine (75) Shortleaf Pine - Oak (76) Slash Pine (84 - dry site variants) Southern Redcedar (73 - dry site variants) Southern Scrub Oak (72) Swamp Chestnut Oak - Cherrybark Oak (91 - dry site variants) Sweetgum - Yellow Poplar (87 - dry site variants) Virginia Pine (79) Virginia Pine - Oak (78)

1 T. T. Huffman and S. W. Forsythe 1981.

2 Forest number, society of American Foresters (1980). Placement of each forest type is based on maximun tolerance to soil moisture. Forest types may therefore be found associated with less frequently inundated sites; however, these associations are not shown. No attempt is made to order forest types within each soil moisture regime due to the high within-group variability.

3 Perennial plant species having an estimated areal coverage per hectare greater than or equal to 30 percent.

specific information about the relative contributions of frequent flooding events, backwater flooding, groundwater discharge, poor drainage, precipitation, evapotranspiration, and other factors that characterize the moisture regime.

HYDROLOGIC SERVICES PROVIDED BY BOTTOMLAND HARDWOOD FOREST-STREAM ECOSYSTEMS

We discuss below four direct hydrologic services of bottomland hardwood ecosystems, which we have chosen to describe as flood stage reduction, flow velocity reduction, base flow augmentation, and groundwater recharge. Surface flood stage reduction is the major recognized hydrologic service of bottomland hardwood systems, but the remaining two surface flow services identified are closely linked. The three services are often referred to in the published literature as "stream flow modification," but we believe it is important to consider them separately. Bottomland hardwood floodplains certainly modify flows, but the utility to humans is generally interpreted by flood reduction (especially downstream), by stream flow velocity reduction, and by base flow augmentation. In addition some bottomland hardwood wetlands appear to act as groundwater recharge sites.

By maintaining bottomland hardwood floodplains the necessity for expensive flood-control structures to protect downstream urban areas can be materially reduced (U.S. Army Corps of Engineers [ACE] 1971). Conversely, levees and channel improvements that isolate a floodplain from its stream and improve stream efficiency result in higher flood stages downstream and more severe flood damage (Belt 1975). Forested floodplain characteristics that contribute to flood reduction are surface area, elevation, and slope, all of which determine storage capacity or volume (Table 2). Floodplains with large water storage capacity experience reduced peak stage both onsite and downstream. Soil saturation, detention storage (permanent or semi-permanent bodies of water on the floodplain), and vegetation cover all reduce storage capacity and decrease flood reduction capacity. Ground surface roughness and vegetation cover decrease the velocity of flow over the floodplain. This delays the return flow to the stream and reduces downstream flood peaks by spreading discharge out over a longer time period.

Table 2. Site characteristics related to services performed by bottomland hardwood ecosystems.

Service	Characteristic	Relation to service
Flood Reduction	Floodplain surface area	+
	Site elevation	-
	Site slopes	-
	Soil saturation	-
	Detention storage	-
	Vegetation cover	+/-
	Surface roughness	+
Velocity Reduction	Floodplain surface area	+
	Width/length ratio	+
	Stream sinuosity	+
	Slope	-
	Vegetation cover	+
	Surface roughness	+
	Debris	+
	Internal drainage	-
Base flow augmentation		
Interflow/groundwater discharge	Area flooded	+
	Hydraulic gradient	+
	Infiltration rate (permeability)	+
	Contact time	+
	Evapotranspiration	-
Surface discharge delay	Storage capacity	+
	Velocity reduction	+
Ground water recharge	Hydraulic gradient	+
	Soil permeability (Infiltration rate)	+
	Detention storage	+
	Evapotranspiration	-

Velocity reduction is closely related to flood reduction. In general, any characteristic of the floodplain that reduces velocity also reduces the flood peak downstream. Since the destructive energy of floods increases rapidly with velocity, bottomland hardwood ecosystem characteristics that reduce velocity generally also reduce erosion and other deleterious effects (Table 2). Stream velocity is reduced by increasing the cross-sectional area across which the water flows. This occurs when water moves out of a stream channel over the floodplain. Its velocity then slows, with the degree of the reduction directly related to the width and surface area of the floodplain. Low gradient floodplains support slower, more sinuous flows than those with higher gradients. Sinuosity decreases channel slope by increasing channel length. Vegetation cover, surface debris, and ground surface roughness all increase the friction of the surface and retard flow. Conversely, an efficient internal drainage network enhances flow by reducing sinuosity and drag.

Base flow refers to the minimum flows maintained in a stream during non-flood periods. It is determined by the magnitude and duration of interflow, which is the shallow subsurface flow from floodplain to stream, and by other factors that delay surface discharge from the flood period into the low flow period. Both interflow and delayed surface discharge increase as a function of the storage capacity of the floodplain. For interflow, a higher storage capacity means that more area will be available through which floodwater can percolate downward into the subsoil. Interflow is also enhanced by factors that reduce velocity, for example, a flat surface water gradient. Generally, characteristics that enhance transfer of surface water to interflow (steep subsurface hydraulic gradients, high permeability and infiltration rates, long contact time) enhance base flow by slowing the releases of water from the floodplain to the stream. However, the same conditions that retard runoff also generally increase evapotranspiration, which under some circumstances can result in a net reduction of base flow (Novitzki 1989).

The details of groundwater recharge and discharge through bottomland hardwood wetlands remain poorly documented (see Taylor et al. Chapter 2). A conceptual model of subsurface flows is proposed in Figure 3 and requires some explanation. We distinguish interflow as generally local and shallow, as compared to regional, deeper groundwater flows. In this chapter we treat sub-surface discharge as an interflow phenomenon and discuss groundwater only in relation to recharge of aquifers from overlying bottomland hardwood forests. There are two reasons for this. First, in this chapter we discuss primarily the hydrology of local bottomland hardwood sites. At this scale most infiltrating water discharges into adjacent streams as interflow. While groundwater discharge may also occur at the site, it does not originate locally, and hence is seldom controlled by local site conditions. Second, although groundwater fluxes may be important because of their timing or location, on a regional basis they appear to account for only a small percentage (trace to 16%) of total flows (Gosselink and Gosselink 1988).

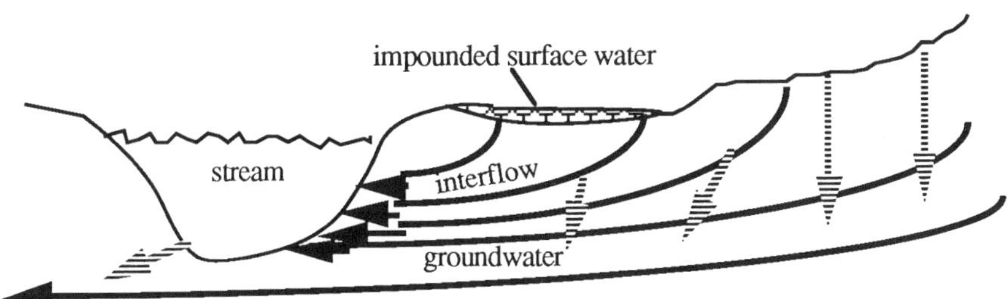

Figure 3. Vertical section across a floodplain and stream, showing local shallow subsurface interflow and regional deeper groundwater flow.

A further clarification is needed with respect to hydraulic gradients. All else being equal, the flux of water increases with the steepness of the hydraulic gradient. For example, water runs off a steep slope faster than a flat one, and groundwater recharge increases with the steepness of the subsurface hydraulic gradient. At the same time, both groundwater recharge and interflow increase when the surface water gradient is flat (for example, in impoundments) because this condition increases contact time for infiltration to occur. This leads to the somewhat surprising conclusion that groundwater recharge is enhanced by both a flat surface water gradient and a steep subsurface hydraulic gradient.

EFFECTS OF HUMAN ACTIVITIES ON BOTTOMLAND HARDWOOD ECOSYSTEM HYDROLOGIC SERVICES

The major types of human activities that affect the delivery of bottomland hardwood ecosystem services are listed in the introduction of this book (Chapter 1), and summarized here in Table 3. There are significant regional differences in the relative importance of these activities. Soybean

Table 3. Dominant human activities in bottomland hardwood forest floodplains, grouped by their generic effect on bottomland hardwood ecosystems.

	Activity	Symbol	Generic Effect on Bottomland Hardwood Ecosystems
Agriculture/forestry	Soybean cultivation	S	Draining (S, P); Clearing (S, P, R, A)
	Pine plantation	P	
	Rice cultivation	R	Impoundment (R, A, I-U, I-O, I-D)
	Aquaculture	A	
Flood control navigation	Impoundment		
	upstream	I-U	
	onsite	I-O	
	downstream	I-D	
	Channelization	C	Channel works
	Levee construction	L	Isolation/constriction

culture, stream channelization, and construction of flood control levees assume the greatest importance in the Mississippi River alluvial floodplain. In contrast, impounding of bottomlands for rice culture has historically been important in South Carolina and Georgia. Channelization is a more recent development along the South Atlantic Coast. These human activities take five basic forms: 1) draining, usually to enhance dry land agriculture and foresting; 2) clearing for crop production, including agricultural, forestry and aquaculture crops; 3) impounding for flood control, reservoirs, and production of flooded crops (rice, catfish); 4) stream channelization to improve stream efficiency for flood control; and 5) floodplain isolation with levees. We focus in this section on the effects of these activities on the ability of forested wetlands to deliver the four primary hydrologic services defined earlier, namely, flood stage reduction, flow velocity reduction, base flow augmentation, and groundwater recharge. We sequentially analyze each in terms of the characteristics listed in Table 2.

Flood Reduction

Floodwater storage and velocity reduction combine to lower flood stages. We consider first human impacts that change flood storage capacity (Table 4a).

The most important characteristic of a bottomland hardwood site affecting flood storage capacity is the surface area of the active floodplain. For a given site elevation and slope, a large floodplain area temporarily stores more floodwater than a small one. Activities that involve construction of levees reduce flood storage by excluding flood flows from the area behind or within the levees. Large mainline levees with crest elevation greater than 2 m above grade remove a bottomland hardwood site from the active floodplain during all but the most severe flood events. Moderate (1 to 2 m) levees associated with aquaculture reduce storage for average floods but are overtopped during severe floods. The small levees associated with rice farming (less than 1 m) appear to have a negligible impact because they are overtopped by almost all flood events. Construction of a flood-control impoundment, in one sense, reduces flood storage of a bottomland hardwood site by partially filling it with the conservation pool. However, the structure allows floodwaters to be held onsite as long as needed, thus increasing the flood storage capability of the area behind the structure.

Flood storage capacity is reduced by the volume of above-ground biomass or vegetation cover. Activities that clear bottomland hardwood forests and replace them with open or agricultural areas (e.g., conversion to rice, soybeans, or aquaculture; clearing associated with channelization) therefore result in a slight increase in flood storage. Clearing associated with conversion to a pine plantation initially increases flood storage capacity somewhat; however, by year 15 in a pulpwood rotation and year 30 in a sawwood rotation, the biomass will return to levels approximating those of bottomland hardwood forests. Clearing associated with levee construction has a negligible effect on flood storage for two reasons. First, the area cleared is fairly small, and second, the levee removes the site from the active floodplain except during extreme flood events.

Microtopographic irregularities or ground surface roughness in a bottomland forest contribute marginally to floodwater storage. Rice farming requires relatively precise leveling of sites for optimum water

Table 4. Impacts of various human activities on the indicated service, as mediated through bottomland hardwood forest characteristics that contribute to the performance of that service.

a. Service: Flood Reduction

Characteristic	R	S	I-U	I-O	I-D	C	L	P	A
					Activity[1]				
Surface area of active floodplain	0	0	0	-	0	0	-	0	-
Vegetation cover	+	+	-	NA	0	+	+	+0	+
Ground surface roughness	-	0	0	NA	0	0	0	0	0
Elevation of site	-	0	0	-	0	0	-	0	-
Slope	0	0	0	NA	0	0	0	0	0
Detention storage	-	-	-	NA	0	0	-	0	0
Soil saturation	-	+	0	NA	+-	+	+	+	-
Overall impact on flood storage capacity	-	+	0	-	0	+	-	+0	-

b. Service: Velocity Reduction

Characteristic	R	S	I-U	I-O	I-D	C	L	P	A
					Activity[1]				
Surface area of active floodplain	0	0	0	NA	0	0	-	0	-
Vegetation cover	-	-	0	NA	0	-	0	-+0	-
Ground surface roughness	-	-	0	NA	0	0	0	+	0
Width/length ratio	0	0	0	NA	0	0	0	0	0
Sinuosity	0	0	0	NA	0	-	0	0	0
Debris	-	-	0	NA	0	0	0	-	-
Slope	+	0	0	NA	0	0	0	0	0
Internal drainage	+	-	0	NA	0	-	0	-	+
Overall impact on velocity reduction	-	+	0	n/a	0	-	-	-+0	-

Table 4. (Continued).

c. Service: Groundwater Recharge[1]

Characteristic	Activity[1] R	S	I-U	I-O	I-D	C	L	P	A
Subsurface hydraulic gradient	+	-	-	+	-	+	-	-	+
Soil permeability (infiltration rate)	-	-	0	0	0	0	0	-	0
Detention storage	+	-	+	+	+	0	+	-	+
Evapotranspiration	?	?	?	?	?	?	?	?	?
Overall impact on ground water discharge modification	-	+	0	n/a	0	-	-	-+0	-

d. Service: base flow augmentation

Characteristic	Activity[1] R	S	I-U	I-O	I-D	C	L	P	A
Enhances Interflow									
Subsurface hydraulic gradient	+	-	-	+	-	+	-	-	+
Soil permeability (infiltration rate)	-	-	0	0	0	0	0	-	0
Area flooded	+	-	-	+	0	-	-	-	+
Surface Discharge Delay									
Storage capacity	0	-	NA	+/-	NA	0	-	0	0
Velocity reduction	+	-	+	+	0	-	-	-	+
Overall impact on base flow augmentation	+	-	+	+	NA	-	-	-	+

[1] A plus indicates that the function is enhances due to the impact of the activity on the characteristic. A minus indicates that the function is impaired. A zero indicated no impact. A combination of signs indicates differing impacts either spatially or temporally. NA means not applicable. R = conversion of site to rice; S = conversion of site to soybeans; I-U = impacts upstream from an impoundment; I-o = impacts from an impoundment onsite; I-D = impacts downstream from an impoundment; C = channelization of stream adjacent to site; L = levee construction onsite; P = conversion of site to pine plantation; A = conversion of site to aquaculture.

control and therefore usually reduces surface roughness. All other activities involve minimal or no leveling and thus have no measurable effect on ground surface roughness and associated flood storage.

Site elevation, that is the height of the ground surface relative to some reference such as mean sea level or average river stage, is affected by activities that impound water on a site. Impoundment that results in pooling creates what amounts to an increase in elevation across the impoundment. Lower elevations within the impoundment are filled with water and flood storage capacity is reduced. Rice farming levees generally impound less than 0.3 m of water from March to August each year. In contrast, aquaculture levees typically impound one or more meters of water for most of the year, and flood-control structures can impound a variable depth conservation pool through the entire year. Construction of a mainline levee that excludes floodwater from the floodplain has a minimal impact on actual site elevation; however, it increases the effective elevation at which floodwaters enter a site and thus has a negative impact on flood storage.

The only activity that affects the slope of a site is conversion to rice. Slope refers to the grade or "tilt" of a bottomland hardwood site along a path perpendicular to the stream channel. However, the associated leveling has no net impact on flood storage capacity because the grading process lowers the highs by depositing sediment at lower elevations.

Depressions on a site (e.g., oxbows) that temporarily hold water following the recession of a flood give rise to detention storage. The stored water eventually flows out through backwater swamps, evaporates, or percolates into the soil. Agricultural conversions to rice or soybeans are not generally considered for sites that require filling of major depressions. Therefore, most agricultural conversions do not significantly affect detention storage. Channelization can greatly increase detention storage if connections to the old channel are blocked; however, these connections are typically left open so that floodwaters filling them drain out as the flood recedes. Construction of mainline levees increases detention storage of precipitation and overland flow from surrounding uplands, and increases detention storage during extreme flooding events. However, the levees prevent average over-bank floodwaters from entering the site in most years and thus are assumed to have an overall negative impact on detention storage.

The volume of water that can be stored in soil pores is determined by the degree of soil saturation, that is the available (air-filled) porosity at the time of flooding. Thus, dry, porous soils provide more storage than saturated or dense soils. Activities that impound surface water (e.g., rice farming, aquaculture) result in saturated soil profiles and thus the loss of this storage capacity. Activities that involve draining a bottomland hardwood site (e.g., conversion to soybeans, channelization, conversion to pine plantation) or that reduce the extent or duration of flows across the site (e.g., levee construction, flood-control impoundment upstream) result in drier soils and more storage capacity.

In summary, construction of large mainline levees causes a significant reduction in flood storage capacity because it removes a bottomland hardwood site from the active floodplain for all but a few extreme flood events. The reduction associated with a conversion to aquaculture is less severe, primarily because the impounding levees are lower. Conversion to rice farming results in a small reduction in flood storage capacity due to field leveling and the need to maintain saturated soils through most of the growing season. Onsite impoundments reduce the flood storage capacity of a site by filling it with a conservation pool of water; however, the structure allows water to be impounded deeper and held longer, thus increasing overall flood control. Conversion to soybeans, channelization, and conversion to pine plantations all result in a slight increase in flood storage capacity due to associated draining and initial vegetation removal. Subsequent tree regrowth in pine plantations offsets the initial increase in flood storage by years 7 to 15 in a pulpwood rotation, and beyond year 30 in a sawlog rotation.

Velocity Reduction

Another aspect of flood reduction involves the extent to which flows are slowed as they flow through a bottomland hardwood site. The most important site characteristics reducing stream velocity include the surface area of the active floodplain, vegetation cover, ground surface roughness, the shape of the site, stream sinuosity, and the amount of debris on the site. Other important characteristics include the slope of the site and the extent of internal drainage channels. The impacts of various activities on these characteristics and the overall impact on the velocity reduction function are presented in Table 4b. Effects of an onsite impoundment were considered not applicable because flow velocities would be controlled by the impounding structure, not by characteristics of the site. It should be pointed out that any activity which increases flood storage capacity (Table 4a) also reduces net velocity by detaining water for a period of time and releasing it later.

When floodwaters overtop stream banks they spread out over the adjacent floodplain and their velocity is reduced as a function of the surface area of active floodplain. The greater the extent of spreading, the greater the velocity reduction. The only activities that affect the size of the active floodplain are those that involve moderate to large levees. The 1 to 2 m levees associated with aquaculture prevent average floods from spreading into the floodplain and thus periodically reduce the active surface area. However, large mainline levees remove a bottomland hardwood site from the active floodplain for all but the most extreme floods and thereby increase velocity in most years.

The extent to which floodwaters are impeded as they flow through a bottomland hardwood site is directly related to the density of vegetation present. Vegetation density is a function of the type of vegetation (e.g., a solid cover of shrubs reduces velocity more than a closed canopy of trees with little undergrowth), the stem density and sizes, and the vertical structure of the plant community. Thus, a pulpwood pine plantation retards flow more than a cypress-tupelo (*Taxodium distichum-Nyssa aquatica*) swamp which retards flow more than a field of soybeans. All activities that involve landclearing result in a decline in the capability of a site to reduce flood flow velocities. The ability of a pine plantation to retard flows gradually increases during the first 7 to 15 years of dense pine growth (1400-2200 stems/ha) following clearing. If the site is managed for pulpwood, clearing at about year 15 would again reduce vegetation cover. If the site is managed for sawlogs, periodic thinnings over the longer rotation period would gradually reduce the ability of the site to retard flood flows to levels approximating those provided by natural bottomland hardwood ecosystems.

Ground surface roughness in inversely related to flow velocity. A bottomland hardwood site with surface irregularities causes additional turbulence in water flowing across the site and thus slows it to some extent. Conversely, very smooth sites allow water to pass through relatively unimpeded. The leveling associated with conversion to rice farming, and to a lesser extent to soybeans, therefore reduces the ability of the site to retard flood flows. However, this reduction is probably minimal when compared to the impacts from associated vegetation clearing and removal of debris. The only other activity that affects surface roughness is conversion to pine plantations. Site preparation typically entails development of a series of ridges and furrows. The ridges provide a slightly higher, and therefore drier, location to plant seedlings, while the furrows provide drainage. Although the ridges and furrows tend to degrade through time, they provide additional surface roughness and thereby reduce flow velocity slightly.

The shape of a bottomland hardwood site, more specifically the width to length ratio, can influence the extent to which floodwaters are retarded. Sites that are wide (along a path perpendicular to the stream channel) provide a greater opportunity for floodwaters to spread out than sites of the same size that run in a narrow band along the stream channel. This characteristic assumes significance when comparing the relative velocity reduction potential of two bottomland hardwood sites; however, it is not affected by any of the activities considered.

Stream sinuosity refers to the extent of meandering of a stream channel. Highly sinuous channels tend to slow down stream flows, while straight channels allow water to flow relatively unimpeded. Channelization and associated debris removal therefore have a major impact on the ability of a bottomland hardwood site to slow stream flow. None of the other activities affect stream sinuosity.

A bottomland hardwood site typically contains an accumulation of fallen trees, broken limbs, and racks of leaf and stem parts. This debris increases the overall roughness of the site and helps to impede flood flows. Extreme accumulations of debris can even block stream channels or impound water in a portion of a bottomland hardwood site. All activities that involve landclearing and debris removal therefore reduce the ability of the site to retard flood flows.

The slope of a site (along a path perpendicular to the stream channel) interacts with surface area and the shape of the site to determine the extent to which floodwaters can spread out. A gentle slope allows more spreading and thus a greater reduction in velocity. The precise leveling associated with a conversion to rice farming can, therefore, have a small positive impact on velocity reduction.

The network of natural slough and small collector channels interspersed within a bottomland hardwood site provides some degree of internal drainage. Sites with extensive networks allow water to drain quickly and therefore provide less retardation of floodwaters than sites with minimal internal drainage. Conversion to rice or aquaculture involves construction of low to moderate size levees to impound water. Floodwaters that overtop the levees are impounded temporarily and their velocities reduced. The improved site drainage associated with channelization or conversion to soybeans or pine plantations reduces the ability of a site to retard floodwaters.

In summary, activities that isolate a bottomland hardwood site from the active floodplain (mainline levees, aquaculture) eliminate the potential of the site to retard flood flows in most years. Activities that improve site drainage (soybean farming, pine plantations, channelization) or remove vegetation (agriculture, silviculture, aquaculture, or channelization) reduce the potential of the site to retard flood flows. This effect is reversed in time in silvicultural operations as the trees regrow.

Groundwater Recharge

Under normal (undisturbed) conditions, on a regional basis and over many years, the groundwater system is in rough equilibrium, and recharge to aquifers equals their discharge. Recharge typically occurs by infiltration through permeable upland soils. Aquifers discharge to lakes, streams, and wetlands. More water is stored in the groundwater system if either recharge increases or discharge decreases. Conversely, less water is stored if either recharge decreases or discharge increases. Groundwater discharge, discussed under "Base Flow Augmentation" below, maintains moist conditions on the floodplain during low water stages and enhances base flow. In this section we consider activities that affect groundwater recharge through their influence on site characteristics (Table 4c).

The subsurface hydraulic gradient is a primary site characteristic influencing the rate at which surface water-groundwater exchanges occur. Under conditions of constant permeability, infiltration is proportional to the steepness of the gradient from surface to water table. Activities that increase the wetness of the site (i.e., that impound water or decrease evapotranspiration) increase the subsurface hydraulic gradient and the possibilities for groundwater recharge. Conversely, activities that reduce the wetness of the site (i.e., that drain the site or increase evapotranspiration) decrease the subsurface gradient and decrease recharge.

Rice farming, aquaculture, and onsite flood-control structures impound surface water on a site, thereby increasing groundwater recharge. Mainline levees isolate bottomland hardwood ecosystems from the active floodplain, preventing river flow from spreading out over the site. This decreases the expectation of recharge through the bottomland hardwood zone, but the increased river stage may increase recharge through the river bed. Conversion to soybeans or pine plantation typically involves ditching and draining which, by enhancing surface runoff, decrease infiltration. Flood-control impoundments and channelization result in less floodplain flooding downstream and consequently decrease recharge at downstream sites.

A fraction of the water falling or flowing on the soil surface percolates into the soil and gradually moves into subsurface layers. The volume of water infiltrating per unit area per unit time is the infiltration rate. This rate depends primarily on soil permeability and to a lesser degree on antecedent soil water conditions, and the uniformity of the soil profile. Permeable sand or gravel soils allow rapid infiltration, whereas clays or tightly compacted organic soils impede infiltration. Alluvial deposits typical of the "back swamp" and other low areas of the bottomland hardwood floodplain away from the natural stream levees are typically fine-grained and relatively impermeable. Permeability is reduced if the surface layers are compacted as by the weight of a levee. The infiltration rate varies with time; it is high in the early stages of infiltration, then asymptotically decreases toward a constant rate that depends on the soil characteristics mentioned above. Activities that add fine-grained sediments (i.e., that improve sediment trapping), enhance the accumulation of organic soils (i.e., increase the duration of soil saturation), or compact the soil surface result in reduced permeability. Conversion to agriculture or pine plantation involves site preparations that compact the soil to varying degrees. The extent of compaction is highly dependent on the soil type, the operator, and the season in which site preparation is conducted.

Any conditions or activities that maintain bottomland hardwood forest surface flooding (e.g., depressions in a bottomland hardwood site that retain water after floods recede) act to increase detention storage and allow more time for water to recharge into the underlying soils. Conversely, any activity that reduces detention storage decreases groundwater recharge. Site preparations associated with a conversion to soybeans or pine plantation reduce detention storage. The low levees constructed for rice farming serve to increase detention storage and increase groundwater recharge.

Evaporation from open water or the soil system and transpiration by plants are important mechanisms that remove near-surface groundwater. Any activity that increases evapotranspiration will therefore decrease infiltration. However, results of recent studies are somewhat contradictory concerning the net effect of vegetation clearing or conversion to agriculture. Bottomland hardwood forests may have high transpiration losses but the micro-climate they produce minimizes evaporation. Clearing these forests or converting them to agricultural fields reduces transpiration but increases evaporation. The net effect of various human activities could not be assessed at the workshop.

In summary, the subsurface hydraulic gradient and soil permeability are the dominant characteristics determining groundwater recharge. Activities that decrease permeability or decrease the subsurface hydraulic gradient reduce recharge. Thus, onsite impoundments, levees, channelization, and conversion to rice or aquaculture increase recharge, while conversion to soybeans or pine plantation and construction of upstream impoundments tend to decrease recharge.

Base Flow Augmentation

A consequence of flood peak reduction is the augmentation of base flow. Even though water is stored on the floodplain and its flow velocity is slowed, the total flux remains unchanged aside from possible changes in evapotranspiration and groundwater recharge. This balance is maintained by enhanced flow during non-flood periods. Base flow is augmented by slowed surface release and by discharge of interflow (i.e., the water percolating from the floodplain through the substrate). As a consequence, physical characteristics of bottomland hardwood sites that enhance base flow are those that increase interflow (steep hydraulic gradient, permeable soils, and large flooded areas), and those that delay surface release (storage capacity and velocity reduction) (see Table 4d).

As indicated in Tables 4c and 4d, the surface hydrologic gradient between the floodplain and the stream is increased by impounding water (rice culture, impoundments, and aquaculture) and by channelizing which lowers the stream surface. Activities like soybean production and clearing for pine plantations accelerate runoff and therefore decrease the hydraulic gradient. During floods levees exclude water from the floodplain and therefore reduce interflow.

Permeability is not affected much by human activities (Table 4c). However, compaction due to site preparation for agriculture and pine production probably reduces permeability.

The area flooded in the bottomland hardwood forest is reduced most by major levees that prevent floodplain inundation under nearly all conditions. To a lesser extent soybean and pine production accelerate runoff, reducing flooding duration if not area. Upstream impoundments generally are operated to reduce peak flows, and therefore also reduce area flooded. Although rice and aquaculture impoundments in the floodplain would be overtopped during major floods, and thus would have no impact on the peak area flooded, they may retain water longer than the natural flood regime and therefore increase the area flooded during low flow periods.

Storage capacity, which delays surface runoff, is determined by the area flooded and the depth of flooding. The capacity is not changed by most activities, although the area actually flooded may be impacted. Impoundments that isolate the floodplain from the river reduce storage capacity if dry but may prolong flooding if inundated, thus enhancing base flow.

Soybean production and pine plantations generally require enhanced drainage, decreasing the velocity reduction service provided by bottomland hardwood forests. Impoundments that release water after peak floods reduce net stream velocity. Levees increase stream velocity by confining the water to the stream bed, increasing its stage. Channelization increases stream efficiency and velocity. To the extent that rice and aquaculture impoundments store water, they reduce flow velocity and augment base flow.

In summary, impoundments and rice production were judged to augment base flow overall because they store floodwaters for later release. Aquaculture also impounds water but does not release it. However, it may increase interflow somewhat. Farming and forestry practices that require good drainage (soybeans, pine plantations) reduce base flow, as do levees and channelization which are designed to increase stream efficiency.

Basin Level Synthesis

In the previous sections we described the influence of human activities on the physical and ecological characteristics of bottomland sites as they influence bottomland hardwood ecosystem services. By focusing on activities that modify the physical characteristics of the bottomland hardwood system, we run the risk of neglecting the overall effects on the surface hydrologic regime of the whole basin. In this section, we discuss flood moderation and desynchronization as determined by changes due to human activities in the timing, frequency, duration, and magnitude of flooding. We also consider the corollary effects on sedimentation. These are basin-level processes.

Flood moderation is the process through which peak flows that enter the floodplain are reduced and delayed in their downslope journey. This moderation results from the interaction of floodwater storage, velocity reduction, and groundwater recharge processes in a number of sub-basins. Flood desynchronization is the process through which simultaneous storage of peak flows in several basins within a watershed and their gradual release in a nonsimultaneous manner result in lower but more persistent downstream flows (Adamus and Stockwell 1983). Impacts of human activities on flood moderation and desynchronization are highly dependent on the configuration of sub-basins within a watershed. While activities at each individual site within a sub-basin may result in impacts (as detailed earlier), here we refer to a cumulative phenomenon that depends on the type and intensity of all activities in the watershed.

The impacts of various activities on the hydrologic regime of the basin are shown in Table 5. These impacts result from the combined modification, at individual sites, of flood storage, velocity reduction, and groundwater discharge. The magnitude of the cumulative impact depends on the intensity of activities on the floodplain and the position in the watershed at which the hydrologic regime is assessed. In general, we assume that impacts occur throughout the length of the watershed, in which case the magnitude of downstream impacts increases in higher order streams as the impacts accumulate.

Table 5. Impacts of various activities on the bottomland hardwood forest downstream hydrologic regime resulting from the combined modification of flood storage, velocity reduction, and ground water discharge.

	Activity[1]						
Hydrologic parameter	R	S	I	C	L	P	A
Seasonality of flooding	n	n	y	n	n	n	n
Flood frequency	+	+	-	+	+	+0	+
Flood duration	?	?	-	?	?	0	?
Flood magnitude (depth)	+	+	-	+	+	+0	+
Soil saturation	?	?	-	?	?	?	?
Sediment trapping	+	+	-	+	+	+-	+
Sediment loading	?	?	+	?	?	?	?

1 A plus indicates that the function is enhanced due to the impact of the activity on the characteristic, a minus indicates impairment, a zero no impact. An n means no change, a y a change in seasonality. A combination of signs indicates differing impacts either spatially or temporally. NA means not applicable. R = conversion of site to rice; S = conversion of site to soybeans; I-U = impacts upstream from an impoundment; I-o = impacts from an impoundment onsite; I-D = impacts downstream from an impoundment; C = channelization of stream adjacent to site; L = levee construction onsite; P = conversion of site to pine plantation; A = conversion of site to aquaculture.

Seasonality of floods is primarily a function of the regional precipitation regime. Floods occur principally during the late winter and early spring in most of the southeast, when precipitation is greatest and evapotranspiration is low. Floods are variable in magnitude and timing on low-order streams since these streams respond to precipitation on small local watersheds, and local precipitation is variable. High-order streams, on the other hand, reflect the coincidence of floods in a number of sub-basins upstream, and thus tend to reflect the regional seasonal pattern of precipitation. For this reason, floods on high-order streams are usually distinctly seasonal, less frequent than on low order streams, and of longer duration.

The only human activity that seriously modifies the natural seasonality of floods is construction of flood control impoundments, which are designed to store water for later release. In practice, the effect of these impoundments is not so much to change flood timing downstream as to "smooth out" the flow by reducing flood peaks and prolonging moderate flows. In this sense they mimic natural floodplains, although at great cost to the taxpayer. Much of the sediment load carried in streams settles out behind flood-control structures. Therefore, water released from the structure has a lower sediment load and a greater capacity to entrain sediments. It therefore tends to scour the stream below the structure. The cumulative effect is a reduction of the sediment load (and hence of sediments available for deposition on the floodplain) of high order streams at the base of the watershed (Koewn et al. 1980, Meade and Parker 1985).

All other activities result in an increase in flood magnitude, since they either decrease storage capacity upstream, or increase the velocity of runoff. As a result floodwaters "stack up" downstream at higher stages than when the upstream floodplain is unmodified. How duration and frequency are affected is not clear. The streamlined flow tends to make streams flashier so that the number of individual flood events might increase (as indicated in Table 5). On the other hand, flood duration may be shorter on severely impacted streams since peak flux is greater, unless the volume of water carried by the stream is significantly increased by less infiltration on the floodplain. The potential for sediment deposition on the floodplain should be increased because more surface area is flooded. It is not clear that sediment loading of the stream by floodplain erosion would be affected.

Several factors must be considered when evaluating the significance of floodplain ecosystems for modifying hydrologic regimes (Adamus and

Stockwell 1983). First, bottomland hardwood sites in headwater areas have less "opportunity" to store floodwaters than similar sites lower in the drainage basin. Second, the configuration of the stream network and floodplain determines the effect of bottomland hardwood forests on flood desynchronization. For example, a number of watershed comparisons between drained and undrained wetlands have strongly suggested a role for wetlands in desynchronizing flood peak flows (Moore and Larson 1979, Novitzki 1979, Brun et al. 1981). However, results are not uniform. For example, one study indicated that flooding was more severe and flashy in a lowland portion of a watershed with extensive forested wetlands (i.e., than in a watershed with less wetland acreage) (Young and Klawitter 1968). Third, downstream hydrologic changes resulting from individual project site modifications are likely to be small unless the project area is large relative to the source area of floodwaters. However, while small projects may not have much effect individually, the cumulative impact of many such projects in a drainage basin can be a significant alteration of the downstream hydrologic regime.

Finally, the economic savings due to maintenance of the flood moderating capacity of natural floodplains depend not only on the size and location of bottomland hardwood ecosystems in the basin but also on the size and frequency of flooding and the proximity and type of dwellings or developments along the river. For example, a reduction of no more than 1% of a hydrograph peak might be significant in many areas of Vermont (E. Swanson, Vermont Department of Water Resources, personal communication, 1982, cited by Adamus and Stockwell 1983). On the other hand, Clark and Clark (1979) indicate that few individual wetland sites are capable of desynchronizing the severe flood flows (e.g., 50- or 100-year probability) that cause most property damage.

A MODEL FOR HYDROLOGIC IMPACT ASSESSMENT

The paucity of available information on most bottomland hardwood sites makes it impossible to develop a quick method for quantitative evaluation of the functional relationships between site characteristics and various measures of the site's hydrologic services (peak flow reduction, etc.). Assessments must be qualitative unless mathematical modeling procedures are applied on a site-specific basis. Furthermore, it became clear during workshop discussions that the extent to which individual site activities modify the hydrologic regime is usually statistically insignificant. Usually, it is only the rather drastic changes in a site (e.g., isolating a site by constructing levees) that result in any measurable modification of stream flows. The cumulative impact of many site modifications can, however, be both measurable and dramatic (see Harris and Gosselink, Chapter 9). It is imperative, therefore, to assess direction and intensity of individual actions in the context of cumulative effects.

The assessment procedure described below produces a qualitative estimation of the direction and intensity of an impact rather than a quantitative determination. The assessment process is designed to be followed in a step-wise approach. The first step or two might be adequate for screening relatively common types of permit requests or in circumstances where time and available staff are limited. For permits that are unusual, controversial, potentially affect endangered species, or involve very large tracts of bottomland hardwood forest, the entire four-step process might be warranted.

The first step is designed to determine if upstream or downstream hydrology would be affected by alteration of a bottomland hardwood site's characteristics (**extent of influence**). From this, it is possible to assess whether the site has basinwide importance from a hydrologic standpoint. Categories of development activities are then identified and the relative impact (high, medium, low) of the activity is determined (**relative impact**). Next, a qualitative description of the direction of potential impacts is given (**direction of impacts**). Finally, if a more detailed analysis is required, procedures for quantifying impacts are described (**quantify impacts**). The model was not developed beyond a conceptual phase. It has the structure of a simple decision tree. The four steps are described below in more detail.

Step I. Extent of Influence

The first step in the hydrologic assessment process is to determine whether the influence of the bottomland hardwood site under consideration extends hydrologically upstream or downstream from the site (Figure 4). Sites that are large relative to the source area of flooding, and sites that are located at hydrologic control points in the basin, are considered to exert a significant influence beyond the immediate vicinity of the site. The workgroup suggested a site area threshold of 10% or more of the upstream basin area, recognizing that there are no data to support a particular threshold (Adamus and Stockwell [1983] used a threshold of 20%). There are well-accepted simulation models (e.g. Chu and Bower 1977) and existing data sets for these models that could be used to evaluate appropriate threshold levels.

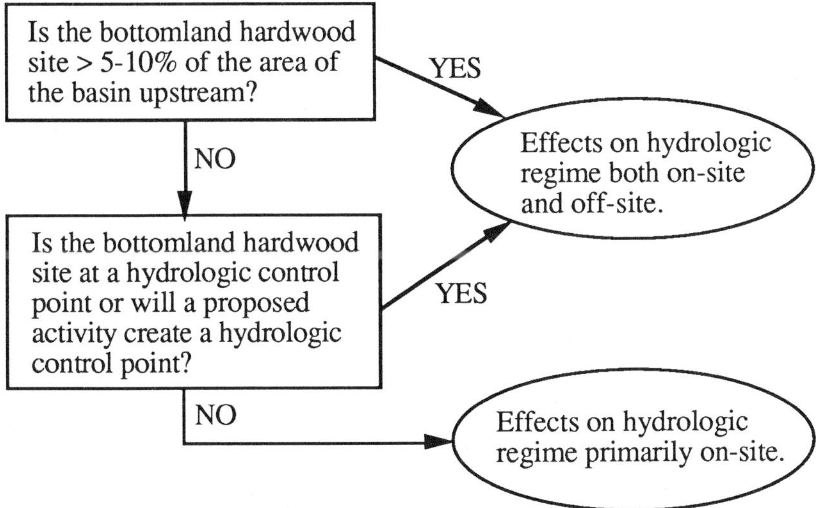

Figure 4. Assessment model for determining the relative ability of a site to influence onsite and offsite hydrologic regime.

If a site is less than 10% of the basin upstream, it may still exert significant influence on basin hydrology if it is at a hydrologic control point where flow is constricted or confined to a narrow channel, or where the proposed activity would create a hydrologic control point. Even if the site is relatively small and not at a hydrologic control point, it is still important to evaluate local hydrologic impacts of planned activities. For example, a low levee constructed around a small bottomland hardwood tract may not significantly affect downstream hydrology but may alter the hydrologic regime of the site to the point where it is no longer available as fish habitat.

Step II. Relative Impact

The second step in the hydrologic assessment process is to determine the generic site modification (see Table 3 for the generic effects) associated with a proposed activity. The relative impact on flood mediation is expressed qualitatively as high, medium, or low. This yields five keys, one for each generic effect (Figure 5). The appropriate keys are used depending on the specific activity being considered. For example, to assess the impact of conversion to soybeans, the portions of the key dealing with clearing and site drainage would be used. The overall level of impact is the highest value obtained from evaluating all relevant characteristics. As an example, assume that a relatively small bottomland hardwood site (less than 10% of the basin upstream) that is not at a hydrologic control point is to be converted to soybeans (clearing the trees and installing some drainage improvements without outlets). The overall relative impact of the conversion is initially high (relative impact due to site roughness modification is high; relative impact due to site drainage is low). However, evaluation of the previous step indicates that impacts are likely to be localized (onsite).

A brief discussion of the keys is appropriate. Site isolation and floodplain constriction result when an area is diked or filled so that floodwaters no longer enter the site. The relative impact depends on the height of the levee as well as the level of flooding relevant to the analysis. For example, an analysis of the effects on downstream flooding of urban areas might focus on the relatively rare but extreme events, while an analysis of the effect on fish habitat might focus on relatively frequent floods. Therefore, a table is incorporated in the Isolation/constriction key, based on the definitions of flooding from the National Soils Handbook (U.S. Soil Conservation Service [SCS] 1972). As an example of use of the key, suppose a levee is constructed so that it prevents frequent flooding. Following construction, over-bank flooding and, thus, storage of floodwaters due to small floods is prevented. Flood flows are confined to the channel and shunted directly downstream with a high level of impact

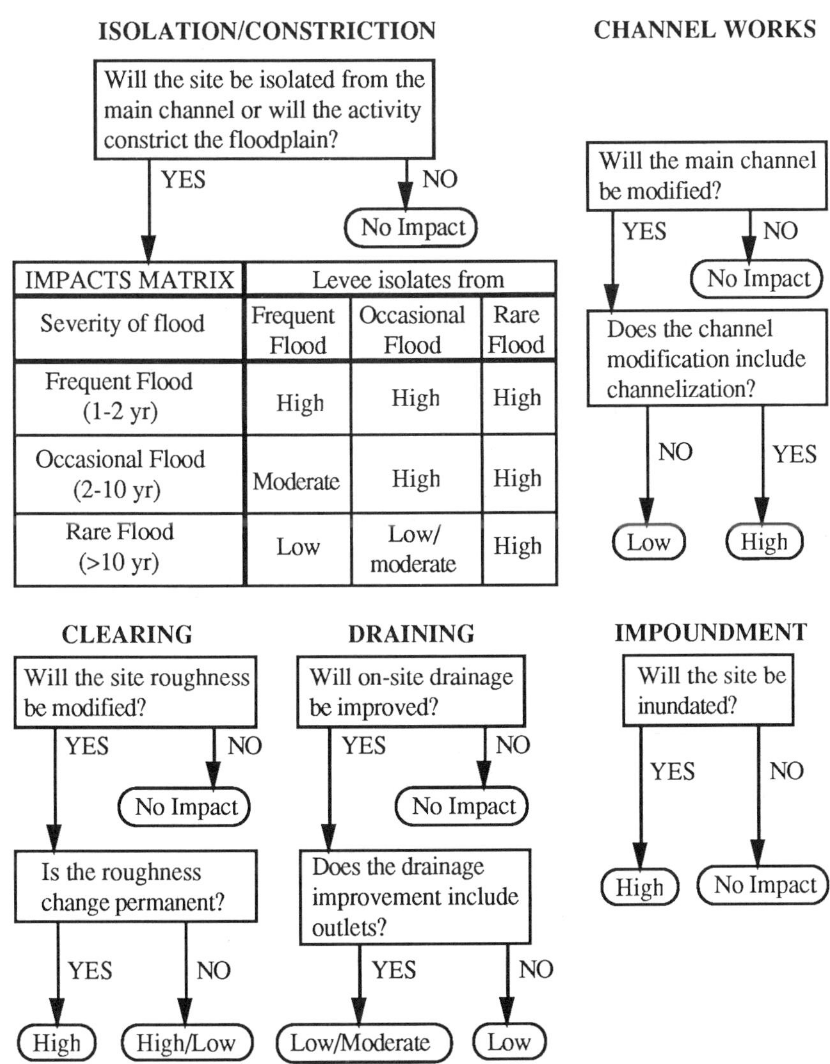

Figure 5. Hydrologic assessment model for determining relative hydrologic impact of generic human activities in bottomland hardwood forests.

on the site and perhaps on downstream hydrology (as determined from step I). However, during floods greater than a 2-year event, waters overtop the levee and the site still provides flood storage much as it did prior to diking. The impact of such levees is moderate.

Channel modification includes a range of activities from clearing snags and debris to channelization. The latter has the highest relative impact; other modifications are considered to have a relatively low impact.

Site roughness is changed by clearing, logging practices, and land leveling. Removal of vegetation may have a major impact on evapotranspiration. Therefore, the overall impact due to a reduction in roughness must account for both flow and water balance changes. If the reduction in roughness is permanent (e.g., soybean production), the relative impact is high. If, on the other hand, the site is allowed to revegetate (e.g., conversion to pine plantation) the high impact is only temporary.

Site drainage involves providing new drainage systems or increasing the efficiency of existing internal natural drains. Rapid drainage decreases the duration of floodwater storage. If the drainage system does not open to the primary channel, the draining water is stored in depressions on site and has little hydrologic impact downstream.

Site impoundment takes two basic forms. Construction of a dam across the primary stream is classified as a modification at a hydrologic control point and results in major hydrologic impacts. The onsite hydrologic regime is changed to permanently flooded, and downstream hydrology depends on the operation of the control structure which is usually designed to reduce flood magnitude. Less severe impacts occur when a site is shallowly flooded for crop or aquatic food production.

Step III. Direction of Impacts

The next step in the hydrologic assessment process is to determine the direction of impacts resulting from the activities considered in the previous step. This is done by reference to Table 6. The combination of Steps II and III yields a direction and relative magnitude of impacts associated with specific human activities in the bottomland hardwood floodplain.

Table 6. Direction of impacts of generic human activities on flood mediation in bottomland hardwood forests.

Hydrologic component	On-site	Impact Offsite Downstream	Upstream
a. Impact:	**Site isolation/constriction**		
Peak flow	Decrease to design capacity. No effect above design capacity.	Decrease (isolation constricts floodplain).	Increase peak stage.
Base flow	Decease up to design capacity.	Decease slightly.	No change.
Flooding freqency	Decrease.	Decrease.	Increase.
Flooding duration	Decrease below design capacity.	Increase.	Increase.
b. Service:	**Channel Modification**		
Peak flow	Decrease	Increase	Decrease
Base flow	Decrease	Decrease	Decrease
Flooding frequency	Decrease	Increase	Decrease
Flooding duration	Decrease	Increase	Decrease
c. Service:	**Clearing (reduce surface roughness)**		
Peak flow	Decrease peak stage slightly.	Increase- extent depends on size and area	Decrease slightly immediately upstream.
Base flow	Decrease.	Decrease.	No change.
Flooding frequency	Slight change depending on size, position in watershed, and hydrograph.	Increase.	No change.
Flooding duration	Increase if vegetation cleared. Decrease if land leveling improves drainage.	Increase.	No change.
d. Service:	**Draining**		
Peak flow	Decrease moderately if outlets built.	Increase - extent depends on size of site and outlets.	Decease very slightly if at all.
Base flow	Increase.	Increase.	No change.
Flooding frequency	Decrease.	Increase.	Decrease slightly if at all.
Flooding duration	Decrease.	Increase.	Decrease.

Step IV. Quantifying Impacts

Normally, the analysis of impacts of human activities to a bottomland hardwood site will not proceed past the first three steps. However, under unusual circumstances (e.g., sites or activities of unusual importance), quantification of impacts may be required. Quantification of hydrologic characteristics should provide data to feed into existing hydrologic models. Although most bottomland hardwood areas are classified as palustrine under current classification schemes (Cowardin et al. 1979), their hydrology is usually controlled by rivers. For this reason, models and techniques for describing the hydrologic characteristics of rivers tend to be more appropriate for bottomland hardwood sites than reservoir models. A major focus of these river models is an accurate description of control points and of the hydraulic characteristics of the channel at these points. Control points are characterized by their cross-sectional areas, channel roughness, slope, length, and other hydraulic parameters, and these should be measured for quantification of impacts. Other measures of importance to river modeling include upstream watershed area, stream length, and floodplain/channel storage.

Data collection requires field staff with some training in basic hydrologic principles and in the purposes of hydrologic modeling. This training need not be extensive and could be provided sufficiently in the form of written materials to accompany an evaluation methodology. Specifically, field staff should have an understanding of watershed processes that affect runoff and movement of water through channels, reservoirs, and across floodplains; and familiarity with common hydrologic methodology. Finally, field staff should know of the most common hydrologic models and have some feel for their capabilities (Table 7).

Simple procedures for estimating hydrologic characteristics are readily available for small watersheds. Included in this category are the SCS method for generating runoff hydrographs (SCS 1972) and the R method for estimating peak flows (Hewlett et al. 1977). These methods can be coupled with standard routing methods (e.g., modified Puhls routing; Chow 1964: 25-38) to create a fairly simple methodology for estimating hydrologic characteristics.

Simple procedures are not as easily devised for large watersheds. Standard hydrologic calculations can, however, still provide a means of screening sites for the purpose of estimating hydrologic importance. Three procedures for use by field personnel for quantifying impacts have been developed: 1) a procedure for quantifying effects of loss of floodplain area at a control point, 2) a procedure for quantifying effects of clearing vegetation at a control point, and 3) a procedure for quantifying effects of loss of floodplain storage when the activity is not at a control point. A step-by-step description of these procedures is given in Roelle et al. 1987, Appendix A. Calculations based on these measures involve several assumptions that restrict their application to a relatively narrow range of situations.

Table 7. Common hydrologic computer programs (a detailed summary of these and other water resources computer programs is given by Chu and Bowers 1977).

Name and primary agency	Description
HEC-1 Corps of Engineers	Computation of hydrographs, optimization of parameters, rainfall and snowfall computation, stream flow routing, and related operations.
HEC-2 Corps of Engineers	Computation and plotting of water surface profiles for river channels of a variety of cross-sections; considers effects of structures such as bridges, weirs, culverts.
TR-20 Soil Conservation Service	Computes surface runoff, develops runoff hydrographs, routes hydrographs.
WSP2 Soil Conservation Service	Computes water surface profiles in open channels, estimates head loss in restricted sections.

For better quantitative understanding of the hydrologic characteristics of a bottomland hardwood site, it is advisable to use analytical models, several of which have already been developed (Table 7; see also Chu and Bowers 1977). If these existing river models were compiled into a central database, a permit evaluator could use them to analyze the effects of a proposed action on that river system. These models require detailed hydrologic, topographic, and climatic information. Appendix A contains a summary of agencies that can provide information required for a hydrologic assessment.

The preliminary hydrologic assessment model described above requires considerable additional work in order to be implemented successfully. This involves:

1) establishing best estimates for the various thresholds in the assessment model;
2) periodically reviewing these thresholds and the need to include additional site or basin characteristics;
3) providing information required to apply the assessment procedure;
4) collecting existing hydrologic models for major river systems and developing a data base so that field personnel can use them to assess hydrologic impacts;
5) providing an institutional memory for a dynamically growing knowledge base and a mechanism for transfer of expertise; and
6) summarizing the results of applying the procedure in a form relevant to policy and operational decisions.

ACKNOWLEDGEMENTS

We thank the hydrology panel participants for the basic material summarized in this report. James Roelle, Greg Auble, Dave Hamilton, R. Johnson, and C. Segelquist summarized the three workgroup reports. We have edited these reports into a single chapter and take responsibility for any misrepresentations or omissions. Paul Kemp's edit of the chapter was extremely helpful.

REFERENCES CITED

Adamus, P. R., and L. T. Stockwell. 1983. A method for wetland functional assessment: volume I. Critical review and evaluation concepts. Office of Research, Development and Technology, Federal Highway Administration, U. S. Department of Transportation, Washington, DC. FHWA-IP-82-23. 176 pp.

Belt, C. B., Jr. 1975. The 1973 flood and man's construction of the Mississippi River. Sci. 189: 681-684.

Brun, L. J., J. L. Richardson, J. W. Enz, and J. K. Larsen. 1981. Stream flow changes in the southern Red River Valley of North Dakota. North Dakota Farm Research 38(5): 11-14.

Chow, V. T., ed. 1964. Handbook of applied hydrology. McGraw Hill, New York. 1418 pp.

Chu, C. S., and C. E. Bower. 1977. Computer programs in water resources. University of Minnesota, Minneapolis. WRRC Bull. 97. 263 pp.

Clark, J. R., and J. E. Clark. 1979. Scientists' report: the national symposium on wetlands, Lake Buena Vista, Florida. National Wetlands Technical Council, The Conservation Foundation, Washington, D.C. 128 pp.

Clark, J. R., and J. Benforado, eds. 1981. Wetlands of bottomland hardwood forests. Elsevier Scientific Publishing Co., New York. 401 pp.

Cowardin, L. M., V. Carter, F. C. Golet, and E. T. LaRoe. 1979. Classification of wetlands and deepwater habitats of the United States. U.S. Department of the Interior, Fish and Wildlife Services, Office of Biological Services, Washington, D. C. FWS/OBS-79/31.

Gosselink, L., and J. G. Gosselink. 1988. Hydrology of wetlands of the southeastern United States. Prepared for National Wetlands Technical Council. Center for Wetlands Resources, Louisiana State University, Baton Rouge. 92 pp.

Hewlett, J. D., G. B. Cunningham, and J. Troendle. 1977. Predicting stormflow and peakflow for small basins in humid areas by the R-index method. Water Res. Bull. 13(2): 231-253.

Huffman, R. T., and S. W. Forsythe. 1981. Bottomland hardwood forest communities and their relation to anaerobic soil conditions. Pages 187-196 *in* J. R. Clark and J. Benforado, eds. Wetlands of bottomland hardwood forests. Elsevier Scientific Publishing Co., New York.

Klimas, C. V. 1988. River regulation effects on floodplain hydrology. Pages 40-49 *in* D. D. Hook, et al. (13 eds.). The ecology and management of wetlands. Vol. 1. Ecology of wetlands. Timber Press, Portland, Oregon.

Koewn, M. P., E. A. Dardeau, and E. M. Causey. 1980. Characterization of the suspended-sediment regime and bed material gradation of the Mississippi River basin. Environmental Laboratory, U.S. Army Corps of Engineers, Vicksburg, Mississippi. Prepared for Army Engineers District, New Orleans, Louisiana. Potamology Invest. Pre 221.

Leitman, H. M. 1984. Forest map and hydrologic conditions, Apalachicola River floodplain, Florida. U.S. Geological Survey Hydrologic Investigations Atlas HA-672.

Leitman, H. M., J. E. Solm, and M. A. Franklin. 1983. Wetland hydrology and tree distribution of the Apalachicola River floodplain. U.S. Geological Survey, Washington, D.C. Water Supply Paper 2196-A.

Meade, R., and R. Parker. 1985. Sediment in rivers of the United States. Pages 49-60 *in* National water summary. U.S. Geological Survey Washington, D.C. Water-Supply Paper 2275.

Mitsch, W. J., C. L. Dorge, and J. R. Wiemhoff. 1977. Forested wetlands for water resource management in southern Illinois. Illinois Water Resources Center, Urbana. Research Report 132. 275 pp.

Moore, I. D., and C. L. Larson. 1979. Effects of drainage projects on surface runoff from small depressional watersheds in the north-central region. Water Resources Research Center, University of Minnesota. Bulletin 99. 225 pp.

Novitzki, R. P. 1979. The hydrologic characteristics of Wisconsin wetlands and their influence on floods, streamflow, and sediment. Pages 377-388 *in* P. E. Greeson, J. R. Clark, and J. E. Clark, eds. Wetland functions and values: the state of our understanding. American Water Resources Association, Minneapolis, Minnesota. 674 pp.

____________. 1989. Wetland hydrology. Pages 47-64 *in* S. K. Majumdar, R. P. Brooks, F. J. Brenner, and R. W. Tiner, Jr., eds. Wetlands ecology and conservation: emphasis in Pennsylvania. Pennsylvania Academy of Science, Easton, PA.

Roelle, J. E., G. T. Auble, D. B. Hamilton, R. L. Johnson, and C. A. Segelquist (eds.) 1987. Results of a workshop concerning assessment of the functions of bottomland hardwoods. U.S. Fish and Wildlife Services, National Ecology Center, Fort Collins, Colorado. NEC-87/16. 173 pp.

Society of American Foresters. 1980. Forest cover types of the United States and Canada. Washington, D.C. 148 pp.

Theriot, R. F. 1988. Flood tolerance indices for palustrine forest. Pages 477-488 *in* D.D. Hook, et al., (13 eds.) The ecology and management of wetlands. Vol. 1: ecology of wetlands. Timber Press, Portland, Oregon. 592 pp.

U.S. Army Corps of Engineers. 1971. Charles River, Massachusetts: main report and attachments. New England Division, Waltham, Massachusetts. 76 pp.

U.S. Soil Conservation Service. 1972. National Engineering Handbook, Section 4, Chapter 16. Washington, D. C.

Young, C. E., and R. A. Klawitter. 1968. Hydrology of wetland forest watersheds. Pages 29-38 *in* Conference proceedings - hydrology in water resources management. Water Resources Institute, Clemson University, South Carolina. Report No. 4.

12. THE USE OF SOIL CLASSES TO DELINEATE TRANSITION ZONES IN BOTTOMLAND HARDWOOD FORESTS: THE REPORT OF THE SOILS WORKGROUP

B. Arville Touchet
U.S. Department of Agriculture, Soil Conservation Service,
Alexandria, LA 71302

with Panel
Stephen Faulkner, Robert Heeren, David Kovacic, William Patrick,
and Charles Segelquist

ABSTRACT

Bottomland hardwoods are forests that occur on floodplains. Definite, identifiable hardwood plant communities related to soil type and flooding regimes occur in different zones in these floodplains. The Soils Workgroup attempted to identify six ecological zones in the floodplains of the Mississippi River in Mississippi and Louisiana, the Savannah River in South Carolina, and Meyer's Branch in South Carolina, to be used as a model for delineating transition zones in similar riverine systems. Soil type, as classified by the National Cooperative Soil Survey, plays an important role in identifying these zones when the hardwood species have been removed.

INTRODUCTION

Historically, the term bottomland hardwoods has been utilized to describe forests that occur on floodplains in the southern and central United States. Under natural conditions, there are definite, identifiable bottomland hardwood plant communities within these floodplains. These systems and upland forests were delineated into six zones (Figure 1) by participants at the Bottomland Hardwood Wetlands Workshop held in

---

Ecological Processes and Cumulative Impacts: Illustrated by Bottomland Hardwood Wetland Ecosystems. Edited by James G. Gosselink, Lyndon C. Lee, and Thomas A. Muir. Chelsea, MI 48118. Printed in USA.

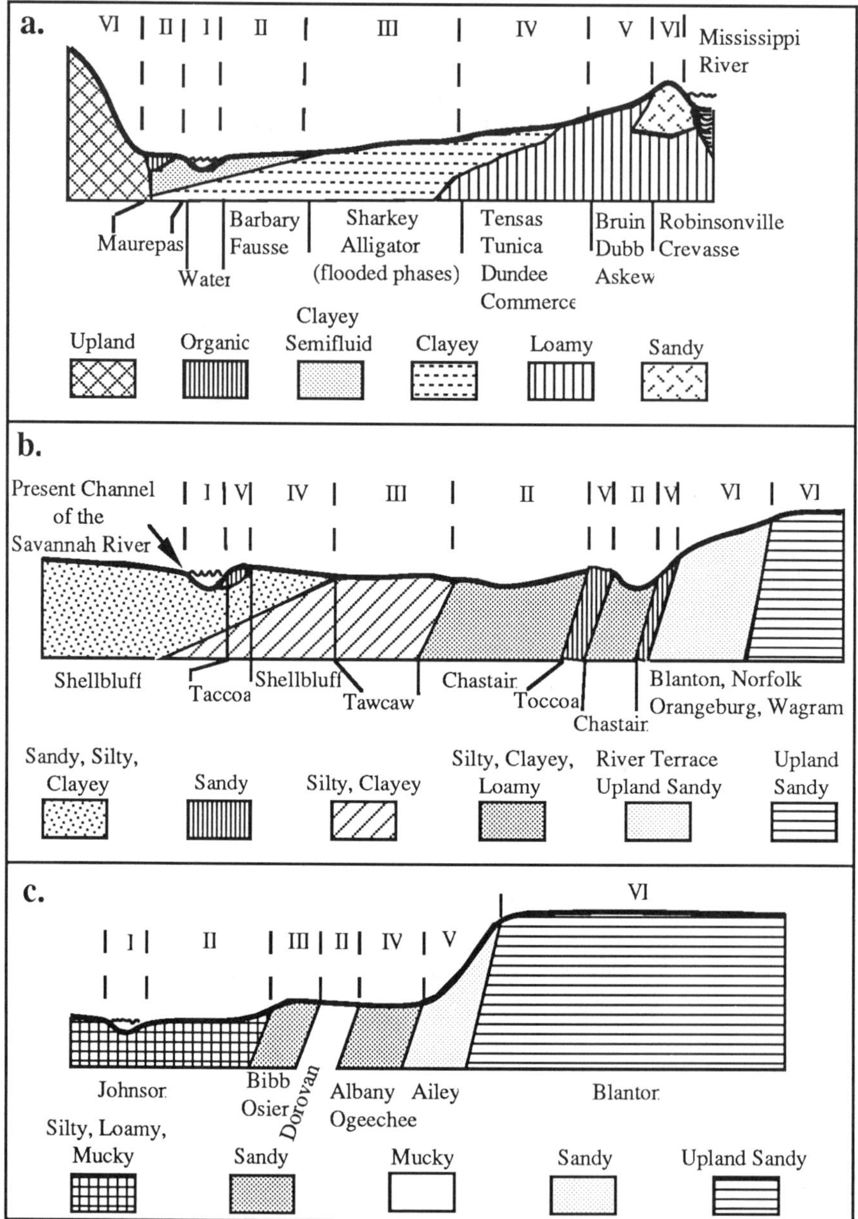

Figure 1. Idealized floodplain to upland profiles from a) the lower Mississippi River floodplain, b) the Savannah River floodplain, and c) Meyer's Branch floodplain.

Lake Lanier, Georgia, in June, 1980 (Clark and Benforado 1981). Soil scientists have long recognized that distinct soils occur in these different bottomland hardwood communities, and soil surveys consistently document these relationships. Even when the indicator plant communities are removed from an area, these sites retain their identity through their soil characteristics. Based on this information, the Soils Workgroup attempted to delineate soils in selected southeastern floodplains into five of the six ecological zones described at Lake Lanier. This delineation was based on soil classification and on various physicochemical characteristics. The three selected river floodplains include the Mississippi River Valley (Rolling Fork, Mississippi; Bordelonville, Louisiana; and Quimby, Louisiana) the Savannah River Valley (Aiken County, South Carolina), and Meyer's Branch Valley (Aiken County, South Carolina). Zone 1 was not treated because it represents permanently inundated aquatic habitat.

The Soils Workgroup found distinct correlations among soil hydrologic regimes for the five zones, as indicated by physicochemical properties, soil types, and vegetation. The redox potential, oxygen content, and water table of hydric soils, for example, indicate their anaerobic nature, and the vegetation on these sites is typically adapted to growing in standing water or in saturated soils. Nonhydric soils, on the other hand, are typically vegetated by species adapted to aerobic soil conditions.

Since soil surveys are recognized and used worldwide for transferring agrotechnology, the Soils Workgroup proposes that soil zonation, as determined by soil series name and supported by physicochemical characteristics shown in this report, be used to delineate transition zones in bottomland hardwood floodplains. An added benefit of using soil series names and map units to delineate these zones is the vast amount of interpretive data, stored in the Soil Conservation Service's computers, that is accessible by soil series name.

PHYSICOCHEMICAL CHARACTERISTICS AND CLASSIFICATION OF SOUTHERN FLOODPLAIN SOILS
(Lower Mississippi River floodplain soils, Savannah River floodplain soils, and Meyer's Branch floodplain soils)

Physicochemical characteristics and soil series classification information for a hypothetical cross section of each of the above-mentioned floodplains are illustrated in Figure 1 and described in

Appendices 1, 2, and 3. A description of these parameters is also presented in considerable detail by zones.

Zone II

Soils are saturated throughout the year most years. Anaerobic conditions prevail causing gray colors in the soil profile. Soil series commonly found in this zone in the different floodplains are as follows:

Mississippi River floodplain - Maurepas, Barbara, and Fausse
Savannah River floodplain - Chastain
Meyer's Branch floodplain - Johnston and Dorovan

Soil oxygenation:	Continuous or nearly continuous anaerobic conditions occur throughout the soil profile throughout the year most years.
Soil drainage and wetness:	Very poorly drained to poorly drained, very wet, hydric soils.
Soil moisture:	Peraquic moisture regime, nearly continually saturated, intermittently exposed.
Soil capability class:	VIw, VIIw, VIIIw.
Hydrology:	Ponded or flooded for very long duration or permanently, surface of soil is exposed during prolonged dry periods, areas lie within the 1-year floodplain or in ponded flats or basins.
Vegetation:	Water tupelo (*Nyssa aquatica*), cypress (*Taxodium distichum*), button bush (*Cephalanthus occidentalis*), swamp privet (*Foresteria acuminata*), black willow (*Salix*

Vegetation (continued): *nigra*), black gum (*Nyssa sylvatica*), and sweet bay (*Magnolia virginiana*).

Texture: Peats, mucks, clays, silt loams, loams, and sandy loams.

Soil pH: Generally ranges from 6.5 to 7.5 in the Mississippi River floodplain and from 3.5 to 6.0 in the Savannah River and Meyer's Branch floodplains.

Zone III

Soils are frequently flooded, ponded, or saturated for very long to long duration. Alternating anaerobic and aerobic conditions produce gray soils with brownish or yellowish mottles. Soil series commonly found in this zone in the different floodplains are as follows:

Mississippi River floodplain - flooded and nonflooded phases of Sharkey and Alligator plus flooded phases of closely similar soils
Savannah River floodplain - flooded phases of the Tawcaw
Meyer's Branch floodplain - Bibb, Osier, and Ogeechee

Soil oxygenation: Anaerobic conditions are prevalent, especially in the lower part of the soil profile, throughout the year most years, alternating with aerobic conditions in the upper part of the profile.

Soil drainage and wetness: Poorly drained to somewhat poorly drained, wet, hydric soils.

Soil moisture: Aquic moisture regime, saturated for long to very long duration.

Soil capability class: III, IV, V, VI.

Hydrology: Frequently flooded or ponded (51 to 100 years per 100 years), to occasionally flooded or ponded (11 to 50 years per 100 years), to rarely flooded or ponded (1 to 10 years per 100 years) for long to very long duration.

Vegetation: Water tupelo, black gum, black willow, pond pine (*Pinus rigida* var. *serotina*), water hickory (*Carya aquatica*), and overcup oak (*Quercus lycata*).

Texture: Mucks, clays, loams, and sands.

Soil pH: Generally ranges from 4.5 to 7.5 in the Mississippi River floodplain and from 4.5 to 6.5 in the Savannah River and Meyer's Branch floodplains.

Zone IV

Soils are rarely flooded or ponded but are saturated in the lower part of the soil profile. Aerobic conditions prevail over anaerobic conditions, producing brownish, yellowish, or reddish mottles in a gray matrix. Soil series commonly found in this zone in the different floodplains are as follows:

Mississippi River floodplain - Commerce, Dundee, Tensas, and Tunica
Savannah River floodplain - Shellbluff
Meyer's Branch floodplain - Albany

Soil oxygenation: Aerobic conditions predominate in the upper part of the profile.

Soil drainage and wetness: Somewhat poorly to moderately well drained, slightly wet, nonhydric soils.

Soil moisture: Aeric aquic to aquic udic moisture regime, commonly not saturated to the surface for long duration.

Soil capability class: IIw, IIIw.

Hydrology: Rarely flooded or ponded (1 to 10 years per 100 years) for long duration; high, fluctuating water table.

Vegetation: Sweet gum (*Liquidambar styraciflua*), sugarberry (*Celtis laevigata*), willow oak (*Quercus phellos*), green ash (*Fraxinus pennsylvanica* var. *lanceolata*), sycamore (*Platanus occidentalis*), yellow poplar, (*Liriodendron tulipifera*), water oak (*Quercus nigra*), cherry bark oak (*Quercus falcata* var. *leucophylla*), scarlet oak (*Quercus occinea*) and black walnut (*Juglans nigra*).

Texture: Silty clay loams, silt loams, loams, and sands.

Soil pH: Generally ranges from 4.5 to 7.2 in the Mississippi River floodplain and from 3.5 to 6.5 in the Savannah River and Meyer's Branch floodplain.

Zone V

Soils are very rarely flooded, ponded, or saturated to the surface. Anaerobic conditions occur, but only in the lower part of the soil profile. Soil colors are generally bright (yellows, browns, or reds) with few, if any, gray mottles. Soil series commonly found in this zone in the different floodplains and terraces are as follows:

Mississippi River floodplain - Askew, Bruin, and Dubbs
Savannah River floodplain and terrace - Toccoa
Meyer's Branch floodplain and terrace - Ailey

Soil oxygenation:	Aerobic conditions predominates nearly all year.
Soil drainage and wetness:	Moderately well drained, not wet, nonhydric soil.
Soil moisture:	Udic moisture regime, very rarely saturated to the surface.
Soil capability class:	I, IIs, IIIs, IIIw, IVs.
Hydrology:	Very rarely flooded (less than 1 year per 100 years), deep to water table.
Vegetation:	Cherry bark oak, swamp chestnut oak (*Quercus prinus*), sassafras (*Sassafras albidum*), American beech (*Fagus grandifolia*), sweet gum, yellow poplar, southern red oak (*Quercus rubra)*, slash pine (*Pinus caribea*) and long leaf pine (*Pinus palustris*).
Texture:	Silt loams, sandy loams, loams, and sands.
Soil pH:	Generally ranges from 4.5 to 6.5.

Zone VI

Soils are not flooded nor saturated and have aerobic conditions prevailing throughout the year resulting in bright soil colors such as yellows, browns, and reds with no gray mottles. Soil series commonly found in this zone in the different floodplains, terraces, and uplands are as follows:

Mississippi River floodplain and terrace - Robinsonville and Crevasse
Savannah River terrace and uplands - Blanton, Norfolk, Orangeburg, and Wagrum
Meyer's Branch terrace and upland - Blanton

Soil oxygenation:	Aerobic conditions prevail throughout the year.
Soil drainage and wetness:	Well to excessively drained, not wet, non-hydric soils.
Soil moisture:	Udic moisture regime, slightly droughty.
Soil capability class:	I, IIs, IIe, IIIs, IIIe.
Hydrology:	Not flooded, ponded, or saturated.
Vegetation:	Cherry bark oak, live oak (*Quercus virginiana*), yellow poplar, white oak (*Quercus alba*), slash pine, long-leaf pine, and red oak.
Texture:	Sandy loams, loams, and sands.
Soil pH:	Generally ranges from 4.5 to 6.5.

MEASUREMENT OF SOIL AERATION

As a result of field data collected from research plots in the lower Mississippi River Valley, generalized diagrams of soil redox potential, soil oxygen content, and water-table cycles have been developed for each of three different sites: dry, intermediate, and wet. These preliminary observations are based on incomplete field research. Final analysis of all the data may lead to modifications of results. However, these initial finds give a better understanding of the nature of the oxidation-reduction processes in bottomland hardwood soils.

Dry Sites

Dry sites are located on high ridges with typical overstory vegetation of *Sassafras albidum*, eastern hophornbeam (*Ostrya virginiana*), and *Fagus grandifolia*. The soils are well to excessively drained, and the water table on these sites stays low throughout the year (Figure 2). The oxygen content generally fluctuates between 15% and 21%, rarely falling below 13%. Redox potential always remains above +300 mv. These sites are not flooded. This is a typical profile for ecological zone VI.

Intermediate Sites

Intermediate sites are generally found between the ridge sites and the true swamps. Dominant overstory plant associations range from the sweetgum-yellow poplar forest type on the higher elevations to the overcup oak-water hickory type on the lower, wetter sites. Soil drainage ranges from well drained to somewhat poorly drained. The poorly drained soils with lingering high water and low oxygen and redox values reflect zone III (Figure 3). The better drained sites, distinguished by lower water tables and higher oxygen and redox values, are indicative of zone V. Zone IV is somewhere between the two extremes. Oxygen content on the drier sites stays above 10%, rising as high as 21% during periods of drought. Soil redox potential stays above +300 mv. The oxidation-reduction cycle on the wetter sites is quite distinct. The water table rises in late winter and early spring with resulting decreases in oxygen content and redox potential. Oxygen content falls below 5% and occasionally to zero. Redox potential remains below +300 mv for a large portion of the year. Oxygen and redox can remain quite low well into the growing season, depending on the hydroperiod.

Wet Sites

Wet sites are found in low landscape elevations and are dominated by saturated soil conditions. Plants are limited to flood-tolerant, obligate wetland species, such as cypress, water tupelo, swamp privet, black willow, and buttonbush. The soils are very poorly drained with

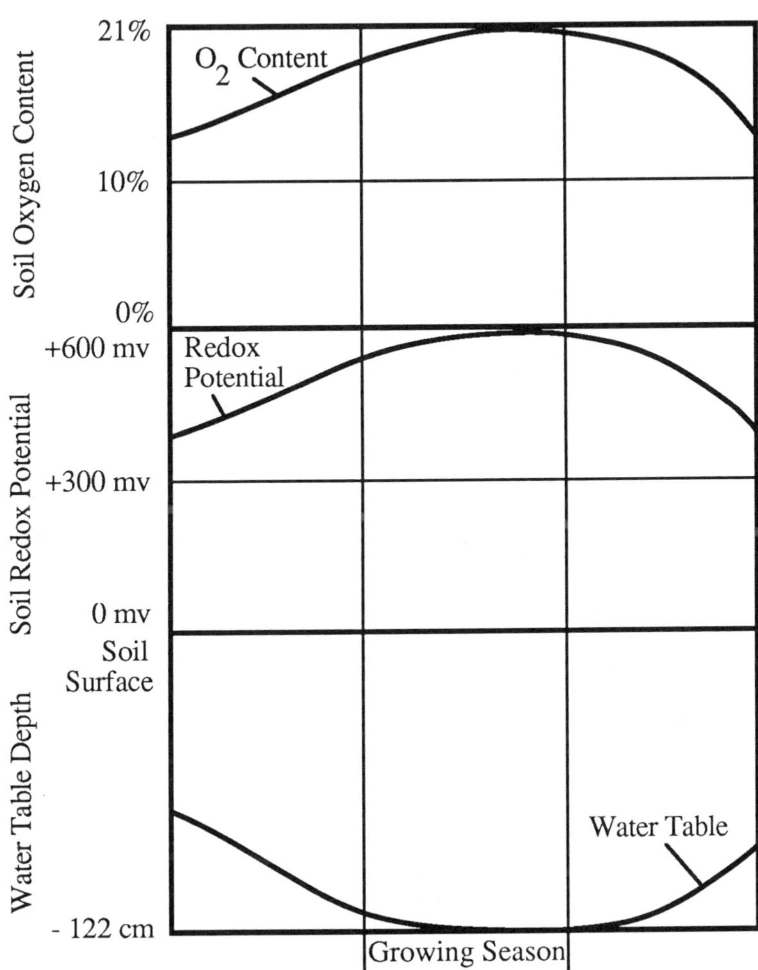

Figure 2. Soil oxygen content, redox potential, and water table cycle for dry sites in the lower Mississippi River Valley.

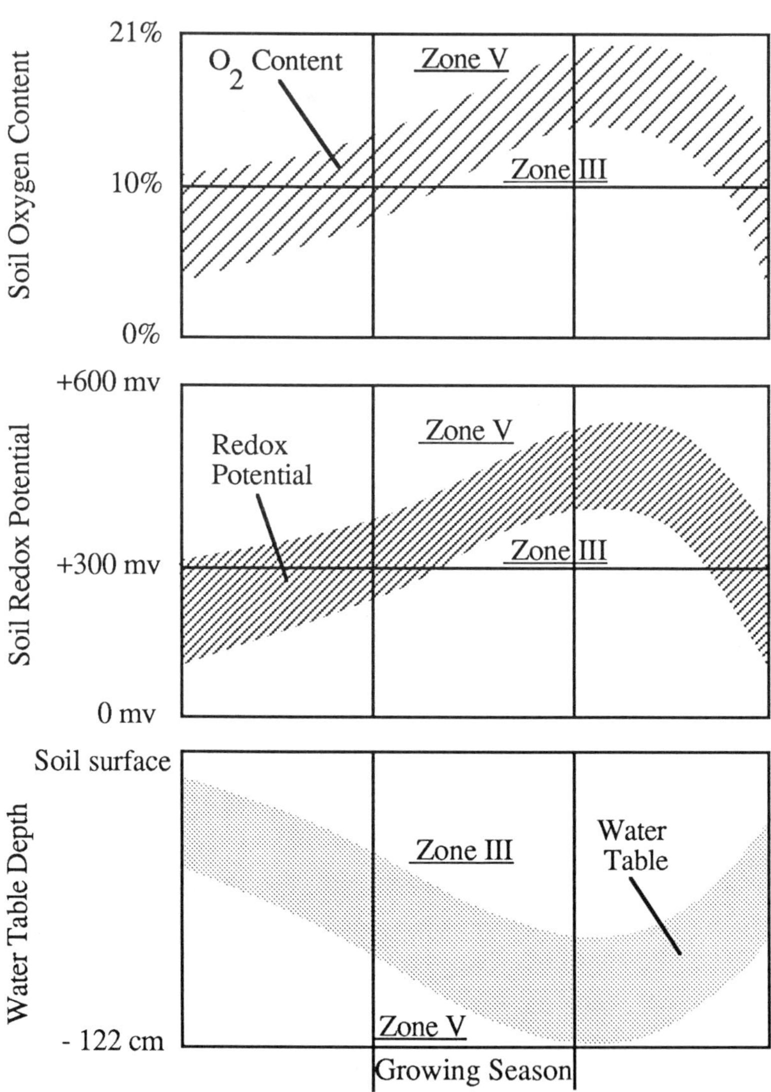

Figure 3. Soil oxygen content, redox potential, and water table cycle for sites with intermediate moisture regimes in the lower Mississippi River Valley.

peraquic moisture regimes. The water table remains at or near the surface for most of the year. Soil oxygen contents are very low (less than 5%) with strong reducing conditions of less than +100 mv (Figure 4). These sites are in zone II.

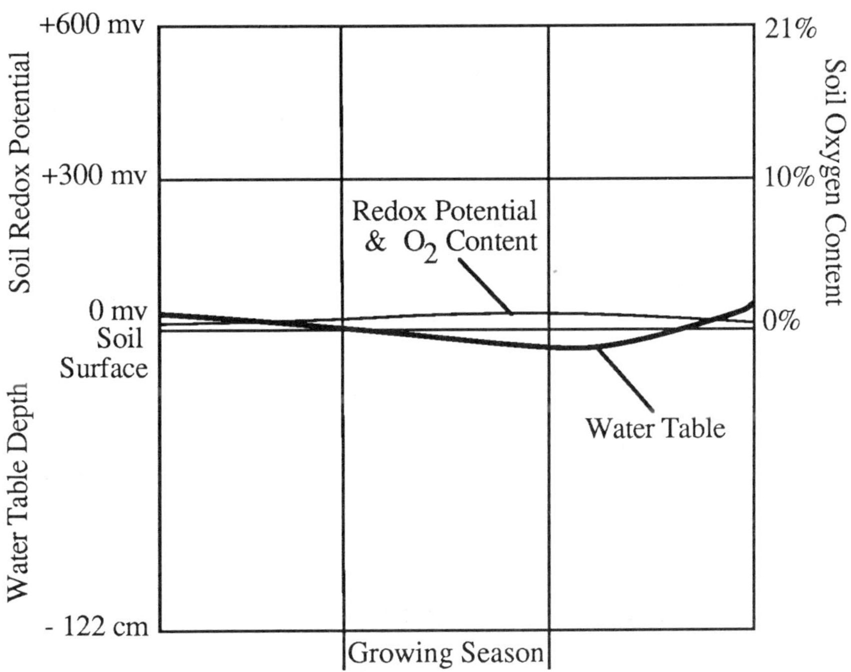

Figure 4. Soil oxygen content, redox potential, and water table cycle for wet sites in the lower Mississippi River Valley.

CONCLUSIONS

The consensus of the Soils Workgroup, based on the preceding analyses, is that there is a direct correlation between soil series mapping units, hydrology, and bottomland hardwood plant communities. The soils are identified by their physicochemical makeup plus vegetation data, when available, but are recognizable even without the vegetation data. The group believes that soils provide a more permanent identification tool for delineating bottomland hardwood sites than vegetation or hydrology. The Soils Workgroup strongly recommends that zonation of bottomland hardwoods be based on soil series information available through soil surveys and validated by multidisciplinary teams of soil scientists, hydrologists, and biologists.

This approach works well in the lower Mississippi River Valley bottomland hardwood area, but additional on-site research, such as that reported here for the lower Mississippi River Valley, needs to be conducted for other bottomland systems to reinforce, refine, and improve soil zonation (soil survey delineation) throughout the Southeast.

It is also recommended that other bottomland hardwood functions; such as water quality improvement, utilization by fish, utilization by animals, and lastly, utilization by humans; be more carefully evaluated through field research and synthesized and integrated into the six zones (corresponding to the five soil zones and the water zone) of bottomland hardwoods.

REFERENCE CITED

Clark, J. R., and J. Benforado, eds. 1981. Wetlands of bottomland hardwood forests. Elsevier Scientific Publishing Company, New York. 401 pp.

APPENDICES

Appendix 1. Physicochemical characteristics of floodplain soils and soil series on the Mississippi River.

	Zone II	Zone III
Soil series (name)	Maurepas, Barbary, Fausse	Sharkey, Alligator (flooded phases) (nonflooded phases)
Soil oxygenation	Anaerobic	Anaerobic to aerobic
Soil drainage and wetness	Very poorly drained, very wet, hydric soils	Poorly drained, wet, hydric soils
Soil moisture regime (soil taxonomy)	Peraquic	Aquic
Soil capability class	VI, VIIw, VIIIw	IIIw, IVw, Vw
Hydrology (duration of saturation during growing season)	~ 100%, except in extreme drought	>25 - 75%
Flooding or ponding	Continually, except in extreme drought	Freq. 11-100yrs/ 100yrs
Vegetation	*Nyssa aquatica, Cephalanthus occidentalis, Foresteria acuminata, Salix nigra*	*Carya aquatica, Salix nigra, Quercus lycata*
Surface texture	Peats, mucks, clays (fluid)	Mucks, clays (firm)
Subsoil texture	Peats, clays (fluid to firm)	Clays (firm)
Soil pH	6.5 - 7.5	4.8 - 7.5
Soil classification (soil taxonomy)	Typic Medisaprists, Typic Hydraquents, Typic Fluvaquents	Vertic Haplaquepts

Appendix 1. (Concluded).

	Zone IV	Zone V	Zone VI
Soil series (name)	Commerce, Dundee, Tensas, Tunica	Askew, Bruin, Dubbs	Robinsonville, Crevasse
Soil oxygenation	Aerobic to anaerobic	Aerobic	Aerobic
Soil drainage and wetness	Somewhat poorly, slightly drained, wet, nonhydric soils	Moderately well drained, not wet, nonhydric soils	Well to excessively drained, droughty, nonhydric soils
Soil moisture regime (soil taxonomy)	Aquic (aeric)	Udic (aquic)	Udic
Soil capability class	IIw, IIIw	I	I, IIs
Hydrology (duration of saturation during growing season)	>12.5 - <25%	>2 - <12.5%	< 2%
Flooding or ponding	Rarely 1-10yrs/100yrs	Very rarely <1yr/100yrs	Not flooded
Vegetation	*Liquidamber styraciflua*, *Celtis laevigata*, *Quercus phellos*, *Quercus nigra*	*Quercus falcata* var. *leucophylla*, *Quercus prinus*	Upland varieties, *Quercus falcata* var. *leucophylla*, *Quercus virginiana*, *Liriodendron tulipifera*
Surface texture	Silty clay loams, loams, silt loams	Silt loams	Sandy loams
Subsoil texture	Silty clay loams, loams, silt loams	Loams	Loams, sandy loams, loamy sands
Soil pH	4.5 - 7.2	4.5 - 6.8	4.5 - 6.5
Soil classification (soil taxonomy)	Aeric Fluvaquents, Aeric Ochraqualfs, Vertic Ochraqualfs, Vertic Haplaquepts	Fluvaquentic Eutrochrepts, Typic Hapludalfs	Typic Udipsamments

Appendix 2. Physicochemical characteristics of floodplain soils and soil series on the Savannah River, Aiken County, South Carolina.

	Zone II	Zone III
Soil series (name)	Chastain	Tawcaw
Soil oxygenation	Anaerobic	Anaerobic to aerobic
Soil drainage and wetness	Poorly drained, very wet, hydric soils	Somewhat poorly drained, wet, hydric soils
Soil moisture regime (soil taxonomy)	Peraquic	Aquic
Soil capability class	VIw	VIw
Hydrology (duration of saturation during growing season)	~ 100% except in extreme drought	>25% - 75%
Flooding or ponding	Continually, except in extreme drought	Freq. 11-100yrs/ 100yrs
Vegetation	*Taxodium distichum, Nyssa aquatica*	*Nyssa aquatica Salix nigra*
Surface texture	Silty clay loams	Silty clay loams
Subsoil texture	Silty clays	Silty clays
Soil pH	4.5 - 6.0	4.5 - 6.5
Soil classification (soil taxonomy)	Typic Fluvaquents	Fluvaquentic Dystrochrepts

Appendix 2. (Concluded).

	Zone IV	Zone V	Zone VI
Soil series (name)	Shellbluff	Toccoa	Blanton, Norfolk, Orangeburg, Wagram
Soil oxygenation	Dominantly aerobic to anaerobic	Aerobic	Aerobic
Soil drainage and wetness	Moderately well to well drained, slightly wet, nonhydric soils	Moderately well to well drained, nonhydric soils	Moderately well to well drained, nonhydric soils
Soil moisture regime (soil Taxonomy)	Udic (aquic)	Udic	Udic
Soil capability class	IIIw	IIIw	I, IIe, IIIe
Hydrology (duration of saturation during growing season)	>12.5% - <25%	>2% - <12.5%	< 2%
Flooding or ponding	Rarely 1-10yrs/100yrs	Very rarely <1yr/100yrs	Not flooded
Vegetation	*Quercus falcata* var. *leucophylla, Quercus cocconea, Quercus nigra, Liquidamber styraciflua, Liriodendron tulipifera*	*Liriodendron tulipifera,Quercus rubra*	*Pinus*, mixed hardwoods
Surface texture	Silty clay loams	Fine sands	Sandy loams
Subsurface texture	Silty clay loams	Loamy	Sandy clay loams
Soil pH	4.5 - 6.5	5.1 - 6.5	4.5 - 6.0
Soil classification (soil taxonomy)	Fluventic Dystrochrepts	Typic Udifluvents	Grossarenic Paleudults, Typic Paleudults, Arenic Paleudults

Appendix 3. Physicochemical characteristics of floodplain soils and soil series on the Meyer's Branch, Aiken County, South Carolina.

	Zone II	Zone III
Soil series (name)	Johnston, Dorovan	Bibb, Osier, Ogeechee
Soil oxygenation	Anaerobic	Dominantly anaerobic to aerobic
Soil drainage and wetness	Very poorly drained, very wet, hydric soils	Poorly drained, wet, hydric soils
Soil moisture regime (soil taxonomy)	Peraquic	Aquic
Soil capability class	VIIw	Vw
Hydrology (duration of saturation during growing season)	~ 100% except in extreme drought	>25% - 75%
Flooding or ponding	Continually, except in extreme drought	Freq. 11-100yrs/ 100yrs
Vegetation	*Nyssa sylvatica*, *Magnolia virginiana*, *Nyssa aquatica*, *Taxodium distichim*	*Nyssa sylvatica*, *Salix nigra*
Surface texture	Silty, loamy, mucky (fluid)	Silty, loamy (firm)
Subsoil texture	Loamy, sandy (fluid to firm)	Loamy, sandy
Soil pH	4.5 - 5.5 3.6 - 4.4	4.5 - 5.5
Soil classification (soil taxonomy)	Cumulic Humaquepts, Typic Medisaprists	Typic Fluvaquents

Appendix 3. (Concluded).

	Zone IV	Zone V	Zone VI
Soil series (name)	Albany	Ailey	Blanton
Soil oxygenation	Dominantly aerobic to anaerobic	Aerobic	Aerobic
Soil drainage and wetness	Somewhat poorly to poorly drained, slightly wet, nonhydric soils	Moderately well to well drained, not wet, nonhydric soils	Moderately well to well drained, not wet, nonhydric soils
Soil moisture regime (soil taxonomy)	Udic (aquic)	Udic	Udic
Soil capability class	IIIw, IVw	IIs, IIIs, IIIa, IVs	IIIs
Hydrology (duration of saturation during growing season)	>12.5% - <25%	>2% - <12.5%	< 2%
Flooding or ponding	Rarely 1-10yrs/100yrs	Very rarely <1yr/100yrs	Not flooded
Vegetation	*Pinus rigida* var. *serotina*, *Liquidamber styraciflua*, *Quercus nigra*	*Pinus caribbea*, *Pinus palustris*, *Quercus rubra*, *Liquidamber styraciflua*	*Pinus caribbea*, *Pinus palustris*, *Quercus rubra*
Surface texture	Sandy	Sandy	Sandy
Subsurface texture	Sandy, loamy	Sandy, loamy	Sandy, loamy
Soil pH	3.6 - 6.5 4.5 - 5.5	4.5 - 6.5	4.5 - 6.0
Soil classification (soil taxonomy)	Grossarenic Paleudults, Typic Paleudults	Arenic Paleudults	Grossarenic Paleudults

13. THE EFFECT OF DEVELOPMENTAL ACTIVITIES ON WATER QUALITY FUNCTIONS OF BOTTOMLAND HARDWOOD ECOSYSTEMS: THE REPORT OF THE WATER QUALITY WORKGROUP

Michael L. Scott
U.S. Fish and Wildlife Service, National Ecology Research Center,
Fort Collins, CO 80525

Barbara A. Kleiss
U.S. Waterways Experiment Station,
Vicksburg, MS 39180

William H. Patrick
Laboratory for Wetland Soils and Sediments,
Center for Wetland Resources, Louisiana State University,
Baton Rouge, LA 70803

Charles A. Segelquist
U.S. Fish and Wildlife Service, National Ecology Research Center,
Fort Collins, CO 80525

with Panel
C. Belin, J. Chowning, L. Glenboski, P. Hatcher, D. Hicks,
H. Howard, E. Hughes, A. Lucas, D. Walker

ABSTRACT

Water-quality functions associated with bottomland hardwood ecosystems include 1) control of sediment detachment and transport, 2) sediment detention, and 3) nutrient and contaminant detention and transformation. Human activities that modify the various physical characteristics of bottomland hardwood ecosystems correspondingly affect the influence of these ecosystems on water quality. In general, water-quality functions associated with bottomland hardwoods are

---

Ecological Processes and Cumulative Impacts: Illustrated by Bottomland Hardwood Wetland Ecosystems. Edited by James G. Gosselink, Lyndon C. Lee, and Thomas A. Muir. Chelsea, MI 48118. Printed in the USA.

adversely affected by all human developmental activities because undeveloped stands provide optimal physical characteristics for improving water quality. Any activity resulting in flood control or conversion of bottomland hardwoods will have a negative effect on water-quality functions, particularly if best management practices or other mitigation measures are not included in the activity.

INTRODUCTION

Forested floodplains of southeastern rivers, or bottomland hardwood ecosystems, provide an environment in which several important ecological functions influence the water quality of associated watersheds. The specific functions identified in this workshop include 1) control of sediment detachment and transport, 2) sediment detention, and 3) nutrient and contaminant detention and transformation. There are a number of physical and chemical properties, or characteristics, of bottomland hardwood ecosystems that are associated with the performance of these identified functions. Human activities that alter these characteristics of bottomland hardwoods also alter the water-quality functions of these ecosystems. Such activities might include conversion of bottomland hardwoods to agriculture, construction of levees, and development activities that eliminate bottomland hardwoods or otherwise isolate the river from its floodplain. Gauging the impacts of particular activities on water quality is a profound challenge to regulatory agencies charged with protecting environmental resources.

In this chapter we present: 1) a general description of the functions and associated characteristics identified with water quality, 2) a description of the impacts of selected human activities on specific characteristics and water-quality functions of bottomland hardwood forests, and 3) a series of response curves designed to illustrate the assumed relationships between functions and associated characteristics of bottomland hardwoods. These are presented from a field assessment perspective, where characteristics are used to assess how well a bottomland hardwood site might perform a particular water-quality function.

FUNCTIONS AND CHARACTERISTICS

Water-quality functions that are performed by bottomland hardwood ecosystems, and characteristics that influence these functions, are identified in Table 1. A brief description of each water-quality function follows.

Control of Sediment Detachment and Transport

Because of the vegetative cover and low gradient of bottomland hardwood floodplains, these areas typically contribute less sediment to adjacent streams than do other cover types, such as an agricultural crop. Using a watershed routing model, Molinas and Auble (1988) estimated that sediment loss, from the delta portion of the Yazoo Basin to the channel, ranged from 0.02 to 0.07 million metric tons per year when cover was nearly completely bottomland hardwood. When bottomland hardwood cover was nominal, this figure rose to 1.5 to 5.5 million metric tons per year. However, these values are small compared to the estimated gross erosion from the uplands of this region, which ranged from 91 million metric tons per year in the 1930's to 25 million metric tons per year in the 1970's (Kolb et al. 1976). In light of this, the role of bottomland hardwoods as sites for sediment detention during overbank flooding is emphasized.

Sediment Detention

Because of their position in the landscape, bottomland hardwoods play an important role in detaining sediments. The relatively level topography, natural ponding, and ground surface roughness of the floodplain create conditions that are favorable for removal of suspended sediment, both organic and mineral, from floodwater moving through the bottomland. The low flow rate of the water and the high surface activity of the organic matter and clay in the soils are major factors in the removal of suspended sediment. Sediment-laden waters entering the bottomland from the channel or adjacent uplands would deposit sediment onto the vegetated, low-gradient floodplain. In a sedimentation study of the Yazoo

Table 1. **Functions and characteristics of bottomland hardwood ecosystems. Characteristics listed under each function are structural or functional attributes that are considered to influence the water-quality functions of bottomland hardwood ecosystems. Selection of characteristics was based on how well the characteristic might predict the function in question. Similar characteristics may be considered differently, depending on the function. Listed under each characteristic are subjective rankings of 1) the ease with which data may be obtained for a particular characteristic, and 2) how important that characteristic is to the associated water-quality function.**

FUNCTIONS / Characteristics	Ease of Obtaining Data[a]	Importance of Characteristic in Determining Function[b]
Control of Sediment Detachment and Transport		
Surface Water Characteristics		
a. duration and frequency of flooding	1-2	2
b. upland runoff and sediment input	2	3
Water Velocity Characteristics		
a. extent of natural ponding	2	1
b. slope	1	3
c. sinuosity	1-2	2
d. water velocity on tract	2	3
Substrate Stability Characteristics		
a. overstory stem density	2-3	2
b. understory density	2-3	3
c. ground surface roughness	2	2
d. soil disturbance	2	2
e. soil clay content	2-3	2
f. soil organic matter content	2-3	2
Sediment Detention		
Sediment Loading Characteristics		
a. stream sediment load	3	3
b. upland runoff sediment load	3	3
c. distance of site from water source	1	2
d. location of site in watershed	1	2
e. sediment particle size	2-3	2
Surface Water Characteristics		
a. duration and frequency of flooding	1-2	3
b. upland runoff	2-3	3
c. distance of site from water source	1	2

Table 1. (Continued).

FUNCTIONS Characteristics	Ease of Obtaining Data[a]	Importance of Characteristic in Determining Function[b]
Water Velocity Characteristics		
a. extent of natural ponding	2	2
b. slope	1	3
c. sinuosity	1	2
d. water velocity on tract	2	3
e. ground surface roughness	2	2
f. overstory stem density	2-3	1
g. understory density	2-3	2
h. distance of site from water source	1-2	2
i. fetch	2	2
Roughness Characteristics		
a. ground surface roughness	2	2
b. overstory stem density	2-3	1
c. understory density	2-3	2
Nutrient and Contaminant Detention and Transformation		
Nutrient and Contaminant Loading Characteristics		
a. instream sediment and nutrient and contaminant load	2-3	3
b. upland runoff and sediment load	2-3	3
Surface Water Characteristics		
a. duration and frequency of flooding	1-2	3
b. distance of site from water source	1	2
Water Velocity Characteristics		
a. extent of natural ponding	2	2
b. slope	1	3
c. sinuosity	1	2
d. ground surface roughness	2	2
e. overstory stem density	2-3	1
f. understory density	2-3	2
g. fetch	2	2
Physical and Chemical Detention Characteristics		
a. clay type and content in soil	1	3
b. organic matter content in soil	1	3
c. overstory stem density	2-3	1
d. understory density	2-3	2
e. soil pH	2	3
Chemical and Biological Transformation Characteristics		
a. clay type and content in soil	1	3
b. organic matter content in soil	1	3
c. soil pH	2	3

Table 1. (Concluded).

[a] Ease Factors
- 1 = requires only office searches including quad sheets, soil maps, and information from applicant.
- 2 = requires a cursory site visit to acquire the data.
- 3 = requires not only a site visit but also considerable off-site data collection.

[b] Importance Factors
- 1 = information that would supplement other data but that could be left out of the data gathering process without affecting the outcome of the evaluation.
- 2 = information that would be of importance in the overall evaluation.
- 3 = an important piece of data that could severely affect the outcome of the evaluation if omitted.

Basin, Simons et al. (1983) developed a model to examine the cumulative sediment yield from Abiaca Creek into an adjacent forested area. Results of the analysis indicated that, initially, 95% of the sediments passing through the forest would be trapped. At the end of a 50-year simulation period, results indicated that trapping efficiency was still 70%. The sediment detention function of bottomland hardwood ecosystems is effective only to the extent that sediment-laden waters are allowed to flow through these systems; sediment detention by bottomland hardwoods in extensively channelized systems would be greatly reduced.

Nutrient and Contaminant Detention and Transformation

Bottomland hardwoods also perform important nutrient transformations that improve water quality. Nitrogen and phosphorus, the nutrients most often implicated in eutrophication, undergo significant reactions after entering a bottomland area that often decrease their concentrations before flowing out of the bottomland (Kitchens et al. 1975). One reaction is through plant uptake by growing plants (Wharton 1970, Kitchens et al. 1975), resulting in either permanent removal or temporary detention of nitrogen and phosphorus. Brinson (1977) demonstrated that immobilization of nitrogen, phosphorus, calcium, and iron on leaf litter provided an important nutrient sink in an alluvial swamp during tree dormancy, a period in which these systems are open to flooding. Denitrification has also been shown to be an important sink for nitrogen in swamp soils of the Mississippi Delta (Patrick and Tusneem 1972, Reddy and Patrick 1975). Bottomland soils are very reactive with phosphate and readily remove it from the water by adsorption and precipitation. Mitsch et al. (1979) found the greatest input of phosphorus to an alluvial swamp forest was due to deposition of phosphate-rich sediments during flooding, and the magnitude of phosphorus deposition was 10 times greater than the outflow of phosphorus to the river.

As with nutrients, sediments are capable of transporting loads of adsorbed pesticides, heavy metals, and other toxins. Deposition of such sediment (e.g., in bottomland hardwood ecosystems) can result in detention of contaminants in an environment where they may be transformed or decomposed (Boto and Patrick 1978). Pionke and Chesters (1973) provide a review of the transport of pesticides by sediments, as well as the rate of decomposition of some pesticides. Included in their review are discussions of factors affecting the distribution and decomposition of pesticides, including pH, soil organic matter content, and clay particle type.

ANALYSIS OF IMPACTS

Throughout the southeastern United States, bottomland hardwood ecosystems have been impacted by human activity in various ways. Consideration of all of these impacts is beyond the scope of this discussion; however, the majority of impacts to bottomland forests have been associated with agricultural conversion and alteration of hydrology. In this section we discuss the impacts of human activities on specific characteristics and water-quality functions of bottomland hardwood ecosystems. The activities considered include 1) conversion to non-wetland dependent agriculture (i.e., soybeans, rice, aquaculture), 2) conversion to pine culture, 3) on-site and off-site channelization, 4) on-site and off-site levee construction, and 5) on-site and off-site construction and operation of impoundments.

Impacts on Characteristics of Bottomland Hardwood Ecosystems

Anticipated impacts that specific human activities will have on some of the bottomland hardwood characteristics considered in this paper are described below. Information contained in the narrative is summarized in Table 2.

Cross-sectional Slope of Tract

The gentle cross-sectional slope of bottomland hardwood sites reduces the velocity of water flowing across the floodplain during overbank flooding. Two of the development activities considered, conversion to rice and aquaculture, serve to reduce the cross-sectional gradient of the floodplain. Conversion to soybeans probably results in steeper gradients due to landforming or landfill activities undertaken to improve drainage. Construction of an on-site impoundment would serve to reduce the cross-sectional slope of the tract, while off-site channelization, levee construction, or the construction and operation of upstream impoundments would likely have little direct impact on the cross-sectional slope of the tract.

Extent of Natural Ponding

Natural ponding contributes to a slowing of overland flows during flooding and provides areas of standing water and sediment deposition following drawdown. Many of the development activities considered in this analysis result in reduced ponding. Conversion to agriculture, pine culture, or aquaculture requires land forming and drainage that reduce natural ponding. On-site levee construction, channelization, and upstream impoundment would limit the extent of ponding, while on-site impoundment would increase the extent of natural ponding. Upstream levee construction and channelization might increase the frequency and extent of overbank flooding at the site, leading to more natural ponding.

Table 2. Summary of the anticipated impacts of specific human activities on selected characteristics of bottomland hardwood ecosystems.

Human Activities

Characteristics of Bottomland Hardwood Ecosystems	Conversion to soybeans	Conversion to rice	Conversion to aquaculture	Conversion to pine culture	On-site channelization	On-site levee construction	On-site impoundment	Upstream channelization	Upstream levee construction	Upstream impoundment
Cross- sectional slope of tract	Negative	Neutral or Unknown	Neutral or Unknown	Negative	Negative	Negative	Positive	Neutral or Unknown	Neutral or Unknown	Neutral or Unknown
Extent of natural ponding	Negative	Negative	Negative	Negative	Negative	Negative	Positive	Positive ?	Positive ?	Negative
Ground surface roughness	Negative	Negative	Negative	Neutral or Unknown	Neutral or Unknown	Neutral or Unknown	Negative	Neutral or Unknown	Neutral or Unknown	Neutral or Unknown
Density of understory	Negative	Negative	Negative	Positive	Positive ?	Positive ?	Negative	Negative	Negative	Positive ?
Density of overstory	Negative	Negative	Negative	Negative ?	Negative ?	Negative ?	Negative	Neutral or Unknown ?	Neutral or Unknown ?	Neutral or Unknown ?
Water velocity on tract	Negative	Negative	Negative	Negative	Neutral or Unknown	Neutral or Unknown	Positive	Negative	Negative	Neutral or Unknown
Duration and frequency of flooding	Negative	Negative	Negative	Negative	Negative	Negative	Positive	Positive ?	Positive ?	Negative
Upland runoff and sediment input to tract	Negative	Negative	Negative	Neutral or Unknown ?	Neutral or Unknown ?	Neutral or Unknown ?	Neutral or Unknown	Neutral or Unknown	Neutral or Unknown	Neutral or Unknown
Fetch and sinuosity	Negative	Negative	Negative	Negative	Negative	Negative	Negative	Neutral or Unknown	Neutral or Unknown	Neutral or Unknown
Percent organic matter in soil	Negative	Negative	Negative	Negative	Negative	Negative	Positive	Positive ?	Positive ?	Negative
Soil disturbance	Negative	Negative	Neutral or Unknown	Neutral or Unknown	Negative	Neutral or Unknown	Neutral or Unknown	Neutral or Unknown	Neutral or Unknown	Neutral or Unknown
Instream sediment and nutrient and contaminant load	Negative	Negative	Negative	Negative	Negative	Negative	Positive	Negative	Negative	Positive

Legend: Negative Impacts on Characteristic (shaded) · Neutral or Unknown Impacts on Characteristic (hatched) · Positive Impacts on Characteristic (blank) · ? Relationship Between Impact & Characteristic Uncertain

Fetch and Sinuosity

These characteristics are influenced by the geomorphic features of bottomland hardwood floodplains which, despite their low topographic relief, are a complex of abandoned river channels and levees, point bar deposits, overflow channels, scour channels, and minibasins (Wharton et al. 1982). We define fetch as the distance or area over which floodwaters interact with a bottomland hardwood floodplain. Sinuosity, or the ratio of stream channel length to downvalley distance (Leopold et al. 1964), refers here to the amount of turning in the travel of water across a site during the rise and recession of flood flows. As the topographic complexity of a site increases, so would the sinuosity and, thus, the fetch of a bottomland hardwood site. Activities such as conversion to agriculture, aquaculture, and pine culture tend to reduce the topographic features of a bottomland hardwood site and would, therefore, reduce the fetch and sinuosity. Similarly, on-site channelization and levee construction would effectively reduce fetch and sinuosity by reducing the extent to which the site would flood. On-site impoundment would inundate many or most topographic features and effectively reduce fetch and sinuosity. Upstream channelization, levee construction, or impoundment would have no obvious direct effect on these characteristics.

Ground Surface Roughness

Ground surface roughness (e.g., irregularities in the soil, fallen logs, stumps, other debris) reduces the velocity of flowing water across a tract of bottomland hardwood trees during overbank flooding. Conversion of these ecosystems to agriculture or aquaculture eliminates most of this surface roughness. Impoundment on-site would decrease ground surface roughness to the extent that floodplain surface features were inundated. On-site channelization, levee construction, or conversion to pine culture results in a temporary reduction in surface roughness of the floodplain; however, long-term effects on surface roughness and subsequent deposition are likely to be relatively minor. Upstream channelization, levee construction, and impoundments would not be expected to impact this characteristic.

Density of Understory

Understory vegetation is similar to ground surface roughness in reducing water velocity across a site. Conversion of bottomland hardwood sites to agriculture or aquaculture removes this understory vegetation and replaces it with a crop. Upstream levee construction or channelization that would potentially increase flooding at the site would be expected to reduce the density of understory vegetation. However, the density of understory vegetation may be increased at sites that have been channelized, converted to pine, or are downstream from impoundments; the increased drainage resulting from these activities may initially reduce tree canopy cover and enhance the growth of a dense, flood-intolerant understory. The long-term effects of increased drainage on understory density are uncertain.

Density of Overstory

Stems of trees, large shrubs, and vines also contribute to the roughness that reduces the velocity of flowing waters across a bottomland hardwood site. As with the understory, clearing and converting bottomland forests to agriculture or aquaculture eliminate all larger woody stems. Clearing and converting to short-rotation pines also results in lower tree basal area over time but may not substantially reduce stem density. On-site impoundment would eventually reduce tree density and basal area because of flooding (Green 1947, Harmes et al. 1980). Improved drainage resulting from on-site channelization and levee construction would be expected to eventually reduce the density and basal area of bottomland hardwood tree species. Upstream channelization, levee construction, or impoundments probably have minimal short-term effects on tree density and basal area. However, the long-term effects of reduced flooding on stand structure are unknown.

Water Velocity on Tract

Water velocity is an important determinant of how much sediment erosion and deposition will occur at a particular site. Conversion activities for rice, soybeans, aquaculture, or pine all tend to speed water flow by various drainage modifications and by the clearing of vegetation.

Upstream impoundment, along with channelization and levee construction on-site, reduces the extent of overbank flooding, while upstream channelization and levees may actually increase water velocity on the tract when flooding does occur. On-site impoundment would decrease water velocity on-site.

Frequency and Duration of Flooding

Frequency and duration of flooding are also strongly associated with the extent of sediment erosion and deposition on a site. All agricultural conversion activities considered here, as well as on-site construction and upstream impoundment, contribute to a reduction in the duration and frequency of flooding. These activities include features that keep floodwaters off the floodplain. On-site impoundment and upstream channelization and levees, on the other hand, could increase the duration and frequency of flooding.

Upland Runoff and Sediment Input to Tract

The runoff from uplands to a site can result in a substantial non-point input of sediments (as well as associated nutrients and toxins) to adjacent bottomland hardwood and aquatic ecosystems. Developmental activities, such as conversion to agriculture or aquaculture, divert upland runoff through ditches and levees into streams and effectively reduce overland flow through a bottomland hardwood site. As a consequence, the volume of upland runoff and transported sediments delivered to a site would decrease. Construction and operation of upstream impoundments, off-site channelization, and levee construction do not directly affect upland flows through a site and thus have no direct effect on this characteristic. On-site channelization, levee construction, and conversion to pine culture would not be expected to have any direct impacts on these characteristics; however, indirect effects are uncertain.

Percentage of Organic Matter in Soil

Generally, soil organic matter exhibits a high capacity to adsorb pesticides (Pionke and Chesters 1973), and metal ions may be complexed by humic and fulvic acids (Boto and Patrick 1978). The soils of bottomland hardwood ecosystems are typically high in organic matter content and, as such, represent an important, chemically reactive soil component capable of detaining certain toxins entering the system. Developmental activities that increase soil drainage also deplete soil organic matter through oxidation. Activities such as agricultural conversion, on-site channelization, and levee construction and upstream impoundment tend to reduce soil saturation and thus reduce the percentage of organic matter in bottomland hardwood soils. On-site impoundment would, in contrast, increase soil organic matter with the creation of anaerobic conditions. Any channelization or levees upstream that would increase the duration and frequency of flooding on the site would maintain or increase the soil organic matter content of the site.

Soil Disturbance

Soil disturbance was considered specifically with regard to the function of sediment detachment and transport control. Clearing and conversion to rice and soybeans create very serious soil disturbance on a site and eliminate many of the structural features of bottomland hardwood ecosystems that help to limit erosion of soil. Aquaculture has less of an impact because disturbance to the soil surface is not as intense or as frequent. Conversion to pine has a short-term impact during clearing and harvesting. On-site levee construction also has a limited short-term impact. Channelization has a severe short-term impact and may also have a long-term adverse effect if channels are not stabilized (Simon and Hupp 1987). There is little or no disturbance of the soil on-site as a result of upstream channelization, levee construction, and operation of impoundments.

Instream Sediment and Nutrient and Contaminant Load

Impacts on these characteristics are discussed from the perspective of the basin or watershed rather than from a site-specific perspective. Developmental activities (e.g., agricultural conversion) on the floodplain and adjacent uplands can greatly increase the gross erosion within a basin and contribute to higher sediment loads in downstream waters (Kolb et al.

1976). Vegetation, through which sediment-laden waters pass, has been shown to effectively trap and detain these sediments (Lee et al. 1976), and simulation studies suggest that sediment trapping efficiency may remain high over the long term (Simons et al. 1983). As the area in bottomland hardwoods within a basin decreases, the corresponding sediment loss within the basin increases (Molinas and Auble 1988). Activities that involve the clearing of bottomland hardwoods, such as agricultural conversion or pine culture, would contribute to increased sediment loads within the basin. On- and off-site channelization and levee construction would also contribute to increased sediment loads by increasing upstream or across-site water velocities. On- or off-site impoundments, however, would act to trap sediments and thus reduce sediment loads within the basin.

The transport and deposition of nutrients and contaminants are closely associated with sediments (Pionke and Chesters 1973, Parr et al. 1974, Mitsch et al. 1979). Given that nutrient and contaminant sources exist within a basin, we consider activities that influence sediment load will also influence the instream load of nutrients and contaminants. Thus, the activities discussed with regard to sediment load would likewise influence the nutrient and contaminant load within a basin.

Overall Impact of Developmental Activities on Soil Detachment and the Transport Control Function

Detachment and transport of soil from natural bottomland hardwood forests are minimal because the typically low elevational gradient of the floodplain reduces water velocity across these sites during overbank flooding. Water velocities are further reduced by ground and vegetative surface roughness features. The soil also is protected by litter on the surface and dense root structure of the vegetation. Precipitation-driven erosion is also minimized by the canopies of overstory and understory vegetation that reduce the kinetic energy of the falling rain.

On-site activities that require the clearing of bottomland hardwood forests and that repeatedly disturb the soil (e.g., conversion to soybeans) can dramatically reduce the effectiveness of a bottomland hardwood site in controlling the pattern and extent of soil detachment and transport. Simulation models have shown that as the area of a basin in bottomland hardwood forest is reduced, sediment losses from the basin increase substantially (Molinas and Auble 1988). Clearing and conversion to rice also increase the potential for soil erosion, but levees and contour farming reduce the process to some extent. Aquaculture would not result in impacts as severe as rice farming because soil disturbance is not repeated as often. To an extent, soil erosion associated with agricultural conversion can be mitigated by use of detention areas, grade control structures, and best-management practices for agriculture, construction, and forestry (Christensen and Wilson 1976, Keown et al. 1977, Goldman et al. 1986).

Channelization (on-site) also contributes to increased and sometimes long-term streambank degradation (Simon and Hupp 1987) that may require some form of structural or non-structural stabilization. However, erosion of soil on the adjacent floodplain may actually be reduced by channelization to the extent that floodwaters are confined to the channel. The effects of construction and operation of upstream impoundments on soil erosion are generally negligible, as overbank flooding is reduced and there is no direct disturbance to the floodplain. Finally, conversion to pine results in intermittent periods of soil disturbance when trees are harvested and planted, but long-term impacts on the control of soil erosion would be relatively small.

Overall Impact of Developmental Activities on the Sediment Detention Function

Sediment-laden waters originating from overbank flooding or from adjacent uplands are trapped and detained by bottomland hardwood forests (Simons et al. 1983). The amount of sediment removed from water flowing through these bottomland forests is dependent upon a number of characteristics typical of bottomland hardwood ecosystems. During overbank flooding, water velocity is slowed considerably because of the level topography and ground surface roughness associated with bottomland hardwoods. Additionally, areas of natural ponding (e.g., old river channels) provide quiescent areas where very fine sediments may be deposited.

Aside from the case of on-site impoundment, the overall effect of development on bottomland hardwoods is the reduction of sediment detention. The primary reason for this is that developmental activities eliminate or reduce the opportunity for sediment-laden floodwaters originating from the stream or adjacent uplands to pass over the floodplain. Developed floodplains that do occasionally flood are no longer as effective in trapping sediments as undisturbed bottomland hardwood sites. This is because the roughness that exists in the form of uneven soil surfaces, stumps, fallen logs, litter, and various layers of vegetation in the undisturbed bottomland hardwood forest is typically reduced or eliminated. The consequence of development on the sediment trapping function in forested floodplains is that sediment in transport is no longer detained, thus leading to a degradation of water quality downstream.

Overall Impact of Developmental Activities on Nutrient and Contaminant Detention and Transformation

Because of level topography, ground surface roughness, and soil properties, bottomland hardwood ecosystems have a unique ability to trap nutrients and contaminants by adsorption or precipitation (Boto and Patrick 1978). Once retained, nutrients may be transformed from inorganic to organic or gaseous forms, while contaminants may be biologically degraded (Pionke and Chesters 1973). Nondegradable contaminants may be bound and deposited long-term within the sediments. Nutrient and contaminant detention and transformation functions are closely linked to the sediment trapping function.

In general, developmental activities have an adverse impact on the capacity of floodplains to retain and transform nutrients and contaminants. The potential of a bottomland hardwood site to trap and transform contaminants from a toxic to a nontoxic form is related to natural overbank flooding, drawdown, and the presence of natural bottomland hardwood vegetation. Factors such as pH, sediment organic matter, and clay content influence the distribution of adsorbed pesticides between sediments and water (Pionke and Chesters 1973). Developmental activities, such as landclearing or cultivation or draining, can result in changes in soil pH, organic matter content, and redox potential that in turn could mobilize nutrients and contaminants detained on-site (Boto and Patrick 1978). Thus, functions that are dependent upon natural conditions are likely to be dramatically altered by development. Furthermore, many agricultural activities implemented on bottomland hardwood areas following clearing require the use of pesticides or fertilizers to control pests or increase productivity. These chemicals may become off-site contaminants that further exacerbate the impact of development on water quality, with the ultimate consequence a deterioration of downstream water quality.

Summary

Water-quality functions of bottomland hardwood ecosystems are adversely impacted by almost all developmental activities, since undeveloped hardwood bottomlands probably optimize the characteristics involved in improving water quality. While direct effects of some developmental activities may be minimal for certain water-quality functions, or may enhance such functions (e.g., sediment detention as related to on-site impoundment), indirect effects are apparently detrimental without exception. For example, in the control of sediment detachment and transport, protection of bottomland hardwood sites from flooding almost always results in conversion activities that require clearing of native vegetation, elimination of ground surface roughness, repeated soil disturbance, and elimination of natural ponding. All of these effects contribute to increased sediment loss from the site and a corresponding deterioration of downstream water quality. Since all water-quality functions described for bottomland hardwood sites require that the floodplain be regularly flooded and that native vegetation be maintained in an undisturbed condition, any activity that results in flood control or conversion of bottomland hardwoods will have a negative impact on the water-quality function of bottomland hardwood ecosystems. This may result in a negative impact on water quality, particularly if best management practices or other mitigation measures are not included in the activity.

ASSESSMENT MODELS

In an effort to make the task of evaluating water-quality functions of a bottomland hardwood site more consistent and complete, a series of flow diagrams and narrative descriptions are presented for each of the two major water-quality functions identified for bottomland hardwood ecosystems. The characteristics discussed above are used to assess how well a particular bottomland hardwood site might perform each function. Generalized response curves were developed for most characteristics to illustrate the presumed nature of their relationship to the function. It should be noted, however, that these functional relationships are speculative in nature, and many are not well-documented in the literature. These diagrams and descriptions differentiate between opportunity and capacity as Adamus and Stockwell (1983) differentiate between opportunity and effectiveness. Some sites may have considerable potential for performing certain water-quality functions but, because of certain constraints, have little opportunity to perform those functions. For example, a densely forested bottomland may be isolated from the stream by a levee and have little opportunity to trap sediment. It seems incorrect to say that such a site has a low capacity to trap sediment; thus, such areas are defined as having low opportunity to trap sediment. The term capacity is used to distinguish how well a site might fulfill a function given the opportunity.

The following narratives and response curves present flow charts designed to evaluate the various functions, explain the rationale used in developing the systems, and discuss the characteristics that should be considered in attempting to answer general questions about a particular bottomland hardwood site. While the characteristics and response curves provide a good framework for the decision-making process, they are not intended to be used in a quantitative sense. For example, a positive response to three out of five response curves does not necessarily imply a "yes" to a question (though it may); judgment based on experience must be employed.

Two basic assumptions were made in presenting this assessment procedure for the water-quality aspects of bottomland hardwoods. First, functional attributes and characteristics were evaluated at the site-specific level. Some effort was made to consider functions and characteristics from the perspective of the entire watershed, and some of the characteristics (e.g., water velocity) selected to evaluate the identified functions do reflect off-site conditions or relationships. However, regardless of their importance, spatial and temporal relationships were given relatively little direct consideration. Second, the assumption was made that a particular bottomland hardwood site has a low overall opportunity to influence water-quality parameters if the integrity of the total watershed is high (i.e., the watershed is undisturbed [Figure 1]). For example, while an undisturbed bottomland hardwood site in a pristine watershed may have the structural capacity to detain sediments, little sediment is produced in an undisturbed watershed; thus, there is little corresponding opportunity for sediment detention. Similarly, a bottomland hardwood site in a disturbed watershed may have a high opportunity to detain sediments, but because of extensive accumulations of sediments, capacity to detain additional sediments may be low. Thus, the opportunity and the capacity of a particular site to detain sediments would be optimized in a watershed with some intermediate level of disturbance.

It should be pointed out that many remaining bottomland hardwood ecosystems exist in disturbed watersheds, particularly in the Lower Mississippi River Valley (MacDonald et al. 1979), emphasizing their opportunity to improve on-site water quality. Because a bottomland hardwood site in a degraded watershed has a greater opportunity to improve water quality than a site in a pristine watershed, it should not be implied that it is of more "importance." Indeed, bottomland hardwood ecosystems in undeveloped settings are the standards against which we measure all others. It is expected that site-specific decisions will be made in the context of present and predicted trends in water quality within a watershed.

Control of Sediment Detachment and Transport

Control of on-site sediment detachment and transport is the prevention of erosion or net removal of soil particles from a site. Because of their forested cover and occurrence on low-gradient floodplains, bottomland hardwood ecosystems are typically sites where erosional

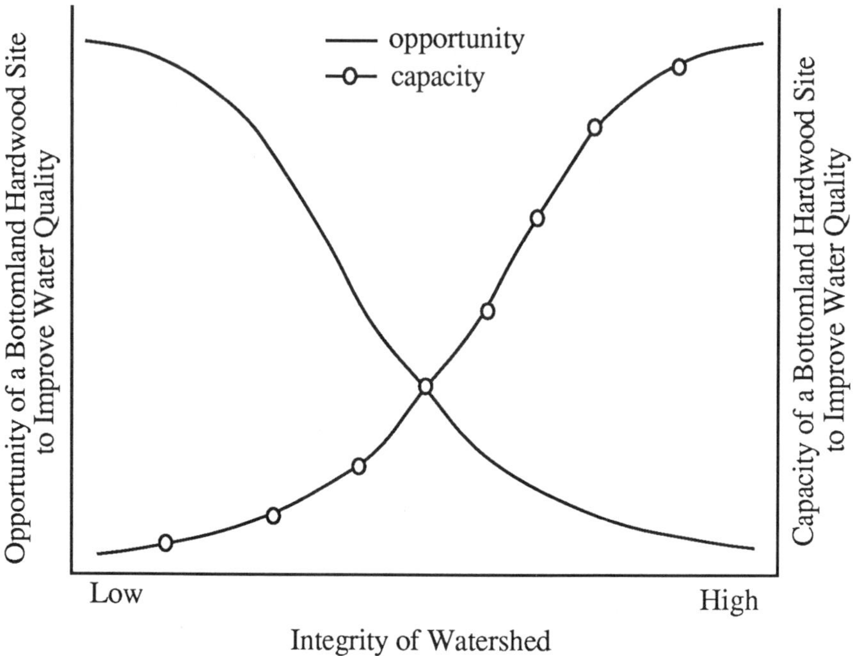

Figure 1. The relationship between capacity and opportunity of a bottomland hardwood site to improve water quality and the integrity of the watershed. The figure illustrates the idea that while a particular bottomland hardwood site in a pristine watershed might have a high structural <u>capacity</u> to detain sediment, the physical <u>opportunity</u> to actually detain sediments would be low because of low stream sediment loads in an undisturbed watershed.

processes are limited. We discuss the function of erosion prevention relative to a typical bottomland hardwood ecosystems capacity to limit or control erosional processes. The flow diagram in Figure 2 describes a set of procedural steps that might be followed in assessing the functional capacity of a bottomland hardwood ecosystem to limit erosion. These steps are based on how the characteristics listed for this function in Table 1 can be used to evaluate that function. The first question posed in this process (Figure 2) is: "Does the flooding regime on the site being evaluated impact the capacity of the site to control erosion?"

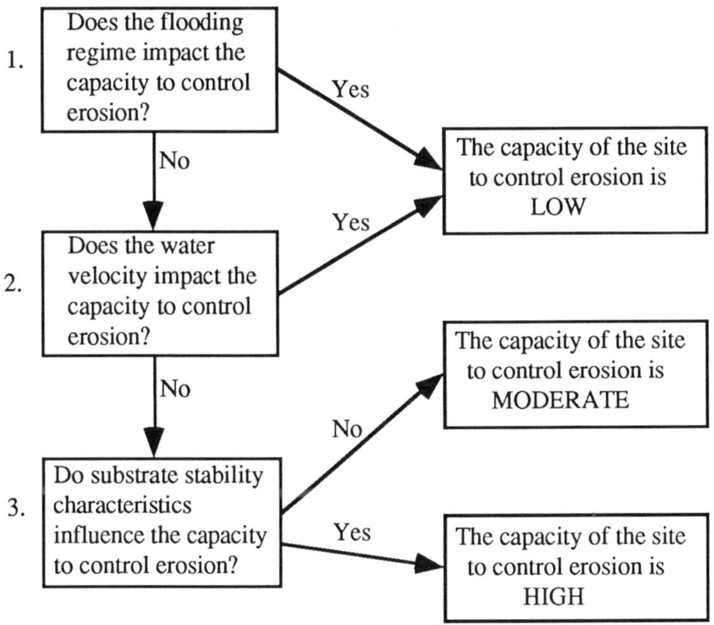

Figure 2. Assessment model for determining the capacity of a site to control erosion.

The relationship of the flooding regime to this function is determined largely by three characteristics: duration of flooding, frequency of flooding, and upland runoff (Figure 3). The duration of flooding is an estimate of the time that a flood event lasts (days, weeks, or months). The longer and more frequently a site is flooded, the more often the site is subjected to erosional forces. Thus, as frequency and duration of flooding increase, the capacity of a site to control erosion decreases (Figure 3a, b). General information on the timing and extent of flooding for a particular site may be obtained from stream gauge data. Additionally, physical evidence of on-site flooding would include the presence of rack lines, scour areas (e.g., areas of exposed soil or roots), and areas of sediment deposition. Upland runoff is another component of this function and, as with flooding frequency, increased flooding from adjacent uplands limits the capacity of a site to control erosion (Figure 3c). For example, adjacent uplands with steep topography and low soil permeabilities could be expected to deliver runoff with considerable erosion potential. Soils and landuse information could provide help in evaluating upland runoff to a bottomland hardwood site. If the flooding regime impacts the capacity to control erosion, then the capacity of the site to control erosion is low (Figure 2).

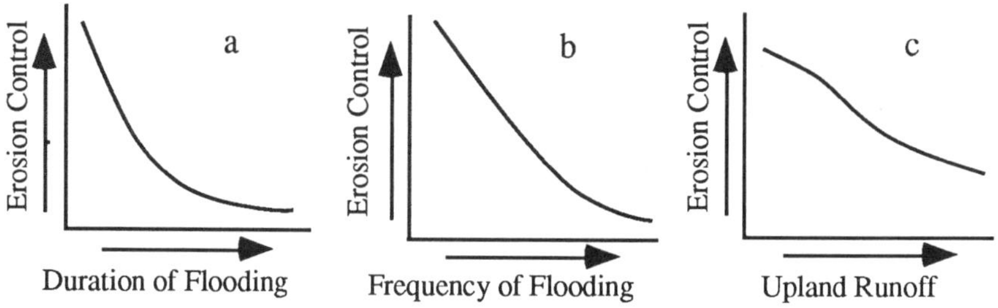

Figure 3. Characteristics that influence the relationship of flooding regime to erosion control.

If flooding does not impact a site's capacity to control erosion, proceed to question 2 (Figure 2): "Does water velocity on the site impact the capacity of a site to control erosion?" High water velocities across the site can detach sediment or soil particles and transport them off-site. This question can be evaluated by direct measurement of water velocity on the site, along with the extent of natural ponding, slope, and sinuosity (Figure 4). The capacity of a site to control erosion decreases with increasing water velocity (Figure 4a) and slope (Figure 4b). Water velocity can be measured directly on-site, or physical evidence may provide clues about the water-velocity regime at a particular site. Extensive litter accumulation and areas of sediment deposition would imply low water velocities across the site, while evidence of erosion would imply at least occasional high water velocities. Slope, or the elevational relief of a site, could be evaluated using a topographic map or by on-site inspection.

The extent of natural ponding and sinuosity is also a characteristic of erosion control (Figure 4c,d). As ponding increases on a site, there is an increase in areas of stagnant or sluggish water flow. This reduction in water velocity across a site is typical of bottomland hardwood forests and characterizes these ecosystems as areas with little or no net erosion. Sinuosity, or the amount of turning or bending in a water course, also influences water velocity. Sinuosity can be quantified and is defined as the ratio of channel length to downvalley distance (Leopold et al. 1964). Increasing the extent of natural ponding and sinuosity decreases water velocity across a site and thus increases the capacity of a site to limit or control erosion.

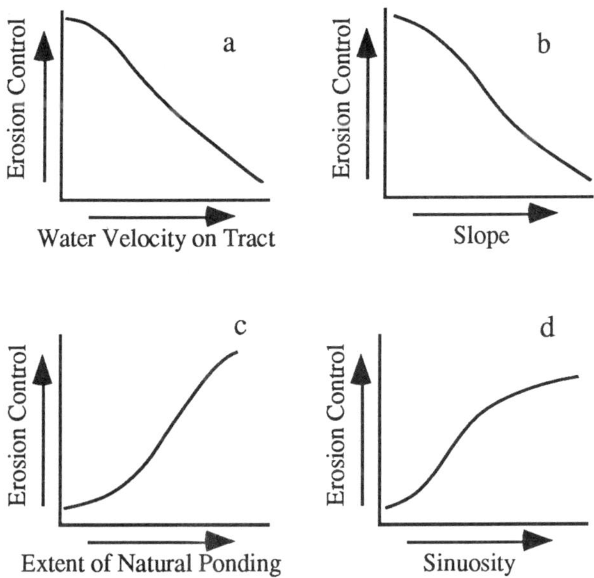

Figure 4. Characteristics that influence the relationship of water velocity to erosion control.

If water velocity and associated characteristics on a site impact erosion control, the capacity of the site to control erosion is low, but if water velocity on the site does not impact this function then question 3 must be evaluated (Figure 2): "Do the substrate stability characteristics influence the capacity of the site to control erosion?" The characteristics considered here are root mass density, overstory stem density and density of understory vegetation, ground surface roughness, soil disturbance, soil clay content, and soil organic matter content (Figure 5). Factors such as overstory density, understory density, and ground surface roughness influence the capacity of a site to limit erosional processes (Figure 5b,c,d), largely as a result of their influence on water velocity across the site.

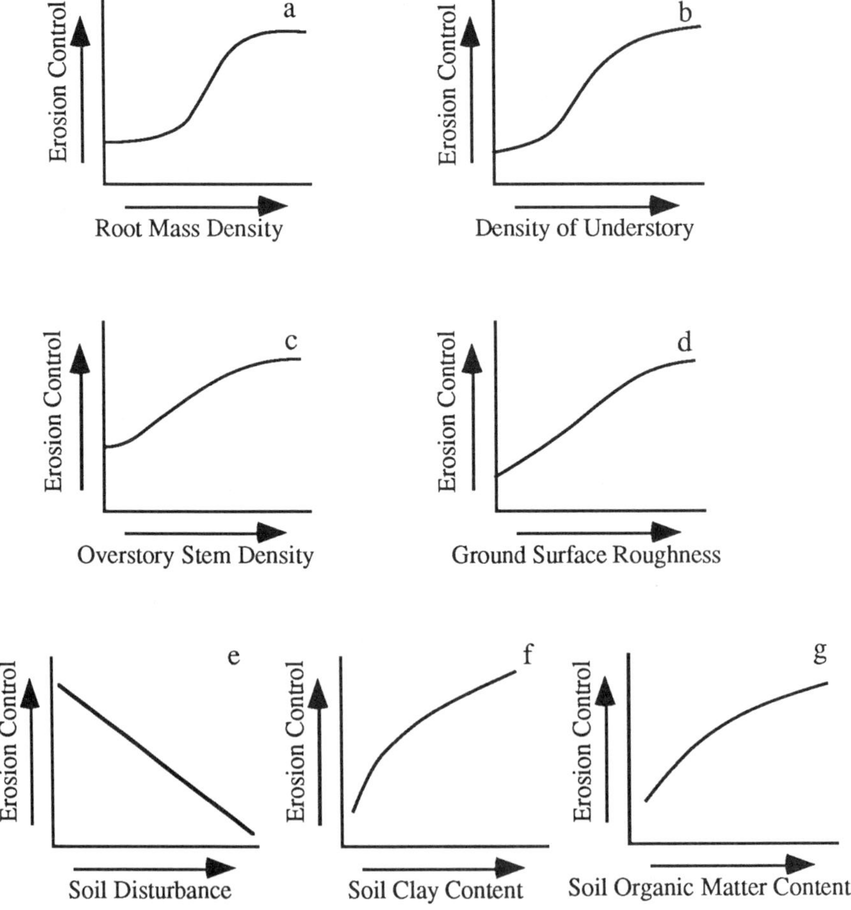

Figure 5. Characteristics that influence the relationship of substrate parameters to erosion control.

As the density of overstory stems and understory vegetation increases and as surface features (e.g., woody debris, hummocks, minibasins) increase, water velocity over the site is reduced, and the capacity of the site to control erosion increases. Whereas it is difficult to evaluate the ground surface and vegetational features of a bottomland hardwood site relative to bottomland hardwoods in general, Wharton et al. (1982) provide a detailed community profile of these ecosystems across the southeastern United States against which a specific site could be compared.

Factors such as soil disturbance, soil root mass, soil clay, and soil organic matter also contribute to the ability of a site to control erosion (Figure 5a,e,f,g). With increasing soil disturbance on a site, the capacity of a site to control erosion decreases. This results primarily from an increase in the extent of erodible surfaces. In contrast, as soil clay and organic matter increase on a site, the capacity of a site to control erosion would increase. Laboratory studies of the resistance of soils to erosion from flowing water have demonstrated that as the percentage of clay or other aggregating materials increases, soils become less erodible (Smeardon and Beasley 1959).

The extent of natural or human-induced soil disturbance could be evaluated by inspection of the site, noting the occurrence of scour channels or areas where vegetation has been disturbed or removed. The soils of bottomland hardwood forests typically form gradients based on elevation which directly influences the frequency and duration of flooding. Wharton et al. (1982) present a generalized example of the distribution of soil types across a bottomland hardwood floodplain and emphasize the variations in soil based on microrelief and the subsurface water table. Because of alluvial sediment inputs, bottomland hardwood ecosystems typically have high soil clay contents, and organic matter contents generally range between those of Pocosins and uplands. A U.S. Soil Conservation Service (SCS) county soils map may provide a general picture of the type of soils present at a particular site. Visual inspection of the amount of surface organic matter and color of subsurface soil would provide additional information on soil organic matter. The technique of making a soil "ribbon" would also give a general metric of the clay content in the soil.

The presence of a dense root system provides an additional structural feature of bottomland forests that helps to limit soil erosion (Figure 5a). Many bottomland ecosystems have surface root mats (see Wharton et al. 1982:28) that are obvious upon inspection. Generally, in assessing subsurface root density, the density of aboveground vegetation would be a good guide, with greater root densities associated with more dense, aboveground vegetation. A series of shallow (30 cm) holes dug into the soil would also provide information on the distribution and extent of subsurface root systems.

If the answer to question 3 (Figure 2) is "no," then the site is considered to have a moderate capacity to control erosion. If the answer to this question is "yes," then the site has a high capacity to control erosion.

Sediment Detention

An important water-quality function of bottomland hardwood floodplain forests is the role they play in temporarily detaining or permanently retaining sediments. This function becomes more important as floodplains and adjacent uplands are altered during development and greater quantities of sediments and commercial, industrial, and agricultural pollutants are introduced into associated aquatic systems.

The first question in this assessment (Figure 6)--"Is there an appreciable sediment load in the stream?"--can be answered by examining the characteristics presented in Figure 7. As the sediment load increases (Figure 7a), the sediment detention potential or opportunity of a site increases until the accumulated sediments decrease trapping efficiency (Simons et al. 1983). The sediment load can be roughly assessed in several ways. A visual inspection of the water can give a general indication, as can turbidity readings when compared to known measurements of other bodies of water in the area. Prior knowledge of upstream inputs, such as from agricultural or mining activities, can also be useful. Evidence of sediment deposition (e.g., buried litter layers, sediment on tree stems) also provides clues about the extent of sedimentation on-site. In addition to inputs from floodwaters from adjacent streams, sediment can be introduced to the site from upland runoff (Figure 7b). The sediment load in runoff can be evaluated by visual inspection and knowledge of the upland drainage area. The SCS soil maps may be useful in this evaluation.

Assessing the relative distance of the site in question from the water source can be beneficial in estimating sediment load (Figure 7c). It can be assumed that if sediment-laden waters travel a considerable distance through a forested or vegetated area to the site, the sediment load will be less than if the site is immediately adjacent to the water.

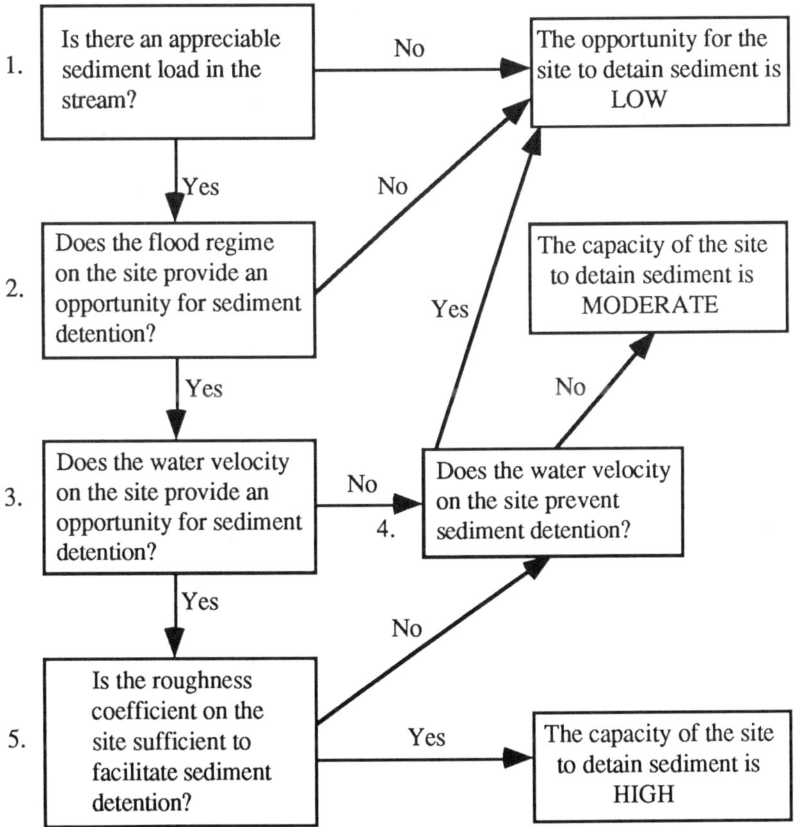

Figure 6. Assessment model for determining the opportunity and capacity of a site to detain sediment.

Frequently, the headwaters, or upstream portions of a watershed, have a lower sediment load than those farther downstream (Figure 7d). Downstream areas tend to receive greater amounts of sediment from the erosion and runoff of a larger area of the watershed. Different soil types within the watershed, different intensities of cultivation, and variation in amount of lateral stream flow all influence this relationship.

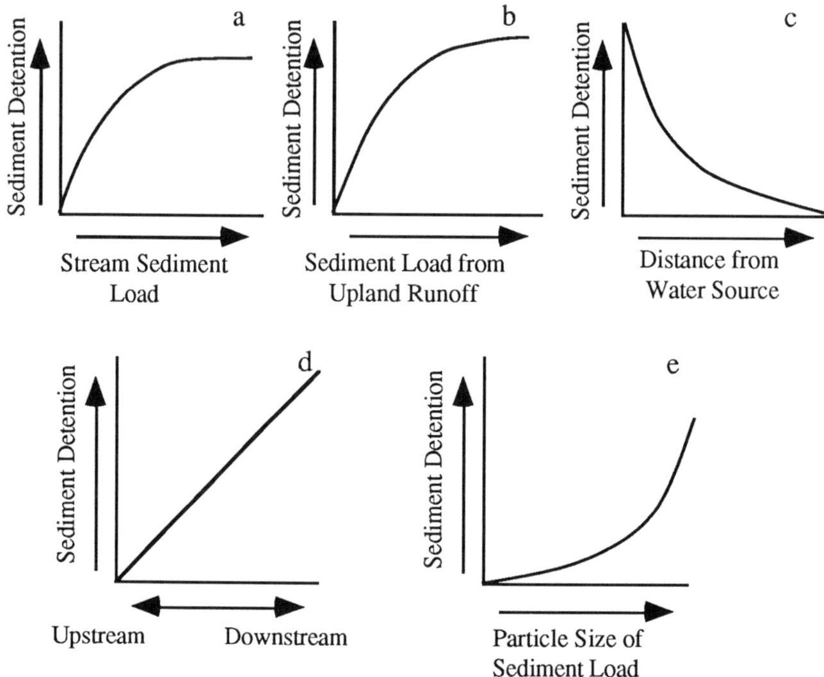

Figure 7. Characteristics that influence the relationship of sediment load to sediment detention.

Another aspect of sediment load important to detention (but difficult to measure) is sediment particle size. Larger, heavier particles "fall out" of the water column more quickly than clays and silts which may stay suspended even after traveling over a considerable distance of forested floodplain (Figure 7e). A knowledge of soil types in the area of upstream inputs can be useful in assessing this characteristic.

If the answer to the first question in Figure 6 is "no," the site has little opportunity to detain sediment. However, if the answer is "yes," then proceed to question 2 (Figure 6): "Does the flood or surface water regime of the site provide an opportunity for sediment detention?" The surface water regime (Figure 8) of a site, as it relates to sediment detention, basically indicates the probability that sediment-laden waters reach the site. Increased duration of floodwaters on the site increases the opportunity for sediment detention to occur, although a point may be reached where little additional sedimentation occurs (Figure 8a).

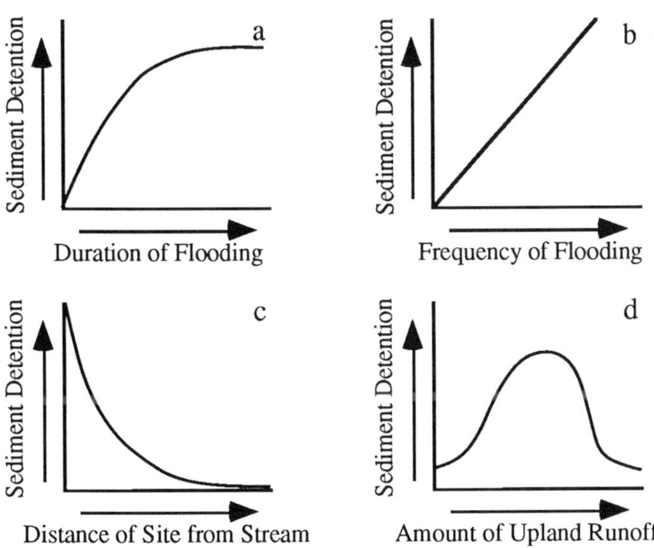

Figure 8. Characteristics that influence the relationship of the surface water regime to sediment detention.

The more frequently a particular site is flooded, the greater the opportunity for sediment detention (Figure 8b). Combining frequency and duration of flooding, the flood regime that best facilitates detention of sediment is many floods of moderate duration. These characteristics can best be determined by examining actual stream gauge data.

A characteristic that may be of some use in approximating flood regimes is the distance between the site and the stream (Figure 8c). Other gross indicators of flooding include the presence of rack lines, water lines, or sediment lines on trees; lichen growth patterns on tree trunks; the presence of objects suspended in shrubs or tree branches by floodwaters; and the absence of leaf litter due to flood scouring.

In addition to flooding, water contributions from upland runoff must be considered. A moderate amount of upland runoff allows for the greatest amount of sediment detention (Figure 8d). If the flooding regime does not provide an opportunity to detain sediments, then the opportunity for sediment detention on-site is low. If the surface water regime does facilitate sediment detention, question 3 from Figure 6, "Does water velocity on the site provide an opportunity for sediment detention?," should be addressed. The amount of sediment held in suspension is directly related to water velocity and to a lesser extent the size of particles being transported (sand, silt, or clay). As water moves through a floodplain and velocities are reduced, heavy, coarse particles settle out initially. Silts and clays settle as velocities are progressively reduced. Conversely, high water velocities can resuspend and transport sediments off-site. During most floods, sediment deposition, resuspension, and transport all occur simultaneously and nonuniformly.

The initial velocity of water flowing into a bottomland hardwood site is determined by off-site factors (e.g., stream or runoff water velocities, resistance to flow between source and site, etc.). Therefore, the position of a bottomland hardwood site in the floodplain, relative to water sources, needs to be considered (Figure 9a). Once sediment-laden waters reach the site, site-specific characteristics can reduce water velocity and increase sediment deposition. Water velocity is directly related to slope; thus, sediment detention is inversely related to slope (Figure 9b).

The surface roughness of a bottomland hardwood site reduces water velocity and increases the potential for sediment detention. A number of features of the site contribute to this characteristic, including ground surface roughness (Figure 9c) (e.g., surface irregularities and ground litter), overstory stem density (measured as the number and size of tree boles) (Figure 9d), and understory density (measured as the densities of saplings, shrubs, and herbs) (Figure 9e).

The kinetic energy of water is dissipated through friction as it moves through channels formed in the floodplain: the greater the sinuosity (Figure 9f), the greater the reduction in velocity. Depressions, sloughs, old levees, or other features that allow water to stand after flooding facilitate settling of sediment (Figure 9g).

The distance over which moving water interacts with the site (fetch) also partially determines the degree to which water velocity is decreased:

the greater the fetch, the greater the water velocity reduction, and the greater the sedimentation (Figure 9h). An analysis of fetch may be quite complex because it entails understanding water direction and pathways at various flood or runoff levels.

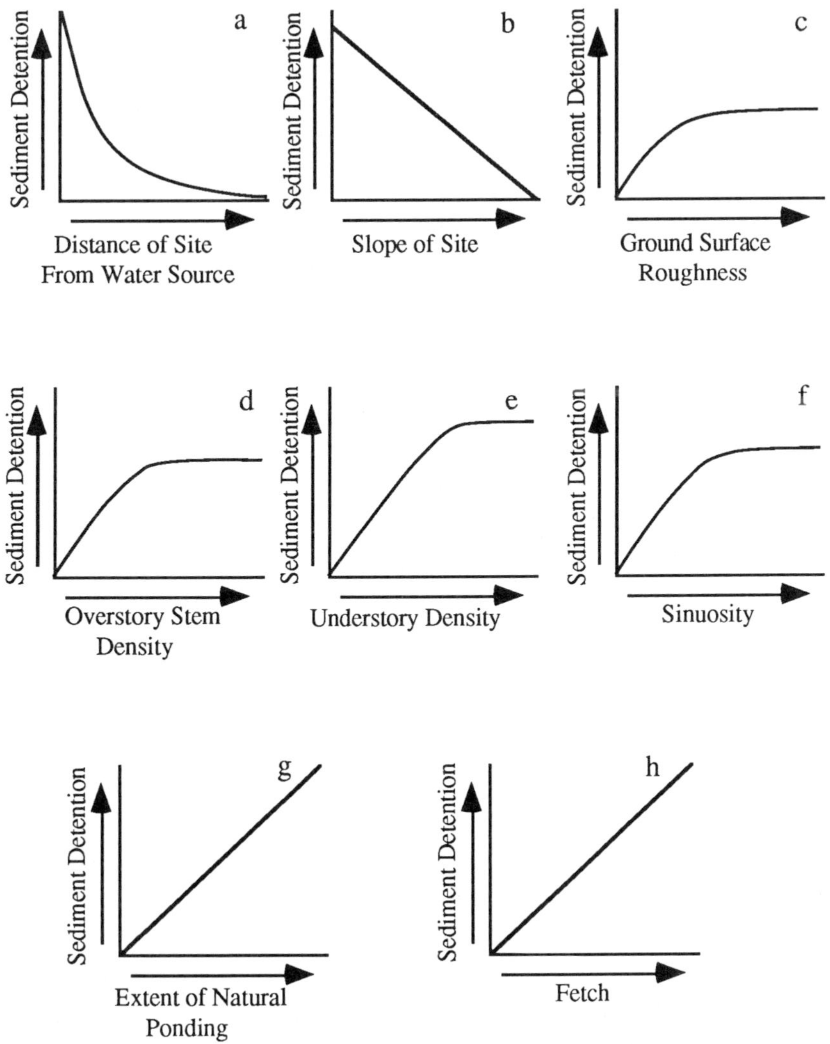

Figure 9. Characteristics that influence the relationship of water velocity to sediment detention.

If the answer to question 3 (Figure 6) is "no," then question 4 should be addressed: "Does water velocity on the site prevent sediment detention?" If water velocity on the site does not preclude sediment detention, the site has a moderate capacity to detain sediment. However, if resuspension and transport are high, then the opportunity for detention is low (Figure 6).

If water velocity on the site allows for deposition, proceed to question 5 (Figure 6): "Is the roughness coefficient of the bottomland hardwood site sufficient to facilitate sediment detention?" Sediment that settles out of the water column may be retained on the site or may be resuspended by subsequent floods and redeposited elsewhere on the floodplain. Characteristics of a bottomland hardwood site that keep water velocities during subsequent flood events below the threshold needed to resuspend sediments, or that trap sediments, are presented in Figure 10.

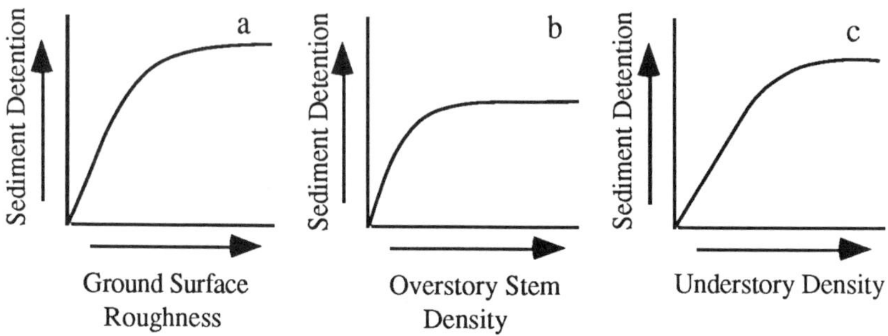

Figure 10. Characteristics that influence the relationship of surface roughness to sediment detention.

Surface irregularities of a bottomland hardwood site (Figure 10a) can trap and detain sediments either temporarily, until subsequent flood events resuspend them, or long-term if the sediments become incorporated into the soil. Ground features that detain sediments include soil fissures, stump holes, and crawfish holes. Leaf or other plant litter that buries sediments also helps detain them on the site.

Stem density of overstory vegetation and density of understory vegetation (Figs. 10b,c) can provide physical niches that trap and detain sediments. If the surface roughness of the site facilitates the detention of sediment, the capacity of the site to detain sediment is high (See Figure 6).

Nutrient and Contaminant Detention and Transformation

Nutrients refer to elements essential to plant and animal growth (e.g., nitrogen, phosphorus, etc.). Contaminants are substances that are toxic or inhibit the growth of plants, animals, or both. In sufficiently high concentrations, some nutrients may be considered contaminants. Detention of nutrients and contaminants relates to the trapping of these substances on a site by processes such as adsorption and deposition or uptake and storage in biomass. Transformation refers to chemical transformations, such as conversions between ferric and ferrous forms of iron, or to biological transformations, such as denitrification by bacteria. Transformation of a contaminant alters its chemical structure but may or may not detoxify the contaminant.

The flow diagram in Figure 11 includes steps that might be followed in assessing the opportunity or capacity of a bottomland hardwood site to retain and transform nutrients and contaminants. The first question in this assessment is: "Does nutrient and contaminant loading provide the opportunity for detention?," or are concentrations of nutrients or contaminants measurable (e.g., do concentrations exceed state or federal guidelines?). Characteristics of bottomland hardwood sites that can be used to answer question 1 are presented in Figure 12(a-c). Sources of nutrients and contaminants from upstream or from adjacent uplands can be ascertained in general from landuse maps of the basin. For example, extensive agricultural or industrial development within the basin might contribute nutrients and contaminants to a site during overbank flooding or directly from upland runoff. Since certain nutrients (Mitsch et al. 1979) and contaminants (Pionke and Chesters 1973) have been shown to be closely associated with sediments, we assume, given a nutrient and contaminant source, that loading and detention of these substances will closely resemble the pattern of sediment loading and detention. If the answer to question 1 is "no" (i.e., no measurable loading of nutrients and contaminants), the site obviously has little opportunity to trap or transform these substances (Figure 11).

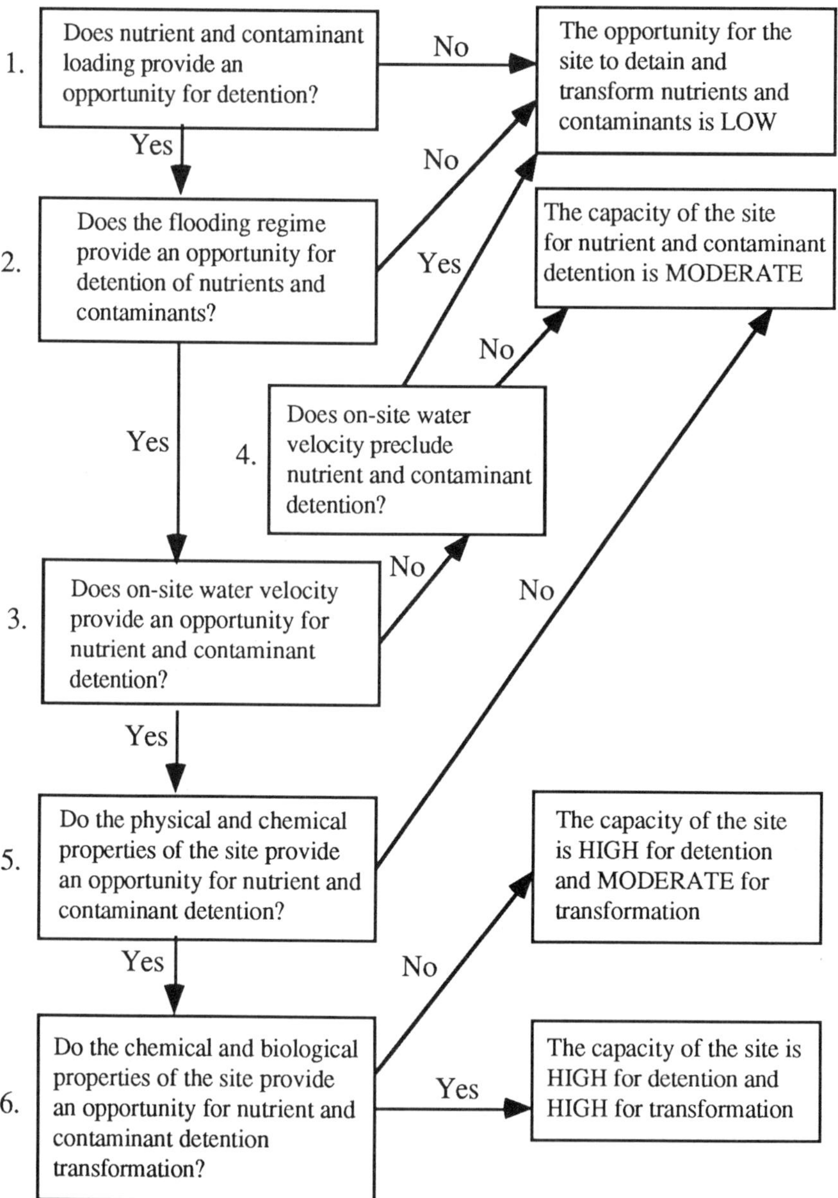

Figure 11. Assessment model for determining the opportunity and capacity of a site to detain and transform nutrients and contaminants.

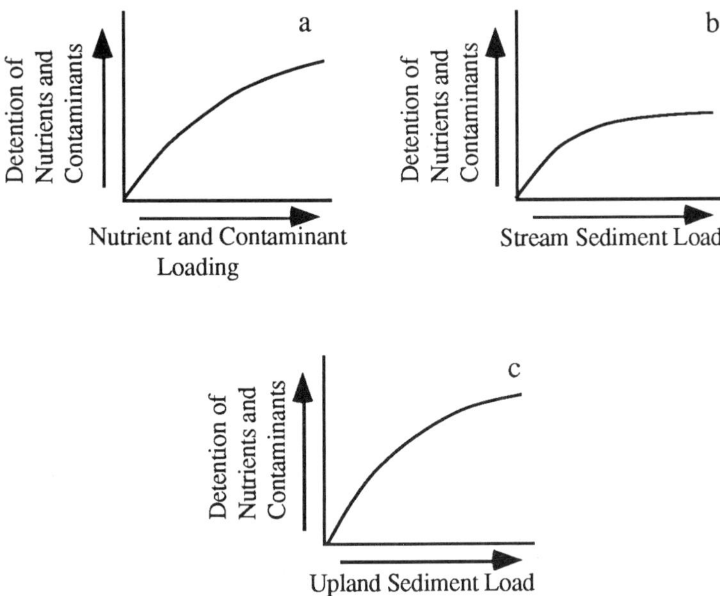

Figure 12. Characteristics that influence the relationship between nutrient and contaminant loading and nutrient and contaminant detention.

If there are appreciable inputs of nutrients or contaminants, proceed to question 2 (Figure 11): "Does the flooding regime provide the opportunity for detention of nutrients and contaminants?" To answer this question it must be determined if the site is subjected to flooding from the river or adjacent uplands. For example, is the site free from structures such as levees that restrict flooding frequency and duration? The response curves in Figure 13 (a-d) illustrate some of these relationships. As with sediment detention, the more frequently a site is flooded and the longer floodwaters remain on the site, the greater the likelihood that sediment-bound nutrients and contaminants will be detained. Field assessment of these characteristics are given above in the "Sediment Detention" section. If the flooding regime on-site does not facilitate the movement of nutrients and contaminants to the site, the opportunity for the site to perform this function is low (Figure 11).

If the flooding regime facilitates the movement of nutrients and contaminants to a site, the next step is to answer question 3 (Figure 11): "Does on-site water velocity provide an opportunity for nutrient and contaminant detention?" That is, do physical aspects of the site help to reduce the velocity of water moving over the site? The same physical characteristics that influence water velocity and subsequent sediment detention (fetch, sinuosity, ground surface roughness, extent of natural ponding, understory density, and overstory stem density) also influence detention of nutrients and contaminants (Figure 14a-f) and can be assessed in the field in the same way.

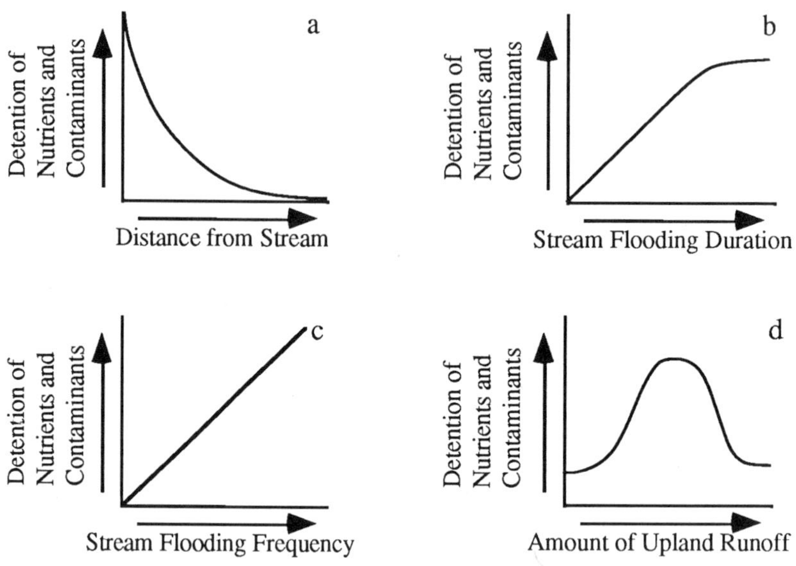

Figure 13. Characteristics that influence the relationship of flooding regime to nutrient and contaminant detention.

If water velocity on the site does not facilitate detention then question 4 (Figure 11) should be answered: "Does water velocity preclude nutrient and contaminant detention?" If water velocity during flooding is such that the site is subjected to scouring or erosion, the opportunity for the site to detain nutrients and contaminants is low. If water velocity does not preclude detention, then the site is considered to have a moderate capacity to detain nutrients (Figure 11).

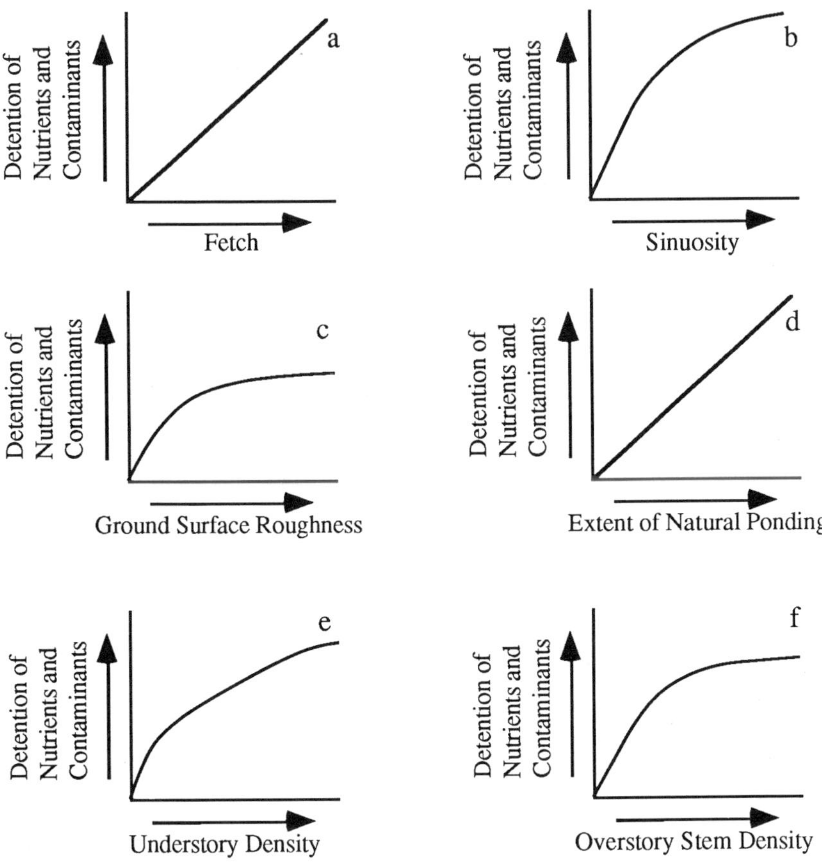

Figure 14. Characteristics that influence the relationship of water velocity to nutrient and contaminant detention.

If water velocity on the site provides an opportunity for detention, then proceed to question 5 (Figure 11): "Do physical and chemical properties of the site provide an opportunity for detention?" That is, is vegetation dense enough (Figure 15a) to provide physical detention? Do soil conditions on the site, such as high soil clay and organic matter content, facilitate the detention of nutrients and contaminants (Figure 15b,c)? In general, soil components with the highest charge and surface areas (e.g., organic matter, certain clays) exhibit high capacity for pesticide adsorption (Pionke and Chesters 1973). Again, field assessments of these characteristics would be the same as for sediment detention.

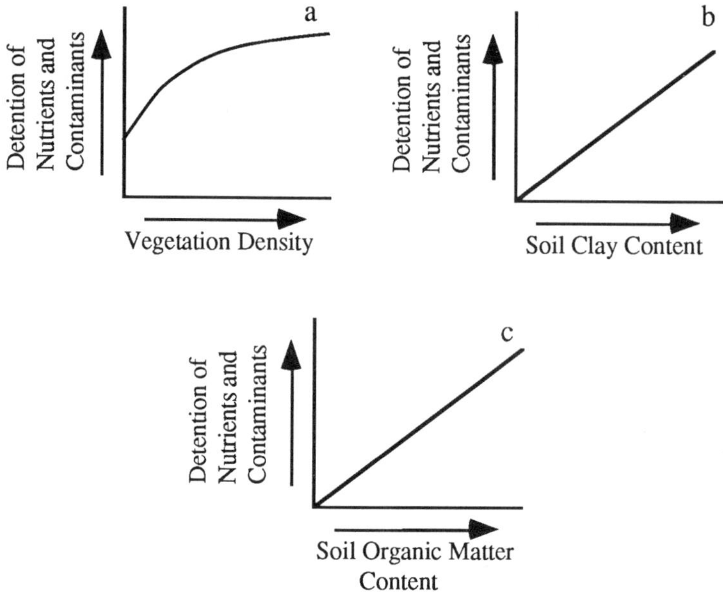

Figure 15. Characteristics that influence the relationship of physical and chemical properties to nutrient and contaminant detention.

If physical and chemical properties do not facilitate detention of nutrients and contaminants, then the capacity of the site to perform this function is moderate. If favorable physical and chemical properties are present, then question 6 (Figure 11) should be answered: "Do the chemical and biological properties provide an opportunity for nutrient and contaminant detention and transformation?"

Characteristics that are associated with the physical, chemical, and biological detention of nutrients and contaminants are similar to those involved in nutrient and contaminant transformations (Figure 16). However, we are concerned with characteristic processes that alter the chemical nature of nutrients and contaminants. In a regional review of the role of wetlands as sources, sinks, and transformers of certain nutrients and metals, Nixon and Lee (1986) discuss data that suggest swamps may remove oxidized, inorganic forms of nitrogen and phosphorus and transform them to reduced, organic forms. While information on metal transformations is even more limited, certain metals may be remobilized from sediments and possibly exported in dissolved or particulate form. We present here the relationships between nutrient and contaminant transformation and vegetation density, soil clay and organic matter content, and soil pH (Figure 16a-d). However, it should be emphasized that the relationship between nutrient and contaminant transformations and chemical and biological processes is known for a very small number of substances under a limited range of environmental conditions.

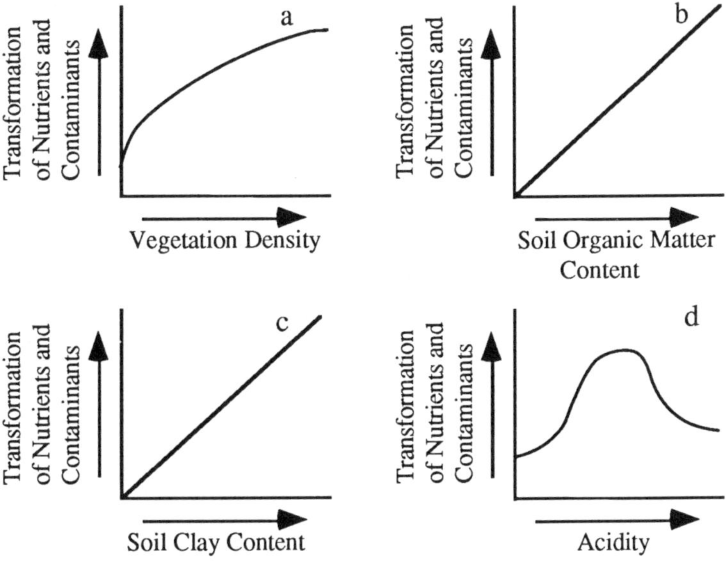

Figure 16. Characteristics that influence the relationship of chemical and biological properties to nutrient and contaminant transformation.

As vegetation density increases, the transformation of nutrients and contaminants should increase to some level (Figure 16a), as the likelihood of plant uptake would increase. Transformation of nutrients and contaminants would also increase as soil clay and organic matter increase (Figure 16 b,c). Numerous pesticides are strongly adsorbed to soil clay and organic material where they may undergo degradation. For example, some organochlorine insecticides (generally the slowest of organic pesticides to degrade) may be transformed rapidly in soils to less toxic compounds (Pionke and Chesters 1973). Finally, the influence of pH on nutrient and contaminant transformation is discussed relative to its specific effect on the biological component of transformation. Microbial degradation of pesticides appears to be the most common transformational pathway (Pionke and Chesters 1973); however, its efficiency is dependent upon soil environmental conditions, including pH. We assume that some transformations take place under pH optima, dependent upon the particular contaminants and array of microbes involved. This relationship is also assumed for biological transformations involving at least some nutrients (e.g., denitrification). Assessment of soil pH could be roughly assessed in the field using relatively simple soil pH kits. Additional soil and vegetation characteristics could be evaluated as they were for the other functions. In general, a bottomland hardwood site that is detaining nutrients and contaminants and exhibits a spatial and temporal range of soil conditions (mineral to organic, aerobic to anaerobic) is likely to provide a range of conditions suitable for the transformation of nutrients and contaminants.

If chemical and biological properties of the site are not conducive to transformation (e.g., little vegetative cover, low soil organic matter, low pH), the capacity of the site to transform nutrients and contaminants is moderate (Figure 11). However, if a site provides the chemical and biological conditions conducive to the transformation of nutrients and contaminants, then the capacity for transformation is high. A site with a moderate or high capacity for nutrient and contaminant transformation is considered to have a high capacity for the detention of nutrients and contaminants (Figure 11).

REFERENCES CITED

Adamus, P. R., and L. T. Stockwell. 1983. A method for wetland functional assessment: Vol. II. FHWA assessment method. Offices of Research, Development, and Technology, Federal Highway Administration, U.S. Department of Transportation, Washington, D.C. FHWA-IP-82-24.

Boto, K. G., and W. H. Patrick. Jr. 1978. Role of wetlands in the removal of suspended sediments. Pages 479-489 *in* P. E. Greeson, J. R. Clark, and J. E. Clark, eds. Wetland functions and values: the state of our understanding. American Water Resources Association. Minneapolis, Minnesota.

Brinson, M. M. 1977. Decomposition and nutrient exchange of litter in an alluvial swamp forest. Ecology 58:601-609.

Christensen, R. G., and C. D. Wilson, eds. 1976. Best management practices for non-point source pollution control. U.S. Environmental Protection Agency, Washington, D.C. EPA-905/9-76-005.

Goldman, S. J., K. Jackson, and T. Bursytynsky. 1986. Erosion and sediment control handbook. McGraw Hill, New York.

Green, W. E. 1947. Effect of water impoundment on tree mortality and growth. J. For. 45:118-120.

Harmes, W. R., H. T. Schreuder, D. D. Hook, and C. L. Brown. 1980. The effects of flooding on the swamp forest in Lake Ocklawaha, Florida. Ecology 61:1412-1421.

Keown, M. P., N. R. Oswalt, E. B. Perry, and E. A. Dardeau, Jr. 1977. Literature survey and preliminary evaluation of streambank protection methods. U.S. Army Engineer Waterways Experiment Station, Environmental Laboratory, Vicksburg, Mississippi. Tech. Rep. H-77-9.

Kitchens, W. M., J. M. Dean, L. H. Stevenson, and J. H. Cooper. 1975. The Santee Swamp as a nutrient sink. Pages 349-366 *in* F. G. Howell, J. B. Gentry, and M. H. Smith, eds. Mineral cycling in southeastern ecosystems. U.S. Government Printing Office, Washington, D.C.

Kolb, C. R., C. R. Bingham, B. K., Colbert, E. J. Clairain, Jr., J. R. Clark, E. Nixon, S. E. Richardson, and F. W. Suggitt. 1976. Data evaluations and recommendations for comprehensive planning for the Yazoo River Basin, Mississippi. Vol. II, Appendix. B. U.S. Army Engineer Waterways Experiment Station, Environmental Laboratory, Vicksburg, Mississippi. Tech. Rep. Y-76-2.

Lee, C. R., R. E. Hoeppel, P. G. Hunt, and C. A. Carlson, 1976. Feasibility of the functional use of vegetation to filter, dewater, and remove contaminants from dredged material. US Army Engineer Waterways Experiment Station, Environmental Laboratory, Vicksburg, Mississippi. Tech. Rep. D-76-4.

Leopold, L. B., M. G. Wolman, and J. P. Miller. 1964. Fluvial processes in geomorphology. W. H. Freeman and Company, San Francisco, California.

MacDonald, P. O., W. E. Frayer, and J. K. Clausee. 1979. Documentation, chronology, and future projections of bottomland hardwood habitat loss in the Lower Mississippi Alluvial Plain, Vols. 1 and 2. U.S. Department of the Interior, U.S. Fish and Wildlife Service, Jackson, Mississippi.

Mitsch, W. J., C. L. Dorge, and J. R. Wiemhoff. 1979. Ecosystem dynamics and a phosphorus budget of an alluvial cypress swamp in southern Illinois. Ecology 60:1116-1124.

Molinas, A., and G. T. Auble. 1988. Application of landscape sediment model. Pages 40-66 *in* A. Molinas, G. T. Auble, and L. S. Ischinger, eds. Assessment of the role of bottomland hardwoods in sediment and erosion control. U.S. Fish and Wildlife Service, National Ecology Research Center, Fort Collins, Colorado. CO NERC-88/11.

Nixon, S. W., and V. Lee. 1986. Wetlands and water quality: a regional review of recent research in the United States on the role of freshwater and saltwater wetlands as sources, sinks, and transformers of nitrogen, phosphorus, and various heavy metals. U.S. Army Engineer Waterways Experiment Station, Environmental Laboratory, Vicksburg, Mississippi. Tech. Rep. Y-86-2.

Parr, J. G., G. H. Willis, L. L. McDowell, C. E. Murphree, and S. Smith. 1974. An automatic pumping sampler for evaluating the transport of pesticides in suspended sediments. J. Environ. Qual. 3:292-294.

Patrick, W. H., Jr., and M. E. Tusneem. 1972. Nitrogen loss from flooded soil. Ecology 53:561-569.

Pionke, H. B., and G. Chesters. 1973. Pesticide-sediment-water interactions. J. Environ. Qual. 2:29-45.

Reddy, K. R., and W. H. Patrick, Jr. 1975. Effect of alternate aerobic and anaerobic conditions on redox potential, organic matter, decomposition, and nitrogen loss in flooded soil. Soil Biol. Biochem. 7:87-94.

Simon, A., and C. R. Hupp. 1987. Geomorphic and vegetative recovery processes along modified Tennessee streams: an interdisciplinary approach to disturbed fluvial systems. Forest hydrology and watershed management. Proceedings of the Vancouver Symposium, August 1987. IAHS-AISH Publ. no. 167.

Simons, D. B., R. M. Li, and G. O. Brown. 1983. Sedimentation study of the Yazoo River Basin: phase II general report. Vols. I and II. Prepared by Colorado State University, Fort Collins, Colorado, for U.S. Army Corps of Engineers, Vicksburg, Mississippi.

Smeardon, E. T., and R. P. Beasley. 1959. The tractive force theory applied to stability of open channels in cohesive soils. Missouri University Agricultural Experiment Station, Columbia, Missouri. Research Bulletin 715.

Wharton, C. H. 1970. The southern river swamp--a multiple-use environment. School of Business Administration, Georgia State University, Atlanta, Georgia.

_____________, W. H. Kitchens, E. C. Pendleton, and T. W. Sipe. 1982. The ecology of bottomland hardwood swamps of the Southeast: a community profile. U.S. Fish and Wildlife Service, Biological Services Program, Washington, D.C. FWS/OBS-81/37.

14. COMPOSITION AND PRODUCTIVITY IN BOTTOMLAND HARDWOOD FOREST ECOSYSTEMS: THE REPORT OF THE VEGETATION WORKGROUP

William H. Conner
Baruch Forest Science Institute,
Georgetown, SC 29442

R. Terry Huffman
Huffman Technologies Company,
San Francisco, CA 94110

Wiley Kitchens
U.S. Fish and Wildlife Service
University of Florida,Gainesville, FL 32611

with Panel
W. Harms, S. Harrison, P. Hatcher, L. Holloway, R. Johnson, L. Lee,
A. Lucas, S. Ray, C. Rhodes, D. Sanders, R. Theriot, J. Wooten

ABSTRACT

Bottomland hardwood forests in the southern United States are transitional between drier terrestrial ecosystems and aquatic ecosystems. Hydrologic and soil moisture regimes have been recognized as the principal factors influencing plant species distribution and productivity. These same factors sometimes make it difficult to clearly separate bottomland forests into distinct communities, especially in areas changing as a result of natural processes or as a result of human intervention. Other than recognizing that bottomland forests are generally more productive than uplands from the same region and that these areas are very open to large fluxes of energy and nutrients, the function of these systems is still poorly understood.

INTRODUCTION

More than 70 commercially important tree species compose the complex and varied hardwood forests of the southern United States

---

Ecological Processes and Cumulative Impacts: Illustrated by Bottomland Hardwood Wetland Ecosystems. Edited by James G. Gosselink, Lyndon C. Lee, and Thomas A. Muir.

(Smith and Linnartz 1980). In 1977, the southern hardwood forests contained an estimated 1 billion cu. m of standing timber growing in bottomland and wetland (Table 1). However, because the rate of clearing of prime bottomland hardwood forests for agriculture is so great, hardwood forest area statistics can be considered only approximations at best.

Table 1. Summary of southern hardwood forest resource data (adapted from Smith and Linnartz 1980).

	Total Commercial Forest[a]		Bottomland and Wetland Forests[b]	
State	Area (thousand ha)	Net Volume, Hardwood Growing Stock (million cu. m)	Area (thousand ha)	Net Volume, Growing Stock (million cu. m)
Alabama	8,633	280	1,254	86
Arkansas	7,368	286	1,265	86
Florida	6,204	119	1,510	119
Georgia	10,041	327	1,276	131
Kentucky	4,817	313	420[c]	22[c]
Louisiana	5,879	231	2,280	148
Mississippi	6,836	214	1,707	106
North Carolina	7,917	436	1,095	108
Oklahoma	1,750	30	172	8
Pennsylvania	7,073	639	85[c]	4[c]
South Carolina	4,928	202	846	103
Tennessee	5,188	289	291	24
Texas	5,065	139	718	40
Virginia	6,451	401	172	15
West Virginia	4,648	370	329	22
Total	92.796	4,275	13,424	1,021

[a]From U.S. Department of Agriculture, Forest Service 1978.

[b]Data for Florida, Georgia, North Carolina, and South Carolina derived from Boyce and Cost (1974); for Virginia, from Cost (1976); for the other states furnished by Renewable Resources Unit, South. Forest Experiment Station, U.S. Department of Agriculture, Forest Service except where noted.

[c]Estimated from most recent state survey and U.S. Department of Agriculture, Forest Service (1978).

Millions of hectares of high-quality hardwood forests on the bottomlands of the Mississippi River and its tributaries have been converted to cropland and pastures because of the economic advantages of growing soybeans and raising beef cattle. At least 20% of the 3 million ha loss of commercial forest land between 1962 and 1970 took place in the Mississippi River Valley (U.S. Department of Agriculture, Forest Service 1974). This conversion from forests to cropland continued into the 1970's and has only recently showed some signs of decreasing (Rosson et al. 1988). In order to understand both the commercial and ecological consequences of this loss, it is important to examine more closely the composition and productivity of the southern bottomland forests.

COMPOSITION

Bottomland forests are transitional between drier terrestrial ecosystems and aquatic ecosystems, and for this reason they have generally been hard for researchers to define and classify. Except for the cottonwood (*Populus* spp.) and willow (*Salix* spp.) stands that colonize newly formed land and the low-diversity cypress-tupelo (*Taxodium distichum-Nyssa aquatica*) swamps, bottomland forests are composed of an extremely heterogeneous mixture of species (Smith and Linnartz 1980). For this reason, there was significant disagreement in the vegetation workgroup concerning the assignment of plant species to particular zones. In some cases, species and communities are distributed in zones such as those identified in previous workshops (Clark and Benforado 1981), but in other cases, the zones and communities are intermingled to the point that they are practically inseparable.

Bottomland hardwood forest areas have been called shallow swamps (Penfound 1952), seasonally flooded basins or flats (Shaw and Fredine 1956), narrow stream margins (U.S. Forest Service 1970), and mixed bottomland hardwood and tupelo-cypress swamps (Stubbs 1973). In this chapter, a bottomland hardwood wetland is defined by Huffman (in Larson et al. 1981:225) as "...a floodplain ecosystem dominated by woody vegetation that has demonstrated ability because of morphological adaptations, physiological adaptations and/or reproductive strategies to perform certain requisite life functions which enable the species to achieve maturity in an environment where the soils within the root zone may be inundated or saturated for various periods during the growing season." The components of these bottomland systems (land, water, plants, animals, atmosphere) are connected by a complex set of physical, chemical, and biological processes with the hydrologic regime strongly affecting their structure and function. Specifically, the hydrology governs the composition and productivity of vegetative communities, the transport and transformation of inorganic and organic materials, and the maintenance of fish and wildlife habitat (Wilkinson et al. 1987).

The vegetation of bottomland hardwood forests is dominated by a large number of trees with a range of commercial value that are adapted to a wide variety of environmental conditions (Table 2), defined principally by the "moisture gradient," or the "anaerobic gradient," as Wharton et al. (1982) calls it. This gradient varies in time and space across the floodplain as a function of water level dynamics. A list of tree and shrub species common to bottomland hardwood forests from wettest to driest conditions is shown in Table 3.

Ecological Zonation

The ecological zones referred to in Table 3 are the result of an interdisciplinary scientific conference held in 1980 (Clark and Benforado 1981). The differentiation of the zones is based on flooding regimes and soil saturation as is graphically depicted in Figure 1. The six zones range from zone I, where the soil is saturated for 100% of the year, to zone VI, where flooding is less than 2% of the growing season. Zone I is considered a strictly open-water ecosystem; zone II, a cypress-tupelo swamp-dominated area; zones III through V, the true bottomland hardwood forest zones; and zone VI, terrestrial, or a transition to an upland area. To better understand the differences among the zones, each one will be examined separately upon the basis of flooding regime, soil conditions, and dominant tree species associations.

The following information on the bottomland hardwood ecological zones is adapted from Slater (1986) who reviewed Larson et al. (1981), McKnight et al. (1981), Fowells (1965), Bahr et al. (1983), Wharton et al. (1982), and Taylor et al. (Chapter 2) unless otherwise referenced.

Table 2. Relative value, growth rate, and tolerance, of common bottomland hardwoods (adapted from Smith and Limmartz 1980).

Relative Value of species	Growth Rate	Tolerance to Competition	Tolerance to Periodic Flooding
Highest Value			
Green ash	Medium	Intolerant	Tolerant
Pumpkin ash	Medium	Intolerant	Tolerant
Eastern cottonwood	Very rapid	Very intolerant	Tolerant
Baldcypress	Slow to medium	Coderately tolerant	Vewry tolerant
Sweetgum	Medium to good	Intolerant	Intermediate
Cherrybark oak	Good to excellent	Intolerant	Very intolerant
Delta post oak	Medium	Moderately intolerant	Very intolerant
Shumard oak	Good to excellent	Intolerant	Very intolerant
Swamp chestnut oak (cow oak)	Medium to good	Moderately intolerant	Intolerant
American sycamore	Good to excellent	Very intolerant	Tolerant
Intermediate Value			
Hackberry	Poor to medium	Very tolerant	Intermediate
Sugarberry	Poor to medium	Very tolerant	Intermediate
Southern magnolia	Medium	Tolerant	Intolerant
Silver maple	Excellent	Intolerant	TolerantNuttall
oak	Good to excellent	Intolerant	Intermediate
Water oak	Good to excellent	Intolerant	Intermediate
Willow oak	Good to excellent	Intolerant	Tolerant
Pecan	Medium to good	Moderately intolerant	Intermediate
Swamp tupelo	Medium	Moderately intolerant	Very tolerant
Water tupelo	Medium	Intolerant	Very tolerant
Black willow	Excellent	Very intolerant	Tolerant
Low Value			
River birch	Good	Intolerant	Intermediate
Swamp cottonwood	Good to excellent	Moderately intolerant	Very tolerant
American elm	Medium	Tolerant	Intermediate
Cedar elm	Poor	Tolerant	Intermediate
Winged elm	Poor to medium	Tolerant	Intolerant
Black tupelo	Poor to medium	Moderately intolerant	Intolerant
Hickory spp.	Poor to good	Very tolerant	Intolerant
Honeylocust	Medium	Intolerant	Intermediate
Waterlocust	Good	Intolerant	Tolerant
Red maple	Medium to good	Tolerant	Tolerant
Red mulberry	Poor to medium	Very tolerant	Intolerant
Laurel oak	Good to excellent	Intolerant	Tolerant
Overcup oak	Poor to medium	Moderately intolerant	Tolerant
Pin oak	Good to excellent	Intolerant	Intermediate

Table 2. (Continued).

Relative Value of species	Growth Rate	Tolerance to Competition	Tolerance to Periodic Flooding
water hickory	Poor	Moderately tolerant	Tolerant
Common persimmon	Poor	Very tolerant	Intermediate
Weed Species			
Boxelder	Excellent	Moderately tolerant	Moderately tolerant
American hornbeam and Eastern hophornbeam	Poor	Very tolerant	Intolerant
Planertree	Poor	Moderately intolerant	Very tolerant
Roughleaf dogwood	Poor	Tolerant	Intermediate
Swamp-privet	Poor	Tolerant	Tolerant
Common buttonbush	Poor	Moderately intolerant	Very tolerant
Hawthorn spp.	Poor	Very tolerant	Intolerant
Possumhaw	poor	Tolerant	Tolerant

Zone I

These are permanently flooded areas. This zone includes river channels, oxbow lakes, and back sloughs. The most shallow parts of this zone are often vegetated by herbaceous aquatic plants.

Zone II

This zone represents the wettest area of flooded forest habitat, and it includes wet flats, sloughs, swales, and back swamps. Flooding and soil saturation is year-round except for periods of extreme droughts. Soils in this area are always highly reduced due to the anaerobic conditions that are caused by continuous flooding. Soils are high in organic matter and contain high levels of nutrients although nutrient availability is severely limited by anoxic conditions. This zone is dominated by bald cypress, pond cypress (*Taxodium ascendens*), water tupelo, and blackgum (*Nyssa sylvatica*) in freshwater areas, or mangrove (*Avicennia* sp.) swamp forests in saltwater areas. Buttonbush (*Cephalanthus occidentalis*) and water-elm (*Planera aquatica*) are common understory species in the freshwater areas.

Table 3. Selected tree and shrub species for bottomland hardwood forests in the southeastern United States (Larson et al. 1981).

Species	Ecological Zone				
	II	III	IV	V	VI
Taxodium distichum (bald cypress)	X	X			
Nyssa aquitica (water tupelo)	X	X			
Cephalanthus occidentalis (buttonbush)	X	X			
Salix nigra (black willow)	X	X			
Planera aquatica (water elm)	X	X			
Forestiera acuminata (swamp privet)	X	X			
Acer rubrum (red maple)		X	X	X	X
Fraxinus caroliniana (water ash)		X	X		
Itea virginica(Virginia willow)		X			
Ulmus americana var. *floridana* (Florida elm)		X	X		
Quercus laurifolia (laurel oak)		X	X	X	
Carya aquatica (bitter pecan)		X	X		
Quercus lyrata (overcup oak)		X	X		
Styrax americana (smooth styrax)		X			
Gleditsia aquatica (water locust)		X	X		
Fraxinus pennsylvanica (green ash)		X	X		
Diospyros virginiana (persimmon)		X	X	X	X
Nyssa sylvatica var. *biflora* (swamp tupelo)		X			
Amorpha fruticosa (lead plant)		X	X		
Betula nigra (river birch)		X	X		
Populus deltoides (eastern cottonwood)		X	X		
Baccharis glomeruliflora (groundsel)			X	X	X
Cornus foemina (stiff dogwood)		X	X		
Virburnum obovatum (black haw)			X		
Celtis laevegata (sugarberry)			X	X	X
Liquidamber styraciflua (sweetgum)			X	X	X
Acer negundo (box elder)			X	X	
Sabal minor Dwarf palmetto)			X	X	
Gleditsia triacanthos (honey locust)			X	X	X
Ilex decidua (possum haw)			X	X	X
Crataequs viridis (green hawthorn)			X		
Quercus phellos (willow oak)			X	X	X
Platanus occidentalis (sycamore)			X	X	X
Alnus serrulata (common alder)			X		
Ulmus crassifolia (cedar elm)			X		
Ulmus alata (winged elm)			X	X	X
Ulmus americana (American elm)			X	X	X
Quercus nuttallii(nuttall oak)			X		
Quercus virginiana (live oak)			X	X	X
Schinus terebinthigolius (Brazilian peppertree)			X	X	

Table 3. Concluded.

Species	Ecological Zone				
	II	III	IV	V	VI
Ascyrum hypericoides (St. Andrews cross)			X	X	X
Bumelia reclinata (bumelia)			X	X	
Carya illinoensis (pecan)			X	X	X
Carpinus cerifera (blue beach)			X	X	
Myrica cerifera (wax myrtle)			X	X	X
Psychotria sulzneri (wild coffee)			X	X	
Psychotria nervosa (wild coffee)			X	X	
Zanthoxylum fragara (wild lime)			X	X	
Morus rubra(red mulberry)			X	X	X
Ximenia americana (hog plum)			X	X	X
Sambucus canadenis (elderberry)			X	X	X
Magnolia virginiana (sweet bay)			X		
Sabal palmetto (cabbage palm)			X	X	
Ligustrum sinense (privet)			X	X	X
Crataegus marshallii (parsley haw)			X	X	
Quercus nigra (water oak)			X	X	X
Quercus michauxii (cow oak)			X	X	
Quercus flacata var. *pagodaefolia* (cherrybark oak)				X	X
Nyssa sylvatica (black gum)				X	X
Pinus taeda (loblolly pine)				X	X
Carya ovata (shagbark hickory)				X	X
Juniperus virginiana (eastern red cedar)				X	X
Callicarpa americana (American beautyberry)				X	X
Asimina triloba (paw paw)				X	
Ilex opaca (American holly)				X	X
Serenoa repens (saw palmetto)				X	X
Prunus serotina (black cherry)				X	X
Fagus grandifolia (American beech)				X	X
Magnolia grandiflora (southern magnolia)				X	X
Ostrya virginiana (eastern hop-hornbeam)				X	X
Sassafras albidum (sassafras)				X	X
Sargeretia minutiflora (sargeretia)				X	X
Quercus alba (white oak)				X	X
Cornus florida (flowering dogwood)				X	X
Tilia caroliniana (basswood)				X	X
Asimina parviflora (dwarf paw paw)				X	X
Euonymus americana (strawberry bush)				X	X
Carya glabra (pignut hickory)				X	X
Ptelea trifoliata (water ash)				X	X

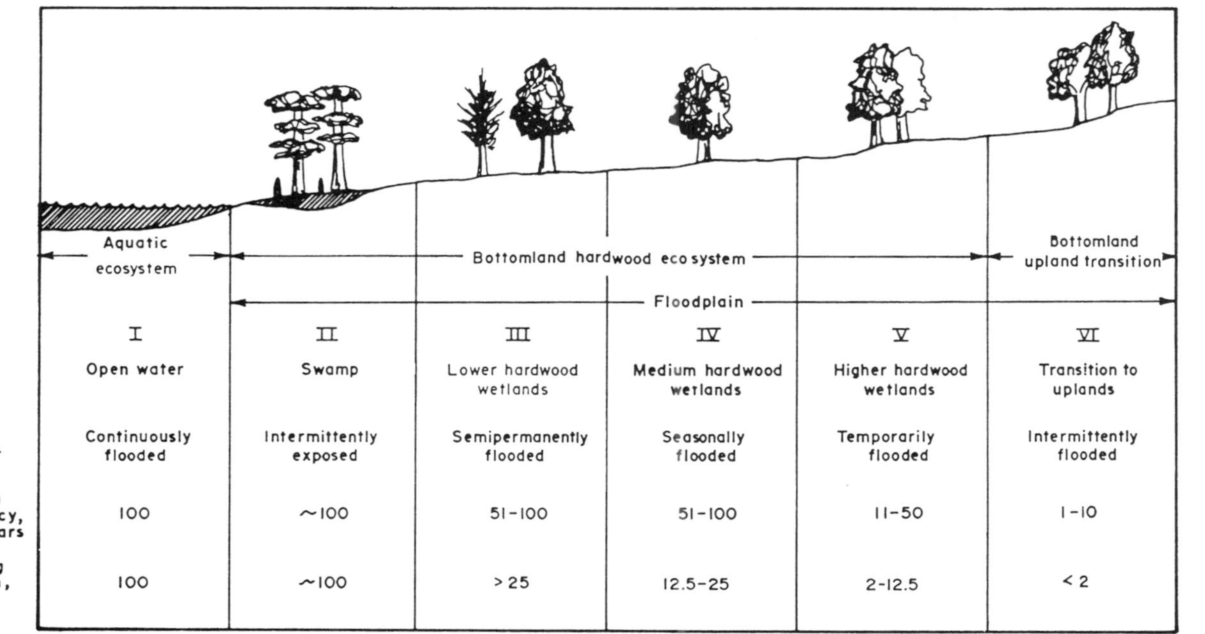

Figure 1. Zonal classification of bottomland forest wetlands showing average hydrologic conditions. (from Mitsch and Gosselink 1986, p.364; Copyright © by Van Nostrand Reinhold, reprinted with permission).

Transition From Zone II to Zone III

The density of bald cypress and tupelo decreases, while the number of water hickory (*Carya aquatica*), ash (*Fraxinus* spp.), and overcup oak (*Quercus lyrata*) increase. It has also been suggested that a drop in the organic matter content to about 5% occurs in this transition.

Zone III

This zone includes wet flats, bank-edge strips, point bars in rivers, and low levees. The probability of annual flooding is is greater than 50% with flooding occurring for extended periods during the growing season. Inundated or saturated soils are prevalent most of the year. However, zone III areas are subject to annual dry-downs when soils become unsaturated. The soils of zone III are usually dominated by clays which are soft at the surface, but firm below 5 cm. Typically, soils are mottled gray, indicative of reduced conditions and a general lack of oxygen movement.

In early stages of succession, this zone supports black willow (*Salix nigra*), red maple (*Acer Rubrum*), and sometimes cottonwood (*Populus heterophylla*). Later, overcup oak, water hickory, and ash dominate the forest canopy. Laurel oak (*Quercus laurifolia*), water locust (*Gleditsia aquatica*), and persimmon (*Diospyros virginiana*) are also common. These trees are shade-intolerant and are slow-growers, therefore, they are quickly replaced by other trees as drainage improves. The seedlings are either flood-tolerant or do not germinate until late in the spring after floodwaters have receded. This zone is characterized by little relief, therefore elevation gradients are difficult to distinguish.

Transition From Zone III to Zone IV

The transition between these two zones is somewhat more abrupt than that between zones II and III. Here, cypress, tupelo, buttonbush, and water-elm (zone II species) are present as scattered individuals along with individuals from zone IV. Most zone III species are present in this transition area, and most actually do appear in zone IV, but their dominance is greatly reduced.

Zone IV

This zone usually forms most of the floodplain area and is found chiefly on flats or terraces of low relief. Two physical features are common in this zone: "washboard" terrain, caused by parallel scour channels, and "hummocky" terrain, where trees either stand above the general soil surface or have large scour channels around them and through the forest. Zone IV is typified by inundation or soil saturation from one to two months of the year, usually during the early part of the growing season. Flooding is usually absent by the end of the growing season. Soils in this zone seasonally alternate between reduced and oxidized conditions. Clays predominate at the surface, but as depth increases the sediments become coarser. Nutrients are more readily available than in Zones II and III because of the extended dry-down periods.

Trees in this zone must be adapted to the sporadic flooding regime. The seeds of the trees germinate earlier in the spring, but because flooding is so variable, they must be able to withstand inundation for short periods of time. The immature trees in this zone are more shade-tolerant than those in the previous zones. The dominant tree species in zone IV are sugarberry (*Celtis laevigata*), green ash (*Fraxinus pennsylvanica*), sweetgum (*Liquidambar styraciflua*), American elm (*Ulmus americana*), and Nuttall oak (*Quercus nuttallii*).

Transition From Zone IV to Zone V

The change from zone IV to zone V is hard to distinguish using plant indicator species due to the large number of zone IV plants found in zone V. Physically, the trend is toward greater topographic relief and less soil saturation.

Zone V

Zone V is characterized by the highest and most well-drained areas.of the floodplain. This zone occurs adjacent to natural levee systems, Pleistocene terraces, and areas adjacent to terrestrial upland systems. Flooding occurs only for brief periods during the growing season. The water table drops far below the soil surface during the remainder of the year. Clay and sandy loam soils dominate. The soil texture and high natural drainage create an ideal soil-water-air relationship for root respiration. Understory species are more prevalent than in wetter sites. The trees are usually flood-intolerant and seedlings and seeds are adversely affected by inundation. Trees in this zone include water oak (*Quercus nigra*), American beech (*Fagus grandifolia*), hickories (*Carya* spp.) and, on the drier sites, live oak (*Quercus virginiana*) and loblolly pine (*Pinus taeda*).

Zone VI

This zone is not considered a "wetland" zone. Bottomland hardwood forest species do exist here, but this area is characterized by trees that are intolerant to soil saturation.

Problems With Zone System

The zonal classification is a practical framework for understanding broad floodplain community patterns (Wharton et al. 1982), but, like all classification schemes, it is flawed. The distribution of plant species in a bottomland hardwood forest community is a response to complex environmental gradients, and plant communities intergrade continuously along environmental gradients rather than clump up in distinct, separate zones. Some bottomland hardwood forest species may occur in more than one soil moisture or hydrologic zone. The occurrence of the same community type across several zones is particularly common in areas that have been significantly altered by flood control or timber management activities.

An example of the complexity of the problem in trying to classify forest vegetation by zones is shown in Figure 2. This is a cross-section of the microtopography of an alluvial floodplain in the southeastern United States. The complex microrelief does not allow a smooth change from one zone to the next. Further, the zonation is not necessarily consistent with the geomorphology of the floodplain. For example, the levee next to the stream, in fact, is often one of the most diverse parts of the floodplain due to fluctuations in its elevation. Plants from Zones II through V often occur on these levees (Wharton et al., 1982). If water levels are raised or lowered, communities will change but require some time to do so.

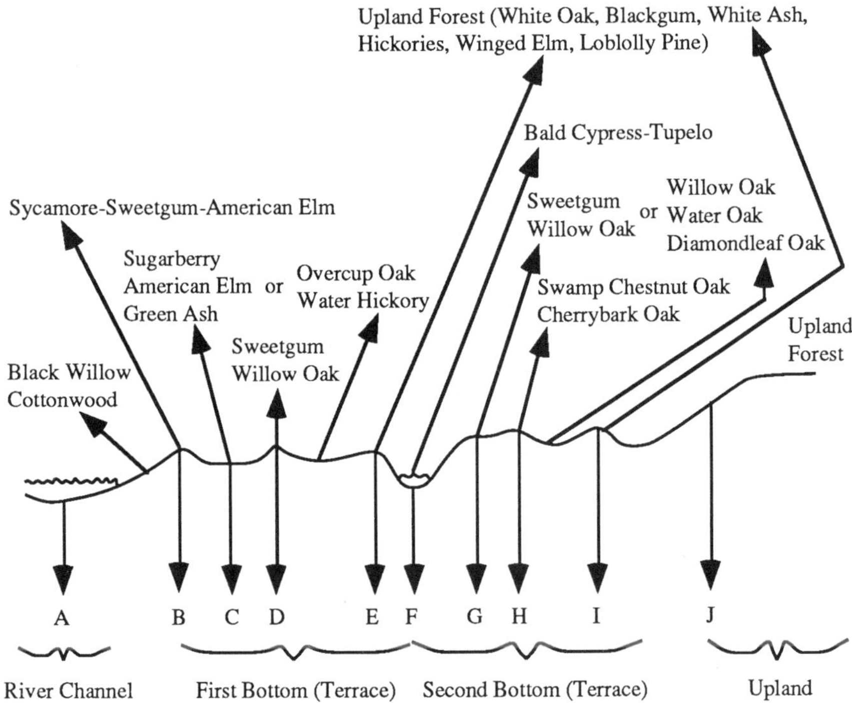

Figure 2. General relationship between vegetation associations and floodplain microtopography. A = river channel; B = natural levee; C = backswamp or first terrace flat; D = low first terrace ridge; E = high first terrace ridge; F = oxbow; G = second terrace flats; H = low second terrace ridge; I = high second terrace ridge; J = upland. (From Wharton et al. 1982)

The narrow, sharply incised riverine systems in the western United States are significantly different from the wide alluvial valleys of the eastern United States. Few species are common to both regions, due to the differences in climate, although species of cottonwood (*Populus* spp.) and willow (*Salix* spp.) are found in western riparian floodplains. One notable feature is the general absence of oak (*Quercus* spp.) in the west (Brinson et al. 1981). Keammerer et al. (1975) and Johnson et al. (1976) described the riparian forest along the Missouri River in North Dakota, where common species included cottonwood, willow, and green ash in the lower terraces, with American elm, box elder (*Acer negundo*), and bur oak (*Quercus macrocarpa*) at the higher elevations. The woody vegetation along the Colorado River in the Grand Canyon region includes salt cedar (*Tamarix chinensis*), mesquite (*Prosopis* spp.) and willow-cottonwood associations at the lower elevations, and willows and alders (*Alnus* spp.) at higher elevations (Johnson 1979). The Rio Grande in Texas and New Mexico has a continuum of riparian vegetation, from a domination of screwbean (*Prosopis pubescens*) in the xeric south to associations of Fremont cottonwood (*Populus fremontii*), Goodding willow (*Salix gooddingii*), Russian olive (*Eleagnus angustifolia*), and salt cedar in the north. The last two species were introduced in the last 50 years and have changed the character and successional characteristics of many riparian woodlands in the arid west (Brinson et al. 1981).

Another factor that must be considered when trying to characterize vegetation zones in forested wetlands is that these areas are very dynamic and in many cases constantly changing as a result of either natural processes or human intervention. For example, the coastal forests of Louisiana have developed as a result of the deposition of sediment from floodwaters of the Mississippi and Atchafalaya Rivers. As the land built up, cypress-tupelo swamps were replaced by bottomland hardwood species. With the completion of flood control levees along these rivers in the 1930's, overbank flooding and sedimentation ceased. In the absence of sedimentation, the regional subsidence is dominant. Thus, the forest are sinking relative to sea level. As a result, flood periods are becoming greater (Conner and Day 1988). Because of this change, plant species indicate one vegetative zone while flooding patterns indicate another. Species regeneration in gaps is more indicative of current hydrology.

PRODUCTIVITY

Bottomland hardwood forests are among the most productive of wetland ecosystems (Conner and Day 1976, Brinson et al. 1981 Brown and Peterson 1983). Bottomland hardwood forests with unaltered seasonal water flow generally have aboveground net primary productivity in excess of 1000 g/m^2/year (Table 4). The high productivity of

Table 4. Biomass and Net Primary Productivity of Deepwater Swamps in Southeastern United States (after Mitsch and Gosselink 1986).

Location/Forest Type	Tree biomass kg/m^2	Litterfall g/m^2/yr	Stem growth g/m^2/yr	NPP[a] g/m^2/yr	Reference
Louisiana					
Bottomland hardwood	16.5[b]	574	800	1,374	Conner and Day (1976)
Cypress-tupelo	37.5[b]	620	500	1,120	ibid
Impounded managed swamp	32.8[b,c]	550	1,230	1,780	Conner et al. 1981
Impoundment stagnant swamp	15.9[b,c]	330	560	890	ibid.
Tupelo stand	36.2[b]	379	–	–	Conner and Day (1982)
Cypress stand	27.8[b]	562	–	–	ibid.
North Carolina					
Tupelo swamp	–	609-677	–	–	Brinson (1977)
Floodplain swamp	26.7[d]	523	585	1,108	Mulholland (1979)
Virginia					
Cedar swamp	22.0[b]	757	–	–	Dabel and Day (1977) Gomez and Day (1982)
Maple gum swamp	19.6[b]	659	–	–	ibid.
Cypress swamp	34.5[b]	678	–	–	ibid.
Mixed hardwood swamp	19.5[b]	652	–	–	ibid.
Georgia					
Nutrient-poor cypress swamp	30.7[e]	328	353	681	Schlesinger (1978)
Illinois					
Cypress-tupelo	45[d]	348	330	678	Mitsch (1979) Dorge, et al. (1984)
Floodplain forest	29.0	–	–	1,250	Johnson and Bell (1976)
Transition forest	14.2	–	–	800	Johnson and Bell (1976)
Floodplain forest	–	491	177	668	Brown and Peterson (1983)
Kentucky					
Floodplain forests	30.3 18.4	420 468	914 812	1,334 1,280	Taylor (1985)

Table 4. (Concluded).

Location/Forest Type	Tree biomass kg/m²	Litterfall g/m²/yr	Stem growth g/m²/yr	NPP[a] g/m²/yr	Reference
Florida					
Cypress-tupelo	19[f]	–	289[f]	760[g]	Mitsch and Ewel (1979)
Cypress-hardwood	15.4[f]	–	336[f]	950[g]	ibid.
Pure cypress stand	9.5[f]	–	154[f]	–	ibid.
Cypress-pine	10.1[f]	–	117[f]	–	ibid.
Floodplain swamp	32.5	521	1,086	1,607	Brown (1978)
Natural dome[h]	21.2	518	451	969	ibid.
Sewage dome[i]	13.3	546	716	1,262	ibid.
Scrub cypress	7.4	250	–	–	Brown and Lugo (1982)
Drained strand	8.9	120	267	387	Carter et al. (1973)
Undrained strand	17.1	485	373	858	ibid.

[a]NPP = net primary productivity = litterfall + stem growth
[b]Trees defined as > 2.54 cm DBH (diameter at breast height)
[c]Cypress, tupelo, ash only
[d]Trees defined as > 10 cm DBH
[e]Trees defined as > 4 cm DBH
[f]Cypress only
[g]Estimated
[h]Average of five natural domes
[i]Average of three domes; domes were receiving high nutrient waste water

bottomland hardwood forests has been attributed, in part, to the alternating wet-dry cycles of major river floodplains. Conner and Day (1976) compared this periodic flushing of bottomland hardwood forests with the tidal flushing of the saline marshes. Several beneficial inputs are offered to the forests because of their location on these floodplains. The inputs include water, particulate and dissolved organic matter, and nutrients in dissolved, particulate, and sediment-adsorbed forms. The yearly input of organic materials and nutrients by rivers and the subtropical temperatures support an enhanced rate of community metabolism. This high metabolism results in:

1) high annual litterfall and nutrient turnover,
2) high detrital decomposition rates,
3) periodic flushing of accumulated detritus and metabolic waste products, and

4) optimal conditions for several microbial conversion processes; i. e., nitrification, ammonification, sulfate reduction, etc. (Wharton and Brinson 1979).

Fluctuating water levels and the flux associated with these water level changes are energy subsidies which generate beneficial variations in hydric conditions, temperature, nutrient levels, and available oxygen. compared with fluctuating water level systems, forested wetlands with stagnant or sluggish waters are usually less productive, and communities in permanently impounded conditions, or on sites with poor drainage leading to continuously high water tables and the accumulation of acidic peat soils, typically have lower productivity. The proximal causes of the productivity decrease are believed to be low nutrient turnover, nitrogen limitations, and low pH (Brown et al. 1979). Similarly, productivity also decreases from seasonally flooded forested wetlands to drained.wetlands. This change in productivity with respect to flooding has been discussed by several authors (Odum 1978, Conner and Day 1976, 1982). These researchers graphically compared the productivity of stagnant, seasonally flooded, and abrasively flooded systems with a regional average of all wetland and upland forests types (Figure 3). Seasonally flooded forested wetlands exhibit the highest productivity.

Another example of the effect of flooding on productivity was shown by Conner et al. (1981). They compared the productivity of wetland forests with different flooding regimes. One area was impounded and permanently flooded; a second was undisturbed and flooded naturally throughout the year; and a third area was a managed crawfish farm which was seasonally flooded in the late fall through late spring and thoroughly drained in the summer. The net primary productivity totals were 887, 1166, and 1780 $g/m^2/yr$, respectively. The authors found that the controlled flooding regime which resulted in the peak production was closest to the natural flooding once common to unleveed riverine systems along the Mississippi River.

The relative distribution of net primary productivity along a generalized floodplain cross section is presented in Figure 4. It appears that bottomland hardwood forest production peaks when flooding occurs once per year during the winter and early spring (Broadfoot 1967, Broadfoot and Williston 1973, Gosselink et al. 1981). This type of

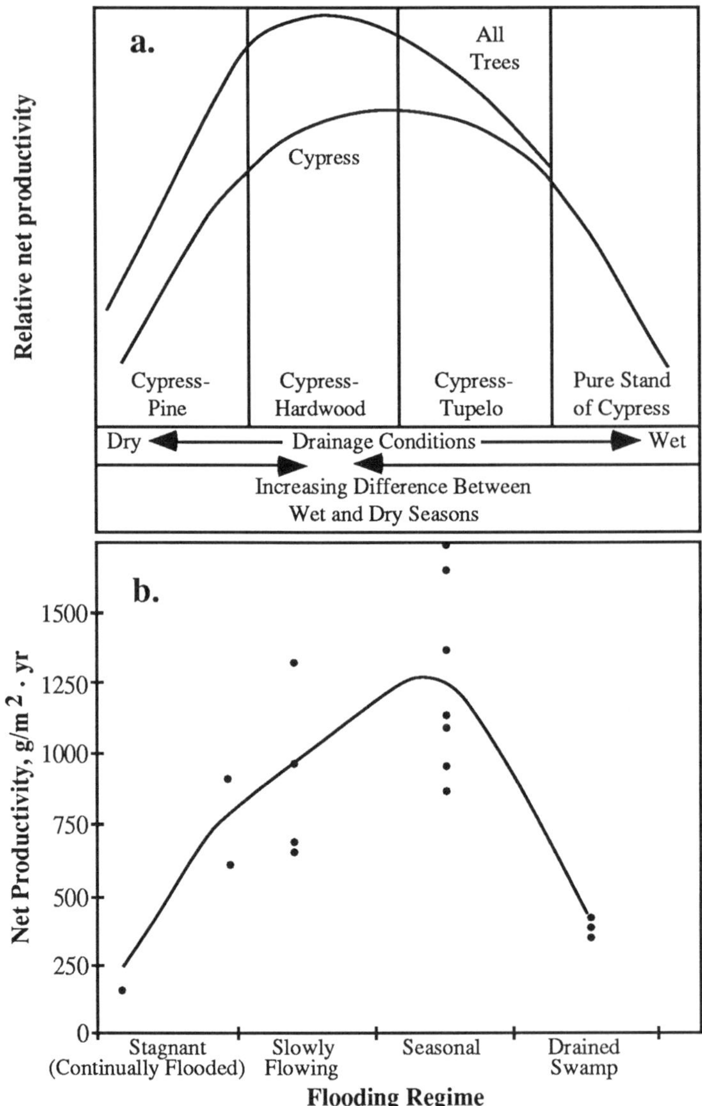

Figure 3. **The relationship of cypress swamp productivity and hydrologic conditions. (a. From Mitsch and Ewel, 1979, p. 424; copyright © 1979 by American Midland Naturalist, reprinted with permission. b. From Conner and Day, 1982, p. 74; copyright © 1982 by International Scientific Publications, reprinted with permission.**

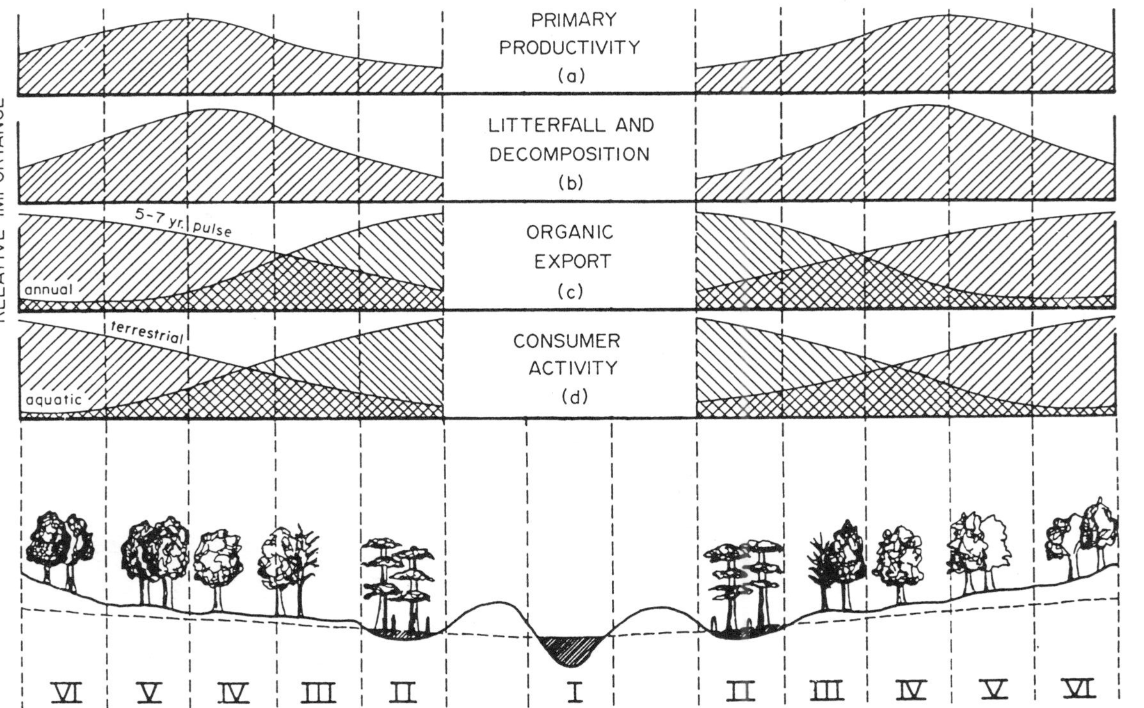

Figure 4. Patterns of community dynamics in bottomland hardwood wetlands (from Mitsch and Gosselink 1986, p.379; Copyright © by Van Nostrand Reinhold, reprinted with permission).

flooding regime offers the best environment for plant growth by assuring an adequate supply of nutrients, adequate soil moisture during the drier times of the year, and possible aerobic soil conditions during the summer leading to inorganic nutrient release from organic debris (Gosselink et al. 1981).

Recently, however, Brown and Peterson (1983) reported that they found above-ground productivity in a flowing-water forest in Illinois to be less than that of a still-water forest, suggesting that the generalization that floodplain forests are most productive is not always true. Also, it is important to note that biomass (that is, stem density) is greater in cypress-tupelo areas than in hardwood communities.

SUMMARY

Bottomland hardwood wetlands, which have a high water table because of their proximity to a river, stream or other body of water, are unique because of their linear distribution along rivers and streams and because they process large fluxes of energy and materials from upstream systems. Major expanses of bottomland hardwood forests are still found in the southeastern United States, although many have been drained and cleared for agriculture. Because bottomland forest composition and productivity depend on hydroperiod, bottomlands have been divided, for convenience,into zones ranging from intermittently exposed to intermittently flooded. Flooding affects soil chemistry by producing anaerobic conditions, importing and removing organic matter, and replenishing mineral nutrients. The plant communities of bottomland hardwood ecosystems are most productive and diverse in areas of seasonal flooding, although recent evidence shows that this relationship does not always hold. Ecosystem function of these systems is still poorly understood, except that primary productivity is generally higher than that in uplands in the same region. The systems are unique in their openness to large fluxes of energy and nutrients. The division of bottomland hardwood forests into distinct zones based on vegetation and productivity is complex, as plant species distributions are a response to complex environmental factors, including rainfall, temperature, nutrients, flooding, and soils.

REFERENCES CITED

Bahr, L. M., Jr., R. Costanza, J. W. Day, Jr., S. E. Bayley, C. Neill, S. G. Leibowitz, and J. Fruci. 1983. Ecological characterization of the Mississippi Deltaic Plain Region: a narrative with management recommendations. U.S. Fish and Wildlife Service, Division of Biological Services, Washington D. C. FWS/OBS-82/69. 189 pp.

Boyce, S. G., and N. D. Cost. 1974. Timber potentials in the wetland hardwoods. Pages 130-151 *in* M. C. Blount, ed. Water resources, utilization, and conservation in the environment. Taylor Printing Co., Reynolds, Georgia.

Brinson, M. M. 1977. Decomposition and nutrient exchange of litter in an alluvial swamp forest. Ecology 58:601-609.

____________, A. E. Lugo, and S. Brown. 1981. Primary productivity, decomposition, and consumer activity in freshwater wetlands. Annu. Rev. Ecol. Syst. 12:123-161.

Broadfoot, W. M. 1967. Shallow water impoundments increases soil moisture and growth of hardwoods. Soil Sci. Soc. Am. Proc. 31:562-564.

______________, and H. L. Williston. 1973. Flooding effects on Southern forest. J. For. 71(9):584-587.

Brown, S. 1978. A comparison of cypress ecosystems in the landscape of Florida. Ph. D. dissertation, University of Florida, Gainesville, Florida. 569 p.

_______, and A. E. Lugo. 1982. A comparison of structural and functional characteristics of saltwater and freshwater forested wetlands. Pages 109-130 *in* B. Gopal, R. E. Turner, R. G. Wetzel, and D. F. Whigham, eds. Wetlands -- ecology and management. National Institute of Ecology and International Scientific Publications, Jaipur, India.

_______, and D. L. Peterson. 1983. Structural characteristics and biomass production of two Illinois bottomland forests. Am. Midl. Nat. 110(1):107-117.

Brown, S., M. M. Brinson, and A. E. Lugo. 1979. Structure and function of riparian wetlands. Pages 17-31 *in* R. R. Johnson and J. F. McCormick, eds. Strategies for protection and management of floodplain wetlands and other riparian systems. U.S. Department of Agriculture, U.S. Forest Service, Washington, D.C. Publication GTR-WO-12.

Carter, M. R., L. A. Burns, T. R. Cavinder, K. R. Dugger, P. L. Fore, D. B. Hicks, H. L. Revells, and T. W. Schmidt. 1973. Ecosystem analysis of the big cypress swamp and estuaries. U.S. Environmental Protection Agency, Region IV, Atlanta, Georgia. EPA 904/9-74-002.

Clark, J. R., and J. Benforado, eds. 1981. Wetlands of bottomland hardwood forests. Elsevier Scientific Publishing Co., Amsterdam. 401 pp.

Conner, W. H., and J. W. Day, Jr. 1976. Productivity and composition of a baldcypress-water tupelo site and a bottomland hardwood site in a Louisiana swamp. Am. J. Bot. 63(10):1354-1364.

______________________________. 1982. The ecology of forested wetlands in the southeastern United States. Pages 69-87 *in* B. Gopal, R. E. Turner, R. G. Wetzel, and D. F. Whigham, eds. Wetlands -- ecology and management. National Institute of Ecology and International Scientific Publications, Jaipur, India.

______________________________. 1988. Rising water levels in coastal Louisiana: implications for two coastal forested wetland areas in Louisiana. J. Coastal Res. 4(4):589-596.

Conner, W. H., J. G. Gosselink, and R. T. Parrondo. 1981. Comparison of the vegetation of three Louisiana swamp sites with different flooding regimes. Am J. Bot. 68(3):320-331.

Cost, N. D. 1976. Forest statistics for the coastal plain of Virginia, 1976. U.S. Department of Agriculture, Forest Service. Resource Bulletin SE-34. 33 pp.

Dabel, C. V., and F. P. Day, Jr. 1977. Structural comparison of four plant communities in the Great Dismal Swamp, Virginia. Torrey Bot. Club Bull. 104:352-360.

Dorge, C. L., W. J. Mitsch, and J. R. Wiemhoff. 1984. Cypress wetlands in southern Illinois. Pages 393-404 *in* K. C. Ewel and H. T. Odum, eds. Cypress Swamps. University Presses of Florida, Gainesville, Florida.

Fowells, H. A. 1965. Silvics of forest trees of the United States. U.S. Department of Agriculture Handbook 271.

Gomez, M. M., and F. P. Day, Jr. 1982. Litter nutrient content and production in the Great Dismal Swamp. Am. J. Bot. 69:1314-1321.

Gosselink, J. G., S. E. Bayley, W. H. Conner, and R. E. Turner. 1981. Ecological factors in the determination of riparian wetland boundaries. Pages 197-219 *in* J. R. Clark and J. Benforado, eds. Wetlands of bottomland hardwood forests. Elsevier Scientific Publishing Co., Amsterdam.

Johnson, F. L., and D. T. Bell. 1976. Plant biomass and net primary production along a flood-frequency gradient in a streamside forest. Castanea 41:156-165.

Johnson, R. L. 1979. Timber harvest from wetlands. Pages 598-605 *in* P. E. Greeson, J. R. Clark, and J. E. Clark, eds. Wetland functions and values: the state of our understanding. American Water Resources Association, Minneapolis, Minnesota.

Johnson, W. C., R. L. Burgess, and W. R. Keammerer. 1976. Forest overstory vegetation on the Missouri River floodplain in North Dakota. Ecol. Monogr. 46:59-84.

Keammerer, W. R., W. C. Johnson, and R. L. Burgess. 1975. Floristic analysis of the Missouri River bottomland forests in North Dakota. Can. Field Nat. 89:5-19.

Larson, J. S., M. S. Bedinger, C. F. Byran, S. Brown, R. T. Huffman, E. L. Miller, D. G. Rhodes, and B. A. Touchet. 1981. Transition from wetlands to uplands in bottomland hardwood forests. Pages 223-275 *in* J. R. Clark and J. Benforado, eds. Wetlands of bottomland hardwood forests. Elsevier Scientific Publishing Co., Amsterdam.

McKnight, J. S., D. D. Hook, O. G. Langdon, and R. L. Johnson. 1981. Flood tolerance and related characteristics of trees of the bottomland forests of the southern United States. Pages 29-71 *in* J. R. Clark and J. Benforado, eds. Wetlands of bottomland hardwood forests. Elsevier Scientific Publishing Co., Amsterdam.

Mitsch, W. J. 1979. Interactions between a riparian swamp and a river in southern Illinois. Pages 63-72 *in* R. R. Johnson and J. F. McCormick, technical coordinators. Strategies for the protection and management of floodplain wetlands and other riparian ecosystems, U.S. Department of Agriculture, Forest Service, Washington, D. C. General Technical Report WO-12.

__________, and K. C. Ewel. 1979. Comparative biomass and growth of cypress in Florida wetlands. Am. Midl. Nat. 101:417-426.

__________, and W. G. Rust. 1984. Tree growth responses to flooding in a bottomland forest in northeastern Illinois. Forest Sci. 30(2):499-510.

__________, and J. G. Gosselink. 1986. Wetlands. Van Nostrand Reinhold Co., N. Y. 539 pp.

Mulholland, P. J. 1979. Organic carbon in a swamp-stream ecosystem and export by streams in eastern North Carolina. PhD dissertation, University of North Carolina, Chapel Hill, North Carolina.

Odum, E. P. 1978. The value of wetlands: a hierarchical approach. Pages 16-25 *in* P. E. Greeson, J. R. Clark, and J. E. Clark, eds. Wetland functions and values: the state of our understanding. American Water Resources Association, Minneapolis, Minnesota.

Penfound, W. T. 1952. Southern swamps and marshes. Bot. Rev. 18(6):413-446.

Rosson, J. F., Jr., W. H. McWilliams, and P. D. Frey. 1988. Forest resources of Louisiana. Southern Forest Experiment Station, New Orleans, Louisiana. Resource Bulletin SO-130. 81 pp.

Schlesinger, W. H. 1978. Community structure, dynamics, and nutrient ecology in the Okefenokee cypress swamp-forest. Ecol. Monogr. 48:43-65.

Shaw, S. P., and C. G. Fredine. 1956. Wetlands of the United States, their extent, and their value for waterfowl and other wildlife. U.S. Department of Interior, Fish and Wildlife Service, Washington, D. C. Circular 39. 67 pp.

Slater, W. R. 1986. Long-term water level changes and bottomland hardwood growth in the Lake Verret basin, Louisiana. M.S. thesis, Louisiana State University, Baton Rouge. 81 pp.

Smith, D. W., and N. E. Linnartz. 1980. The southern hardwood region. Pages 145-230 *in* J.W. Barret, ed. Regional Silviculture of the United States. John Wiley and Sons, New York.

Stubbs, J. 1973. Atlantic oak-gum-cypress. Pages 89-93 *in* Silvicultural systems for the major forest types of the United States. U.S. Department of Agriculture, Washington, D.C. Handbook 445.

Taylor, J. R. 1985. Community structure and primary productivity in forested wetlands in western Kentucky. PhD dissertation, University of Louisville, Louisville, Kentucky. 139 pp.

U.S. Department of Agriculture. 1974. The outlook for timber in the United States. Forest Service, Washington, D.C. Forest Resources Report No. 20. 374 pp.

__________________________. 1978. Forest statistics of the U.S., 1977. U.S. Government Printing Office, Washington, D. C. 133 pp.

U.S. Forest Service. 1970. Florida's timber. U.S. Department of Agriculture, Forest Service. Resource Bulletin SE-20. Feb. 1971.

Wharton, C. H., and M. M. Brinson. 1979. Characteristics of southeastern river systems. Pages 32-40 *in* R. R. Johnson and J. F. McCormick, eds. Strategies for protection and management of floodplain wetlands and other riparian systems. U.S. Department of Agriculture, Forest Service, Washington D. C. GTR-WO-12.

_____________, W. M. Kitchens, E. C. Pendleton, and T. W. Sipe. 1982. The ecology of bottomland hardwood swamps of the Southeast: a community profile. U.S. Fish and Wildlife Service, Biological Science Program, Washington D. C. FWS/OBS-81/37. 133 pp.

Wilkinson, D. L., K. Schneller-McDonald, R. W. Olson, and G. T. Auble. 1987. Synopsis of wetland functions and values: bottomland hardwoods with special emphasis on eastern Texas and Oklahoma. U.S. Fish and Wildlife Service, Washington, D.C. Biological Report 87(12). 132 pp.

15. THE ECOLOGICAL SIGNIFICANCE TO FISHERIES OF BOTTOMLAND HARDWOOD SYSTEMS: VALUES, DETRIMENTAL IMPACTS, AND ASSESSMENT: THE REPORT OF THE FISHERIES WORKGROUP

H. Dale Hall
U.S. Fish and Wildlife Service,
Washington, D.C. 20240

Victor W. Lambou
U.S. Environmental Protection Agency,
Las Vegas, NV 89193

with Panel
Paul Adamus, James Brown, C. Fred Bryan, Ellis Clairain,
Fred Dunham, Gerry Horak, Joseph Jacob, Richard Johnson,
Albert Korgi, William Kruczynski, and Edward Smith

ABSTRACT

This paper examines the relative values of bottomland hardwood wetland zones in supporting finfish and shellfish (crawfish) populations; the probable impacts of various developmental activities on finfish and shellfish habitat; and the characteristics of a site that determine its ability to provide food, cover, or reproductive habitat for finfish or shellfish. It was determined that the floodplain habitat in bottomland hardwood systems cannot realistically be separated from the permanent waterbodies when assessing fisheries productivity, but an increase in values in the higher bottomland zones was predicted when flooded. All potential activities to convert bottomland hardwood wetlands to other uses were found to have overall negative impacts to fisheries functional values. A practical field model to determine the value of a particular bottomland hardwood site for fisheries productivity was also developed.

Ecological Processes and Cumulative Impacts, as illustrated by Bottomland Hardwood Wetland Ecosystems. Edited by James G. Gosselink, Lyndon C. Lee, and Thomas A. Muir.

INTRODUCTION

Bottomland hardwood ecosystems provide a wide variety of natural resources, including extremely significant fish and fishery resources. The material presented here represents the findings of the Fisheries Workgroups at the three workshops sponsored by the U.S. Environmental Protection Agency (EPA) in order to elicit answers to key questions concerning bottomland hardwood wetlands based on the best scientific and technical information currently available.

The material provided here is subdivided into three main sections which correspond to the organization and objectives of the three workshops. The section "Values of Bottomland Hardwood Zones" examines the relative values of bottomland hardwood zones in supporting finfish and shellfish (crawfish) populations covered at the first workshop, while the section "The Relationship of Characteristics to Human Activities in Bottomland Hardwood Systems" examines the probable impacts of various developmental activities on finfish and shellfish habitat covered at the second workshop. The third workshop attempted to consolidate the characteristics that are determinants of the ability of a site to provide food, cover, or reproductive habitat for finfish or shellfish; identify numerical criteria; and develop an approach that might be used as a tool in the field for evaluating fisheries' habitat quality on a bottomland hardwood site. This is covered in the section "Field Analysis."

This paper represents the collective findings of the participants at the three Fisheries Workgroups listed in the panel at the beginning of this report. We, the primary authors, have attempted to present the decisions made by the Workgroup, which included us as active participants, and not our own personal opinions and biases.

VALUES OF BOTTOMLAND HARDWOOD ZONES

Introduction

Bottomland hardwood ecosystems provide a wide variety of natural resource benefits. Uses such as timber production, water quality enhancement, flood storage, nutrient recharge, and waterfowl habitat are well-established values. The high-quality benefits provided to subtropical fishery resources for reproduction, protection from predation, food availability, and harvest for human consumption and recreation are extremely significant. Sport and commercial fishes dependent upon bottomland hardwoods support a large economic base that is one of the most prolific in the United States (Lambou, Chapter 3). Overall, fish harvests of 9000 kg/km^2/year have been documented.

Harvest of the fisheries resources is accomplished by both commercial and sport interests. The bulk of the commercial catch in bottomland hardwood systems consists of buffalofish (*Catostomidae*), catfish (*Ictalurus* spp.), carp (*Cyprinus carpio*), drum (*Aplodinotus grunniens*), suckers (*Catostomidae*), and crawfish (*Procambarus* spp.); bass (*Micropterus* spp.), crappie (*Pomoxis* spp.), sunfish (*Lepomis* spp.), catfish, and crawfish dominate the sport catch. All these species spend at least a portion of their life cycle in flooded bottomland hardwood forests or derive benefits in nutrition from the flood cycle.

In June, 1980, a workshop was held at Lake Lanier, Georgia, to assess the status of knowledge regarding bottomland hardwood ecosystems and their functional values. The results of that workshop were published in 1981 (Clark and Benforado) and were used as a base of departure for our analyses. At that workshop, Larson et al. (1981) divided the forested portion of bottomland hardwood ecosystems into "ecological zones" based on "soil-moisture/hydrologic habitat conditions" and described the forest types associated with the various zones (Figure 1). We were asked to review these zones with particular regard to intrinsic fishery resource values and, specifically, to evaluate each zone relative to the life requisites of indigenous sport and commercial fishes.

We accepted this charge knowing that the literature does not contain a wealth of information directly assessing functional fisheries values, but does contain associated findings that can be applied through extrapolation. To facilitate analyses, we established basic assumptions that guided our discussion of relative values of the various zones.

First, it was assumed that all areas inundated by a given frequency flood are connected by hydrologic avenues for finfish movement. This assumption alleviated the need for differentiation of flooded bottomland hardwoods relative to stream overflow or ponded rainfall origin (Hall 1986).

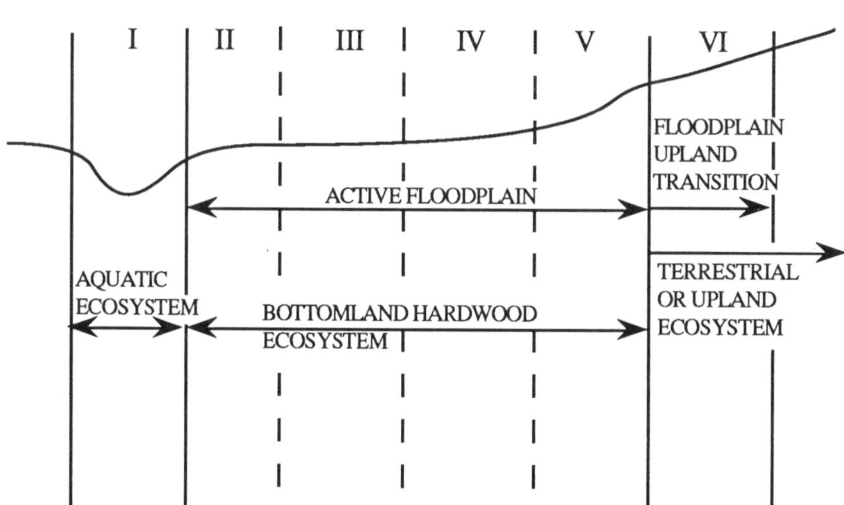

Figure 1. Generalized profile of ecosystems and transitions associated with floodplains of the southeastern United States (after Larson et al. 1981).

Second, we realized that yearly and/or seasonal variations in flooding events could cause a shift in the importance of the zones due to timing, duration, and/or absence of inundation. To alleviate this variable, the 10-year frequency event was selected for geographic extent of inundation. Aerial extent of inundation across the zones and predicted duration were taken from Larson et al. (1981) for this flood event. For purposes of comparison, however, Figure 2 illustrates the relative value of the zones for fish use averaged over many years. Zone I would be permanently inundated and would, therefore, have the highest value to the fishery resource. The value of the zones would gradually decrease in an upslope direction because the frequency and duration of inundation, and consequently the available habitat, would decrease. However, Figure 2 does not reflect the relative importance of the higher zones for a single flood event that crosses all zones in the floodplain. Analyses of the importance of zones during such a single event would elevate the importance of the higher, more shallowly flooded zones for particular life requisites of aquatic species.

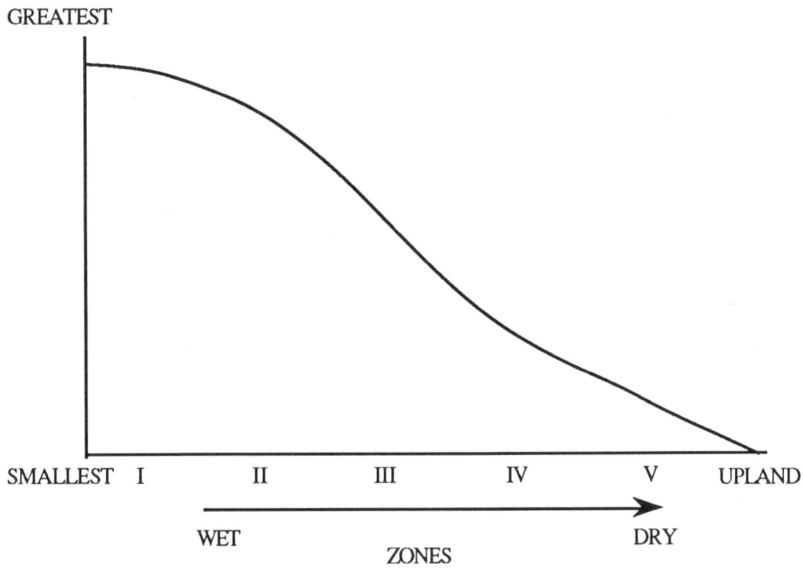

Figure 2. Relative importance of bottomland hardwood zones to fisheries resources averaged over many years.

Finally, we assumed sufficient quantities of phytoplankton, micro- and macrozooplankton, aufwuchs, detritus, and terrestrial insects to be available as food organisms during the early life stages of finfishes and during flood events for crawfish.

Even though Figure 1 shows the "aquatic ecosystem" as being restricted to zone I, we believe that this ecosystem encompasses the total floodplain, including the "bottomland hardwood floodplain" areas and "permanent waterbodies." In our analyses, the interdependency of rivers and their floodplains became a constant reminder that one cannot realistically separate the "adjacent permanent waterbody" from the "bottomland hardwood floodplain" when discussing the viability of fishery resources. However, we made a concerted effort to explore the functional interrelationships between bottomland hardwood zones for fishery populations. The remainder of this section elucidates this effort and our conclusions.

Approach

Bottomland hardwood ecosystems are extremely diverse and dynamic. Hydrologic conditions vary dramatically from permanently flooded (zone I) to infrequently flooded (zone V). These ecosystems support a wide diversity of aquatic biota during all life stages, as was amply demonstrated in the Lake Lanier workshop (Wharton et al. 1981) and in this workgroup (Lambou, Chapter 3).

However, a detailed evaluation of the utilization of all zones by all life stages for all species was beyond the scope of our effort. Therefore, we identified target organisms and evaluation criteria for analyses. This approach was not intended to ignore the importance of flooding to the life cycles of other aquatic biota (e.g., macroinvertebrates, phytoplankton, or zooplankton) or other life requisites, but rather to permit a manageable scope for this evaluation. A brief description of the life forms and analysis criteria is presented below.

1) Fisheries Functions. Two broad categories of organisms were examined: finfishes and crawfishes. Finfishes were further sub-divided into adults and young-of-the-year because of substantial differences in the life requirements of these groups. Young-of-the-year were considered to be all eggs, larvae, and juvenile fishes that have not completed their first year of life nor reached marketable status. Adults were considered to be all finfishes that have completed their first year of life. The value of the zones to each of these groups of organisms is amplified in the discussion below.
2) Functional Characteristics. The basic life requisites of food availability, living space/protective and predatory cover, and reproductive substrate were examined for each life form (crawfishes, adult finfishes, and young-of-the-year finfishes) by zone.

Results and Discussion

The yield of fish to man from bottomland hardwood floodplains is largely dependent upon the maximum area flooded (Lambou, Chapter 3; Welcomme 1979). In the case of a mixed fishery (finfishes and crawfishes) in the Atchafalaya Basin, Louisiana, a harvest of 8797 kg/km^2/year over the maximum area flooded has been documented (Lambou, Chapter 3), while in the case of tropical finfish it has been estimated that the yield varies from 4000 to 6000 kg/km^2/year over the area flooded for normally exploited systems (Welcomme 1979).

The relationship of yield to area flooded, however, is complex and depends upon present and past flooding cycles, timing and duration of flooding, amount of permanent water during the low-water season, the kind and types of finfishes found, and the characteristics of the fishery harvest. The maximum amount of fishes is present during the flood event and the least just prior to onset of flooding. Reproductive strategies of many riverine species are cued to spring floods. Thus, the excess produced during a flood must be lost during the low-water period to bring the population within the carrying capacity of the permanent habitat.

Simulation models of floodplain finfish populations by Welcomme and Hagborg (1977) indicate that differences in high water, or flood regime, from year to year generate great differences in ichthyomass produced. However, the carry-over success of the community is largely dependent upon the amount of water remaining during the dry seasons of the year (zone I). These simulations reflected that the more water remaining during the low-water period, the more efficient the transmittal of population increases during the high-water years. Production and biomass were maximized during high flood years and production ranged from 24,100 to 56,400 kg/km^2/year of the mean area flooded (Welcomme and Hagborg 1977).

Alternate exposure and flooding of the land is of great importance in maintaining the viability and productivity of fishery populations occurring in streams that drain bottomland hardwood ecosystems. The effects depend more upon the timing, area, and duration of flooding than on the degree of fluctuation as measured in vertical meters. This alternating wet/dry cycle has important effects on fish population dynamics and controls the amount of forage available. Food availability, in turn, sets upper bounds on the population size any species of finfish may attain. Thus, one can begin to understand the complex interdependence that exists between the permanent waterbodies and their associated floodplains. The effects of fluctuating water levels on fish populations are discussed by Wood (1951), Lambou (1959), Wood and Pfitzer (1958), and Hall (1979).

The vast majority of southern freshwater finfishes spawn in water less than 1.2 meters in depth (Breeder and Rosen 1966) and are oviparous (egg-laying), with advanced embryological development taking place post-hatching. For the first few days after emerging from the egg, the larva receives nutrition from a yolk sac while primitive organs of digestion are developing. Within 3 to 7 days, the yolk material may be exhausted, and the larva makes the transition from endogenous to exogenous nutrition. The main factors involved in the success of this transition are 1) the density of food organisms available, and 2) the capability of the larva to capture its food (Hall 1979 after Hempel 1965, Blaxter 1974, Lasker 1965).

Some fishes, such as bass and sunfish, build nests and guard the eggs and young while fanning the nest to keep the eggs aerated and free from sediment. Many others, such as shad (*Clupeideae*), buffalo, and carp, merely broadcast (or spray) their eggs over vegetation. In the latter, the eggs possess an adhesive coating that causes them to cling to vegetation or bottom detritus. This spawning behavior enhances survival; the eggs are kept free from siltation and aerated by water movement. After hatching, the vegetation becomes cover for protection against predation (Balon 1975, Hall 1979).

The volume of water in which a larva can search for food is measured in liters or less for the first few days of life. In the search for food and "avoidance" (albeit more or less passive) of predators, the larval fish is dependent on the whims of currents to position it near food and away from enemies. The larva's centers of olfaction and sight are so primitive that, from their dietary items, one can only conclude that it is an opportunistic feeder. It will apparently swallow almost any suspension material and will, therefore, survive to the next mouthful only if the first one proved nutritious. In view of the fact that the gape (mouth) of most new hatchlings is measured in fractions of millimeters, those suspensoids (algae, detritus, protozoans, rotifers, copepod nauplii [not adults], small cladocerans, etc.) must be smaller than the gape in order for the larva to swallow them.

While fish populations can exist and maintain themselves in areas of permanent inundation, recruitment and production are fostered during the flood cycle. The end result is a continuum of fisheries production that expands with the migrating waters' edge. Concomitantly, limitations in access to floodplain habitat may limit fishery productivity. Therefore, the viability of fish populations in permanent waterbodies located in bottomland hardwood ecosystems cannot be isolated from the productivity of adjacent floodplains.

As floodwaters rise and recede, an edge of shallow water migrates across the floodplain. It is this shallow zone of the shifting water's edge that houses the highest fishery production at any given time (Lambou, Chapter 3). Therefore, the values assessed in our analysis must be considered transient between zones as the water crests and recedes, and these can vary from year to year with different flood events. As will be seen, however, we believe the productivity of shallow-water areas increases with inundation of zones IV and V. Also, the inundation of zones IV and V is normally associated with longer-duration flood events which increases the production and recruitment of fish.

Adult Finfishes

The values attributed to space/cover, food availability, and reproduction habitat were assessed for each zone during the 10-year flood event (Table 1). During this flood, zones I through V are expected to be inundated for 30 days or more during the growing season (Larson et al. 1981).

Cover/Living Space. The amount of cover in each zone varies with the amount of water present (flood event). Cover is an impediment to feeding by predator species and important as protection for prey species. A weighted value to reflect the overall considerations of cover was assigned to each zone (Table 1). Cover in zone I was considered of medium value to adults before the flood, low value during a flood, and medium-to-high value afterwards. This was based upon the difference in location of fishes in the floodplain and the increase in depth of zone I during a flood, as well as the concentration of fishes following recession of floodwaters. Even though zone I was considered by the workgroups to be of medium value to adult finfish before the flood and of medium-to-high value after the flood (primarily based on the need for cover), zone I is essential for providing living space to carry over adult finfishes between flood events. In this sense, i.e., of providing living space for adult finfishes during low-water periods, zone I should be considered of high importance.

Table 1. Relative importance of bottomland hardwood zones in meeting certain life requisites for adult finfish, young-of-the-year finfish, and crawfish (L = low, M = medium, H = high).

Life requisites	Zone						
	Ibf[1]	Idf[2]	Iaf[3]	II	III	IV	V
Adult finfish							
cover/living space	M	L	M/H	L	M/H	H	H
food availability	L	M	H	L	M/H	H	H
spawning substrate	L	L	M	L	M	H	H
Young-of-the-year							
cover/living space	H	L	H	M	M/H	H	H
food availability	M	L	H	L	M/H	H	H
Crawfish							
food				M	H	H	H
burrow utilization				H	M	L	L

[1] bf is before the flood in zone I
[2] df is during the flood in zone I
[3] af is after the flood in zone I

The values of functional characteristics for zones II through V were considered, by definition, to be present only when the water exceeds bank full. Cover was considered to have low-to-no value (importance) during the 10-year event in zone II due to concentrations of fishes at the shallower depths of the floodplain. Zone III was assigned high-to-medium values for cover based upon an increase in ground cover vegetation and the increased use by adult finfishes. Zones IV and V were considered to provide the greatest amount of cover due to a great increase in annual, perennial, and shrub vegetation. These zones were assigned high values for cover and utilization due to concentrations of finfishes in the shallow edge area.

Food Availability. Food was considered to be of low value to adult finfishes in zone I prior to flooding due to reduced metabolism in response to low-water temperatures. Adult fishes feed only infrequently during the cold winter months. During the flood, which typically occurs in spring, the water temperature rises and metabolic processes increase, thus elevating to medium the value of food in zone I. Following recession of floodwaters, a concentration of new life forms is washed into zone I at a time when metabolic processes are rapidly increasing. Therefore, food attains a high value to adult finfishes following the flood. Zone II was considered to have low value as a food area during the flood analyzed. However, this could change for less frequent events. For zone III, food concentrations increase due to shallower depth and heavier organic material (detritus). The value for food in zone III was therefore considered to be medium-to-high. Zones IV and V provide the highest concentrations of food organisms, both invertebrate and vertebrate, and were considered to be of high value.

Reproduction Substrate. Spawning substrate is of low importance prior to the flood. Late winter and early spring spawners use either mainstem floodplain bayous, bend-way cutoffs, or meander lakes in which there are usually sufficient substrates for spawning sites. After the flood, late spawners (June, July, August, and even September) may find suitable substrates only in zone I; thus they are extremely important to those species (high). Spawning substrates in zone II are less important during our assumed flood than sites near the water's edge in zones III, IV, and V, but perhaps of similar importance to those substrates in zone I for the same reasons listed above (low during the flood).

As the floodwaters rise to the levels of zones III, IV, and V, the availability (or proximity) of spawning substrates becomes increasingly important because we envision that it is at these levels that the water has arrived in a timely fashion for the bulk of the spawning populations of fishes. Therefore, we assigned values of medium for zone III and high for zones IV and V.

Young-of-the-Year Finfishes

Cover/Living Space. Cover in zone I (permanent water) is of low importance to larval and juvenile finfishes during a flood, but has high value before (for carry over of juveniles) and after the flood (for concentration of current hatchlings; Table 1). Zone II provides medium value due to importance for young as the waters recede down to zone I. Medium-to-high values are found for cover in zone III due to increased vegetation, with zones IV and V providing high values.

Food Availability. As previously discussed, food is of paramount importance to larval and juvenile finfish survival. Zone I provides some food for young-of-the-year before the flood (medium) in the form of microbial and invertebrate colonization of plant stems and clay banks. If the cover is of suitable surface, bacterial flora, protozoans, algae, rotifers, and other small organisms are available. In this zone, there is little contribution of food to young-of-the-year finfishes during the flood due to fish concentration in the higher zones, but high contributions of food following the flood when newly produced prey organisms are concentrated in zone I. Zone II was rated low during the flood due to the water depth of a 10-year event in that zone. Zone III was rated medium-to-high due to increased vegetative (organic) material. Zones IV and V were considered the optimum habitat for production of needed food organisms (high). This value was attributed to heavy plankton and aquatic invertebrate production along with availability of terrestrial insects provided by the higher sites.

Crawfishes

It was the opinion of the group that zone I receives insignificant use by crawfish populations (Lambou, Chapter 3). Therefore, our analyses were limited for the most part to zones II through V.

Cover/Living Space. Most individual members of crawfish populations (e.g., red swamp crawfish [*Procambarus clarkii*]) spend their whole life cycle in the forested area of the floodplain. Their life cycle strategy is flexible, which allows them to take advantage of the timing and duration of the flood event. When floodwaters overtop the burrow, overwintering crawfishes and their young are released from the protected depths of the groundwater table. The crawfishes, vegetative/detritus feeders, concentrate in the shallow water edge of the flood, moving with the water as stages rise and fall.

Because of abundant detrital material available on the forest floor, crawfish densities as high as 46.9 /m^2, with biomass as high as 241.9 g/m^2, have been estimated in shallow floodwaters of the Atchafalaya River, Louisiana (Lambou, Chapter 3). In the whole flooded, forested study area, a mean number of 3.4 /m^2 and a biomass of 29.5 g/m^2 were estimated for the month of June, with production of 69,717 kg/km^2.

In the Atchafalaya Basin Floodway, crawfishes normally burrow from late June through July. The particular bottomland hardwood zones in which this occurs depend upon the frequency flood present; however, in one area studied, the majority of the burrows were in zones II and III, while most of the production occurred in zones IV and V.

The abundance of burrows at a particular location in the overflow area appears to be dependent upon the water level at the time of burrowing, which can vary from year to year. The higher numbers of burrows in zones II and III may, however, reflect evolutionary adaptations for the higher frequency floods. Because of the massive number of individuals that usually occur in flooded hardwood wetlands, predation does not appear to be a significant factor for survival and was not considered for analysis of crawfishes.

Food Availability. The amount of food present in zone II was assessed to have medium value to crawfish populations (Table 1). This value was assumed due to the marginal ability of crawfishes to digest cellulose, including the needle leaves of cypress (*Taxodium distichum*). Zones III, IV, and V were assigned high values for food due to increased organic production and concentration of crawfish populations in these zones during our flood of analysis.

Burrow Utilization. Because crawfishes carry the eggs and young attached to their ventrum until release in the spring, burrow utilization was examined here to replace spawning substrate. Zone II has very high burrow activity and is an important area for this life requisite. Zone III was rated as medium value based on numbers observed in field analyses. Zones IV and V were considered to have low values for burrow utilization due to less frequent flooding and, therefore, less long-term probability for repetitive survival opportunity.

Conclusions

There are trends in the importance of "ecological zones" in bottomland hardwoods when evaluation criteria are applied to life requisites of fishes. Adult finfishes have fluctuating values in zone I, depending upon stages in the flood cycle (Figure 3), but zone I certainly provides a place for carry over until the next flood. Zones II and III reflect gradual increases in values further into the floodplain, with optimum values in zones IV and V. Young-of-the-year finfishes nearly mirror value increases across the zones (Figure 4) that are seen for adult finfishes. Zones IV and V provide optimum values to larval and juvenile finfish species. Crawfish populations reflect a marked increase from zone II to III and optimum values for food in zones IV and V (Figure 5). Burrow utilization is inversely related to values for food production in the zones. Zones II and III function for crawfishes as zone I functions for finfishes.

Therefore, when one examines the overall functions of bottomland hardwood zones (Figure 6), a trend of increasing values is seen across the zones for all fishery aspects considered except for crawfish burrow utilization. Zones II and III appear to provide lesser overall value to the fish communities, while zones IV and V appear to be optimal for spawning, rearing, and growth.

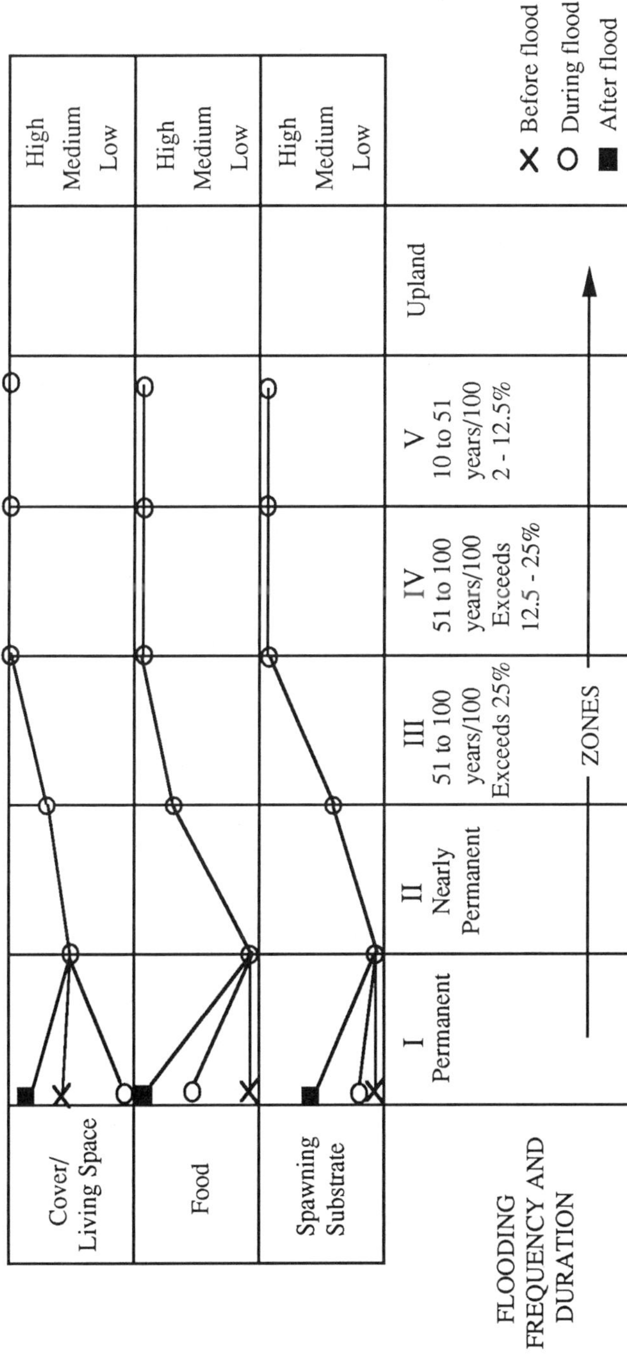

Figure 3. Relative value of bottomland hardwood zones in meeting the life requirements of adult finfishes in a 10-year flood event.

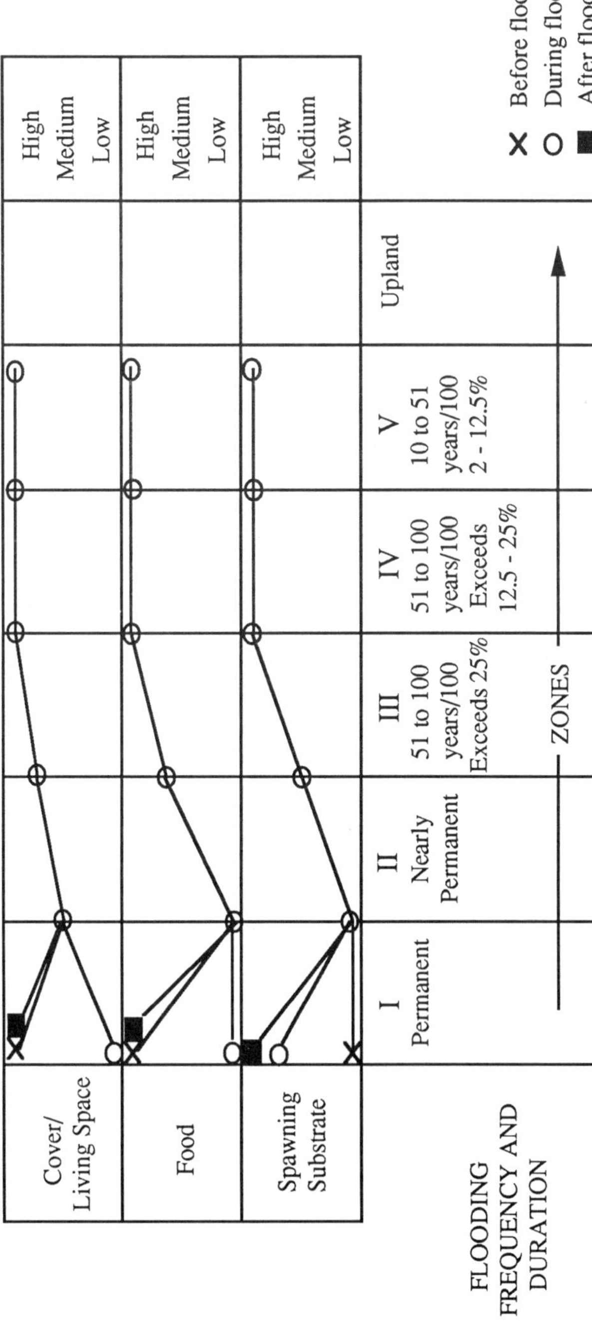

Figure 4. Relative value of bottomland hardwood zones in meeting the life requirements of young-of-the-year finfishes in a 10-year flood event.

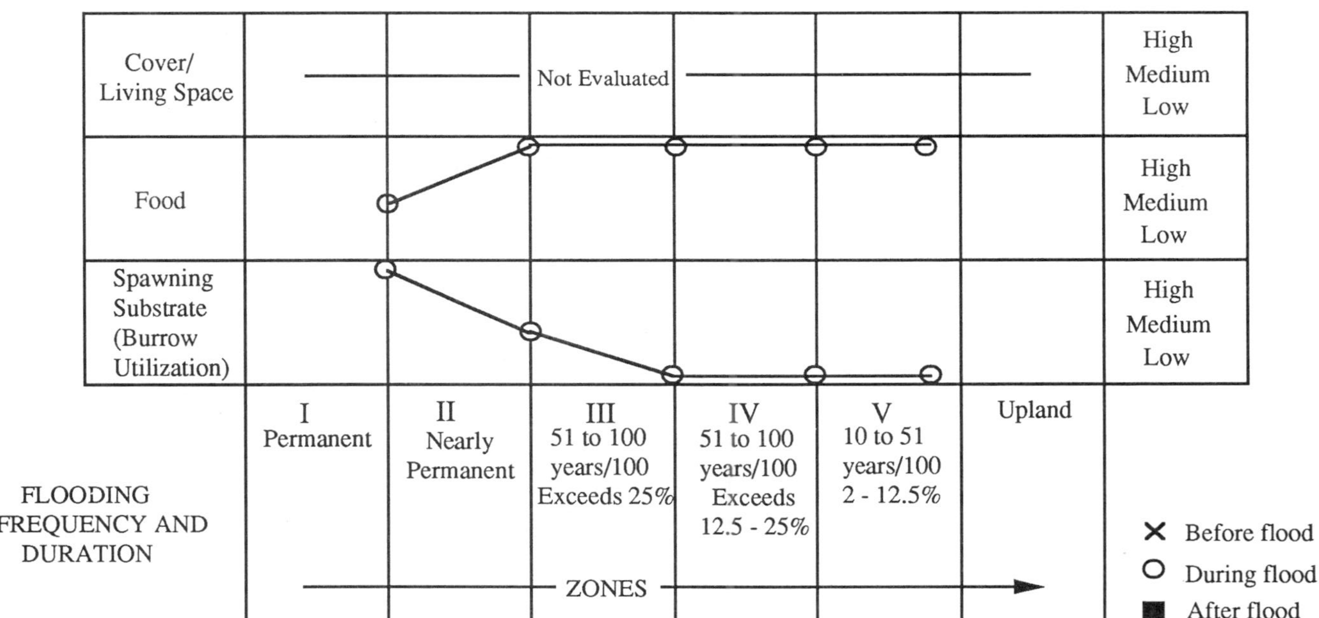

Figure 5. Relative value of bottomland hardwood zones in meeting the life requirements of crawfishes in a 10-year flood event.

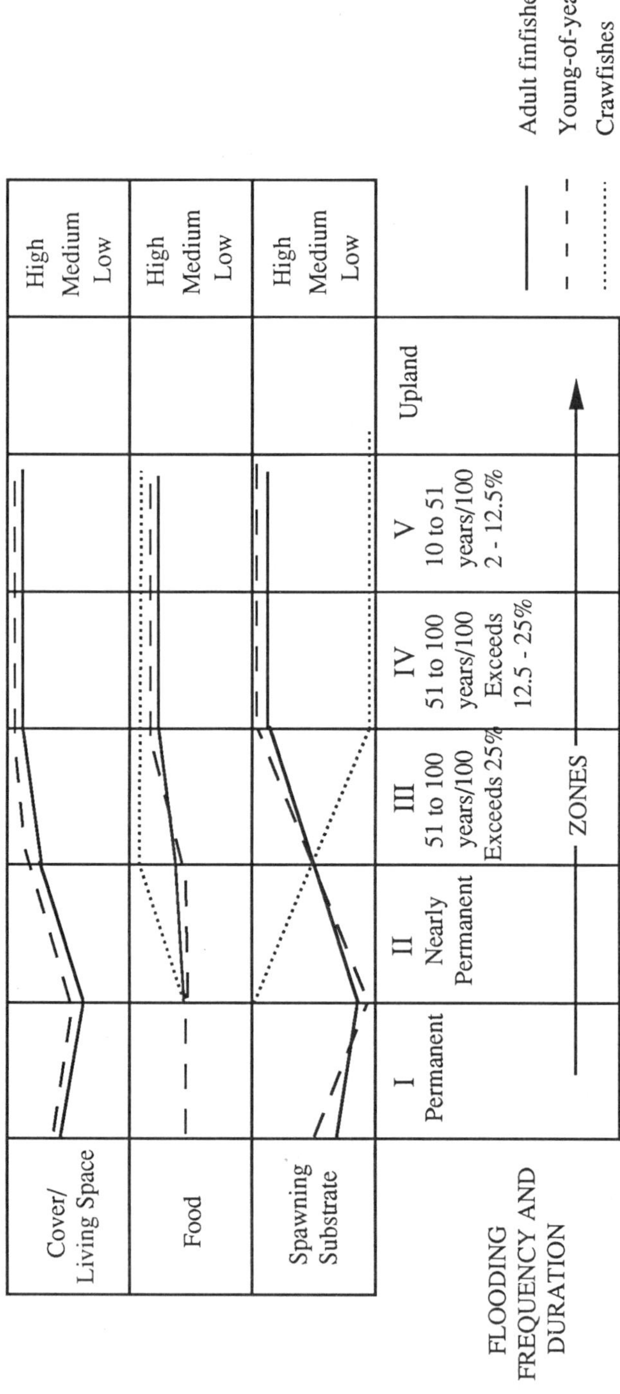

Figure 6. Relative value of bottomland hardwood zones in meeting the life requirements of all fishes in a 10-year flood event.

FISHERIES PRODUCTIVITY IN BOTTOMLAND HARDWOOD SYSTEMS: IMPACTS FROM COMMON HUMAN DISTURBANCES

Introduction

Our previous discussions have focused on the overall function of fisheries production in bottomland hardwood wetlands as related to adult and young-of-the-year finfishes and crawfish sub-functions. Basic characteristics associated with certain elevations (zones) of the floodplain that contribute directly to fisheries production were also discussed and assigned values along the slope of the bottomland hardwood system. The next logical step in the architecture of an evaluation is to introduce normal or potential activities carried out by humans in bottomland hardwood habitats that might create impacts, positive or negative, to the fisheries function and the characteristics that support it.

Seven activities (Table 2, see Chapter 1) with their associated actions were used by all workgroups for evaluation of impacts. In addition to the activities that the workgroup was instructed to consider, the impacts of oil and gas development, aquaculture, and commercial pine tree plantations on finfish and shellfish habitat were examined. Oil and gas development in bottomland hardwood forests includes those activities associated with access to drilling sites, disposal of pollutants (e.g., drilling muds and brine), and transport of the product. Access to drilling sites includes dredging of canals, deposition of dredged materials adjacent to canals, and construction of roads through filling and ditching. Disposal of pollutants is either into holding ponds or directly into waterways. The transport of products via pipeline requires clearing, dredging, and filling. Pine plantations and aquaculture include leveling of land, ditching, and leveeing.

Finfish Habitat

In order to more specifically address the impacts to fisheries of man's activities in bottomland hardwood wetlands, the general characteristics of cover/living spaces, food availability, and reproductive substrate were sub-divided to create nine characteristics of measurement. These are areal/stage relationship, natural floodplain vegetation, gentle topography, stream sinuosity, snags and instream cover, access to permanent water, presence of toxicants, oxygen level, and sedimentation. Table 2 presents the impacts of the activities on these characteristics.

Table 2. Impacts of various activities on habitat for finfish as mediated through certain characteristics that contribute to the performance of this function.

Characteristic	Relationship of characteristic to function[1]	Activity[2]									
		R	S	I-U	I-O	I-D	C	L	P	A	O
Areal/stage relation	+	-	-	0	-	-	-	-	-	-	-
Natural floodplain vegetation	+	-	-	0	-	-	-	-	-	-	-
Gentle topography	+	-	-	0	0	0	-	-	-	-	-
Stream sinuosity	+	0	0	0	-	-	-	0	0	0	-
Snags and instream cover	+	-	0	0	-	-	-	0	0	0	-
Access to permanent water	+	-	0	-	+	-	-	-	-	-	-
Presence of toxicants	-	-	-	0	-	0	-	0	-	0	-
Oxygen level	+	-	-	0	+	-	-	-	0	0	-
Sedimentation	-	-	-	0	-	-	-	-	-	0	-
Overall impact on habitat for finfish		-	-	0	-	-	-	-	-	-	-

[1] A plus indicates a positive correlation between the characteristic and the function. A minus indicates a negative correlation.

[2] A plus indicates that the function is enhanced due to the impact of the activity on the characteristic. A minus indicates that the function is impaired. A zero indicates no impact. R = conversion of site to rice, S = conversion of site to soybeans, I-U = impacts upstream from an impoundment, I-O = impacts from an impoundment on-site, I-D = impacts downstream from an impoundment, C = channelization of stream adjacent to site, L = levee construction on-site, P = conversion of site to pine plantation, A = conversion of site to aquaculture, O = oil and gas development on-site.

Areal/Stage Relationship

The largest yields of finfish tend to be supported by those bottomland hardwood areas where flooding occurs for the longest time (and during a season critical to fish) and over the largest area of floodplain. The seasonal duration of flooding is difficult to determine precisely and varies greatly from year to year, but a general idea may be gained from interviewing knowledgeable local citizens. Existing gage data generally report only the mainstem conditions rather than flooding duration in seasonally flooded zones. During unflooded conditions, the areal extent of flooding may be determined by presence of watermarks, drift lines, discolored leaf litter, and sediment deposits on vegetation; examination of aerial photos and topographic contour maps (paired with stream gage data); and interviews with local sources (Hall 1986). From these data the flooded area may be delineated and planimetered. The absence of a positive indicator of flooding from any of these items should not be assumed to be conclusive evidence of lack of flooding.

The ratio of mainstem area (or volume) to flooded area (or volume) during various frequency floods is one parameter used to measure the areal/stage relationship. We believe that data are insufficient to indicate whether a 1:1 ratio of these habitats, or any other ratio, is significant to fisheries.

In the impact analysis it was assumed that an adverse impact on areal/stage relationship is one that reduces the extent or duration of overbank flooding or alters the season of its occurrence. A positive impact results from an activity that increases the extent or duration of overbank flooding. No impact results from an activity having no effect on the natural flood regime.

Except for the impacts upstream from an impoundment, the activities were judged to reduce areal/stage relationships and thus impair finfish habitat. Conversion to rice, channelization of stream adjacent to site, levee construction on-site, and aquaculture (through the construction of levees or berms) reduce the area to be flooded and thus reduce available area for the production of food, cover, and reproduction sites for fish. The areal/stage relationship is diminished by conversion to soybeans due to land clearing, leveling, and ditching, which result in improved drainage and shorter duration of flooding. Conversion to pine plantations also reduces flooding events through increased drainage (ditching) and blockage of sheet flow by road construction. Although an impoundment on-site may flood a larger acreage than normally occurs in the bottomland hardwood area, the water level fluctuations are not seasonal. Thus, the floodplain finfishery is converted to a lake finfishery.

Finfish habitat is adversely affected by the construction of a dam upstream due to the alteration of the natural areal/stage relationship in a controlled tailwater volume. The result is a loss of spawning and nursery habitat in a reduced floodplain. Oil and gas development reduces areal/stage relationships by the construction of canals, ditches, and roadways.

Natural Floodplain Vegetation

The largest yields of finfish and crawfish tend to be supported by those bottomland hardwood areas where vegetation has not been altered by man. Under such conditions, diversity of vegetational layers, species richness, and annual leaf-fall are at their maximum, and availability of both terrestrial and aquatic invertebrates is consequently highest. Also, flood flows may be retarded by vegetation on the floodplain, and detrital material may be transported to the channel and downstream areas. Diversity of vegetation conditions is easily recognized by individuals with local knowledge of typical species and their usual densities within each zone. Specific parameters, such as vertical layer structure and canopy closure, may be measured. However, the workgroup identified no criteria for these parameters nor rapid methods for their assessment. In the impact analysis, the group assumed that any deviation from natural conditions would constitute an adverse effect, while preservation of natural conditions would constitute no effect.

Natural floodplain vegetation is affected by conversion to rice, soybeans, and pine plantations; impoundment on-site; and aquaculture through the removal of all vegetation by land clearing. Although floodwaters can still reach the converted agricultural fields and pine plantations, the resulting vegetation and detrital material are not as valuable to finfish. Bottomland hardwood areas downstream from impoundments are impacted by the leveling of seasonal flows from the dam. The natural flows are critical to the fertilization and hydrologic environment of the natural floodplain vegetation. Thus, the altered flows can encourage species composition changes of natural vegetation to more of an upland vegetation community. The ultimate result is the elimination of crawfish and finfish habitat.

Likewise, channelization and oil and gas development reduce and alter the flooding of bottomland hardwood areas, resulting in drier conditions that are favorable to more mesic species. Backswamp areas on the protected side of levee systems or areas that have undergone ditching result in plant community changes due to spatial and temporal alterations of flooding. Thus, the supply and extent of dissolved and particulate organic matter are lessened, which reduces the food base for finfish populations. Again, bottomland hardwood areas upstream from impoundments would not generally be affected by this characteristic.

Gentle Topography

The largest populations of finfish tend to be supported by those bottomland hardwood areas where the topography is gentle and mildly complex. Such an area would have an average gradient from the mainstem up through zone V that is very slight, but complex both vertically and horizontally in the sense of having small mounds, mild undulations, and backwaters. Gentle, complex topography is important because it allows water to be trapped for long periods over a wide area with a large edge effect between upland and flooded soils. Specific criteria for identification of this characteristic will be discussed in the final section. Adverse impacts were assumed to be those associated with leveling of the irregularities in the topography or creation of other unnatural conditions (e.g., steep mounds).

Topography is affected by most activities through leveling and/or levee or berm construction. Levees drastically reduce the area of shallow water during flood events, and it is in these shallow areas where the maximum production of finfish and shellfish occurs in bottomland hardwood forests. However, impoundments were judged to have an insignificant effect on topography.

Stream Sinuosity

Bottomland hardwood areas adjacent to permanent water channels that are highly meandering or sinuous support the largest yields of finfish. Sinuosity is important because it increases the availability of cover for finfish inhabiting the permanent water during seasons when the floodplain is dry and offers hydrologic constraints that favor overbank flooding. Meandering or sinuous streams also provide more total water area during low-water periods per length of floodplain than non-meandering streams and thus provide more living space for carry over of finfish between flood events. Recent topographic maps or aerial photos can be used to measure sinuosity by determining the ratio of direct distance to thalweg distance between two stream points. Negative impacts were assumed to be those resulting in straightening of a channel and positive impacts to be those resulting in an increase in its sinuosity.

Only four of the activities evaluated affect stream sinuosity. Impoundments on-site totally inundate the channel and remove meandering characteristics. Stream channels associated with bottomland hardwood areas downstream from impoundments are typically straightened either mechanically or by the fact that seasonal fluctuations are reduced. The result is that water-channel finfish habitat contains less cover on a year-round basis. Although habitat space may be somewhat greater during the normal dry season, this does not compensate for loss of cover typically associated with channel meandering. Channelization of the stream adjacent to a site and on-site oil and gas development can drastically reduce stream or permanent water sinuosity.

Snags and Instream Cover

A direct relationship exists between finfish yields and bottomland hardwood areas where the floodplain, and especially the mainstem, have moderate to high accumulations of large organic debris (e.g., tree trunks and roots). Snags in the channel provide cover for finfish as well as serving as a stable substrate for high densities of invertebrate food organisms. Site visits would be required to determine the occurrence of snags. Adverse impacts result from removal of instream cover, while positive impacts occur when instream cover is increased.

Bottomland hardwood areas downstream from impoundments receive lower peak seasonal flows. This reduces the opportunity for woody debris to be washed out of the floodplain and into the channel. Moreover, the dam physically cuts off the supply of large debris from upstream areas. Thus, the reduced instream cover in this tailwater area can adversely affect finfish. Snags and in-stream cover are totally inundated by an impoundment on-site and are unavailable because of seasonal stratification. Channelization removes snags and other debris.

Access to Permanent Water

Bottomland hardwood wetlands can only provide fisheries values when they are connected directly (i.e., a "water bridge" is present) to the permanent water during flooding. Such water bridges are essential because they allow adult fish to have access during flooding to fertile floodplain areas which are critical sites for reproduction, food, and cover. At the same time, presence of water bridges allows fish to follow receding floodwaters back into the main channel or other permanent waterbodies.

The presence of water bridges acts as a modifier of the areal/stage relationship; i.e., when calculating a ratio of mainstem acreage to floodplain acreage, only those parts of the floodplain that would be accessible to fish during flooding should be considered. Such accessible areas may be identified by use of topographic maps; field inspections during flooding; interviews with knowledgeable sources; or by scour marks on the forest floor, relative lack of leaf litter build-up, or presence of adventitious (above-ground) roots on trees (Hall 1986). Absence of any of these indicators should not be taken as *prima facie* evidence of lack of a water bridge. On the other hand, watermarks and discolored leaves do not necessarily indicate presence of a water bridge, as their appearance may be due to dried-out precipitation pools or water table highs. An adverse impact was judged to occur when fish ingress or egress is blocked from the floodplain and a positive impact when a greater acreage is accessible to fish.

The construction of levees negatively affects access to permanent water. The levees act as a barrier to finfish movement and may prevent the retreat of finfish to permanent-water areas as the floodwaters recede. Bottomland hardwood areas downstream from impoundments receive reduced peak seasonal flows which prohibit finfish from getting into many floodplain areas needed for feeding and spawning. On-site impoundments make some backwater areas accessible that previously were not, and the acreage of permanent water is obviously enlarged. However, dams physically restrict natural dispersal upstream and downstream, so finfish populations can suffer.

It should be remembered that areas maintaining good natural vegetation and topography conducive to ponding rainfall for extended periods during the winter and spring may not be affected for crawfish production, even if access to permanent water is absent.

Presence of Toxicants

Low finfish yields may reflect bottomland hardwood areas where fish are exposed to excessive levels of toxic substances or conditions, such as certain pesticides, heavy metals, acidity, brine, and oil. These may be directly toxic or they may indirectly affect fish through impacts on food and vegetative cover. The upper zones of bottomland hardwood areas are especially vulnerable to pesticide runoff. To assess the occurrence of toxicants, field staff should begin by reviewing locations of point sources and agricultural fields.

The Fisheries Workgroup assumed that an adverse impact would result from any increase in toxic substances within an environment where fishes might be exposed. An "O" rating was assigned to activities that caused no change in natural background levels.

Insecticides, herbicides, and fertilizers are used on agricultural fields and pine plantations. Therefore, toxicants increase in kind and quantity due to these activities and result in a negative impact to fishery habitat. Also, on pine plantations decomposition of pine needles acidifies the soil, reduces bacterial action, and can produce a condition too acidic for other lower forms of life that are found in the food chain of fishes. Impoundments on-site concentrate sediment with attached toxic substances. Under certain mixing conditions, the substances can be mobilized within the impoundment. Spoil disposal from oil and gas development negatively influences the amount of spawning and nursery habitat as well as the quality of available food.

Oxygen Levels

Bottomland hardwood areas where dissolved oxygen (DO) levels in the water column are adequate tend to be associated with large yields of fishes. Although specific criteria were not assigned, it was assumed that levels of approximately 3 mg/l are adequate. The workgroup assumed that an adverse impact would result from any decrease in DO levels, and that a positive impact would result from increases in DO levels up to about 4 mg/l, beyond which DO assumes less importance as a limiting factor to bottomland hardwood fishes.

Conversions to rice and soybeans lower oxygen levels by increased water runoff of higher temperatures and lower oxygen detention potential. Oxygen levels within an on-site impoundment are, in most cases, less stressful than the levels that previously existed in backwaters. Channelization and levees alter and reduce the exchange of water between the bottomland hardwood area and main channel. Because of the decomposition of plant litter on the forest floor during flood events, areas devoid of water movement may have significantly lower oxygen contents and thus be less suitable habitat for finfishes.

Sedimentation

Where levels of suspended sediment are high or the depth of annually deposited sediment is excessive, yields of fishes in bottomland hardwood areas are small. Suspended sediment's direct adverse effects on fish are well known, although moderate turbidity may provide protective cover from predators in the case of some young fish. The role of deposited sediment is less certain. Excessive deposited sediment may destroy invertebrate foods and smother vegetation vital as cover, while moderate amounts may essentially fertilize the floodplain vegetation, ultimately providing better cover (see Scott et al. Chapter 13). An increase in sediment was assumed to result more often in an adverse impact than in a positive one. Conversely, a positive impact would be any effect that results in reduced levels of incoming sediment.

Sediments increase from conversion to rice, soybeans, and pine plantations through land clearing, leveling, seedbed preparation, and drainage. Although impoundments upstream of bottomland hardwood areas reduce turbidity in the tailwaters, the predominant effect on finfish is adverse due to reduced peak seasonal flows. Seasonal flows normally fertilize backwater areas with moderate amounts of sediment. Channelization causes bank erosion, longer periods of sediment suspension, and deposition of sediment downstream. Levees prevent the export of sediments (and associated nutrients) from main channel areas to the forest floor, which over time may reduce primary productivity and decrease the food base for finfish.

Shellfish Habitat

Six characteristics of shellfish (mainly crawfish) habitat were evaluated for eight activities. The characteristics are areal/stage relationship, natural floodplain vegetation, gentle topography, presence of toxicants, oxygen level, and sedimentation. Table 3 summarizes the relationships between activities, characteristics, and the shellfish habitat function. The Fisheries Workgroup concluded that the impacts on the function from the relationship of the activities and characteristics are similar to those described above for finfish.

Cumulative Impacts

During the analyses of the ten perturbations identified for evaluation, members of the Fisheries Workgroup realized that development activities seldom occur in isolation. For example, the acreage of soybean production present in the Lower Mississippi Valley would be substantially lower today without the massive flood-control efforts of the Federal government.

Flood-control projects offer a degree of floodplain protection locally, increase flood problems downstream, and induce additional flood-control measures to control the downstream problems. Through this process, an expectation is spawned and cultivated among the public that flood protection will be provided simply because it was provided upstream. This expectation, in addition to federal financial subsidies, encourages landowners to convert valuable bottomland hardwood wetlands to farmland. When crop damage occurs as a result of flooding,

Table 3. Impacts of various activities on habitat for shellfish (mainly crawfish) as mediated through certain characteristics that contribute to the performance of this function.

Characteristic	Relationship of characteristic to function[1]	Activity[2]									
		R	S	I-U	I-O	I-D	C	L	P	A	O
Areal/stage relation	+	-	-	0	-	-	-	-	-	-	-
Natural floodplain vegetation	+	-	-	0	-	-	-	-	-	-	-
Gentle topography	+	-	-	0	0	0	-	-	-	-	-
Presence of toxicants	-	-	-	0	-	0	-	0	-	0	-
Oxygen level	+	-	-	0	+	-	-	-	0	0	-
Sedimentation	-	-	-	0	-	-	-	-	-	0	-
Overall impact on habitat for shellfish		-	-	0	-	-	-	-	-	-	-

[1] A plus indicates a positive correlation between the characteristic and the function. A minus indicates a negative correlation.

[2] A plus indicates that the function is enhanced due to the impact of the activity on the characteristic. A minus indicates that the function is impaired. A zero indicates no impact. R = conversion of site to rice, S = conversion of site to soybeans, I-U = impacts upstream from an impoundment, I-O = impacts from an impoundment on-site, I-D = impacts downstream from an impoundment, C = channelization of stream adjacent to site, L = levee construction on-site, P = conversion of site to pine plantation, A = conversion of site to aquaculture, O = oil and gas development on-site.

levee construction, channelization, or upstream impoundments, additional flood protection measures are demanded. A technique for illustrating this contention is to use a basin case history.

The Yazoo Basin is located in the northwestern portion of Mississippi and is composed of approximately 1.6 million ha of delta (old Mississippi River floodplain) and 1.8 million ha of hill area (Figure 7).

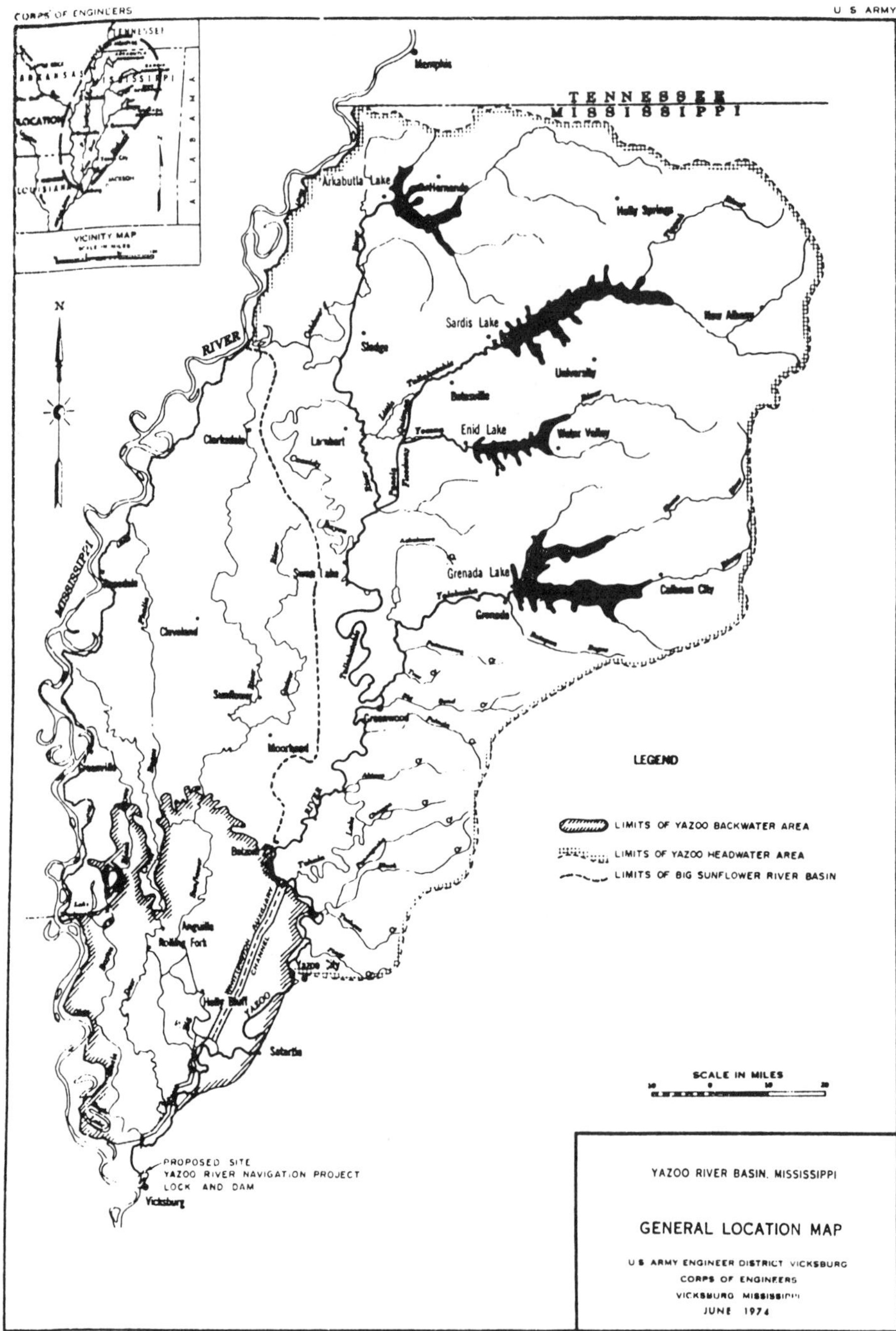

Figure 7. General location map of the Yazoo Basin, Mississippi.

When James J. Audubon visited this delta in its predevelopment days, he was explicit in his accounts of streams filled with fishes and covered with ducks and geese (U.S. Fish and Wildlife Service [FWS] 1979).

Flood control in the area began with private construction of levees along the Mississippi River. After the great floods of 1927 and 1928, the federal government became involved in major flood-control efforts for the first time (for an in-depth discussion see McCabe et al. 1982). Federal action began with the construction of flood-control reservoirs by the U.S. Army Corps of Engineers (USACE) and small retention reservoirs by the U.S. Soil Conservation Service (SCS), both in the hill area to control headwater flooding. With these efforts came increased conversion of tributary floodplains to agriculture and the resultant need for additional protection in the form of levees and channelization.

These flood-control measures allowed the waters of the hill region to move quickly down to the delta region with larger volumes and flows than historically existed. This resulted in a series of flood-control actions that would culminate in more than one billion dollars in federal expenditures. Levees and channel improvements were accomplished on the Coldwater, Tallahatchie, Yalobusha, and Yazoo Rivers in the upper region of the delta. These actions to protect property and crops from flooding also made it possible to convert massive bottomland hardwood areas, once believed to be too wet to regularly sustain crops, to agriculture. When this occurred, on-farm drainage of the cleared areas overloaded the downstream capabilities to contain the flows within the channels. Thus, additional actions were performed to channelize, levee, and build overland canals to move the floodwaters to the lower portion of the delta more expeditiously.

The pattern of federal flood-control measures moving down the Basin was well recognized by landowners as they "waited their turn." Extensive areas of productive bottomland hardwood wetlands were converted to marginal farmlands with the expectation that flood-control measures would upgrade them to prime farmlands. The same process of pushing the once stored floodwaters downstream occurred as it had in the upper reaches of the Basin. By the late 1970's, the only area in the Yazoo Delta that had not received complete flood control and the ensuing devastation of the bottomland hardwood resources was the area just above Vicksburg known as the "Backwater Area." It was so named because before construction of the Yazoo Backwater Levee the Mississippi River regularly flooded the area during spring flows. With creation of the backwater levee, the entire Yazoo Delta became "ring leveed" by the hill area to the east, mainline Mississippi River levees to the west, and the backwater levee connecting the two at the southern tip of the Yazoo Delta (Figure 7).

The vast majority of these flood-control measures were justified by "agricultural intensification" on areas previously supporting bottomland hardwood wetlands. Not only were these valuable wetlands lost, but they were replaced by farmland that exhibited 10 to 20 times the previous erosion rate with extremely high concentrations of agricultural chemicals. In the 1960's, pesticide input into natural waters forced the closing of lakes, such as Mossy and Wolf, to fishing in the central Yazoo Delta. These chemically "hot" waterbodies were, unfortunately, not the exception but the rule. Analyses of the pesticide concentrations in fish flesh at the outfall stream (Yazoo River) of the Delta revealed agri-chemicals in concentrations 320 times the accepted EPA level. Because of the loss of bottomland hardwood habitat, extremely high erosion rates after conversion (up to 44.8 metric tons/ha/year), and dangerous levels of pesticides, only 20% of the Yazoo Basin stream area is capable of sustaining a fish population today (FWS 1979).

While more than one billion federal dollars have been spent in the Yazoo Basin for "flood control," nothing has been spent solely for fish and wildlife conservation. Indeed, no consideration has ever been given to the cumulative and self-perpetuating effects of impoundments, levees, channelization, soybeans, rice production, aquaculture, and oil and gas development in the Yazoo Delta. Yet the overall effects of all perturbations considered in this workshop have eliminated 80% of the stream habitat and 90% of the spawning and nursery habitat for Yazoo Delta fishes. The individual impacts of any of the perturbations analyzed above are certainly important. However, without full consideration of those forces that cause the perturbations and other associated activities, the extent of long-term fisheries impacts cannot be determined. The Fisheries Workgroup believes it is a primary responsibility of the Section

404 permit program to ensure that these cumulative impacts are fully considered.

FIELD ANALYSIS

Introduction

Thus far we have examined the relative values of bottomland hardwood sites in supporting finfish and shellfish (crawfish) populations and the probable impact of various developmental activities on finfish and shellfish habitat. We will now begin the process of integrating the information into an evaluation system that might be used by field personnel to assess the relative ability of a site to provide finfish and shellfish habitat. By transposing the impacts of specific activities in bottomland hardwood systems, a qualitative analysis can be performed regarding potential impacts from development on a particular bottomland hardwood site. Through identification of field characteristics that establish a bottomland hardwood site as high-quality habitat for fishes, one also can identify potential remedial or mitigatory measures to restore or enhance a bottomland hardwood site for fisheries utilization.

During initial discussions, workgroup members decided that practicality and ease of application should be the litmus tests of any methodology developed. Therefore, an attempt was made to modify indices of the characteristics into an even more pragmatic yardstick that could serve as overall assessment criteria. For example, flood regime was used to represent areal/stage relationship, flood duration, and gentle topography. Table 4 illustrates the results of this exercise. By identifying certain assessment criteria, the foundation is laid for a pragmatic series of questions that would lead the field analyst to a sound conclusion regarding fisheries habitat values of a given bottomland hardwood site.

Once the task of consolidation had been accomplished, an attempt was made to complete all of the cells in Table 4 for shellfish, adult finfish, and young-of-the-year finfish. However, if all applicable measurements were performed for all characteristics, the field reviewer would become bogged down in an evaluation that was too complicated. Therefore, an exercise in establishing priorities was undertaken to identify the most critical elements of concern for each sub-function (Table 4).

Table 4. Consolidated characteristics that are critical determinants of finfish and shellfish habitat.

Characteristic Index	Finfish			Shellfish	
	Food	Cover	Reproduction	Food	Cover
Instream habitat					
A. Snags and instream cover		young			
B. Stream morphology		adult			
Access					
A. Access to permanent water			adult		
Flood regime					
A. Areal/stage	young	young		crawfish	crawfish
relationship	young	young	adult	crawfish	crawfish
Water quality					
A. Dissolved oxygen level			adult		
B. Presence of toxicants			adult		crawfish
Floodplain habitat					
A. Natural vegetation	young	young	adult	crawfish	
B. Sedimentation	young		adult	crawfish	

While workgroup members supported the conclusion from the first workshop that permanent-water areas and adjacent floodplains constitute a single aquatic system, the model consists of a series of logical questions leading from the stream to the floodplain. This approach provides the necessary insight for assessing fishery habitat quality and potential impacts. However, no explicit attempt was made to incorporate specific analyses of impacts from the second workshop. To the extent that the characteristics and criteria in Table 4 reflect a healthy environment, they can be used by field personnel to estimate changes resulting from various project perturbations. This approach is consistent with the assumption that the question of response to a particular permit application or field analysis requires considerable judgement on a case-by-case basis and should be left in the hands of field personnel.

Assessment Model

The five major elements listed in Table 4 were organized into a flow-chart (model) to assist field personnel in assessing the ability of a bottomland hardwood site to provide habitat for finfish and shellfish (Figure 8). A 'yes or no' decision is required for each of the major elements, and the final result is a ranking of high, moderate, or low for the site. The following section describes the basic rationale for the assessment model. Subsequent sections describe various factors that should be considered in arriving at a 'yes or no' decision for each of the five major elements.

Finfish

Instream Habitat. The first major element of the model is intended to determine the occurrence of finfish in a permanent waterbody adjacent to a bottomland hardwood site or, in the absence of explicit information concerning the presence of finfish, to estimate the likelihood that the permanent waterbody provides habitat suitable for finfish. If the permanent waterbody is known not to support finfish or does not exhibit characteristics necessary for finfish habitat, the value of the adjacent bottomland hardwood site is assumed to be low for finfish resources. If the permanent waterbody supports finfish or exhibits characteristics of good finfish habitat, the model proceeds to the question of access.

Access. In order for a bottomland hardwood site to provide habitat for finfish, individuals in the adjacent waterbody must have access to the site. If they do not, the value of the bottomland hardwood site is assumed to be low. If they do, there is an opportunity for the bottomland hardwood site to support finfish. However, simply having a source population with access to a site does not guarantee that the site is important to finfish. Additional, site-specific criteria must be examined.

Flood Regime. Certain hydrologic conditions are required to meet minimum spawning and nursery requirements. It was assumed that a period of about 20 days is required to complete the processes of

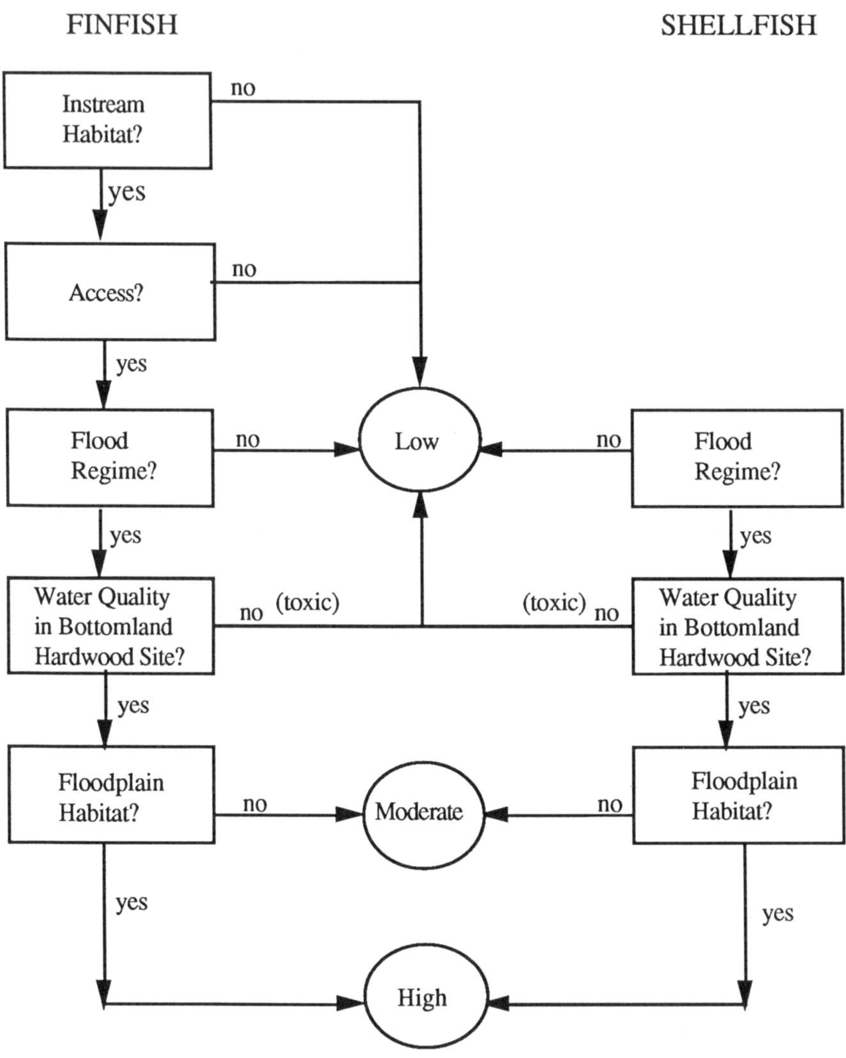

Figure 8. Flow chart for assessing aquatic habitat value of a bottomland hardwood forest site.

spawning, egg development, and larval development to produce juvenile fish. Sites flooded for 20 days or more would thus be of greatest value to finfish. Sites flooded for more than 14 days would be of some value, but young produced on such sites would be forced into the permanent waterbody prior to completion of larval development and thus would be subject to higher predation by adult fish. However, it is difficult to estimate the flooding regime on a particular site with such specificity. It was therefore assumed that any site inundated long enough to exhibit evidence of flooding is flooded for a minimum of 20 days. In such cases, the assessment model proceeds to a consideration of water quality. If there is no evidence of flooding, the value of the site is assumed to be low (see Hall 1986).

Water Quality. A variety of water-quality constituents may be either acutely or chronically toxic to finfish. If evidence of acute toxicity (e.g., fish kills) is present, the value of the site is assumed to be low. Appropriate state agencies responsible for monitoring water quality should be notified when such conditions are found, or consulted as a source of information. If no evidence of acute toxicity is available, the assessment model continues. However, users of the model must be aware that chronic stress on a fish population may still limit the utility of a site.

Floodplain Habitat. If each of the above criteria is satisfied, the quality of the habitat on the bottomland hardwood site itself is used as the final component in the assessment model. Primarily, this component deals with whether or not the structure and distribution of vegetation are suitable for spawning and rearing. If they are, the site receives a high ranking. If they are not, the ranking of the site is moderate.

Shellfish

The basic assessment model for shellfish (mainly crawfish) follows the same logic as that for finfish, except that instream habitat and access are not considered. This is because the majority of shellfish of interest, particularly crawfish, are capable of completing their entire life cycle on the floodplain itself.

Assessment Model Components

In the following section, each of the five model components is addressed in greater detail. A word model utilizing a question format is employed to yield the habitat quality estimates required in Figure 8. For each component, a general question is posed, followed by an explanation of its importance. More detailed questions are then asked to provide an answer to the general question. Appendix A provides an indepth listing of sources of information for bottomland hardwood ecosystems.

Instream Habitat

Is there reason to expect that the instream habitat supports a population of finfish that would use the bottomland hardwood site?

Adequate permanent water is essential to support healthy fish populations in bottomland hardwood systems. During low-water seasons, the permanent-water areas support fish until the next flood, at which time the bottomland hardwood site can again provide habitat. Fish populations in a bottomland hardwood systems are likely to be at their lowest during these low-water periods; the presence of adequate habitat is thus critical for the maintenance of viable populations. When considered over many years, permanent-water habitat is perhaps the most critical component in a bottomland hardwood system, although bottomland hardwood sites away from the permanent water may be extremely important during individual flood events. The quality of permanent water habitat is determined by a variety of factors, including stream morphology, snags and aquatic vegetation, and water quality, which are treated in the following questions.

Is there a commercial or recreational fishery?

In most areas, fishing pressure is sufficiently heavy that existing populations will be exploited. In many cases the existence of a fishery can be established from published accounts, unpublished surveys by state conservation agencies, personal observations of local resource managers, or contacts with local residents. Surveys of these sources will often produce adequate information concerning the existence of a fishery. Additional evidence in the form of footpaths used by fishermen, snagged fishing line and lures, and discarded bait containers may also be obtained during site visits. If none of these factors indicate the presence of a sport or commercial fishery, it is likely that fish populations are too low to warrant fishing effort, and the site should be rated as having low value as fishery habitat. If this is the case, the assessment need not proceed further. If the evidence points to the existence of a sport or commercial fishery, additional evaluation factors should be examined.

Are there sufficient snags, logs, and other forms of instream cover?

An adequate number of snags, logs, and other forms of cover in the permanent-water areas is essential to provide habitat for fish inhabiting bottomland hardwood systems. Young-of-the-year fish use these as cover to escape predation and as shelter from fast stream currents. Adult fish use them as hiding places from which they can feed on passing prey. Smaller fish, such as minnows, feed on bacteria and fungi growing on the underwater surfaces. The presence of these forms of cover can best be determined during the low-water season. Whirlpools and eddies may provide evidence of their existence during periods of higher water. Guidelines on clearing and snagging published by the American Fisheries Society (1983) may give quantitative estimates of the amount of cover needed to support a viable fishery.

Does the stream meander to provide a variety of habitats?

In a healthy bottomland hardwood system, the stream exhibits a characteristic meandering form. This meandering produces a variety of instream habitats, including deep holes with undercut banks on the outsides of meanders and mud or sandbars on the insides. This stream morphology, especially when combined with instream snags and logs, provides excellent habitat for finfish. The deep holes and undercut banks provide cover for both adults and young-of-the-year, and may be the only habitat available during periods of extremely low water. Mud and sandbars support aquatic vegetation and provide spawning habitat for species such as large-mouth bass and other sunfish. A stream lacking this characteristic morphology should generally not be rated as high for finfish habitat as one that does.

Are there slackwater areas that support vascular vegetation?

Emergent aquatic vegetation (e.g., water hyacinths [*Eichhornia*], water lilies [*Nymphaea*]) provides excellent habitat for finfish. Such vegetation provides cover for both young-of-the-year and adults. Fungi and bacteria growing on vegetation provide food for minnows. Rich populations of snails, insects, and other invertebrates provide food for larger fishes. Vegetation also provides shade during the hot summer months. If other cover is present in sufficient quantity, a permanent-water area may still provide high-quality finfish habitat in the absence of emergent aquatic vegetation. However, the presence of such vegetation provides additional evidence that a site has high value as fishery habitat.

Is the permanent water of adequate quality to maintain a healthy finfish population?

In order for permanent water to support a healthy finfish population, it must be of adequate quality. In general, three parameters may be sufficient indicators of water quality: DO levels, presence or absence of contaminants, and turbidity levels. In order to support healthy finfish populations, DO levels should regularly exceed 3 mg/1. The low-water period in late summer is usually the most critical because high water temperatures reduce the oxygen-holding capacity of the water and low streamflows reduce oxygen uptake. In addition, biological oxygen demand is often high at this time. If DO levels regularly fall below 3 mg/1, only those species that are tolerant of low levels (e.g., gar [*Lepisosteus* spp.], bowfin [*Amia calva*]) are likely to be present. Published and unpublished information on DO levels can often be obtained from state water-quality agencies and local resource managers.

Contaminants, such as agricultural chemicals (pesticides) and industrial chemicals, can also stress fish populations and other aquatic organisms. Again, published and unpublished information on such contaminants may be available from state water-quality agencies and local resource managers. In the absence of such information, field investigations of the potential for agricultural or industrial runoff should be conducted. If the watershed has extensive agricultural lands with substantial sediment runoff, agricultural chemicals are probably present in the stream. Discarded chemical containers may provide an indication of industrial contamination. In general, unless a severe problem (e.g., acute toxicity) is known to exist, chemical contaminants in a stream system should be viewed as a stress on aquatic life, even though other signs may point to healthy populations.

Turbidity is frequently a water-quality problem, particularly in row crop agricultural areas. Turbidity reduces light penetration and results in deposition of fine sediments on aquatic vegetation and other structures in the streambed. Sedimentation reduces photosynthesis and covers the food supply of small grazing species. State water-quality standards should be consulted to determine acceptable turbidity levels.

During low-water periods in late summer, high water temperatures may also affect finfish and other aquatic organisms, especially in concert with low DO levels. In general, adequate streamside vegetation (e.g., overhanging trees) and emergent aquatic vegetation are important factors in maintaining appropriate water temperatures.

Access

Are there access points that would allow establishment of a water bridge for movement of finfish from permanent water to the bottomland hardwood site?

This question is basic to the establishment of the importance of bottomland hardwood sites for finfish spawning and feeding. Water must be capable of flowing from permanent-water areas to the bottomland hardwood site in order for fish to move to the site. Many bottomland hardwood sites that were historically connected to permanent water by water bridges are presently isolated by dikes, levees, drainage structures, water-control structures, or other man-made or natural features. Such sites may often retain their wetland character through ponding of rainwater or runoff. However, no matter how productive or diverse the habitat may be for crawfish, it has no direct value to finfish in the absence of access routes.

The presence of water corridors from permanent waterbodies to seasonally flooded bottomland hardwood sites may be established through examination of existing maps, aerial photographs, and other data. Field verification of access points and indicators of flooding (e.g., watermarks, debris lines, scouring) may be necessary to prove water access conclusively. The following questions can be used as a guide to establishing whether or not water access to a given site exists.

Do existing photographs taken during flood periods show water on the site, and are flooded corridors evident?

Aerial photography is available for large portions of the United States and represents a valuable resource for natural resource managers. Inventories of existing photography from the U.S. Geological Survey's (USGS) National High-Altitude Survey, from the EPA, and from other sources is available through EPA's Aerial Imagery Centers in Vent Hill, Virginia, and Las Vegas, Nevada. Aerial photographs that show water on a site prove conclusively that the site floods; however, flooding could be the result of ponding of rainwater, runoff, or backwater flooding from a permanent source. Thus, photographs must also be examined for conveyance channels, breaks in levees, or other physical features that provide evidence of a water bridge.

Absence of water on a site should not be construed as proof that water bridges do not exist, because flood events vary in timing and extent. Likewise, the presence of potential conveyance corridors should not be regarded as conclusive proof of flooding, because flood elevation cannot be ascertained from aerial photography alone. However, the existence of tributary channels or contiguous vegetation types with permanently flooded areas is strong presumptive evidence that the bottomland hardwood site is sometimes flooded.

Do existing stream stage and topographic data indicate flood elevations sufficient to inundate the bottomland hardwood site?

Correlation of elevations at the bottomland hardwood site with stage records for the adjacent stream, along with verification of an existing water movement corridor, can be used to demonstrate conclusively that water access exists and that a site has potential to support finfish. Stage data are available from many sources, including the USGS, the USACE, and various state and local agencies. Topographic maps are also available for most areas of the United States. In addition, some agencies (e.g., the Federal Emergency Management Agency) may have maps that integrate stage and topographic data (e.g., maps of the 100-year floodplain). In general, the scale of topographic maps is sufficient to show main points of water access, such as tributaries and major drainages. However, such maps may not indicate the presence of minor tributaries or other access points. It is also important to note the date on topographic maps. Older maps may not reflect current conditions.

Soil maps showing distribution of soil types may also provide useful information. Continuous, uninterrupted distributions of hydric soils between a permanent waterbody and the bottomland hardwood site may provide additional evidence of a historic water connection. Again, however, knowledge of current conditions is essential to determine conclusively that such a connection still exists.

Are there other indications of the existence of a water connection between the bottomland hardwood site and the permanent waterbody?

Personal contacts with local residents, resource managers, and researchers can also provide information on the flooding regime of a site. Fishermen, trappers, and farmers are often particularly good sources. In addition, a site visit may produce useful information concerning access. Flooding indicators such as watermarks and debris lines show that water can get to the site, but they do not provide much information about its source. Fishing gear lost or abandoned on the floodplain itself (e.g., trot lines, lures, floats) provides strong evidence that the site is of value to finfish.

Flood Regime

Is the flood regime sufficient to provide finfish and shellfish habitat on the bottomland hardwood site?

If a bottomland hardwood site is to provide habitat for finfish, the areal extent and duration of flooding must be sufficient to permit spawning, hatching, larval development, and juvenile development. The value of a site to shellfish is likely to be proportional to the area and duration of flooding, because flooding provides access to food and a physical environment suitable for reproduction.

Is there field evidence of flooding on the bottomland hardwood site?

In the absence of actual water on the site, field signs might include any of the following:

Watermarks. Watermarks are commonly observed as stains on the bark of woody vegetation. If staining has occurred, it is likely that floodwater was present long enough to permit spawning and egg hatching. Watermarks can also be used in combination with topography to estimate the areal extent of flooding. The highest watermarks reflect the maximum extent of the greatest inundation event.

Drift lines. Drift lines consist of debris deposited in a line on the ground surface or entangled in above-ground vegetation. They provide an estimate of the minimum area inundated during a flood event.

Drainage patterns. Soil erosion patterns, debris piled against vegetation, absence of leaf litter, and scouring around the base or roots of perennial vegetation all provide information on the drainage of a site. Topographic position is important in assessing whether water originated from upland runoff or from flooding.

Shallow root zones. Woody vegetation having roots at or near the surface is likely to have developed under flooded conditions or a high groundwater table.

Adventitious roots. Woody vegetation often develops adventitious, or above-ground, roots in response to flooding or a high groundwater table.

Soil water. Free water at or near the soil surface is often a good indicator of the flood regime on a site. Ponded water may be observed under some conditions, or a small hole may be dug in the soil and the presence or absence of water noted.

Silt deposition. Silt deposits on understory vegetation provide information on both the areal extent and duration of flooding, because fine sediments take a relatively long time to settle out of the water column.

If none of the above field indicators provide evidence of flooding, flood-stage data, aerial photography, topographic maps, and other information sources should be examined before drawing firm conclusions. If examination of these sources leads to the conclusion that there is no flooding, the bottomland hardwood site is likely to be of low value to finfish and shellfish. If evidence of ·flooding exists, the assessment should proceed to the next question.

Are the timing and duration of flooding appropriate for spawning and egg hatching?

Spawning in some species begins as early as February and in others can occur throughout the summer (Hall 1979). For most species, spawning occurs between mid-March and mid-May. Egg hatching requires 7-14 days for most species. If flooding does not occur at the appropriate time or is not of sufficient duration for spawning and egg hatching to take place, the bottomland hardwood site has a low value for finfish and shellfish. If flood timing and duration are appropriate, the assessment should proceed.

Are the timing and duration of flooding appropriate for larval and juvenile development?

For most species, larval development takes a minimum of 20 days. The most accurate information on flood duration is obtained from flood stage data in combination with topographic data and aerial photography. Verification of this information by local experts is also useful.

If this information indicates that the duration of flooding is insufficient to allow larval and the beginning of juvenile development, then the bottomland hardwood site is of low to moderate value to finfish. Additional periods of ponded water during the winter and spring would indicate high values for shellfish.

When assessing the suitability of the flood regime at a particular site, it is important to remember that water depth is also an important criterion. Thus, higher areas that are flooded infrequently may be extremely important at the times they are flooded, because they may provide the only habitats with suitable water depth for finfish spawning (generally less than 1.2 m).

Water Quality

Is the water quality on the site adequate for spawning and nursery habitat?

Water quality is important both to finfish and shellfish in various life stages. It is probably most critical during reproduction, because eggs and young fish are generally more sensitive to water quality. Poor water quality during the initial stages of development can result in high mortality or improper development of eggs and larvae.

Numerous water-quality constituents can have either chronic or acute effects on finfish and shellfish. In addition, new pollutants are being introduced into the environment at a rapid rate, and combinations of constituents can often have synergistic effects. Isolating one or a few constituents of concern is therefore difficult. Current, specific information should be obtained from state and federal regulatory agencies. In general, however, water-quality concerns can be grouped into three categories.

Is the DO level regularly at or above 3 mg/1?

The DO level is an important factor in maintaining a suitable growth medium for finfish and shellfish. Both have metabolic rates that require an adequate oxygen supply; 3 mg/1 is usually sufficient for most species that utilize bottomland hardwood sites. A variety of factors can affect DO levels. Temperature is one of the most important, because warmer water can hold less oxygen. Removal of woody species from a floodplain site or from a streambank reduces the amount of shading and thus leads to higher temperatures and lower DO levels. In streams, the most critical time is during low-water periods when fish are restricted with respect to available habitat.

If DO levels regularly drop below 3 mg/1, the value of the site to finfish and shellfish is low. If they do not, the assessment continues with the next question.

Are there point sources of industrial pollutants upstream that are close enough to affect the bottomland hardwood site?

Industrial sites such as paper mills, battery factories, sewage treatment plants, and toxic waste disposal areas can often discharge materials that are harmful to finfish and shellfish. Information on locations of such sources and water-quality standards for various pollutants can be obtained from the state and federal regulatory agencies. Licensing information on particular chemicals can also provide data on the potential effects on aquatic life. Direct evidence of potential problems may be obtained from observations of fish kills and studies of benthic populations. If any of these sources indicate lethal levels of pollutants, the value of the bottomland hardwood site is likely to be low for finfish and shellfish. If no evidence of lethal concentrations is found, the assessment should proceed, but with the recognition that sublethal concentrations of many pollutants can still affect growth and development of aquatic organisms.

Is there evidence that sediments from nearby agricultural fields reach the bottomland hardwood site?

Soil particles washed from agricultural fields often have high concentrations of agricultural chemicals associated with them, and sediment deposition can retard organic decomposition. Irregular distributions of organic carbon in soil levels near the surface may reflect high levels of turbidity and thus negative impacts to fish eggs and larvae.

Floodplain Habitat

Is the bottomland hardwood site habitat of sufficient quality to provide spawning and nursery habitat for finfish and/or food sources for shellfish?

The primary objective of this question is to evaluate the potential for 1) spawning sites for adult finfish, 2) cover and food availability for young-of-the-year, and 3) food availability for crawfish. It is important to recognize these values as the evaluation targets so a logical application of field observations can be made. Simply stated, it is not the trees, detritus, or bushes in bottomland hardwood systems that the Clean Water Act sets out to protect, but the values from these components to the fisheries resources.

We identified the importance of vegetational dominance in bottomland hardwood systems to fishes for sites of egg deposition, cover for larval and juvenile fishes, production of plankton for nutrition of the young and detrital material for the base link of the food chain and a direct food source for crawfish. Hall (1979) and Gosselink et al. (Chapter 17) pointed out the importance of detrital export to the sustenance of stream fisheries during the non-flood season. The basic premise is that in a bottomland hardwood system, detritus is the source of nutrient introduction that feeds the aquatic ecosystem. Not only does detritus provide the base link of the food chain, it also aids in erosion control through buffering the "splash" effects of rainfall on the soil. This, in turn, contributes to good water quality that is important to sensitive egg, larval, and juvenile fishes.

Vegetation is also very important in providing spawning sites for adult finfishes. Balon (1975) indicates that spawning sites can be characterized by reproductive guilds or methods that are common to several species. In bottomland hardwood systems, the two primary guilds are nest builders and broadcast spawners. Nest builders select a site, excavate or build a nest, and then guard the eggs and young. Breeder and Rosen (1966) identified the preferred substrate for many of the bottomland hardwood system fishes. Common preferences include excavated nests surrounded by detritus, nests constructed under logs or in exposed tree roots, and cavity areas such as inside hollow cypress trees. Therefore, the abundance of these structures is an important field character that should be considered.

Broadcast spawners release the eggs in the upper water column which allows drift to deposit eggs on submerged vegetation. These eggs are initially covered by an adhesive coating that allows attachment to physical structures in the water. After attachment, the chorion of the egg hardens and embryological development continues. Bushes, logs, bases of trees, low branches, heavy detritus, and other such potential egg deposition sites are, therefore, very important considerations in assessing the value of a bottomland hardwood site for finfish spawning. Day et al. (1977) found that a bottomland hardwood system in Louisiana had an above ground net productivity of 1584 g dry weight $\cdot$ m^{-2} $\cdot$ yr^{-1}. This indicates a heavy layer of detrital material and infers a high percentage of canopy cover. Finally, the information derived from detrital production can be very helpful in assessing the quantity of food available for emerging crawfish.

What percentage of the bottomland hardwood site is covered by exposed root systems, twigs, bushes and herbaceous vegetation (both living and dead) to provide spawning sites, nursery values and good production for young-of-the-year fishes and crawfishes?

From field experience and information considered from the literature the following scale may be used to answer this question.

% Cover = /______/______/______/
0 Low Value 40 Moderate Value 70 High Value 100

Using this scale, the evaluator can estimate a reasonable value for a bottomland hardwood site to the fisheries resources from an available habitat standpoint. It should be remembered that the time of year a site is visited may reflect different values. The evaluator should attempt to apply a value that would reflect habitat conditions during the flood event, when the bottomland hardwood site is most important. The most critical aspect of any evaluation for fisheries use in bottomland hardwood wetlands is the understanding that a complete analysis should be performed prior to forming a conclusion.

ACKNOWLEDGMENTS

The primary authors wish to gratefully acknowledge the remainder of the workgroup participants at each of the three workshops. These individuals were active contributors to each workshop report, which in turn has been condensed and edited as presented here. The expertise provided by each of these professionals is highly respected and appreciated. Each workgroup was chaired by H. Dale Hall. The authors also wish to express a sincere appreciation to the editors, James G. Gosselink, Lyndon C. Lee, Tom A. Muir, and Susan C. Hamilton for their understanding and editorial assistance.

REFERENCES CITED

American Fisheries Society. 1983. Stream obstruction removal guidelines. Stream Renovations Guidelines Committee. A joint report of the Wildlife Society, American Fisheries Society, and International Association of Fish and Wildlife Agencies. American Fisheries Society, Bethesda, Maryland. 9 pp.

Balon, E. K. 1975. Reproductive guilds of fishes: a proposal and definition. J. Fish Res. Board Can. 32(6): 821-864.

Blaxter, J. H. S. 1974. ed. Introduction. Pages III-IV*in* The early life history of fish. Proceedings International Symposium, Scottish Marine Biological Association, Oban, Scotland. May 17-23, 1973.

Breeder, C. M., Jr., and D. E. Rosen. 1966. Modes of reproduction in fishes. Natural History Press, Garden City, New Jersey. 941 pp.

Clark, J. R., and J. Benforado, eds. 1981. Wetlands of bottomland hardwood forests. Elsevier Scientific Publishing Company, New York. 401 pp.

Day, J. W., T. V. Butler, and W. H. Conner. 1977. Productivity and nutrient export studies in a cypress swamp and lake system in Louisiana. Pages 255-269 *in* M. V. Wiley, ed. Estuarine processes. Vol. 2. Circulation, sediments and transfer of materials in the estuary. Academic Press, Inc., London.

Hall, H. D. 1979. The temporal and spatial distribution of ichthyoplankton of the upper Atchafalaya Basin. Unpublished M.S. thesis, Louisiana State University, Baton Rouge. 60 pp.

__________. 1986. Transposing wetlands characteristics to wetlands values: the 404(b)(1) analysis. Pages 128-131 *in* J. A. Kusler and P. Riexinger, eds. Proceedings, National Wetlands Assessment Symposium. June 17-20, 1985. Association of State Wetland Managers, Chester, Vermont. Technical Report 1.

Hempel, G. 1965. On the importance of larval survival for the population dynamics of marine food fish. Calif. Coop. Oceanic Fish. Invest. Vol. X:13-23.

Lambou, V. W. 1959. Fish populations of backwater lakes in Louisiana. Trans. Am. Fish. Soc. 88(1): 7-15.

Larson, J. S., M. S. Bedinger, C. F. Bryan, S. Brown, R. T. Huffman, E. L. Miller, D. G. Rhodes, and B. A. Touchet. 1981. Transition from wetlands to uplands in southeastern bottomland hardwood forests. Pages 225-273 *in* J R. Clark and J. Benforado, eds. Wetlands of bottomland hardwood forests. Elsevier Scientific Publishing Co., New York. 401 pp.

Lasker, R. 1965. Field criteria for survival of anchovy larvae: the relation between inshore chlorophyll maximum layers and successful first feeding. Fish. Bull. 73(3):453-462.

McCabe, C. A., H. D. Hall, and R. C. Barkley. 1982. Yazoo area pump study, Yazoo backwater project, Mississippi. U.S. Fish and Wildlife Service, Fish and Wildlife Coord. Act Report. Vicksburg, Mississippi. 181 pp.

U.S. Fish and Wildlife Service. 1979. The Yazoo Basin: an environmental overview. Jackson Area Office, Jackson, Mississippi. 39 pp.

Welcomme, R.L. 1979. Fisheries ecology of floodplain rivers. Longman, Inc., New York. 317 pp.

______________, and D. Hagborg. 1977. Towards a model of a floodplain fish population and its fishery. Environ. Biol. Fishes. 2(1):7-24.

Wharton, C. H., W. M. Kitchens, E. C. Pendleton, and T. W. Snipe. 1982. The ecology of bottomland hardwood swamps of the southeast: a community profile. U.S. Fish Wildl. Serv., Biol. Serv. Program, Washington, D.C. FWS/OBS-81/37.

______________, V. W. Lambou, J. Newsom, P. V. Winger, L. L. Gaddy, and R. Mancke. 1981. The fauna of bottomland hardwoods in southeastern United States. Pages 87-160 *in* J. R. Clark and J. Benforado, eds. Wetlands of bottomland hardwood forests. Elsevier Scientific Publishing Co., New York. 401 pp.

Wood, R. 1951. The significance of managed water levels in developing the fisheries of large impoundments. J. Tenn. Acad. Sci. 26(3): 214-235.

_________, and D. W. Pfitzer. 1958. Some effects of water-level fluctuations on the fisheries of large impoundments. International Union for Conservation of Nature and Natural Research, Seventh Technical Meeting. 11-19 September 1958. Brussels, Belgium.

16. THE RELATIONSHIP OF HUMAN ACTIVITIES TO THE WILDLIFE FUNCTION OF BOTTOMLAND HARDWOOD FORESTS: THE REPORT OF THE WILDLIFE WORKGROUP

Steven W. Forsythe
U.S. Fish and Wildlife Service,
Tulsa, OK 74127

James E. Roelle
National Ecology Research Center,
U.S. Fish and Wildlife Service,
Fort Collins, CO 80525

with Panel
R. Banks, R. Boner, F. Dunham, T. Glatzel, L. Harris, J. Hefner, D. Lofton, J. Neal, C. Newling, T. Pullen, D. Sanders, H. Sather, B. Tomlinson, and T. Welborn

ABSTRACT

Wildlife workgroup discussions, held as part of three bottomland hardwood workshops, are summarized. The wildlife function is defined as provision of habitat for wildlife species; such a function cannot appropriately be described in terms of zones. The relationship of human (developmental) activities to the wildlife habitat function of bottomland hardwoods is generally negative because impacts of the activities decrease (to varying degrees) the value of the function. Characteristics of bottomland hardwoods that contribute to their value for wildlife are identified, and a tentative model (index) is developed to assess the occurrence of such characteristics. However, information constraints prevented development of a full complement of indices for the wildlife habitat function. Management goals for providing habitat for wildlife in bottomland hardwoods are not stated, but they would likely be based on amount and distribution of the forest at a regional or continental scale.

Ecological Processes and Cumulative Impacts: Illustrated by Bottomland Hardwood Wetland Ecosystems. Edited by James G. Gosselink, Lyndon C. Lee, and Thomas A. Muir. Chelsea, MI 48118. Printed in USA.

INTRODUCTION

The importance of bottomland hardwood forests as habitat for a broad assemblage of wildlife species has been well documented. Results of these studies have been discussed and summarized extensively in the last decade by Fredrickson (1979), Klimas et al. (1981), Wharton et al. (1982), and Wilkinson et al. (1987). Complexity, diversity, and productivity are recurrent themes in these descriptions of the value of bottomland forests as habitat for wildlife. The basis for these themes is the combination of water, organic matter, sediment, and nutrients interacting with environmental gradients, such as hydroperiod and elevation, to stimulate diverse floras. These floras, in turn, support complex wildlife communities.

Despite the validity of this generalization, it is also true that bottomland hardwood sites differ in their ability to support wildlife. There is considerable natural variability in bottomland hardwood forests, both regionally (e.g., the Atlantic Coastal Plain versus the Lower Mississippi Valley) and locally (e.g., backwater swamps subject to floods of long duration versus narrow, riparian woodlands subject to frequent floods of short duration). This natural variability, in combination with the impacts of human developments (e.g., reservoirs, levees) and activities (e.g., timber management), results in bottomland hardwood sites that differ substantially in their value as wildlife habitat.

Recognition of this variability was one of the prime factors that motivated the workshops described in this book and the wildlife workgroup discussions summarized in this chapter (Roelle et al. 1987a,b,c). Generally, the objectives of the workshops were to:

1) examine the functions performed by bottomland hardwood communities and the relationship of these functions to the zonation concept described by Larson et al. (1981);
2) examine the characteristics that determine the extent to which a particular bottomland hardwood site performs the functions and, using these characteristics, assess the probable impacts of several development activities on the functions; and
3) describe indices that might be used for field assessment of the magnitude or importance of a particular function and, where

possible, target values or management goals for these indices or for the functions themselves.

The workshops can thus be viewed as an attempt to understand and characterize the natural and man-induced variability of bottomland hardwood communities, variability that was a constant source of concern in the wildlife workgroups. In order to accomplish the workshop tasks in the limited time available, participants felt compelled to think in terms of "typical" bottomland hardwood sites, yet realized that such sites rarely exist.

DESCRIPTION OF FUNCTIONS AND CHARACTERISTICS

The following definitions have been adopted for the purposes of this report:

> Function - an ecological process, generally identified because it has social value.
>
> Characteristic - a property of an ecosystem that confers, leads to, or is in part responsible for a function. A characteristic can be considered an independent variable for a dependent function.

While these definitions are more specific than any provided at the workshops, they can still be interpreted at several levels of detail with respect to wildlife (Table 1). The difficulty in precisely defining functions and characteristics was the source of some confusion during the workshops, especially since a characteristic at one level of detail can be thought of as a function at another level. However, the common themes throughout the series of workshops were that: 1) bottomland forests are important to wildlife because of the resources they provide, 2) these resources can be generally categorized and described, and 3) bottomland hardwood sites differ in their value to wildlife because of differences in their ability to provide these resources. Taking into consideration the various workshop discussions, for the purpose of this chapter we define the wildlife function as the provision of habitat for the diverse assemblage of wildlife species.

Table 1. Examples of possible levels of detail concerning functions and characteristics related to wildlife.

Example Functions	Example Characteristics
1. Provide habitat for wildlife	Size of tract
	Connectivity with other habitats
	Diversity
	Geographic location
2. Provide reproductive sites (or food, or cover) for wildlife	Abundance of tree cavities
	Abundance of ground-level cavities
3. Provide tree cavities (or any other specific resource) for reproductive sites	Abundance of tree cavities
	Abundance of trees prone to produce cavities
4. Provide habitat for wood ducks (or any other "important" species)	Abundance of tree cavities for nesting
	Abundance of swamps, ponds, and sloughs for brood habitat
	Abundance of hard mast for food

RELATIONSHIP BETWEEN HABITAT FUNCTION AND ZONATION

The primary objective of the wildlife workgroup at the first workshop (Roelle et al. 1987a) was to evaluate the relationship between the wildlife habitat function of bottomland hardwoods and the zonation concept described by Larson et al. (1981). The wildlife workgroup concluded it would be difficult to describe a general wildlife support function (provision of habitat) in terms of zones in bottomland hardwood forests Furthermore, it seemed inappropriate to characterize one zone as being more important than another zone. The reasons for these conclusions were as follows:

1) The various wildlife species supported by bottomland hardwoods have vastly different habitat requirements--many are capable of using several zones to meet their life requisite.
2) Many wildlife species are extremely mobile and thus take advantage of certain habitat components of particular zones as the necessity occurs. An example of this mobility would be migratory waterfowl that only use bottomland hardwoods during the winter flooding for some life requisites.

In addition, it is important to remember that on-site inspections of bottomland forests usually result in observation of only a small fraction of the wildlife species present. Investigators must rely instead on features such as hydrology, soils, and vegetation as indicators of wildlife habitat. This requires a broad base of training and experience possessed by a limited number of individuals. Assumptions concerning fixed relationships between zones and wildlife habitat, if used in the regulatory process, would tend to minimize the importance of professional experience and judgment in determining the actual habitat values of a particular site.

CHARACTERISTICS THAT DETERMINE THE VALUE OF BOTTOMLAND HARDWOODS AS WILDLIFE HABITAT

The wildlife workgroups continually struggled to identify a set of characteristics that could be "assessed" or "measured" to determine the value or worth of a particular bottomland hardwood site as wildlife habitat. At the first workshop (Roelle et al. 1987a), these efforts concentrated on the vegetative zones (see above). During the second workshop (Roelle et al. 1987b), the workgroup identified functions of bottomland hardwood sites that provide habitat for representative species or groups. The third wildlife workgroup (Roelle et al. 1987c) considered life requisites of a broad range of wildlife species as "subfunctions" to the overall wildlife function. These "subfunctions" seem to be more appropriately defined as characteristics, shown in Table 2.

Table 2. Characteristics of the wildlife habitat function of bottomland hardwoods, and examples of wildlife species or groups that use the habitat characteristics.

Characteristics	Example Species or Groups
Production of hard and soft mast, berries, and fruit	Waterfowl, deer, turkeys, bear, squirrels, small mammals, furbearers, passerine birds
Production of browse	Deer, rabbits, small mammals
Production of invertebrates (e.g., crayfish, snails, worms, insects)	Waterfowl, passerine birds, small mammals, furbearers, amphibians, reptiles
Production of vertebrates (e.g., small mammals, reptiles, amphibians, fish)	Raptors, furbearers, bear, wading birds
Presence of arboreal cavities	Wood ducks, squirrels, passerine birds, woodpeckers, furbearers
Presence of ground-level cavities (in trees and logs)	Furbearers, bear, rabbits, reptiles, amphibians
Presence of dense, ground-level vegetation	Deer, rabbits, small mammals, bear, waterfowl, reptiles
Presence of downed timber	Deer, rabbits, small mammals
Variable topography (interspersion of swamps and higher ground)	Deer, bear, rabbits, furbearers

Additionally, during the series of workshops, participants identified several characteristics that would probably be associated with sites of high value to wildlife, including the following:

1) **Size of the tract.** Large blocks of contiguous habitat may be important in providing for the needs of large, mobile species, as well as providing diverse habitats for smaller species.

2) **Connectivity with other habitats.** Forested corridors, especially those along watercourses and those connecting upland habitats with bottomland forests, may be particularly important in allowing movement of wildlife species. Movements between blocks of habitat are important for several reasons, including repopulating vacant areas, exchanging genetic material, and allowing large, mobile species access to a variety of resources. Many species tend to follow watercourses in these movements. Corridors connecting bottomlands to upland habitats may be important as escape routes during floods and in providing alternate food sources.
3) **Diversity.** Tracts with high diversity (i.e., interspersion of forest types, vertical vegetation structure, and flooding regime) may have the highest probability of meeting the needs of a variety of wildlife species.
4) **Geographic location, both local and regional.** From a local perspective, factors such as proximity to permanent waterbodies and proximity to other areas of known value to wildlife may be important determinants of habitat values. Similarly, tracts immediately outside mainline levees and floodways may be important in providing refuge during floods. Regionally, the location and pattern of bottomland hardwood sites with respect to flyways and other movement corridors may be as important to migrating species as specific characteristics of the sites. For example, tracts along the Gulf Coast may be especially important to species that migrate across the Gulf of Mexico.

RELATIONSHIP OF WILDLIFE HABITAT FUNCTION TO HUMAN ACTIVITIES

The relationship of the wildlife habitat function of bottomland forests to human (developmental) activities was the primary focus of the second workshop (Roelle et al. 1987b). The wildlife workgroup analyzed the impact of human activities (Table 3) on five wildlife species or species groups. This information indicates that specific actions analyzed by the workgroup have two common impacts: 1) removal or substantial modification of natural vegetation, and 2) alteration of the hydrologic regime (most frequently a reduction in wetness). These two impacts vary in intensity, depending on the specific design of a particular activity, but an activity such as land clearing for agriculture virtually removes the characteristics responsible for the wildlife habitat function.

Table 3. Human activities and associated specific actions analyzed by the wildlife workgroup.

Activity	Actions	
Conversion to rice	Land clearing Leveling Levee construction Flooding	Drainage or drying Fertilization Pesticide application Seed bed preparation
Conversion to soybeans	Land clearing Leveling Ditching	Fertilization Pesticide application Seed bed preparation
Impoundment construction (upstream of site, on-site, downstream of site)	Land clearing (on-site) Dredging	Filling
Channelization	Trapezoidal cut Land clearing (both sides)	Dredging Filling
Levee construction (e.g., a mainline river levee)	Land clearing Dredging	Filling
Conversion to pine plantation	Clearcutting Land clearing Site preparation	Ditching Herbicide application Pesticide application
Conversion to aquaculture	Land clearing Leveling Levee construction	Ditching Irrigation

Thus, even if the hydrologic regime remains unchanged (as often occurs with agricultural conversion), the wildlife habitat function is substantially diminished. If the development also includes reducing the hydrologic regime (i.e., "drying out" the site), the wildlife habitat characteristics are essentially eliminated.

There are some developmental activities that only modify the hydrologic regime (i.e., reduce flooding with a levee or increase flooding with a dam). In these instances, forest vegetation remains but usually undergoes some change in value to wildlife, and these changes are often difficult to predict.

The conclusion of this exercise was that all of the activities generally have negative impacts on the species or groups considered, and that the extent of the impacts depends on factors such as the type of habitat that is being replaced or altered, the geographic location of the site, the geographic extent of the activity, the specific purpose of the activity, habitat conditions in the surrounding area, construction techniques, future management of the site, and specific wildlife management objectives for the area.

Another noteworthy finding of the second workgroup, one often overlooked in impact analyses, is the need for consideration of more than specific, on-site characteristics of the wildlife habitat function. To elaborate, indices of bottomland hardwood characteristics (discussed later) are often measured along a transect or by some other accepted sampling technique. This approach tends to place considerable emphasis on the site-specificity of wildlife habitat and, in the process, overlooks other important characteristics (e.g., tract size, connectivity) that contribute to the overall wildlife habitat function. Such considerations may seem obvious to experienced field biologists, but an assessment can easily become site-specific. Furthermore, one of the tasks of the workgroup was to identify the relationship of characteristics to activities, primarily for an audience of "decision-makers" unfamiliar with the wildlife habitat function of bottomland hardwood forests.

Finally, the second workgroup recognized that the secondary and cumulative impacts of most development activities, while difficult to predict or evaluate, are also likely to be negative. Gosselink and Lee

(1989) describe a methodology that may be useful in assessing cumulative impacts in bottomland hardwoods.

INDICES OF WILDLIFE HABITAT FUNCTION

The third wildlife workgroup (Roelle et al. 1987c) attempted to develop indices, or models, to assess the magnitude of wildlife habitat function in bottomland forests. The group used the characteristics listed in Table 2, which appear to provide reasonable coverage of the attributes of bottomland forests for wildlife habitat function. The workgroup then developed tentative models, essentially indices, to assess the occurrence, or potential for occurrence, of the characteristics.

An example for the ability of a site to produce acorns, an important component of hard mast in bottomland hardwoods, is shown in Figure 1. The model consists of a branching key that produces a high, medium, or low ranking based on the species composition, size, and density of oaks (*Quercus* spp.). While workgroup members believed that the approach illustrated in Figure 1 may be workable, they also emphasized several difficulties, including the following:

1) Developing the evidence to support such models would not be an easy task. Data pertaining to important vegetation parameters (e.g., abundance and species diversity of oaks) may be available, but are likely to be widely scattered in the literature. Correlating these data with levels of wildlife use would be very difficult.
2) Verifying such models would also be a difficult job. Specific studies could probably be designed to examine relationships between wildlife use and model parameters. However, such studies would be intensive and relatively expensive, and their cost-effectiveness would likely be questionable. In the absence of such studies, a less desirable level of verification might be obtained by comparing model results with expert judgment.

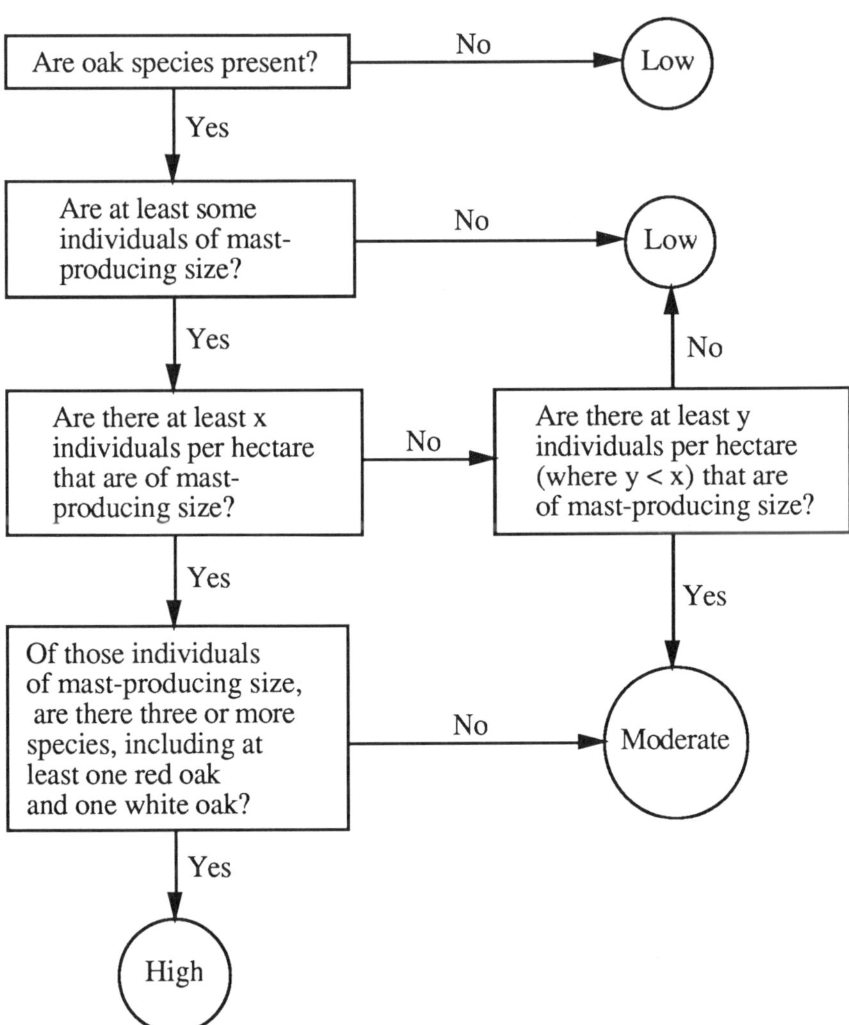

Figure 1. **Model to assess the ability of a site to produce acorns.**

3) The utility of such models would be highly dependent on the level of aggregation necessary in the regulatory process. Combining results from several models to produce an aggregate assessment of a site would introduce greater subjectivity, and would make verification even more difficult.
4) Finally, workgroup members believed that even relatively good models, if they could be developed, would still be reliable only when applied by trained biologists with experience in wildlife habitat evaluation. Bottomland hardwood forests vary geographically, and experience with local conditions would still be important in producing a reliable assessment.

To further elaborate on the subject of an index for the wildlife habitat function, it may be useful to discuss use of the Habitat Evaluation Procedures (U.S. Fish and Wildlife Service 1980). The procedures measure certain life requisite habitat variables for a particular wildlife species and are frequently used to determine the quantitative value of a bottomland hardwood site for certain species. This baseline information is then used to develop projections of impacts and, in many instances, to develop mitigative measures. It must be remembered that the emphasis of these procedures is on species habitat function, <u>not</u> the holistic wildlife habitat function of bottomland forests that is the subject of this chapter. At this point in our understanding of bottomland hardwood ecosystems, an assessment technique for the wildlife habitat function has yet to be developed. The development of such a technique should include assessment of the "nonsite-specific" characteristics of a bottomland hardwood tract, as discussed earlier in this chapter.

ESTABLISHMENT OF TARGETS AND GOALS

An additional task of the third workgroup was to develop "target values" or "management goals" for the wildlife habitat function of bottomland hardwoods. The basic idea was to develop relationships between characteristics and functions, and thus identify points at which perturbations might begin to reduce performance of the functions. Time constraints prevented explicit consideration of this task. However, even with additional time, the job would have been difficult.

At the level of a single, general function (i.e., to provide habitat for wildlife), target values or management goals would likely be stated in terms of amounts and distribution of bottomland hardwood forests that should be maintained at a regional or continental scale. Such goals would depend on factors ranging from habitat requirements of individual species, to agency objectives and missions, to economic considerations. As such, establishment of these goals was well beyond the capabilities of a small workgroup.

REFERENCES CITED

Frederickson, L. H. 1979. Lowland hardwood wetlands: current status and habitat values for wildlife. Pages 296-306 *in* P. E. Greeson, J. R. Clark, and J. E. Clark, eds. Wetland functions and values: the state of our understanding. American Water Resources Association, Minneapolis, Minnesota.

Gosselink, J. G., and L. C. Lee. 1989. Cumulative impact assessment in bottomland hardwood forests. Wetlands 9:83-174.

Klimas, C. V., C. O. Martin, and J. W. Teaford. 1981. Impacts of flooding regime modification on wildlife habitats of bottomland hardwood forests in the Lower Mississippi Valley. U.S. Army Corps of Engineers, Waterways Experiment Station, Vicksburg, Mississippi. Tech. Rep. EL-81-13. 137 pp.

Larson, J. S., M. S. Bedinger, C. F. Bryan, S. Brown, R. T. Huffman, E. L. Miller, D. G. Rhodes, and B. A. Touchet. 1981. Transition from wetlands to uplands in southeastern bottomland hardwood forests. Pages 225-273 *in* J. R. Clark and J. Benforado, eds. Wetlands of bottomland hardwood forests. Elsevier Scientific Publishers, New York.

Roelle, J. E., G. T. Auble, D. B. Hamilton, R. L. Johnson, and C. A. Segelquist, eds. 1987a. Results of a workshop concerning ecological zonation in bottomland hardwoods. U.S. Fish and Wildlife Service, National Ecology Center, Ft. Collins, Colorado. NEC-87/14. 141 pp.

Roelle, J. E., G. T. Auble, D. B. Hamilton, G. C. Horak, R. L. Johnson, and C. A. Segelquist, eds. 1987b. Results of a workshop concerning impacts of various activities on the functions of bottomland hardwoods. U.S. Fish and Wildlife Service, National Ecology Center, Ft. Collins, Colorado. NEC-87/15. 171 pp.

____________________________________, R. L. Johnson, and C. A. Segelquist, eds. 1987c. Results of a workshop concerning assessment of the functions of bottomland hardwoods. U.S. Fish and Wildlife Service, National Ecology Center, Ft. Collins, Colorado. NEC-87/16. 173 pp.

U.S. Fish and Wildlife Service. 1980. The Habitat Evaluation Procedures. U.S. Fish and Wildlife Service, Ecological Services Manual 102. 124 pp.

Wharton, C. H., W. M. Kitchens, E. C. Pendleton, and T. W. Sipe. 1982. The ecology of bottomland hardwood swamps of the southeast: a community profile. National Coastal Ecosystems Team. U.S. Fish Wildl. Serv., Biol. Serv. Program, Washington, D.C. FWS/OBS-81/37. 133 pp.

Wilkinson, D. L., K. Schneller-McDonald, R. W. Olson, and G. T. Auble. 1987. Synopsis of wetland functions and values: bottomland hardwoods with special emphasis on eastern Texas and Oklahoma. U.S. Fish Wildl. Serv., Biol. Serv. Program, Washington, D.C. Biol. Rep. 87(12).

17. HUMAN ACTIVITIES AND ECOLOGICAL PROCESSES IN BOTTOMLAND HARDWOOD ECOSYSTEMS: THE REPORT OF THE ECOSYSTEM WORKGROUP

James G. Gosselink
Center for Wetland Resources,
Louisiana State University, Baton Rouge, LA 70803

Mark M. Brinson
Biology Department, East Carolina University, Greenville, NC 27834

Lyndon C. Lee
Savannah River Ecology Laboratory,
University of Georgia, Aiken, GA 30365

Gregor T. Auble
National Ecology Research Center,
U.S. Fish and Wildlife Service,
Fort Collins, CO 80525

with Panel
J. Day, J. Dowhan, L. Harris, P. Hatcher, B. Keeler, C. Kiker,
W. Kruczynski, R. Lea,W. Mitsch, L. Pearlstine, J. Pomponio,
J. Safley, D. Sanders, M. Scott, L. Tebo

ABSTRACT

Bottomland hardwood forest ecosystems are familiar parts of integrated landscapes that include the watershed and the adjacent streams. We identify ecological principles relevant to their functional integrity. Naturally functioning forested bottomlands provide services to man by mediating floods, improving water quality, providing primary production and food web support for animal populations, and maintaining natural biotic diversity. The level of these services depends on the magnitude of a number of interrelated attributes of intact floodplain forests. We consider

---

Ecological Processes and Cumulative Impacts: Illustrated by Bottomland Hardwood Wetland Ecosystems. Edited by James G. Gosselink, Lyndon C. Lee, and Thomas A. Muir. Chelsea, MI 48118. Printed in USA.

eight common human activities in floodplains that influence the level of services by their effects on these ecosystem attributes.

While the impact of individual actions can be assessed at the local level, the cumulative effect of many such actions provokes a landscape-level response. Regulation of cumulative impacts therefore requires regulation of landscapes. A broad outline of the technical and regulatory process requirements for such regulation is presented, and some specific suggestions are provided for cumulative impact assessment criteria.

INTRODUCTION

Ecosystems respond to man's intrusions at two distinct levels. In the first response level, discussed in foregoing chapters, either internal processes, individual components, or both react to a specific human action. At the second response level, the entire ecosystem shows a collective reaction to many discrete alterations.

The Ecosystem Workgroup focused on the second level response by considering five basic questions.

1) How does a bottomland hardwood ecosystem function in an integrated landscape?
2) What services do humans derive from bottomland hardwood ecosystems?
3) What attributes of bottomland hardwood ecosystems determine these benefits?
4) How do human activities modify these attributes and subsequently affect the benefits?
5) How can the cumulative impact of human activities in bottomland hardwood ecosystems be regulated?

This report begins with some important definitions, then progresses through a discussion of 10 ecological principles and their relevance to the structure and function of bottomland hardwood ecosystems and landscapes. In the next section, four services of bottomland hardwood systems to man are identified along with the ecological attributes responsible for them. A discussion follows on how common human activities affect bottomland hardwood forest services through their impact on relevant attributes. The chapter concludes by considering the cumulative impacts of these human activities and the regulatory mechanisms by which they might be appropriately managed.

DEFINITIONS

It was essential to the workgroup's discussions that the questions be understood on the basis of agreed-upon, standardized terminology. After much deliberation, the workgroup adopted the following terms and definitions.

Ecosystem - a self-maintaining, functional system that includes the organisms of a natural community together with their environment. As traditionally used, this term focuses on the environmental driving forces, the trophic components, and their interactions through time. Often the system, as defined, is considered spatially homogeneous in order to focus on trophic structure and functional processes; but this is not a necessary part of the definition. As used in this chapter, the areal extent of a bottomland hardwood ecosystem encompasses the entire spatial gradient from river to upland.

Landscape - a heterogeneous land area composed of a cluster of interacting ecosystems that are repeated in similar form throughout (Forman and Godron 1986). Landscape ecology deals with large, naturally defined areas and emphasizes spatial patterns and spatial interactions. This contrasts with the term "ecosystem" which is used to emphasize functional interactions. There is considerable overlap in the two terms, since spatial pattern and ecological process are inseparable. Both are influenced by the scale of inquiry. This workgroup did not focus on the spatial pattern and process of trees in a bottomland hardwood forest but rather on patterns of different forest types and extent as determined by geomorphic and hydrologic processes acting over thousands of years. In the context of cumulative impacts as defined below, landscapes are large areas outlined by natural boundaries, for example, watersheds, drainage basins.

Cumulative Impacts - "the impact on the environment which results from the incremental impact of the action when added to other past, present, and reasonably foreseeable future actions .... Cumulative impacts can result from individually minor but collectively significant actions taking place over a period of time" (Regulations of the Council on Environmental Quality, 29 November 1978, 40 C.F.R. Parts 1508.7 and 1508.8). A technical hindrance to protecting bottomland hardwood forest wetlands has been the difficulty of managing cumulative impacts (Lee and Gosselink (1988). The Section 404 permit process of the Clean Water Act, codified at 33 C.F.R.§§ 320-330) concerns the impact of a proposed activity at an individual wetland permit site. In contrast, cumulative impacts are landscape-level phenomena that result from decisions at many individual permit sites (Gosselink and Lee, 1989). A number of issues make cumulative impact assessment and regulation in bottomland hardwood forests difficult to achieve (Horak et al. 1983, Canadian Environmental Assessment Research Council 1986).

1) The present framework for regulation of dredge and fill activities in wetlands (Section 404 of the Clean Water Act) and the attendant programmatic focus on review of individual permit sites is inadequate to address cumulative impacts. Therefore, individual permit requests must be considered in the context of the landscape and management goals for that landscape. The analysis must be anticipatory rather than reactive.
2) Cumulative impacts are incremental. An individual action seldom makes a "significant" or detectable impact on the whole landscape. Impacts may exist, but they are not obvious because of natural variability within landscapes. However, the sum of all incremental impacts can have, and often does have, extremely serious environmental consequences. Environmental toxicology and epidemiology provide an accurate analogy. Often a strong statistical relationship exists between exposure to an environmental toxin and the incidence of related disease. However, on an individual basis it is impossible to relate a marginal increment in exposure to contracting the disease. Furthermore, an absolute, specific threshold rarely separates harmful from harmless levels. Nevertheless,

management of such impacts does proceed without detailed knowledge of the marginal impacts of each additional increment.

3) Previous actions should influence present decisions. From the cumulative impact perspective, local actions must be compared with past actions in the same ecosystem and landscape. Local action must support total ecosystem management not just site-specific management.
4) Cumulative impact assessment requires landscape-level reference points from which to evaluate both the status of the system and cumulative deviations from desired standards. Effective management of cumulative impacts requires that these reference standards be clear, measurable, defensible, and shared in the sense that they represent enough of a consensus to support regulation of individual actions.

Ecosystem services - the term "services" is used throughout this chapter in an anthropocentric sense to denote ecosystem products and processes that benefit humans.

Ecological attributes - key structural or functional ecological properties of ecosystems related to self-maintenance. Under normal conditions the presence of these attributes leads to services to humans. For example, provision of diverse flora and fauna depends on such ecological attributes as forested area, linear continuity of forest, and habitat heterogeneity.

Ecological or landscape indices - quantitative, measurable parameters by which the magnitude of ecosystem services can be estimated. These indices are necessary because the services provided by bottomland hardwood ecosystems are complex and difficult or impossible to measure directly. For example, the magnitude of the service "maintaining natural biotic diversity" can be estimated by the index "species richness."

ECOLOGICAL PRINCIPLES APPLIED TO BOTTOMLAND HARDWOOD ECOSYSTEMS

Ecological principles governing bottomland hardwood ecosystems must be understood before sound management guidelines can be developed. The Ecosystem Workgroup identified 10 principles organized around 6 topics: time scales, hydrology, spatial patterns, productivity, transformation and transport of materials, and landscapes. These principles in some cases are offered as provocative statements and may not have been demonstrated explicitly in bottomland hardwood ecosystems. As with other generalizations about bottomland hardwood ecosystems, they need further testing to determine the limits of their validity. However, the workgroup felt comfortable using them to represent the ecosystem-level processes judged to occur in most bottomland hardwood systems.

Time Scales

Principle 1. Components of the bottomland hardwood ecosystem respond in different ways and at different rates to temporal variations in the major driving variables of sun, hydrology, nutrients, and sediments.

Long-term processes dating back to the Pleistocene, acting through hydrology, were responsible for the large-scale features of floodplains that support bottomland hardwood ecosystems. An example is the creation of recognizable topographic features, such as Deweyville Terrace, which continue to influence ecosystem processes in floodplains. On a shorter time scale, variations exist in rainfall and river flow, with low-frequency, high-intensity events superimposed on a dominant annual cycle. The availability of sediments and nutrients, in part a consequence of hydrology, has a strong influence on surface drainage, biotic populations, and biogeochemical processes and is basically a reflection of long-term sedimentation and diagenetic processes. Solar radiation is relatively uniform throughout the Southeast. Its variability occurs principally in seasonal and daily time-scales although longer-term climatic variation has also been reported (Stahle et al. 1988).

Time scales relevant to bottomland hardwood ecosystem processes thus range from thousands of years to hours. Sediment deposition and erosion, occurring over thousands of years, produce the major geomorphic features of bottomland hardwood systems, including soil type, ridge and swale topography, point bars, floodplains, and natural levees. Some of these processes can be rapid and episodic (Gagliano and Van Beek 1975), but most are generally much slower and have a more permanent effect than biotic ones. Distinct vegetation communities (Wharton et al. 1982) develop over a scale of hundreds of years as a function of soil type, topography, and hydrologic regime. Precipitation and river flow go through irregular wet and dry cycles in scales of a few to tens of years (Stahle et al. 1988). Seedling establishment is often a response to these relatively short-term cycles; the frequent occurrence of even-aged stands of a few tree species in bottomland hardwood forests reflects the occurrence of unusual weather conditions that allowed the germination and survival of seedlings where they ordinarily would not be the sole dominant. Similarly, long-lived animals become more or less abundant in response to year class dominance, habitat changes, and variability in environmental parameters (Hellgren and Vaughan 1988).

The annual cycle is the dominant time scale, reflecting the temperate climate and the seasonality of the hydrologic regime in most southeastern bottomland hardwood ecosystems. These factors drive the annual cycle of primary production, litter decomposition, and secondary biotic activity. For example, Lambou (Chapter 4) illustrated the movement of young red swamp crawfish (*Procambarus clarkii*) into and out of bottomland hardwood areas during the annual spring flood. Harris and Gosselink (Chapter 9) discussed the complementary temporal distribution of fruit and seed sources used by wildlife over the year. They also cited the annual migrations of birds through bottomland hardwood forests. These spring migrations are timed to the availability of such foods as emerging insects.

Many processes, such as the reproduction of small animals and herbaceous plants, operate on time scales of less than a year. At one extreme, bacteria may replicate in less than a day. Some chemical processes are even faster. The timing of these processes and the activities of organisms may be correlated with temperature, a particular stage in the hydrologic cycle, rainfall, daylength, or other physical-chemical factors.

In general, the generation time (or turnover time) of organisms and physical structures is related to size (Figure 1). Usually the larger a component is, the more it contributes to overall ecosystem structure. Small components are often more mobile and more ubiquitous, and tend to couple process to structure. Using a "principle" from hierarchical theory, the behavior of a system at any level in a hierarchy is explained in terms of the level below; and its significance is found at the level above (Webster 1979). For example, the pattern of vegetation provides an understanding of bottomland hardwood forests functioning on a scale of approximately 100 years. Tree growth can be understood in terms of processes that occur on shorter time scales, but the vegetation pattern is a response to geomorphic and hydrologic events, and anthropogenic influences, that have occurred over longer time scales.

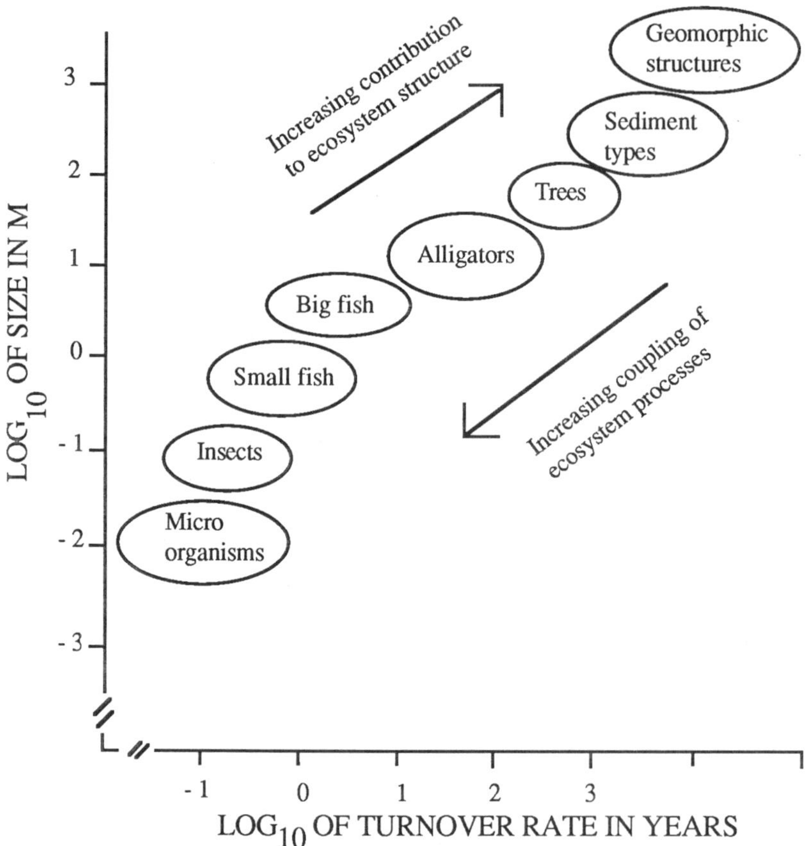

Figure 1. The generalized relationship between structural size and temporal scale of associated processes.

Hydrology

Principle 2. Hydroperiod, or the timing and duration of inundation, is the principal driving force for ecological processes and vegetation structure in bottomland hardwood ecosystems.

Bottomlands receive water from precipitation, runoff, interflow, groundwater, and tidal flow. The presence, movement, quality, and quantity of water strongly influence the physical, chemical, and biotic processes that characterize a bottomland hardwood ecosystem (Clark and Clark 1979). Water motion shapes and maintains floodplains by transporting and redistributing sediments and nutrients. The alternating wet-dry hydrologic cycle represents an energy subsidy that maintains high productivity (Odum 1979, Gosselink et al. 1981). Because the timing and length of periodic inundation and the direction of flow change with elevation within the system (i.e., along a gradient from the stream to upland), bottomland hardwood ecosystems contain a mosaic of microenvironments and species assemblages able to withstand varying lengths of soil saturation.

Climate, topography, channel slope, and soils in the ecosystem are among the more important factors that determine the hydroperiod and other flooding characteristics of bottomland hardwood forests. Small watersheds with steep slopes and dense clay soils experience fast runoff and infrequent flooding for short durations (Wharton et al. 1982). Large, flat bottomlands of some southeastern rivers are inundated from 18% to 40% of the year (Bedinger 1981). Their heavily forested floodplains slow water movement, allowing for increased infiltration and sediment deposition and for protection of downstream areas from flooding. Floodwater allows movement of fish and shellfish into the floodplain (Hall and Lambou, Chapter 15), dispersal of seeds (Huenneke and Sharitz 1986), and transport of sediments and nutrients. Runoff from adjacent uplands flows across bottomlands, carrying nutrients and sediments. These are typically modified in their passage, usually with a net reduction of entrained sediment and a conversion of nutrients from dissolved inorganic to particulate organic form (Asmussen et al. 1979, Cooper and Gilliam 1987).

Saturated floodplains may be directly connected to water table aquifers (Gosselink et al., Chapter 11). Discharge or recharge can occur depending on the local topography, soils, precipitation, and river stage. Seeps, which may occur when relatively steep upland slopes abut floodplains, may maintain soils in saturated condition during summer months when river flow is low and evapotranspiration is high (Porcher 1981). Seeps are important in maintaining populations of terrestrial and amphibious species on the floodplain during these periods. Conversely, aquifer recharge may occur in these same areas during periods of high water.

Principle 3. At the landscape level bottomland hardwood ecosystems receive, absorb, and dissipate hydrologic energy generated elsewhere in the watershed. The patterns of flood intensity, frequency, and depth determine patterns of ecosystem structure and function.

Bottomland forest floodplains function as units to absorb the energy of floods, ranging from relatively high-frequency, low-energy events to low-frequency, high-energy occurrences (e.g., 100-year flood). For a given elevation or site within a floodplain, the total hydrologic energy received over time is the accumulation of all flood events of varying magnitude (Figure 2). The whole spectrum of flooding events contributes to the hydrodynamics at low elevations or near the stream channel. But only high-energy, low-frequency events reach higher elevations, and most of the energy of these unusual floods is dissipated at lower elevations in the floodplain.

Although the exact relationship between elevation and kinetic energy (Figure 2) is not known, and undoubtedly varies from one bottomland hardwood ecosystem to another, the workgroup hypothesized an exponentially decreasing hydrologic energy with increasing elevation. At the bottomland hardwood-upland transition, at highest elevations, upland runoff and seeps represent an additional vector of hydrologic energy, thus causing the total curve to slope upward at both ends. This suggests that the lowest energy regime in the bottomland hardwood forests lies somewhere intermediate between the river channel and the upland.

In terms of hydrologic function, the whole bottomland represents a continuum of flood events, from nearly continuous inundation to unflooded upland. On this continuum, the vegetation changes, often forming distinct zones of different dominant plant species. When the lower zones are flooded, they absorb hydrologic energy through the frictional resistance provided by sediment and vegetation surfaces. When larger floods occur, the area of energy dissipation increases. Flooded areas provide habitat for fish and shellfish, the dry zones for terrestrial animals. The interface of water and land is especially heavily used by wildlife. As floods wax and wane, the aquatic and terrestrial habitats change in size and the interface moves up and down the vegetation zones. Thus, functionally, the bottomland hardwood forest provides a continuum of flooding conditions that move across the vegetation zones through time rather than a static series of inundation zones.

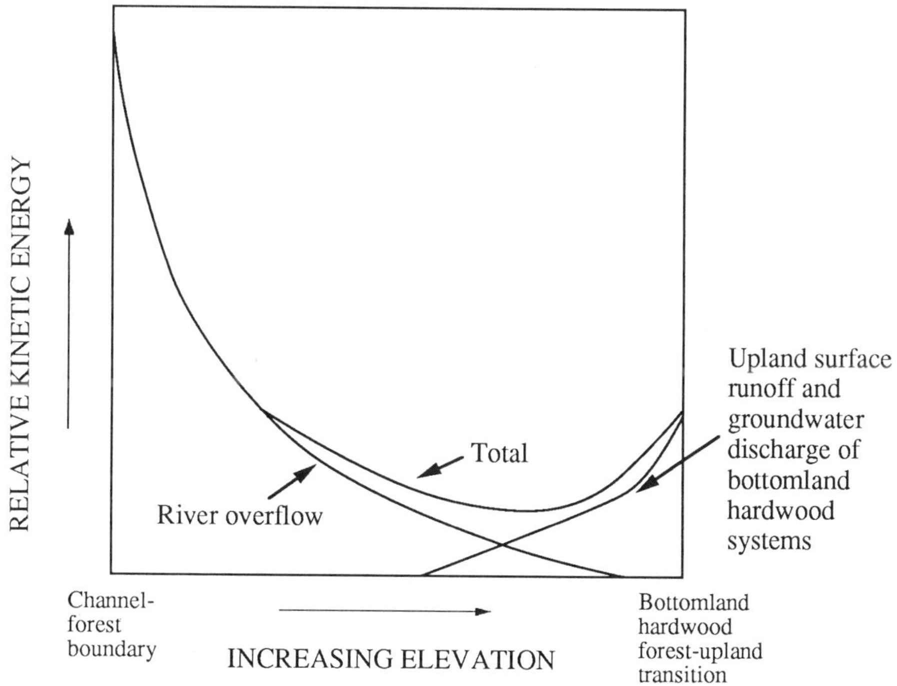

Figure 2. Distribution of kinetic energy associated with different riverine flooding and upland runoff events along a generalized bottomland hardwood forest elevation gradient. (Total kinetic energy is the sum of overflow derived from the river channel and runoff or seepage originating in the upland.)

Principle 4. Bottomland hardwood ecosystems are mosaics. The patterns of these mosaics (i.e., complexity and degrees of interspersion) have significant effects on ecological processes.

Geomorphic processes operating on long time scales have produced highly varied patterns of physical surfaces within many forest floodplains, ranging from simple gradients characteristic of first-, second-, and third-order streams (Figure 3a), to much more heterogeneous cross-sections composed of backswamp, channel fill, oxbow lake, ridge, swale, point bar, and hummock features (Figure 3b). The flood pattern of this complex topographic surface determines the vegetation patterns, with low sites occupied by water tupelo (*Nyssa aquatica*), bald cypress (*Taxodium distichum*), and aquatic macrophytes interspersed with a variety of different plant associations characteristic of drier sites, across fairly abrupt ecotones. Because of this interspersion, activities of animals in dry sites may be more directly coupled to those in the wet sites than is generally assumed.

The complex spatial pattern within the bottomland hardwood ecosystem probably has important effects on ecological processes although the relationships of many have not been explicitly studied. For example, as spatial complexity and interspersion increase one might expect that the diverse topographic pattern will create a complex pattern of flooding, drainage, sediment deposition, and soil formation. As a result, the future physical development of the floodplain is shaped and modified by the present structure.

Because of the complex topographic pattern, transport of waterborne materials is complex and difficult to predict. The variable flooding and drainage patterns will lead to variable flushing of organic and inorganic materials from the floodplain as floodwaters recede. The dispersal of vegetation through waterborne seeds may be channeled to localized sites, impeded by poor drainage, or otherwise modified by the complex topographic pattern (Huenneke and Sharitz 1986).

In conjunction with the topographic mosaic, ecotones (gradients between habitats) will proliferate. Ecotones are usually characterized by increased plant and animal diversity compared to either adjacent habitat. This may favor eurytopic animal species over stenotopic species. Other possible consequences are the genetic isolation of populations and the provision of refugia in swales for aquatic species between large floods, and on ridges and hummocks for small terrestrial animals during moderate floods. Extirpation of local populations (e.g., aquatic species under dry

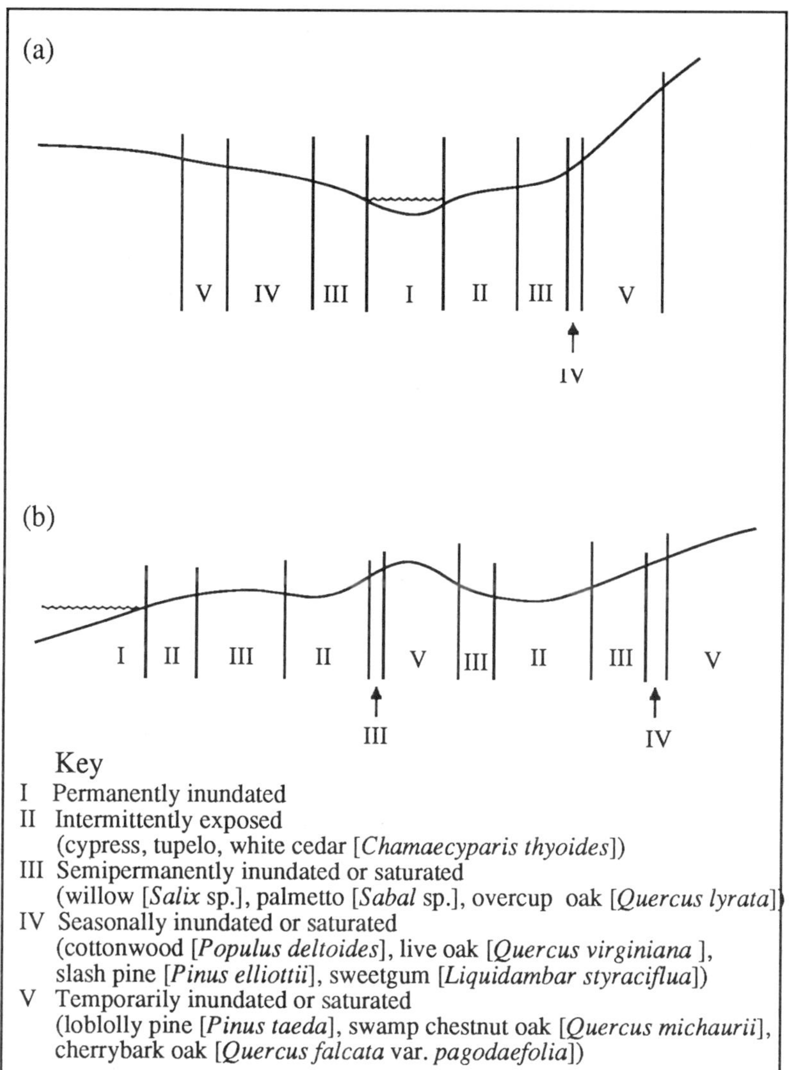

Figure 3. Cross-section of bottomland hardwood forests with (a) simple spatial patterns of regularly increasing elevation and regularly decreasing hydroperiod, and (b) more complex spatial pattern of elevation and hydroperiod (modified from Clark and Benforado 1981).

conditions and terrestrial species during wet conditions) will increase to the extent that the total suitable habitat is broken into isolated patches. Finally, a mosaic of habitats may provide a richer, more continuous food source for large, mobile fauna than that provided by a single habitat.

Productivity

Principle 5. Net primary production of bottomland hardwood ecosystems is generally as high or higher than that of mesic upland ecosystems in the same geographic region.

Foresters and ecologists generally accept that biomass production in bottomland hardwood ecosystems is as high or higher than that of other forested ecosystems in similar geographic locations. Some reasons for high productivity have been suggested. First, bottomland hardwood ecosystems are unlikely to experience moisture deficits except during the most extreme droughts (Bell and Johnson 1974), and essential nutrients are unlikely to be limiting to growth in fertile floodplain sediments (Peterson and Rolfe 1982a). Litterfall rates tend to be as high or higher in bottomland hardwood forests than in upland forests (Brinson et al. 1980, Peterson and Rolfe 1982b). Net primary production tends to be greatest at sites with intermediate hydroperiods (Conner and Day 1982) and decreases when a wetland forest is drained (Burns 1978). Net primary production also may vary across specific bottomland hardwood forest sites as a function of stand age.

Trophic Structure

Principle 6. Bottomland hardwood ecosystems support a multifaceted assemblage of consumers because the ecosystem alternates between terrestrial and aquatic conditions.

Bottomland hardwood ecosystems provide a dynamic interface between aquatic and terrestrial habitats. Use of bottomland hardwood ecosystems by both aquatic and terrestrial consumers at different stages in the wet-dry cycle results in consumer communities and food webs that are more structurally and temporally complex than communities in adjacent upland and aquatic sites. The presence of consumers from both environments at different times of the year may result in a more complete transfer of energy to higher trophic levels in both grazing and detrital food webs (Harris and Gosselink, Chapter 9).

Periods of wetting and drying in bottomland hardwood ecosystems provide peaks in habitat availability and consumer production. Flooding increases the habitat available to fish for migration, feeding, and spawning (Lambou, Chapter 4). In contrast, dry-downs concentrate food sources for various bird species such as the wood stork (*Mycteria americana*) (Kahl 1964). Still other species such as fingernail clams (*Corbiculacea pisidiidae*), tree frogs (*Hyla aurea*), and some salamanders (*Salamandroidea*) appear to benefit most from periods of partial inundation (Wharton et al. 1982). Because bottomlands are interfaces linking stream to upland, their destruction may have unexpectedly large effect on river and upland fauna as well as the more predictable effects on inhabitants of the floodplain itself (Wharton et al. 1982).

Transformation of Materials

Principle 7. Bottomland forest ecosystems are open systems that transform energy and materials derived from inflows. Their outflows differ quantitatively and qualitatively from inflows.

Studies have variously demonstrated that bottomland hardwood ecosystems are net sources for some elements, sinks for others, and/or transformers among forms of the same element (Nixon and Lee 1986). Generally, the bottomland hardwood floodplain is a depositional environment that accumulates material (see Scott et al. Chapter 13). In addition to their high primary productivity, bottomland hardwood ecosystems provide biological energy for transformations of many elements to more reduced forms. Some of these reduced forms (e.g., N_2, N_2O, H_2S, CH_4) are volatile and leave the ecosystem as gases.

Studies of the transformation of materials brought onto bottomland hardwood floodplains from upland runoff or stream overflow generally show a high potential for microbial transformation, root uptake, and at least short-term immobilization (Yarbro 1979; Brinson et et al. 1980, 1983; Lowrance et al. 1984; Peterjohn and Correll 1984). Leaf litter acts as a site for strong, short-term immobilization of nitrogen (Qualls 1984). These studies illustrate the stabilizing effect of recycling in the face of hydrologic forces that otherwise would favor downstream movement of nutrients. Furthermore, sequestering and immobilization of heavy metals and organic toxins (and their degradation products) have been demonstrated in wetland and aquatic sediments (Pionke and Chesters 1973, Schlesinger 1979).

In large-scale comparisons of similar watersheds with or without bottomland hardwood forests, Kuenzler et al. (1977) and Peterjohn and Correll (1984) demonstrated that channelized streams which were prevented from flooding overbank into bottomland hardwood forests export more nitrogen and phosphorus than natural streams with wetland exchanges. They interpreted this as evidence that the bottomland hardwood forest along natural streams provides the mechanisms for biogeochemical processes such as 1) sorption of soluble phosphorus and ammonium to inorganic soil particles, 2) removal of nitrate and ammonium through the nitrification-denitrification pathway, and 3) uptake of soluble forms of nitrogen and phosphorus by immobilization in decomposition and algal growth (short-term removal) and through uptake and recycling by vascular plants, especially trees (long-term removal). Bottomland hardwood systems generally transform inorganic inputs to organic outputs. For example, an upstream-downstream study of the lower Apalachicola River (Elder 1985), which passes through an extensive bottomland hardwood ecosystem, showed only small differences in the total amount of incoming and outgoing nitrogen and phosphorus, but significant transformation of dissolved, inorganic inputs to particulate organic material in the outflow.

Export of Materials

Principle 8. Episodic flooding events serve two roles: to transport accumulated net primary production from bottomland hardwood sites to aquatic consumers and food webs in other sites, and to provide a medium for mobile aquatic consumers such as fish to exploit large floodplain areas.

Net primary production (NPP), its distribution into various components (roots, stems wood, leaf litter, fruit), and its processing and decomposition rates determine the organic matter available for export. The timing and duration of flooding and the velocity of floodwaters determine how much of that NPP is actually exported. A number of studies show the importance of exported organic material. For example, feeding and spawning of fish populations in Barataria Bay, Louisiana, are correlated with annual pulses of organic export from an upstream alluvial swamp (Day et al. 1977). Major infrequent flushes of organic matter (occurring at a frequency of 5- to 7-year intervals) are related to peaks in commercial fishing harvests in Apalachicola Bay, Florida (Livingston et al. 1976).

As with regularly flooded tidal marshes, export of detritus from wetlands no longer need be invoked as an exclusive or even dominant mechanism of food web support (Rozas and Odum 1987). Mobile organisms such as fish simply move to recently flooded areas of wetland surfaces to feed. Even in very small floodplains of second-order streams, fish populations have available to them much greater areas for feeding when river flow exceeds channel capacity (Walker 1984).

Ratio of Recycling to Throughput

Principle 9. Mineral cycling in bottomland hardwood forests has a lateral vector (throughflow by floodwaters) and a vertical vector (recycling by litterfall decomposition, and plant uptake). The lateral vector is a feature that causes bottomland hardwood forests to depart from other ecosystems with regard to nutrient cycling.

Mineral cycling in bottomland hardwood ecosystems has both a recycling component and an import-export (throughflow) component. The ratio of these two, called the cycling index (CI= recycling/throughflow), increases across the gradient of increasing elevation perpendicular to the stream bed. The cycling index (Finn 1976) is affected principally by the system throughflow because mineral recycling probably differs by only a factor of about two or three across the elevational gradient, while throughflow may change by several orders of magnitude. Litterfall and decomposition rates (measures of recycling), determined by total NPP and species composition, varied by less than a factor of three for five bottomland hardwood forest communities on the Apalachicola River floodplain (Elder and Cairns 1982). In contrast, throughflow is a product of water discharge and frequency of flooding for a given elevation in the floodplain. It has at least three components: lateral, bidirectional transport between stream and floodplain; flow on the floodplain parallel to the stream; and downslope transport from uplands to floodplain. Because flooding frequencies at different elevations on the floodplain vary between once per year and once per 100 years, throughflow within bottomland hardwood forest communities varies by at least a factor of 100.

The relationship between recycling, throughflow, and CI are shown in Figure 4. With increasing elevation CI increases to a maximum then falls again as upland runoff contributes kinetic energy to throughflow at the bottomland hardwood forest-upland transition (Lowrance et al. 1984, Peterjohn and Correll 1984). Consequently, the CI may reach a peak toward the upland-floodplain transition.

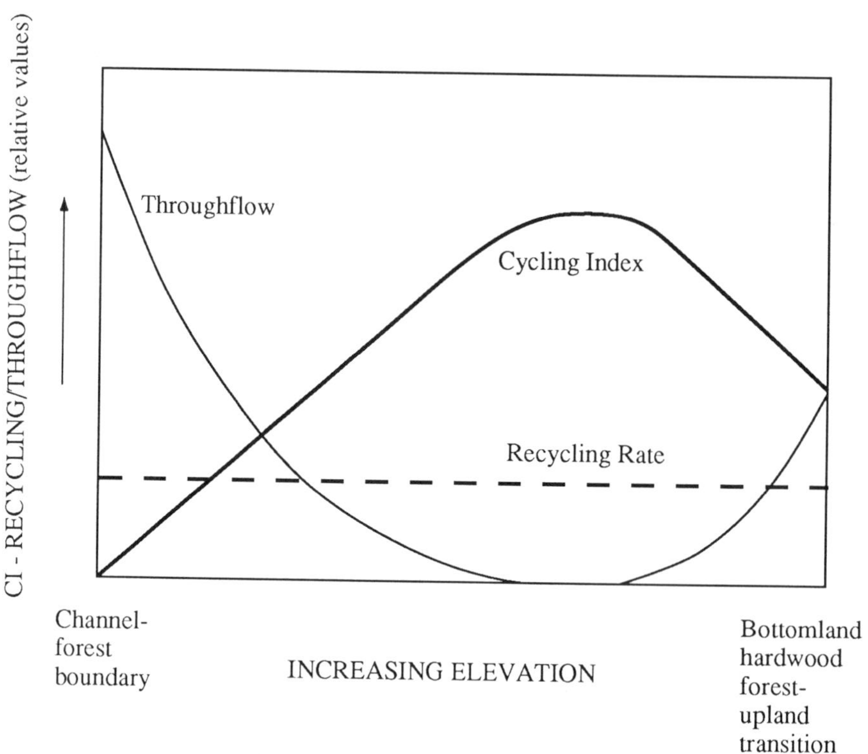

Figure 4. Cycling index (CI) as a function of generalized elevational gradient from stream channel to upland in a bottomland hardwood ecosystem. See Figure 3 for rationale for shape of throughflow curve.

Landscapes

Principle 10. The bottomland hardwood system (floodplain and river) acts as a functional unit within the landscape.

The bottomland hardwood forest with its river is a complete unit with gradients of water, soil types, and vegetation as parts of a unified whole. Zones or areas of different elevation and vegetation within the bottomland hardwood ecosystem do not function independently (Principles 3, 4, 6, 7, 8, 9). Flows across the floodplain connect the zones; to modify one zone is to induce changes in the adjacent zone. In addition, some overall properties of bottomland hardwood ecosystems are determined by the spatial and temporal configuration of multiple zones. A simple example is the wetted edge which moves across the zones as floodwaters rise and recede. The ecological processes occurring at this edge also move from zone to zone. Floodwater provides access for fish and shellfish species, many of which require the floodplain habitats for completion of their life cycles (Bryan et al. 1975; Lambou, Chapter 5). Terrestrial animals also may move back and forth across the floodplain following the water's edge. These animals take advantage of the seasonal change in food supplies provided by the different bottomland hardwood forest zones (Wharton et al. 1982). Thus, it is misleading to assume that any zones in the floodplain remain unaffected when other zones are removed or altered or when riverflow is modified.

Within the larger landscape the river-floodplain is an open system responding to events upstream and in the adjacent upland and influencing downstream ecosystems. This intimate relationship is illustrated by several examples. First, the size and characteristics of both the stream channel and the floodplain are clearly functions of both streamflow volume and velocity. Figure 5 illustrates that for a large number of rivers across the nation, regardless of size, bankfull depth is exceeded on the average of once every 1.5 years. Second, reducing the floodplain area increases flood stages downstream (Belt 1975, Zinn and Copeland 1982), a fact that influenced the U.S. Army Corps of Engineers (USACE) to purchase the Charles River floodplain rather than construct expensive flood-control structures to protect the city of Boston (USACE 1971). Third, the floodplain responds to changes in the adjacent upland watershed. Generally, surface and groundwater moving downslope from upland to stream are reduced in volume through infiltration and evapotranspiration, and in pollutant concentrations through removal in the floodplain. When the upland ecosystem is forested, runoff is minimal. But when the upland area is agricultural, water, eroded sediments, inorganic fertilizer nutrients, and pesticides may run off in quantity. Under these circumstances the bottomland hardwood forest is particularly effective in buffering the stream from these accelerated flows.

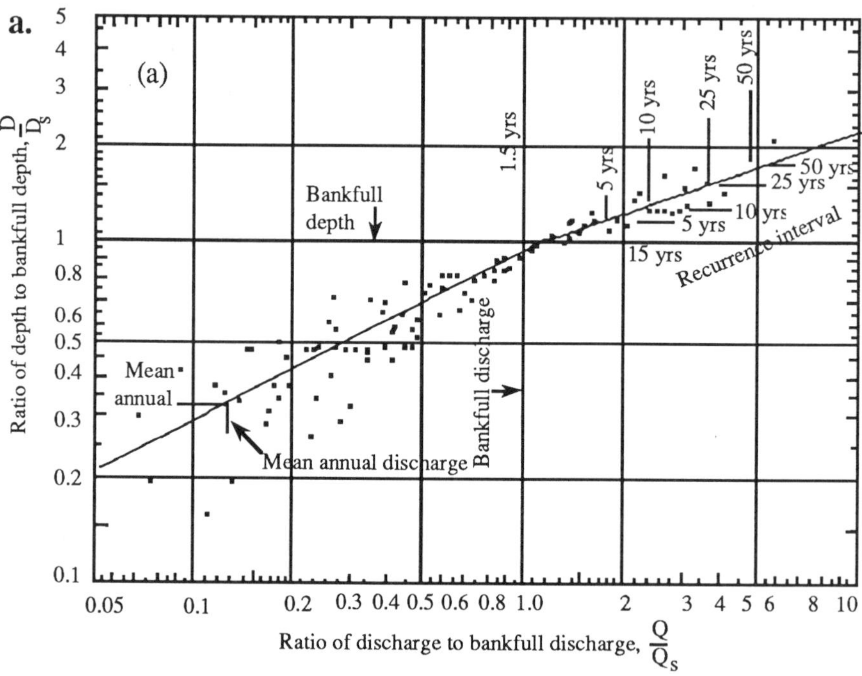

Figure 5. Recurrence interval for bankfull flooding on many different rivers in the U.S. (Leopold et al. 1964, p.219; Copyright © 1964, by W. H. Freeman, reprinted with permission).

SERVICES OF BOTTOMLAND HARDWOOD FORESTS

Ecological processes in bottomland hardwood forests, operating under the principles discussed in the previous section, provide valuable services to human society. Four of these services are described in this section: maintenance of natural biotic diversity, primary production and food web support, streamflow mediation, and water-quality modification.

Ecosystem characteristics that contribute to the services are also discussed. Because services are difficult to measure directly, indices of those services are also identified. Each results from ecological attributes of bottomland hardwood ecosystems. Shown in Table 1 are the indices by which effectiveness is measured as well as the ecological attributes responsible for the service. These indices represent the service though they are not synonymous with it. For example, the natural biotic diversity index is native species richness. The discussion in this section lays a foundation for the later consideration of how human activities modify ecological services by altering the controlling attributes of bottomland hardwood forest landscapes.

Maintenance of Natural Biotic Diversity

Biotic diversity serves human society in many ways although the direct market implications are often obscure. The World Conservation Strategy has correctly identified the preservation of genetic diversity as "a matter of insurance and investment--necessary to sustain and improve agricultural, forestry, and fisheries production; to keep open future options as a buffer against environmental change; and as the raw material for much scientific and industrial innovation and a matter of moral principle as well." (U.S. Department of State 1982:27-28).

The maintenance of natural, or native, biotic diversity is specifically mandated by several federal and state statutes (e.g., the Endangered Species Act of 1973). Diversity refers to the native flora and fauna, excluding exotics which may have been introduced to the area. Numerous species of native plants and animals are either endemic to bottomland hardwood forests or depend on them for one or more of their survival needs at various seasons or life stages. Large tracts are probably the key ecological attribute of the landscape for maintenance of natural biotic diversity for the following reasons:

1) Large tracts generally provide a diverse array of sites that have the potential to support a balanced, indigenous population of plants and animals (Harris 1984, Diamond 1975, Soulé and Wilcox 1980).

Table 1. Services provided by bottomland hardwood forest ecosystems.

SERVICE			
Maintenance of Natural Biotic Diversity	Primary Production and Food Web Support	Streamflow Mediation	Water Quality Modification
INDEX			
Native species richness	Species composition and condition of forest community	Downstream flood stage/ standard flood	Nutrient concentration
ECOLOGICAL ATTRIBUTES			
Effective size of forest patch Effective breadth of forested floodplain	Net primary production quantity quality	Percentage of watershed area that floods freely	Percentage of watershed in bottomland hardwood forests versus other cover types
Forest linear continuity	Organic matter processing	Hydrologic detention time	Percentage of stream bordered by bottomland hardwood forests
Percentage of regional landscape forested	Access/transport vectors		
Ecologic heterogeneity (number and area of different community types and ecological zones)	Landscape mosaic		
Edge between community types			
Structural diversity of vegetation			
Mast production and diversity			
Nesting and escape cover			

2) For vegetation, large tracts maintain natural patch-dynamic processes whereby shade-intolerant species can occur in all life stages (Forman and Godron 1986). These patch dynamics are especially important in producing regeneration sites in close proximity to undisturbed stands with mature, seed-bearing individuals.
3) Large tracts, because they generally incorporate intra-patch habitat heterogeneity, preserve genetic diversity within a species. In contrast, habitat fragmentation and resulting small populations are likely to suffer from problems related to random environmental and demographic fluctuation, inbreeding, or genetic drift (Noss 1987).
4) Large tracts allow wide ranging, terrestrial organisms to occur in populations sufficiently large to forestall or prevent inbreeding, regional extirpation, or extinction (Terborgh and Winter 1980).

<u>Native species richness</u> is a quantitative index of natural biotic diversity. Natural biotic diversity includes, in addition to the total number of native species (species richness), the relative importance or dominance of those species. This is much more difficult to measure than species richness, and information of this kind is available from few sites.

Several structural attributes of the landscape related to large tract size influence natural biotic diversity (Table 1). <u>Effective size</u> measures the extent of bottomland forest tracts relative to species' habitat requirements (Eisenberg 1980). <u>Breadth</u> refers to the width of the forested floodplain. The full set of ecological zones (Clark and Benforado 1981) is needed to support the entire range of bottomland hardwood species. Bottomland hardwood forests of higher elevational zones provide refugia for terrestrial species during floods; slightly lower elevational zones provide shallow, inundated areas for aquatic species during the same events; and adjacent streams provide year-around habitat for migratory fish and shellfish. <u>Linear continuity</u> refers to continuous dispersal corridors both in the forested portion of the floodplain and in the adjacent stream (Diamond 1975, Noss 1987). <u>Percentage of regional landscape in bottomland hardwood forests</u> is a measure of the integrity of the total ecosystem. As this parameter declines, the ecological integrity of remaining isolated patches of bottomland hardwood forest is reduced. Remaining patches are increasingly vulnerable to complete change by weather phenomena and to invasion by weedy species. At some point patches drop below critical size for the support of interior forest species. Ecological heterogeneity refers to the diversity of vegetation associations, community types, and ecological zones characteristic of many topographically complex bottomland landscapes that contribute to the maintenance of species diversity (Lack 1976).

At the local level, additional ecological attributes contribute to biotic diversity. The amount of edge between community types is a slightly different measure of ecological diversity (beta diversity) which reflects additional aspects of the site's spatial mosaic. An increase in the amount of edge has both positive and negative effects on biotic diversity. Some species respond positively to edge whereas when edge is gained by fragmenting large, continuous forest stands, the resulting small patches may not support other stenotopic species. Structural diversity of vegetation within a stand refers to the number of effective layers of vegetation. Several measures of structural diversity (e.g., foliage height diversity) are positively correlated with faunal diversity, especially for birds. Mast production and diversity (in type and time) are important characteristics related to the amount and seasonal availability of food for wildlife. Nesting, thermal, and escape cover refers to such structures as snags, logs, and butt cavities that provide an important aspect of overall structural diversity often not included in measures of vegetative layers.

Primary Production and Food Web Support

The harvested products of bottomland hardwood ecosytems--timber, fish, and wildlife--depend directly on the broad services implied in the term "primary production (NPP) and food web support." The ability of a bottomland hardwood ecosystem to support animals and to provide organic energy for other ecosystem processes, such as decomposition, depends on the magnitude of NPP, the provision of this primary production in an appropriate form (organic matter processing, Table 1), and accessibility to the products either by movement of the organic matter to the consumers or vice versa (access/transport vectors). Although the emphasis in this service is on energetic support of trophic structures, the configuration of the landscape mosaic also determines the availability of different food sources at different times of the year.

Three aspects of NPP are relevant to food web support: the quantity per unit area, the total net production (area times production), and the quality of that production in terms of its chemical composition, location, and timing. Distribution and spatial configuration of various plant associations are also critical at the landscape level. Species in bottomland hardwood forest food chains must be able to satisfy all life requisites (not just food) and thus may require certain configurations of plant associations in order to meet requirements such as for cover or nesting habitat. Many species use different habitats at different times and thus require certain distributions and interspersions of various community types.

Although leaves, fruits, flowers, and seeds may be eaten directly by herbivores, most of the primary production is processed through a complex detrital food chain in which abiotic and microbial processes assume critical importance. This organic matter processing provides a rich and diverse distribution of organic products. Given this diverse food supply, flooding provides the primary medium both for transport of processed detrital material to downstream aquatic foodchains and for access of aquatic consumers to bottomland hardwood forests. Flooding also determines the accessibility by terrestrial animals to bottomland sites.

The direct measurement of NPP is an index of this ecological service (NPP and food web support) since NPP determines the maximum potential for the size and complexity of the food web. From a practical standpoint, NPP for a specific site may be estimated from knowledge of stand age and species composition (densities, sizes, and age distributions by species; see Mengel and Lea, Chapter 3).

Streamflow Mediation

The ability of a floodplain wetland to store and slow water provides a valuable service in that downstream flood stages are reduced compared to those on comparable streams with no floodplains (Taylor et al. Chapter 2). Bottomland sites are often subject to stream overflow from some combination of headwater and backwater flooding. The water stored on the floodplain does not drain as long as the river stage is as high as that of the floodplain waters. Consequently, storage of floodwaters on bottomlands reduces and retards downstream flood peaks and prolongs the flooding event at lower stages. In other words, the floodplain desynchronizes potential downstream floods. Conversely, when storage capacity is lost and water is confined to the river channel, downstream flood peaks are increased. Furthermore, as the stage versus discharge relationships of the river change, the channel may become unstable with consequent problems of scouring and filling (Belt 1975).

Two related bottomland hardwood forest characteristics that affect streamflow are the percentage of freely flooding area in the watershed, and the hydrologic detention time (Table 1). The former has two components. One directly relates to the natural flood storage capacity of the floodplain, e.g., the extent and depth of flooding. The second is determined by the vegetation cover which modifies the frictional drag (Manning's N) of the flooded surface. The hydrologic detention time (stored volume/discharge rate) is a slightly different characteristic that relates the storage capacity of the floodplain to the discharge rate. The larger the storage capacity relative to discharge the greater will be the flood mediation downstream.

An index of streamflow mediation in response to some alteration of the floodplain is the resultant change in the downstream hydrograph, measured as a change in the ratio of stage to discharge. An increase in the ratio is an indication of reduced bottomland hardwood forest ability to desynchronize floods. This is often accompanied by increased instability of the index, indicating stream instability (Belt 1975).

Water-Quality Modification

An important service of bottomland hardwood forests, a consequence of their transforming and filtering capacity, is their ability to maintain and improve water quality (Taylor et al., Chapter 2; Lowrance et al. 1984; Peterjohn and Correll 1984). Erosion is one important aspect governing water quality, which is also often closely correlated with nutrient loading. One way of describing various patches of the landscape, therefore, is to classify them according to various "soil loss types," reflecting a combination of cover, management practices, slope, and soil characteristics.

Erosion and nutrient loading increase dramatically with forest clearing (which increases soil loss). The location of the remaining erosion-resistant, forested soil loss type is important because of the capability of bottomland hardwood forest along stream margins to remove suspended silt and nutrients from water flowing across it. In this context key attributes that influence the degree to which bottomland hardwood ecosystems perform transformation and filtering functions are the percentage of the watershed in bottomland hardwood forests versus other soil loss types; and location and configuration of various soil loss types, specifically the buffer configuration of bottomland hardwood forests calculated by the percentage of stream bordered by bottomland hardwood forests (Table 1).

Convenient indices of water-quality modification are the concentrations of selected nutrients, toxins, or suspended sediments in waters of the bottomland hardwood forest stream, referenced to independent water-quality standards. The selection of the specific index or indices should be governed by the conditions of the watershed, especially by the kinds of human impacts. Phosphorus, for example, is a good indicator for agricultural activity because it has no volatile phase, is tightly bound to sediment particles (hence is closely related to erosion levels), and is a universal constituent of agricultural fertilizers.

HUMAN ACTIVITIES AND THEIR IMPACTS ON BOTTOMLAND HARDWOOD ECOSYSTEMS

The previous section identified four types of ecological services provided by bottomland hardwood ecosystems and listed ecosystem attributes that contribute to or determine their effectiveness. The impact of human activities can be gauged either directly by changes in the indices of these services, or indirectly by changes in the related ecosystem attributes. This section analyzes the impact of common human activities on the ecological services and their attributes. It ends with a general discussion of the intensity and permanence of the major kinds of human activities in bottomland hardwood forests.

The Impact of Human Activities on Ecological Services

The indices and attributes of each ecological service (from Table 1) are shown in Tables 2 through 5 in the left-hand column with human activities across the top. The matrix of pluses, minuses, and zeroes in these tables reflects our best judgment about the influence of each activity (positive, negative, or neutral) on each index or ecosystem attribute. The eight common human activities, grouped as agricultural conversion, flood control, forestry, and aquaculture, were discussed in the Introduction (Chapter 1).. Most of the attributes are negatively affected by the human activities common in bottomlands, and the overwhelming influence on the services is also negative.

Maintenance of Natural Biotic Diversity

Human activities often fragment the landscape, resulting in smaller tracts of bottomland hardwood habitat (Table 2; Krummel et al. 1987, Turner 1987, Gosselink and Lee 1989). Human disturbance tends to occur first at higher elevations in the bottomlands, eliminating or reducing the full range of elevational ecotypes and natural ecological diversity. It tends to break the continuity of the bottomland hardwood corridor, reducing the free movement of both terrestrial and aquatic organisms. On the other hand, overall ecological diversity may be increased at local sites by the formation of agricultural fields and increased edge diversity. This kind of change generally favors eurytopic edge species, often exotics, over interior, stenotopic species that suffer from habitat loss. In general, well-managed greentree reservoirs (forested sites within which water levels are managed, usually for waterfowl) are the least intrusive developments which, in the best circumstances, may have little or no measurable effect on the bottomland hardwood system.

Primary Production and Food Chain Support

Agricultural production may, in some instances, increase the total above-ground NPP of a bottomland hardwood forest site (Table 3). However, a site in agricultural production no longer functions as a bottomland hardwood forest, and the yield does not reflect the energetic

Table 2. Probable impacts of various human activities on maintenance of natural biotic diversity, as indicated by ecological attributes and indices related to the performance of this service.

	Activity[1]							
	Agriculture		Flood Control			Forestry		Aqua-Culture
	R	S	I	C	L	P	G	A
Index								
Native Species richness	-	-	-	-	-	-	-	-
Ecological Attributes								
Effective size of forest patch	-	-	-	-	-	-	0	-
Effective breadth of forested floodplain	-	-	-0	-0	-0	-	0	-
Forest linear continuity	-	-	-	-	-	-	-0	-
% of regional landscape forested	-	-	-	-	-0	-	0	-
Ecologic heterogeneity (number and area of different community types and ecological zones)	-+	-+	-+	-+	-+	-+	-+	-+
Edge between community types	-	-	-+	-+	-+	-	-0	-0
Structural diversity of vegetation	-	-	-	-+	-+	-	-	-
Mast production and diversity	-	-	-	-+	-+	-	-+	-
Nesting and escape cover	-	-	-	-+	-+	-	-+	-

[1] A plus indicates that the attribute or index is enhanced by the human activity, a minus indicates impairment, and zero indicates no impact. A combination of signs indicates differing impacts depending on initial conditions, multiple causal pathways, or interactions. R = conversion of site to rice, S = conversion of site to soybeans, I = impoundment impacts in general, C = channelization impacts in general, L = levee construction impacts in general, P = conversion of site to pine plantation, G = conversion of site to greentree reservoir, and A = conversion of site to aquaculture.

Table 3. Impacts of various activities on primary production and food web support as indicated by ecological attributes related to performance of this service.

	Activity[1]							
	Agriculture		Flood Control			Forestry		Aqua-Culture
	R	S	I	C	L	P	G	A
Ecological attributes								
Net primary production								
Quantity	+	+	-+	-+	-+	0+	-0	-
Quality	-	-	-	-	-	-	-0	-
Organic matter processing	-	-	-	-	-	-	-0	-
Access/transport vectors	-	-	-	-	-	-	-0	-
Landscape mosaic	-+	-+	-+	-+	-+	-+	-+	-+

[1] A plus indicates that the attribute or index is enhanced by the human activity, a minus indicates impairment, and zero indicates no impact. A combination of signs indicates differing impacts depending on initial conditions, multiple causal pathways, or interactions. R = conversion of site to rice, S = conversion of site to soybeans, I = impoundment impacts in general, C = channelization impacts in general, L = levee construction impacts in general, P = conversion of site to pine plantation, G = conversion of site to greentree reservoir, and A = conversion of site to aquaculture.

cost of machinery, fertilizer, and fuel. Conversion to agricultural production involves a qualitative change in NPP that affects this service negatively, since primary production is maximized for human use at the expense of secondary production and food web support. Flood control, to the extent that it provides a better drained environment for plant growth, may also increase the quantity of NPP for flood-intolerant plant species within the ecosystem (Conner et al. 1981) The long-term effect of the change in the flood regime will be a qualitative shift in plant composition

Table 4. Impacts of various human activities on streamflow mediation as indicated by ecological attributes and indices related to performance of this service.

	Activity[1]							
	Agriculture		Flood Control			Forestry		Aqua-Culture
	R	S	I	C	L	P	G	A
<u>Index</u>								
Downstream flood stage/standard flood	0	0	+	-+	+	0	0	+
<u>Ecological attributes</u>								
% of watershed that floods freely	-	0	-+	-0	-	-	-	-
Hydrologic detention time	0+	-	+	-	-	-	0+	0+

[1] A plus indicates that the attribute or index is enhanced by the human activity, a minus indicates impairment, and zero indicates no impact. A combination of signs indicates differing impacts depending on initial conditions, multiple causal pathways, or interactions. R = conversion of site to rice, S = conversion of site to soybeans, I = impoundment impacts in general, C = channelization impacts in general, L = levee construction impacts in general, P = conversion of site to pine plantation, G = conversion of site to greentree reservoir, and A = conversion of site to aquaculture.

and related food webs. The forest mosaic pattern in the landscape is altered by all human activities considered, and the net effect may be either positive or negative in providing food web support, depending on the species of concern and the pattern of development. For some animals agricultural fields provide a food subsidy not available in pristine bottomland hardwood forests. These tend to be eurytopic species rather than the interior stenotopic ones. With the exceptions noted, other human activities were all considered to be detrimental to bottomland hardwood ecosystem food web support.

Table 5. Impacts of various activities on water-quality modification indicated by ecological attributes and indices related to performance of this service.

	Activity[1]							
	Agriculture		Flood Control			Forestry		Aqua-Culture
	R	S	I	C	L	P	G	A
Index								
Nutrient Concentration	+	+	+	+	+	+	0	+
Ecological attributes								
% of watershed in bottomland hardwood forests versus other cover types	-	-	-+	-	-+	-	-0	-
% of stream bordered by bottomland hardwood forests	-	-	-+	0	-	-	-0	-

[1] A plus indicates that the attribute or index is enhanced by the human activity, a minus indicates impairment, and zero indicates no impact. A combination of signs indicates differing impacts depending on initial conditions, multiple causal pathways, or interactions. R = conversion of site to rice, S = conversion of site to soybeans, I = impoundment impacts in general, C = channelization impacts in general, L = levee construction impacts in general, P = conversion of site to pine plantation, G = conversion of site to greentree reservoir, and A = conversion of site to aquaculture.

Streamflow Mediation

Human activities that prevent free flooding of bottomlands alter streamflow mediation (Table 4). Therefore, impoundments required for rice and aquaculture were generally considered negative. Flood-control impoundments which are designed to hold excess floodwaters may reduce flood peaks downstream. Although they may increase flooding upstream, they decouple streamflow from natural flood events. The most common activity in bottomland hardwood forests, conversion to row crops, does not necessarily change the flood storage capacity of the system although it could affect desynchronization by increasing the run-off rate. Subsequent pressure for flood-control construction to make farming economically viable often leads to secondary detrimental impacts (Stavins 1987). To the extent that the flood storage capacity of a bottomland ecosystem is reduced by human activities, water detention time is also reduced further. All activities that impact these two attributes negatively tend to raise downstream flood levels and to decrease stream stability.

Water-Quality Modification

Disturbance of bottomland hardwood ecosystems can directly affect water-quality constituents (e.g., total suspended solids, dissolved oxygen, phosphorus) either by causing direct loading of adjacent waters from eroding bottomland soils or by allowing untransformed or poorly filtered nutrients, organic matter, and contaminants to reach adjacent waters (Table 5; Richardson 1985). The transforming capacity of bottomland hardwood ecosystems seems to be directly related to their ecological condition and to the rate at which materials are transferred through them. Disturbing the natural structure through various landuse or water-regulation activities reduces the ecosystem's efficiency as a filter and therefore its function to maintain or improve water quality.

The Overall Effects of Human Activities on Bottomland Hardwood Forest Ecosystems

We have attempted to document specific effects of activities on ecological attributes in Tables 2 through 5. However, seldom are these interactions clearly separated in the real world. Furthermore, it is difficult to display relative importance, such as intensity and permanence of these activities, through single factor analyses such as this. We have attempted a multi-factor synthesis in Figure 6. In this figure the total effect of all the activities addressed in Tables 2-5 is estimated, plus mining, on the area disturbed and the intensity and permanence of disturbance. The activities are simply ranked from one to nine on each axis and no attempt was made to weight their relative differences more accurately. Nevertheless, the ordination shows some interesting features that should be important considerations in bottomland hardwood forest regulation. The activities can be considered in four clusters.

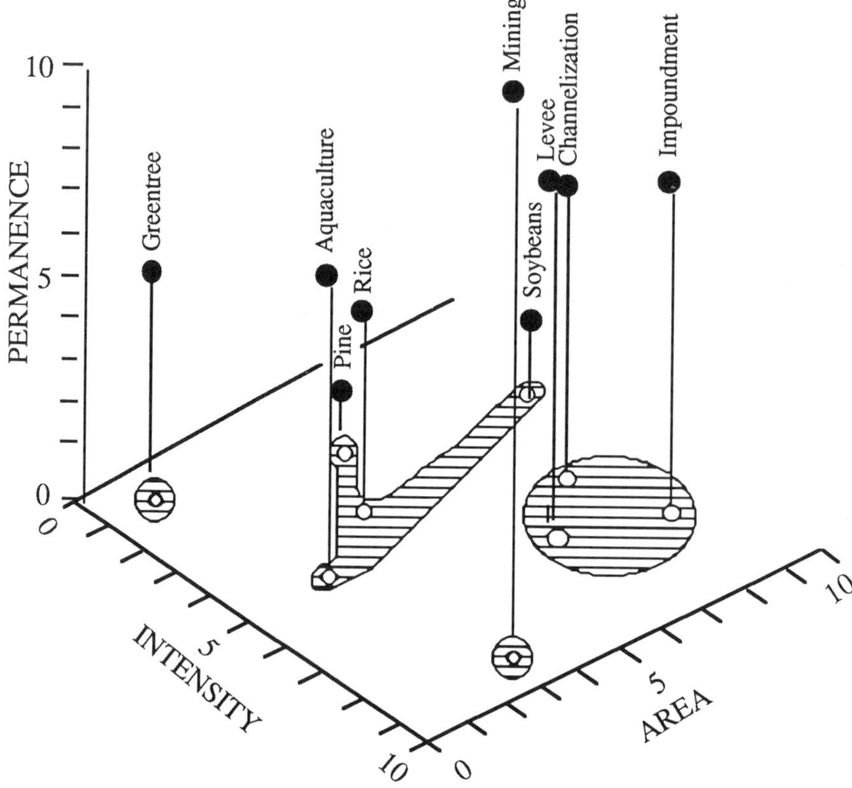

Figure 6. Simple ordination of human activities by the importance of their cumulative impacts (Lee and Gosselink 1988, p.596; Copyright © 1988 by Springer-Verlag New York, Inc, reprinted with permission).

1) Water resource development activities (impoundment construction, channelization, and levee construction) were judged to cause serious cumulative effects because they combine relatively intense and pervasive impacts with a high degree of permanence or irreversibility. Many of these projects permanently modify the flooding regime with long-term effects on the bottomland hardwood forest.
2) Strip-mining was ranked as the most severe effect in terms of intensity and permanence, despite a relatively small areal

extent. The surface extraction of overburden constitutes a major and permanent landform alteration. Petroleum and sulfur extraction also have surface effects that may be severe. They often involve the water resource development activities discussed above through their requirements for access canals.

3) Conversions to soybean and rice agriculture, pine plantations, and aquaculture, were grouped in a broad cluster of intermediate to low intensity and permanence. Conversion to soybean production was at an extreme edge of this cluster because of its large areal extent. Conversion to pine plantations was at another edge due to its relatively low intensity (i.e., a woody plant community is maintained). These activities are similar in that they all involve primarily alteration of the vegetation with only localized modification of the flow and circulation of water on bottomland hardwood sites. They are probably reversible on a scale of 70-100 years needed for regeneration and maturation of bottomland hardwood forests.
4) Conversion to greentree reservoirs was judged as the least severe activity on all dimensions.

THE REGULATION OF CUMULATIVE HUMAN ACTIVITIES IN BOTTOMLAND HARDWOOD FORESTS

Cumulative impacts have long been understood to degrade natural resources; but their regulation has not generally been attempted and has been, on the whole, ineffective where tried. Cumulative impacts in bottomland hardwood forests can be approached only at the landscape level rather than at the local scale of individual permit applications, which is the present regulatory approach. Landscape-level regulation requires the ability to characterize a system at this level, that is, to evaluate its ecological integrity as a landscape. This section identifies a set of landscape structure and function indices based on the ecological attributes of ecosystem services. Finally, the conceptual problems of cumulative impacts are discussed, and a suggested procedure for cumulative impact regulation is proposed.

Cumulative impacts are landscape-level phenomena. By definition they cannot be observed at the local level because they involve the incremental accumulation of many small, local impacts on the whole system. Cumulative impact analysis and regulation must therefore be conducted at large scales (e.g., basin, watershed, region). The discipline of landscape ecology is new and evolving, so few principles and techniques are available to guide regulators. This fact makes the task of cumulative impact assessment doubly difficult.

The Characterization of Landscapes

A first step in any environmental assessment is an ecological characterization. One of the reasons cumulative impact regulation has been difficult to accomplish is the absence of generally accepted procedures and standards for characterization of landscape integrity. This workgroup focused on bottomland hardwood landscapes as large, integrated systems. Consequently, the services identified, and most of the attributes that affect those services are viewed as properties of the whole bottomland hardwood landscape, including its stream and watershed. In the foregoing discussion (see Table 1) we identified measurable indices of the four services of bottomland hardwood forests. In this discussion we identify a small additional set of indices that relate broadly to bottomland hardwood forest structure (and thus indirectly to its services).

As can be seen in Table 1 certain structural features of landscapes control several different services. These attributes are consistently related to area and pattern of bottomland hardwood forests. This same information about attributes is collapsed in Table 6 to forest area, forest patch size distribution, connectivity of bottomland hardwood patches, and contiguity of bottomland hardwood forests to uplands and stream. These key structural features are not rigidly defined. Such definition depends on the susceptibility of various possible indices to measurements. As an aid, therefore, several different variables are listed for each index. Data sources that might be helpful in the analysis are also identified.

The relationship of total bottomland hardwood forest area to other structural indices is not necessarily linear. For example, for a given area of bottomland hardwood forest, one large tract will give different values

Table 6. Indices of landscape spatial structure.

Index	Variable	Units	Tools	Nomenclature
Fraction of bottomland hardwood area remaining	1. Current/historical potential 2. Current/recoverable 3. Recoverable/historical potential	Dimensionless (area/area)	Current wetland (e.g., NWI) cover maps; aerial imagery; historical vegetation potential maps (e.g., Bailey 1980 and Kuchler 1964; FEMA floodplain maps; SCS soil surveys)	Historical potential refers to estimates of pre-development vegetation potential. Recoverable refers to areas that would revert to bottomland hardwood areas if the specific sites were left alone (e.g., agricultural fields that would grow back to bottomland hardwood areas without major changes in the current hydrologic regime of the system as a whole).
Forest patch size distribution	Number of bottomland hardwood patches in various size classes frequency histogram	Number by size class	Current wetland (e.g., NWI) cover maps; aerial imagery	
Connectivity: bottomland hardwood forest-bottomland hardwood forest	1. Number of intrusions 2. Size of intrusions 3. Contrast of intrusions	1. Number 2. Area of intrusions or intrusion area/bottomland hardwood area 3. Differences between intrusions and bottomland hardwood area	Current wetland (e.g., NWI) cover maps; aerial imagery	Contrast of intrusion refers to qualitative or quantitative difference between the intrusion and adjacent bottomland hardwood area. This might be measured with respect to habitat requirements of a set of species or with respect to hydrologic parameters such as a roughness coefficient.

Table 6. (Concluded).

Characteristic	Variable	Units	Tools	Nomenclature
Contiguity: bottomland hardwood area-water	1. Distance from bottomland hardwood area to channel reach 2. Water detention (see Table 7) 3. Streamside buffer ratio: length of channel side with adjacent bottomland hardwood area/2 x total length	1. Linear distance 2. (see Table 7) 3. Dimensionless (length/length)	Current wetland (e.g., NWI) cover maps; aerial imagery; hydrologic feature maps	Channel refers to both natural and man-made. Functioning, adjacent bottomland hardwood forest requires adequate width (15-70 m?) and natural or semi-natural flow (i.e., not bypassed by feeder channels).
Connectivity: bottomland hardwood area-upland	Streamside buffer ratio (see contiguity: bottomland hardwood area-water above)			

of continuity and contiguity than many smaller tracts. These relationships have not been worked out and require investigation. Norms or reference standards of desired structure should be developed based on the relationship of structure to the services performed by bottomland hardwood ecosystems. Several additional points related to how these structural indices might be used require further consideration.

1) Nationally, the remaining bottomland hardwood forest area is a relatively small fraction of historical potential. Relatively few large, continuous bottomland hardwood forest tracts remain.
2) The shape of the functional curve relating bottomland hardwood forest area to services is not clear. Losses of bottomland hardwood forests that produce a greater than proportional loss of function are clearly of highest concern. In Figure 7 two such areas are identified on a generalized response curve. The first potential crisis occurs when nearly 100% of the bottomland hardwood forest is still intact but where initial declines in area may dramatically change the connectivity characteristics by fragmenting large forest tracts. The second potential crisis is farther down the loss curve where area of bottomland hardwood forests remaining may drop below some minimum amount required to perform certain services (e.g., there is no longer enough connected habitat to support a viable population of stenotopic bird species).
3) A distinction needs to be made between the level of performance of some ecological process ("function" curve in Figure 7) and the value to society of those processes ("value" curve in Figure 7). For example, the last fractions of bottomland hardwood forest remaining in a watershed may be very highly valued (e.g., as relict, rare stands) while actually performing ecological processes poorly (e.g., they no longer provide stable, suitable, long-term habitat).

The approach illustrated in Table 7 emphasizes the indicators of bottomland hardwood forest services listed in Table 1: natural biotic diversity, water-quality modification, and streamflow mediation. Several alternate indices are identified for each service, and possible data sources are identified. One of the most important and difficult services to quantify

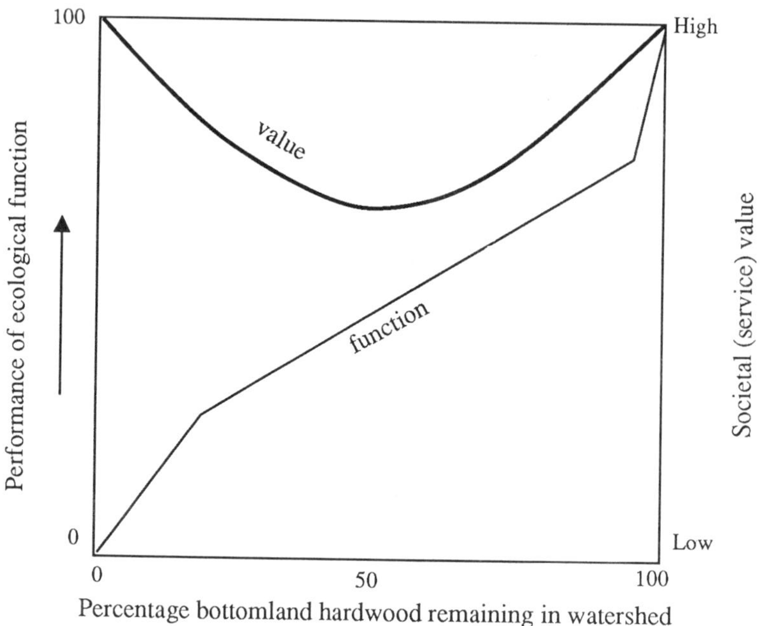

Figure 7. Generalized response curve relating bottomland hardwood forest area to ecological process and societal (service) value (adapted from Lee and Gosselink 1988).

is natural biotic diversity of maintenance of "balanced, indigenous populations" of plants and animals as stipulated in the Clean Water Act. Whole disciplines are devoted to the problems of defining, measuring, and predicting the species and densities of flora and fauna that occupy a given area. Nevertheless, the species present and their relative dominance are powerful indicators, not only of the integrity of any indigenous communities, but also of the general "health" of the whole system.

Table 7. Indices of landscape services

Service	Index	Units	Data sources	Nomenclature
Water quality modification	1. Selected constituents with reference to established standards	1. Concentration of selected constituents (e.g., Total Suspended Solids, forms of Nitrogen and Phosphorus)	1. Water quality observations at selected monitoring stations	1. Established standards refers to specific acceptable concentrations such as those developed under Section 401 of the Clean Water Act.
Stream flow mediation				
Water detention	1. Residence time 2. Duration and extent of inundation	1. Time 2. Inundated area of bottomland hardwood forest	Hydrologic records and models; aerial imagery	
Stage-discharge	1. Annual stage-discharge rating curve 2. Maximum and minimum annual stages and discharges	1. Stage (height) versus discharge (volume/time) at selected monitoring stations on channel	Hydrologic records	1. Annual stage-discharge curve is developed from plots of observed stages and discharges for a year.
Natural biotic diversity	1. Species composition 2. Community balance 3. Diversity indices 4. Vegetation vigor	1. Species present 2. Histogram of dominance by species 3. Various units 4. Primary productivity leaf area index, regeneration (presence of propagules and small size classes	Regional surveys and field measurements at selected sites	Species composition refers to the species present, whereas community balance refers to both composition and the relative ranking of those species in dominance (e.g., has a common species become rare?).

Cumulative Impact Regulation

Elements of a Cumulative Impact Regulation Framework

Assessing cumulative impacts for any given area requires quantifiable indicators, independent performance standards for each indicator, and a meaningful application of those indicators. The relationship between measured index and standard is then used to set limits to further change. The necessary steps are discussed in more detail below.

1) Establish workable, landscape-level ecological indicators of system structure and function and set reference criteria specifying acceptable ranges for them. Tables 6 and 7 identify potential indicators that should be quantifiable and relevant to levels of bottomland hardwood forest disturbance. Establishing standards or norms, however, is a difficult process that goes beyond strict scientific criteria to include the social values within which the standards are set. In order to be effective, the desired reference state must be reasonably specific, measurable, defensible, and shared in the sense of being accepted in a democratic society. Establishing such goals is extremely difficult, but precedents exist in the water-quality standards set as a result of the Clean Water Act. The most common response to this problem for threatened habitat is to adopt the default goals of either historical or existing conditions, that is, to restore the system to a state presumed to exist before significant human disturbance or to adopt a policy of no further ecosystem degradation. Either of these goals involves assigning a high value to the currently existing resource, as has essentially occurred with *Spartina* marsh. Many workgroup participants strongly believed that remaining bottomland hardwood forests should be assigned this level of value by society.
2) Assess the cumulative deviation of the system under study by comparing the current state with the desired state or range of states. Several multivariate analysis methods are available for making such comparisons in the general case. In practice, the comparison may be based on one or two simple indicators that

reflect and/or integrate many bottomland hardwood ecosystem processes (e.g., remaining bottomland hardwood forest area).

3) Assess the direction of the proposed action on the ecological indicators; that is, will the proposed activity move the system farther from the desired goal or closer to it? As discussed above, at the cumulative impact level it may not be possible or necessary to evaluate the absolute magnitude of the individual impact.
4) Differentially regulate individual actions based on the cumulative deviation (3 above) from the desired state (2 above). Two issues are important here. First, differential regulation implies a graded or "directional" response in which approval of individual actions becomes more rare as the system degrades toward a minimum acceptable standard; or, moving in the other direction, regulation becomes more liberal as the landscape system improves. Second, differential regulation poses real problems of equity. The consideration of new individual actions would be influenced by prior actions. Actions previously acceptable might no longer be acceptable because of accumulated changes in the overall system. In one sense this would be unfair to individuals seeking permits to develop floodplains. In another sense this type of differential treatment is the essence of consideration of cumulative as opposed to individual impacts. Again, referring to water-quality standards, the concept of differential regulation has been accepted as part of the regulatory process governing point-source discharges to waters of the United States.
5) Develop and maintain mechanisms for "institutional memory" so that individual regulatory actions fit into a cumulative program of regulation just as individual impacts fit into a pattern of cumulative impacts. This is an especially important point. The accumulation of management experience and information concerning bottomland hardwood forests needs to be systematically organized and retained in order to maximize its use and to protect against the disruptions caused by shifting personnel. Much information needs to be synthesized to quantify the type of large-scale characteristics necessary to

assess cumulative impacts. This synthesis needs to be accomplished in forms that allow anticipatory management in the context of existing regulatory approaches rather than quantification of cumulative impacts without application.

Additional Information Needs

The ecosystem workgroup made considerable progress in identifying a workable set of indicators of landscape structure and services. Nevertheless the set needs continued refinement and evaluation to enhance its utility, and to develop standards and norms based on measurable indices. Basic information is needed on the function of typical reference bottomland hardwood sites, especially in light of the need to develop standards for landscape indicators. Likewise, much more information is needed on the response of bottomland hardwood landscapes to various forms of disturbance. For example, more comparisons of the structure and function of altered sites with reference sites are needed. Techniques must be developed for synthesizing and integrating landscape overlays which describe the history and current state of landscape systems supporting bottomland hardwood forests. Sound decisions for regulating cumulative impacts require a process that allows easy access to relevant information and involvement of individuals with diverse specialized expertise at multiple organizational levels. Advances in both the scientific/technical area of cumulative impact assessment and the political/social area of problem-solving and decision-making are needed for effective regulation of cumulative impacts.

REFERENCES CITED

Asmussen, L., J. M. Sheridan, and C. V. Booram, Jr. 1979. Nutrient movement in streamflow from agricultural watersheds in the Georgia Coastal Plain. Trans. Am. Soc. Agric. Eng. 22:809-815,821.

Bailey, R. G. 1980. Description of the ecoregions of the United States. U.S. Department of Agriculture. Miscellaneous Publication Number 1391.

Bedinger, M. S. 1981. Hydrology of bottomland hardwood forests of the Mississippi embayment. Pages 161-176 *in* J. R. Clark and J. Benforado, eds. Wetlands of bottomland hardwood forests. Elsevier Scientific Publishing Company, New York.

Bell, D. T., and F. L. Johnson. 1974. Flood caused tree mortality around Illinois reservoirs. Trans. Ill. State Acad. Sci. 67 (1): 28-37.

Belt, C. B. Jr. 1975. The 1973 flood and man's constriction of the Mississippi River. Science 189:681-684.

Brinson, M. M., H. D. Bradshaw, R. N. Holmes, and J. B. Elkins. 1980. Litterfall, stemflow, and throughfall nutrient fluxes in an alluvial swamp forest. Ecology 60:827-835.

Brinson, M. M., H. D. Bradshaw, and R. N. Holmes. 1983. Significance of floodplain sediments in nutrient exchange between a stream and its floodplain. Pages 199-221 *in* T. D. Fontaine and S. M. Bartell, eds. Dynamics of biotic ecosystems. Ann Arbor Science Publishers, Ann Arbor, Michigan.

Bryan, C. F., F. M. Truesdale, and D. S. Sabins. 1975. A limnological survey of the Atchafalaya Basin, annual report. Louisiana Cooperative Fisheries Unit, Louisiana State University, Baton Rouge. 203 pp.

Burns, L. A. 1978. Productivity, biomass, and water relations in a Florida cypress forest. Ph.D. dissertation. University of North Carolina, Chapel Hill.

Canadian Environmental Assessment Research Council and United States National Research Council. 1986. Cumulative environmental effects: a binational perspective. Minister of Supply and Services, Canada. 174 pp.

Clark, J. R., and J. Benforado, eds. 1981. Wetlands of bottomland hardwood forests. Elsevier Scientific Publishing Company, New York. 401 pp.

Clark, J. R., and J. E. Clark, eds. 1979. Scientist's report. National Wetlands Technical Council Report, Washington, D.C. 128 pp.

Conner, W. H., J. G. Gosselink, and R. T. Parrondo. 1981. Comparison of the vegetation of three Louisiana swamp sites with different flooding regimes. Amer. J. Bot. 68:320-331.

Conner, W. H., and J. W. Day, Jr. 1982. The ecology of forested wetlands in the southeastern United States. Pages 67-87 *in* B. Gopal, R. E. Turner, R. Wetzel, and D. Whigham, eds., Wetlands: ecology and management. International Scientific Publishers, Jaipur, India.

Cooper, J. R., and J. W. Gilliam. 1987. Phosphorus redistribution from cultivated fields into riparian areas. J. Soil Sci. Soc. Am. 51:1600-1604.

Day, J. W., Jr., T. J. Butler, and W. H. Conner. 1977. Production and nutrient export studies in a cypress swamp and lake system in Louisiana. Pages 255-269 *in* M. Wiley, ed. Estuarine processes, Vol. 2. Academic Press, New York.

Diamond, J. M. 1975. The island dilemma: lessons of modern biogeographic studies for the design of natural reserves. Biol. Conservation. 7:129-146.

Eisenberg, J. F. 1980. The density and biomass of tropical mammals. Pages 35-55 *in* M. E. Soulé and B. A. Wilcox, eds. Conservation biology. Sinauer Association, Inc., Sunderland, Maryland.

Elder, J. F. 1985. Nitrogen and phosphorus speciation and flux in a large Florida river wetland system. Water Resource Res. 21: 724-732.

_________, and D. J. Cairns. 1982. Production and decomposition of litterfall on the Apalachicola River floodplain, Florida. U.S. Geological Survey, Tallahassee, Florida. Openfile Report 82-252, 73 pp.

Finn, J. T. 1976. Measures of ecosystem structure and function derived from analysis of flows. J. Theor. Bio. 56:363.

Forman, R. T. T., and M. Godron. 1986. Landscape ecology. John Wiley and Sons, New York, New York. 619 pp.

Gagliano, S. M., and H. Van Beek. 1975. Environmental base and management study, Atchafalaya Basin, LA. U.S. Environmental Protection Agency, Washington, D.C. EPA-600/5-75-006.

Gosselink, J. G., S. E. Bayley, W. H. Conner, and R. E. Turner. 1981. Ecological factors in the determination of riparian wetland boundaries. Pages 197-219 *in* J. R. Clark and J. Benforado, eds. Wetlands of bottomland hardwood forests. Elsevier Scientific Publishing Company, New York.

______________, and L. C. Lee. 1989. Cumulative impact assessment in bottomland hardwood forests. Wetlands 9:83-174.

Harris, L. D. 1984. The fragmented forest. University of Chicago Press, Chicago, Illinois. 211 pp.

Hellgren, E. C., and M.R. Vaughan. 1988. Home range and movements of winter-active black bears in the Great Dismal Swamp of Virginia and North Carolina. Proc. Int. Bear Conf. Res. and Management 7:227-234.

Horak, G. C., E. C. Vlachos, and E. W. Cline. 1983. Methodological guidance for assessing cumulative impacts on fish and wildlife. Prepared for Eastern Energy and Land Use Team, U.S. Fish and Wildlife Service. Prepared by Dynamic Corporation, Fort Collins, Colorado. Unpublished Report.

Huenneke, L. F., and R. R. Sharitz. 1986. Microsite abundance and distribution of woody seedlings in a South Carolina cypress-tupelo swamp. American Midland Naturalist 115:328-335.

Kahl, M. P. 1964. Food ecology of the wood stork (*Mycteria americana*) in Florida. Ecol. Monogr. 34:97-117.

Krummel, J. R., R. H. Gardner, G. Sugihava, and R. V. O'Neill. 1987. Landscape patterns in a disturbed environment. Oikos 48:321-324.

Kuchler, A. W. 1964. Potential and natural vegetation of the conterminous United States. Manual to accompany the map. American Geographical Society, Washington, D.C. Special Publ. 36. 116 p. Revised edition of ;map, 1965.

Kuenzler, E. J., P. J. Mulholland, L. A. Ruley, and R. P. Sniffen. 1977. Water quality in North Carolina coastal plain streams--effects of channelization. Water Resources Research Institute, University of North Carolina, Raleigh, North Carolina. Report 127. 160 pp.

Lack, D. 1976. Island biogeography illustrated by the birds of Jamaica. University of California Press, Berkeley.

Lee, L. C., and J. G. Gosselink. 1988. Cumulative impacts on wetlands: linking scientific assessments and regulatory alternatives. Environ. Manage. 12(5):591-600.

Leopold, L. B., M. G. Wolman, and J. P. Miller. 1964. Fluvial processes in geomorphology. W. H. Freeman, San Francisco. 522 pp.

Livingston, R. J., R. L. Iverson, and D. C. White. 1976. Energy relations and the productivity of Apalachicola Bay. Final Report. Florida Sea Grant College, Tallahassee, Florida. 437 pp.

Lowrance, R., R. Todd, J. Fail, O. Hendrickson, R. Leonard, and L. Asmussen. 1984. Riparian forests as nutrient filters in agricultural watersheds. Bioscience 34:374-377.

Nixon, S. W., and V. Lee. 1986. Wetlands and water quality: a regional review of recent research in the United States on the role of freshwater and saltwater wetlands as sources, sinks, and transformers of nitrogen, phosphorus, and various heavy metals. Prepared by the University of Rhode Island for the U.S. Army Engineer Waterways Experiment Station, Vicksburg, Mississippi. Technical Report Y-86-2.

Noss, R. F. 1987. Protecting natural areas in fragmented landscapes. Natural Areas Journal 7 (1):2-13.

Odum, E. P. 1979. Ecological importance of the riparian zone. Pages 2-4 *in* R. R. Johnson and J. F. McCormick, tech. coord. Strategies for protection and management of floodplain wetlands and other riparian ecosystems. U.S. Forest Service, Washington, D.C. General Technical Report WO-12.

Peterjohn, W. T., and D. L. Correll. 1984. Nutrient dynamics in an agricultural watershed: observations on the role of a riparian forest. Ecology 65:1466-1475.

Peterson, D. L., and G. L. Rolfe. 1982a. Seasonal variation in nutrients of floodplain and upland forest soils of central Illinois. J. Soil Sci. Soc. Am. 46:1310-1315.

Peterson, D.L., and G. L. Rolfe. 1982b. Nutrient dynamics and decomposition of litterfall in floodplain and upland forests of central Illinois. Forest Science 28:667-681.

Pionke, H. B., and G. Chesters. 1973. Pesticide-sediment-water interactions. J. Environ. Qual. 2:29-45.

Porcher, R. D. 1981. The vascular flora of the Francis Beidler forest in Four Holes Swamp, Berkeley and Dorchester Counties, South Carolina. Castanea 46:248-280.

Qualls, R. G. 1984. The role of leaf litter nitrogen immobilization in the nitrogen budget of a swamp stream. J. Environ. Qual. 13:640-644.

Richardson, C. J. 1985. Mechanisms controlling phosphorous retention capacity in freshwater wetlands. Science 228:1424-1427.

Rozas, L. P., and W. E. Odum. 1987. Use of tidal freshwater marshes by fishes and macrofaunal crustaceans along a marsh stream-order gradient. Estuaries 10:36-43.

Schlesinger, R. B. 1979. Natural removal mechanisms for chemical pollutants in the environment. Bioscience 29:95-101.

Soulé, M. E., and B. A. Wilcox, eds. 1980. Conservation biology. Sinauer Assoc., Inc., Sunderland, Maryland. 395 pp.

Stahle, D. W., M. K. Cleveland, and J. G. Hehr. 1988. North Carolina climate changes reconstructed from tree rings: A.D. 372 to 1985. Science 240:1517-1519.

Stavins, R. 1987. Conversion of forested wetlands to agricultural uses: an econometric analysis of the impact of Federal programs on wetland depletion in the Lower Mississippi Alluvial Plain 1935-1984. Environmental Defense Fund, Inc., Consultant's Report, New York. 291 pp.

Terborgh, J., and B. Winter. 1980. Some causes of extinction. Pages 119-133 *in* M. E. Soulé and B. A. Wilcox, eds. 1980. Conservation biology. Sinauer Assoc., Inc., Sunderland, Maryland.

Turner, M. G. 1987. Spatial simulation of landscape changes in Georgia: a comparison of three transition models. Landscape Ecol. 1:29-36.

U.S. Army Corps of Engineers. 1971. Charles River, Massachusetts: Main report and attachments. New England Division, Waltham, Maryland. 76 pp.

U.S. Department of State. 1982. Proceedings of the U.S. strategy conference on biological diversity. Department of State, Washington, D. C. Publication 9262. 126 pp.

Walker, M.D. 1984. Fish utilization of an inundated swamp-stream floodplain. M.S. Thesis. East Carolina University, Greenville, North Carolina.

Webster, J. R. 1979. Hierarchical organization of ecosystems. Pages 119-129 *in* E. Halfon, ed. Theoretical systems ecology: advances and case studies. Academic Press, New York, New York.

Wharton, C. H., W. M. Kitchens, E. C. Pendleton, and T. W. Sipe. 1982. The ecology of bottomland hardwood swamps of the southeast: a community profile. U.S. Fish and Wildlife Service, Biological Services Program, Washington, D.C. FWS/OBS-81/37.

Yarbro, L. A. 1979. Phosphorus cycling in the Creeping Swamp floodplain ecosystem and exports from the Creeping Swamp watershed. Ph.D. dissertation, University of North Carolina, Chapel Hill. 231 pp.

Zinn, J. A., and C. Copeland. 1982. Wetland management. Congr. Res. Serv. Rep. Ser. 97-11. 149 pp.

18. ANALYSIS OF IMPACTS OF SITE-ALTERING ACTIVITIES ON BOTTOMLAND HARDWOOD FOREST USES: THE REPORT OF THE CULTURAL/RECREATIONAL/ECONOMIC WORKGROUP

Clyde F. Kiker
Food and Resource Economics Department,
University of Florida, Gainesville, FL 32611

with Panel[1]

ABSTRACT

Human activities in bottomland hardwood forests are identified as: 1) those that are nonobtrusive to natural forest functions, and 2) those that inherently alter the forest and its functions. The regulations that define the public's involvement in wetland use decisions suggest consideration of multiple factors. While economic evaluation of activities is seen as one such factor, other nonmonetary valuations should be considered. Results of the Cultural/Recreational/Economic Workgroup's evaluation of forest-altering activities on four natural forest uses--sensory experience, recreation, information storage, and renewable harvesting--are presented. Although the results are qualitative, they give insight into the direction of impacts, positive or negative.

INTRODUCTION

Many human activities can take place within a bottomland hardwood forest or at the site where a forest once grew. These activities range from the occasional interloping sportsman enjoying the flora and fauna, to

[1] Section 3 of this chapter, Analysis of Impacts, was prepared by the Cultural/Recreational/Economics Workgroup made up of Robert Davis (Chairman), Richard Johnson (Recorder), Robert Heeren, Max Reed, John Stierna, and Clyde Kiker with special contribution by Richard Smardon.

Ecological Processes and Cumulative Impacts: Illustrated by Bottomland Hardwood Wetland Ecosystems. Edited by James G. Gosselink, Lyndon C. Lee, and Thomas A. Muir. Chelsea, MI 48118. Printed in USA.

agricultural production where all remnants of the forest are removed. All human activities have impacts on the forest, but some are far more serious than others. The sportsman leaves only footprints, whereas clearing and drainage for agriculture or other commercial purposes change virtually all natural functions of the forest and can create impacts away from the site itself. Impacts on hydrologic regimes, water quality, and wildlife are examples.

This chapter focuses on the relationship between various human activities and the functions provided by bottomland hardwood forests. The perspective is one of viewing the forest as an entity and then asking the question, "What impacts do activities that substantially change the nature of the forest have on human use of the forest in its relatively natural state?"

The objectives of this chapter are to identify the impacts of forest-altering activities and venture opinion on the likely direction of impacts, positive or negative. The chapter consists of four sections. The first deals with bottomland hardwood forest-using activities and the public decision process administered by the federal government. The potential for explicitly using economic valuation in decision processes is discussed in the second section. The third section covers the major activities undertaken by the Cultural/Recreational/Economic (CRE) Workgroup at the July 1985 U.S. Environmental Protection Agency (EPA) workshop dealing with impacts of various activities on bottomland hardwood forests. The workgroup identified four fundamental uses that are the basis of many human activities associated with bottomland hardwood forests. Linkages between site-altering activities and forest uses were identified and were the basis of the impact analysis. A brief list of suggestions from the workgroup is presented in the concluding section.

BOTTOMLAND HARDWOOD FOREST-USING ACTIVITIES AND DECISIONS

It is useful to place human activities in bottomland hardwood forests in one of two categories: ones that are nonobtrusive, in that they have little impact on natural functions, and others that inherently alter forest functions. The activities of the first type (listed under "Limited

Disruption" in Table 1) can be undertaken with minimal disruption of bottomland hardwood forest functions. Hunting and fishing, along with careful natural product harvesting, can be undertaken while maintaining the integrity of the forest. Other activities (listed under "Substantial Disruption" in Table 1) require significant modification of the bottomland hardwood forest, and many natural functions are affected. In each case, the parcel of land to be used has much of the vegetation removed and hydrologic characteristics changed, thereby affecting a broad range of functions. Crop production, silvicultural plantations, rights-of-way, resource extraction, and most industrial and commercial enterprises result in dramatic change of many forest characteristics and ultimately loss of ecological functions.

Figure 1 portrays bottomland hardwood forests in terms of use values. The perspective is specifically anthropocentric. Inherent is the view that all uses depend upon either the natural processes of the forest itself or characteristics of a specific parcel to provide products or services that increase people's well-being in the present or immediate future. Path (a) represents a holistic view of the forest. Natural ecological and hydrological functions are recognized as providing service flows to a broad range of nonobtrusive uses of the forest from which humans derive value. In order to identify the effects of obtrusive uses of an individual forest parcel, it is useful to consider the parcel as an integral part of the bottomland hardwood forest. Path (a_p) reflects the perspective that an individual parcel is a subset of the forest as a whole, and its natural functions are integrated with the forest functions, again, viewed as a whole. The services and use values associated with the parcel in this unaltered state are also viewed as a subset of those provided by the forest itself.

Path (b), on the other hand, identifies the potential the parcel can have when it is substantially altered for production of a service or product. Now, value results not from the forest but from site characteristics. (All the activities listed in Table 1 under "Substantial Disruption" are of this type.) The impacts activities on altered parcels can have on the natural functions of the surrounding bottomland hardwood forest are often overlooked. Drainage of a parcel, for example, can cause loss of the forest's potential for ameliorating flooding and water-quality problems downstream as well as loss of on-site ecological function. These types of

Table 1. Activities and their impact on the functions of bottomland hardwood forests.

ACTIVITY	COMMENTS
LIMITED DISRUPTION	
Commercial fishing	
Sport fishing	
Timber harvesting	
Water supply	
Hunting	
Fur harvest	
Recreation	
Leave it alone use	
Historical/archaeological	
Preservation	
Aesthetic/passive recreation	
SUBSTANTIAL DISRUPTION	
Renewable with transformation	
Cropland Rice Other crops Aquaculture Silvicultural plantations Water resource development Includes flood control, impoundments, release schedules, channelization, levees, dredge/spoil areas	Require various combinations of clearing, brush control, draining, flood control, agricultural chemicals, impoundments, use of heavy machinery
Linear disruptions	
Powerline corridors Oil pipelines Coal slurry pipelines Water pipelines Highways and roads Drilling access Recreation trails	All require landclearing, landgrading and alteration, and chemical conditioning
Resource extraction	
Peat extraction	Removes resource
Oil and gas development	Subsidence
Borrow pits Top soil, sand, gravel, phosphate, lignite coal	Remove resources, change land level

Table 1. (Concluded).

ACTIVITY	COMMENTS
Waste treatment and industrial processing	
Waste-water treatment	All remove surface cover and involve site preparation and some landforming
Cooling ponds	
Solid waste disposal	
Manufacturing and industrial operations	
Water-based transportation	
Waste-water renovation	
Commercial/residential/military use	
Broad range of activities that utilize the location	Require removal of cover, site preparation, and possible filling

impacts are illustrated in Figure 1 by the linkage between the use level of the parcel (path b) and the bottomland hardwood forest function (path a). Although impacts of this type are difficult to identify and quantify, they do exist and can ultimately reduce the overall value of bottomland hardwood forests. In evaluating potential forest-altering landuses and activities these impacts should be considered.

Figure 2 illustrates a broader view of individual valuation of bottomland hardwood forests. Use values as considered in Figure 1 are given in the left-hand set. In the present and immediate future, direct and indirect use of the bottomland hardwood forest can occur. The right-hand set represents non-use values. People may, in addition, value the bottomland hardwood forest for its existence with no expectation of ever receiving service from a direct or indirect use. Bequest value is closely related. Here individuals value the potential of leaving the bottomland hardwood forest as part of a heritage. The intersection of the two sets is a subset that identifies option value. Where the forest remains unaltered, the option to make use of it or to leave it in its natural state continues. Individuals may express a value for maintaining the options.

Under the Rivers and Harbors Act, the Clean Water Act, and other federal laws, the U.S. Army Corps of Engineers (USACE); in consultation with the EPA, U.S. Fish and Wildlife Service, and other

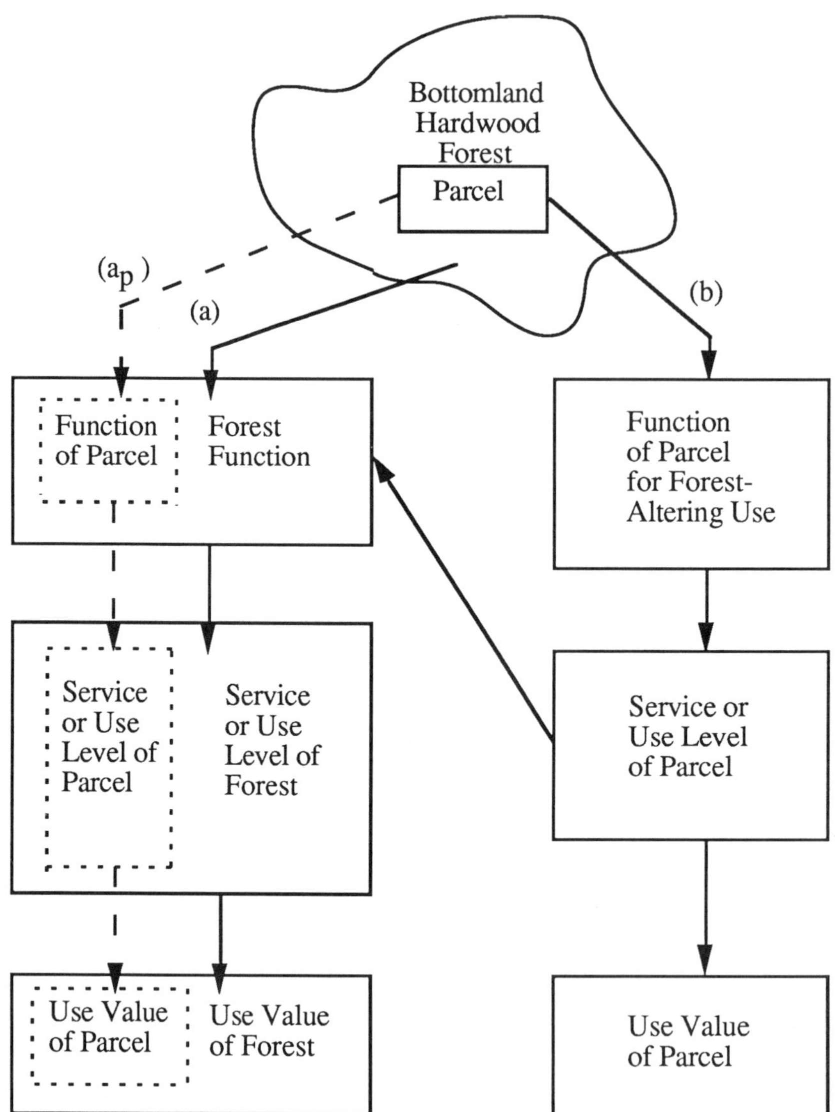

Figure 1. Relationship of functions and uses to value of unaltered and altered bottomland hardwood forests.

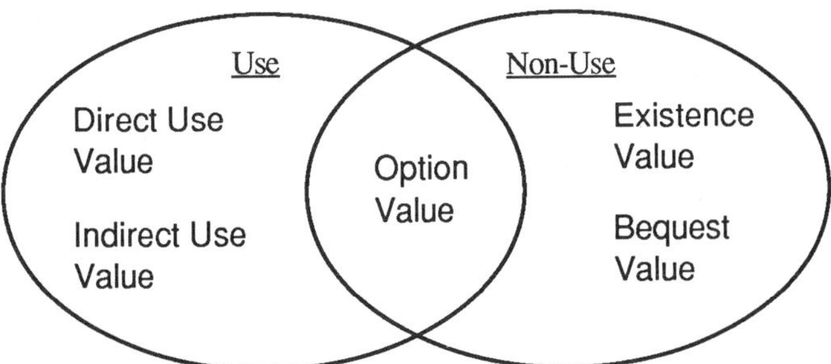

Figure 2. Components of individual valuation that underlie economic valuation of bottomland hardwood forests.

federal agencies; is responsible for representing the public's many values through a permitting process. Proposed activities in wetland areas are evaluated, and a permit is either awarded or denied. The following, taken from the General Regulatory Policy, gives an overview of the factors considered in the process.

> The decision whether to issue a permit will be based on an evaluation of the probable impacts, including cumulative impacts, of the proposed activity and its intended use on the public interest. Evaluation of the probable impact which the proposed activity may have on the public interest requires a careful weighing of all those factors which become relevant in each particular case. The benefits which reasonably may be expected to accrue from the proposal must be balanced against its reasonably foreseeable detriments. The decision whether to authorize a proposal, and if so, the conditions under which it will be allowed to occur, are therefore determined by the outcome of this general balancing process. That decision should reflect the national concern for both protection and utilization of important resources. All factors which may be relevant to the proposal must be considered including cumulative effects thereof: among those are conservation, economics, aesthetics, general environmental concerns, wetlands, historic properties, fish and wildlife values, flood hazards, floodplain values, land use,

> navigation, shore erosion and accretion, recreation, water supply and conservation, water quality, energy needs, safety, food and fibre production, mineral needs, consideration of property ownership and, in general, the needs and welfare of the people (51 Fed. Reg. 219, 41223 [1986] Section 320.4).

It is clear from this policy statement that a balance of benefits against reasonably foreseeable detriments is required, and that a number of factors, including cumulative impacts, must be considered. The implication is that a multi-dimensional evaluation approach is necessary.

However, many responsible for administering the wetland permit process desire a defensible, empirical measurement tool that allows comparison of values derived from alternative uses of wetlands. The ultimate hope is for a conceptually consistent method of contrasting the benefits and detriments of allowing or disallowing an activity on a specific parcel of wetlands.

POTENTIAL FOR ECONOMIC VALUATION

Under previous interpretations of property entitlements, the public, through government, had little to say about the way in which a landowner chose to use wetlands. The landowner judged whether the expected net value of the use service flow from a parcel in an altered state was greater than the expected net value of the use service flow from the parcel in an unaltered state. If it was, it would be reasonable from a financial perspective for the owner to make the decision to change the parcel. In earlier times, technologies available to the owner were limited. Thus, the cost of altering the wetland could be quite high, resulting in a low expected net value. In the last fifty years, however, technologies have allowed substantial alteration of wetlands at much lower costs, and landowners have moved to alter wetlands.

The General Regulatory Policy as reflected in the above quotation implies a perspective broader than that of the individual landowner's. Factors of importance to the citizenry as a whole are to be included. Expressed in economic terms, the criteria for the public are similar to the landowner's, except now the loss of use value beyond the parcel boundary, as well as the non-use value precluded by the alteration, are involved. Recall the link between path (b) and path (a) that introduces the loss of ecological and hydrologic functions in the forest, as well as in systems occurring beyond the immediate region. Inclusion of the effects site alterations have on wetland service flows is the difference between the landowner's position and that of society. These positions can be written as follows:

Landowner's position:

Value of altering the parcel	=	Net use value of parcel in altered state	-	Net use value of parcel in unaltered state

Society's position:

Value of altering the parcel	=	Net use value of parcel in altered state	-	Net use value of parcel in unaltered state	-	Use value of forest function precluded by alteration	-	Non-use value of forest precluded by alteration.

Clearly, the estimated values from the landowner's and society's perspectives can differ. The owner's position can be positive, while society's can be negative.

The problem in using solely an economic framework for decision-making stems from the difficulty in identifying and quantifying the items in the boxes in Figure 1 and relating these to the terms in the two equations. Traditionally, economic research has focused on establishing the net use value of sites in altered states. That is, much effort has gone into determining the value of wetlands for agriculture, silviculture, aquaculture, and other commercial purposes. The products and services of these activities are traded in markets. Similarly, the resources necessary to conduct the activities are traded in markets. These markets provide a forum for people to express their willingness to pay for the products and services. With market information it is possible to make monetary estimates of the value of the wetland as a site for these wetland-altering activities.

The majority of goods and services stemming directly from the forest, on the other hand, do not enter the market place. The lack of market-established values does not mean that these wetland-provided goods and services do not have economic value; it means that there is no systematic forum for people to express their willingness to pay for such goods and services. Without such a forum for expressing values it becomes much more difficult to express the service flows from wetland functions in monetary terms.

It has been suggested that the benefit/detriment balance be extrapolated into an economic criterion of decision making. Shabman and Batie (1988:14) express the idea: "The wetland permit process should seek to ensure that the higher value option, e.g., wetland alteration (permit granted) versus wetland preservation (permit denied) be pursued. In general terms, denial of a wetland alteration permit requires an analysis documenting that the benefits to human users of maintaining natural wetlands exceed the costs, measured as foregone development use values." While it is recognized that this quote can be read in terms broader than monetary values, there is presently substantial pressure to use these values.

The problem with such a limited view is that the wetlands would be altered unless there was economic evidence of greater detriment resulting from lost wetland functions. Because of the intricacies of wetland processes (as documented in other chapters of this book), it is difficult to specify many of these natural processes and establish their economic value. Under this criterion, one could conclude that the wetlands will be altered not because people place a low value on the services and goods they provide, but because there is uncertainty as to how wetland systems actually work. The implication is that alteration of the wetland will not have detrimental effects because we cannot identify wetland functions and give monetary values for lost services and goods.

Unfortunately, little information exists on the economic value of natural wetland services. Shabman and Batie (1988), in a review of the literature on the socioeconomic values of wetlands make this clear in their conclusions: "The literature search . . . found relatively few articles which provide estimates of the value of wetlands. Of those articles a small number employ conceptually valid approaches to valuation. Even those [analysts] familiar with the correct application of the concepts often report results that were, in their own view, less than satisfactory. To a large degree, this can be attributed to inadequate data and poorly documented linkages between wetland areas and wetland services in the scientific literature" (1988:50). They also conclude that even the few studies that are conceptually correct have "...site-specific results [that] are not transferable beyond their case study areas" (1988:51).

It is not currently possible to express cost benefits and detriments associated with altering a specific parcel of wetlands. Were the permit process to use a decision rule based upon such a measure, the result would inevitably be alteration of wetland parcels that the public might, if there were a forum for consistently expressing values, choose to leave in a natural state. To illustrate the type of information that can presently be provided to the decision process, the CRE Workgroup analyzed a range of activities that can take place in bottomland hardwood forests. The next section presents the results of the analysis.

ANALYSIS OF IMPACTS

The CRE Workgroup focused on uses that stem from ecological and hydrologic functions of the forest and the impact bottomland hardwood forest-altering activities have on these uses. The workgroup identified four broad, generic uses that stem from fundamental natural forest processes and that are the basis of the service flows from the forest. The four uses--sensory experience, recreational experience, information storage, and renewable harvesting--arise from aesthetic, cultural, and material values. Certain bottomland hardwood forest characteristics are associated with the four generic use categories. Tables 2 and 3 present the forest characteristics, generic uses, and specific activities.

The majority of the workgroup's time was spent analyzing landuse activities that substantially change bottomland hardwood forest characteristics, both on-site and off-site, thereby affecting the potential for the four broad categories of use. Disruptive landuse activities selected as representative are: conversion of the site to rice production, soybean production, pine plantations, aquaculture, or water impoundment; levee construction; and stream channelization. The analysis that follows identifies links between these land-altering activities and natural bottomland hardwood forest uses and conjectures whether the resulting impacts are likely to be positive or negative.

Table 2. Categories of values, uses, and non-disruptive activities.

Values	Uses	Non-Disruptive Activities
Passive aesthetic	Appreciation/ Sensory experience	Boating Sightseeing
Active aesthetic	Recreational experience	Boating Sightseeing Photography Wetland art Bird watching Canoeing, kayaking ATV use Sport fishing Sport hunting
Cultural	Information storage	Nature study Education, research Wetland art Photography Literary work Archeology/historic
Material	Renewable harvesting	Natural food gathering Commercial fishing Fur harvesting Herbaceous harvesting Timber and fiber harvesting

Table 3. Characteristics of bottomland hardwood forests that underlie fundamental human uses.

	Use			
Characteristics	Sensory Experience	Recreational Experience	Information Storage	Renewable Harvesting
Visual access	x			
Physical access	x	x		
Vegetation diversity	x			
Water quality	x	x		
Species richness	x			
Fauna	x	x		x
Flora	x	x		x
Water flow velocity		x		
Water depth minimum		x		
Stream pattern complexity		x		
Contiguity and size		x		
Historical/archaeological			x	
Representative ecosystems			x	
Relict stands			x	
Existing documentation			x	
Presence of food species				x
Presence of finfish, shellfish				x
Presence of wildlife				x
Presence of fiber species				x

Sensory Experience

The quality of sensory experience in the landscape is related to a complex range of cultural, psychological, and environmental factors (Smarden 1978, 1983, 1988a, Lee 1983). Sensory experiences in bottomland hardwoods involve nonconsumptive contacts between people and the environment. These experiences depend on the physical, biological, and hydrologic characteristics of the bottomland hardwood ecosystem, the mode of experiencing it, perceptions of value by users, and the information gained by users. Newby (1971) described human perception of the environment as incorporating "the interaction of man's senses into a system whereby he is able to adapt to a world of constantly changing environmental conditions" (Lee 1983:43). Bottomland hardwood forest characteristics shown in Table 4 are important in determining the quality of sensory experiences that may be produced by bottomland hardwood forests and that may be affected by various changes in landuse (disruptive activities). These characteristics include visual access, physical access, vegetative diversity, water quality, species diversity, fauna, and flora.

Visual Access

Visual access means that vistas are open to different environments. Four factors are identified by Lee (1983) as being important to visual preferences in the environments. Legibility involves the clarity or coherence of a scene and aids in individual recognition of visual elements. Spatial definition primarily concerns the arrangement of three-dimensional space within the visual array. It affects orientation and has a definite influence on individual perception and preference. Complexity relates to the number and relative distribution of landscape elements. Finally, mystery involves the promise of additional visual information and encourages an individual to enter a visual display in order to seek this additional visual data (see Smarden 1983 for additional information).

As access to visual information increases, the sensory experience use of bottomland forests also increases initially. The relationship is positive, and may turn negative only when visitation demand enters the relationship. Crowding, due to the quality of visual access, may produce a negative impact on an individual's sensory experience.

Table 4. Sensory experience as affected by the impact of disruptive activities on bottomland hardwood forest characteristics.

Bottomland hardwood forest characteristic	Relationship of characteristic to use1	Activity2								
		R	S	I-U	I-0	I-D	C	L	P	A
Visual access	+-	-	-	0	-	0	-	+	-	-
Physical access	+-	-	-	0	-	0	+-	+-	+	-
Vegetation diversity	+	-	-	0	-	0	-	-+	-	-
Water quality	+	-	-	0	-	+	+-	0	0	-
Species richness	+	-	-	0	-	0	-	-+	-	-
Fauna	+	+-	+-	0	-	0	+-	-+	+-	-
Flora	+	-	-	0	-	0	+-	-+	+-	-
Overall impact on sensory experience		-	-	0	-	0	-	-+	-	-

1 A plus indicates a positive correlation between the characteristics and the use. A minus indicates a negative correlation. A combination of signs indicates that the relationship may differ depending on specific circumstances.

2 A plus indicates that the use is enhanced due to the impact of the activity on the characteristic. A minus indicates that the use is impaired. A zero indicates no impact. A combination of signs indicates differing impacts depending on specific circumstances. R = conversion of site to rice. S = conversion of site to soybeans. I-U = impacts upstream from an impoundment. I-O = impacts from an impoundment on-site. I-D = impacts downstream from an impoundment. C = channelization of stream adjacent to site. L = levee construction on-site. P = conversion of site to pine plantation. A = conversion of site to aquaculture.

All site-changing activities considered at the workshop were estimated to have a negative impact on sensory experience through visual access except water impoundments and levees (Table 4). Conversion to agricultural crops (rice or soybeans) destroys the natural scenic environment, and conversion to pine plantations drastically changes the scenic environment. Channelization mars the environment and restricts visual access to views up and down the channelized stream. Aquaculture and on-site impoundments destroy natural scenes, but there is no visual effect upstream or downstream unless the hydrologic regime is drastically altered. Only levees have a potential positive effect. They improve viewer elevation and increase visual access to more legible, spatially defined, and complex environmental vistas. However, they can also block views if the observer is standing on either side of the levee.

Physical Access

Physical access indicates the degree of access to or within bottomland hardwoods by trail, road, or boat (Smardon 1983). Improved physical access has a positive effect on sensory experiences unless it becomes so convenient that individual enjoyment is impaired by crowding. Conversion of bottomland forests to rice, soybeans, aquaculture, or on-site impoundments destroys the natural environments that create sensory experiences. Locations upstream or downstream from impoundments and aquaculture are affected minimally by these developments. Conversion to pine plantations improves access to surrounding bottomland forests as well as to the new plantations, while levees and channelization improve water-based access but may block foot access from one area to another.

Vegetative Diversity

Vegetative diversity generally interacts positively with sensory experience. This is especially true as variable-height trees increase legibility, spatial definition, complexity, and mystery at wetland or waterbody edges. However, vegetation diversity may be inversely related to visual access if growth becomes too dense and pervasive to permit clear views of varied environmental vistas.

Conversion of bottomland hardwood forests to rice, soybean production, pine plantations, aquaculture, and on-site impoundments drastically changes vegetation, resulting in severe reductions in sensory experience. Vegetative diversity is not affected upstream from an impoundment, and is not likely to be affected downstream unless drastic changes in the water regime are initiated. Vegetative diversity is reduced by channelization as flora outside the channel shifts to more mesic species, and flora within the channel shifts to aquatic plants. Levees have a negative impact on vegetative diversity, and thereby sensory perception, at the exact location of the levee. They may have either a positive or negative effect on lands influenced by the levee. Species mixes normally change as a result of a levee.

Water Quality

Sensory experiences of visitors to bottomland hardwood resources are affected by the quality of water encountered during their stay. Threshold values for tactile, olfactory, and visual attributes of water are all important determinants of the sensory experience. Once the threshold levels have been achieved, further improvements in these water-quality attributes go unnoticed.

Conversion of bottomland hardwood forests to rice, soybeans, or aquaculture is likely to negatively affect sensory experiences through reductions in water quality caused by increased sediment loads and chemical concentrations. The visual quality of water in shallow, windswept impoundments may be reduced because of increased turbidity. Water upstream from an impoundment is likely to carry reduced sediment loads.

The quality of water in pine plantations that have been converted from bottomland hardwood tree species is expected to be unchanged because the land will still be forested after an initial period for stand establishment. Levees are also expected to be neutral in their effect on water quality, while channelization will have a negative effect in the short term due to increased turbidity and a positive effect later because of reductions in water stagnation.

Species Diversity

The quality of sensory experience is estimated to improve as animal species diversity increases. Diversity of species, and thereby the sensory experience, is reduced drastically when bottomland forests are converted to rice, soybeans, aquaculture, and on-site impoundments because the nature of landuse and productivity is changed totally. Conversion to pine plantations also reduces species diversity through replacement of bottomland hardwood habitat with a monoculture pine forest. Areas upstream and downstream from an impoundment are not likely to be affected.

Channelization reduces animal species diversity both in and outside the channel. Levees result in the elimination of many animal species at the levee site and changes in species mix in areas influenced by the levee. Whether species diversity increases or decreases in those areas is uncertain.

Fauna

A positive relationship is expected between total fauna and sensory experience. Aquaculture and on-site impoundments affect these relationships negatively. Potential changes in fauna upstream or downstream from impoundments are not expected to have noticeable effects on sensory experience.

Conversion of bottomland hardwood forests to either rice or soybeans has a drastic effect on faunal species mix, but the effect on total fauna is uncertain. Conversion to pine forests also affects species mixes and total numbers in uncertain ways. Channelization causes short-term reductions in fauna and a long-term shift in species, with uncertain implications for total populations long-term. Animal populations are drastically reduced at levee sites and affected uncertainly in areas influenced by the levee.

Flora

Sensory experience is estimated to vary directly with vegetation growth. Conversion of bottomland forests to rice, soybeans, aquaculture, or on-site impoundments results in drastic changes in plant species and reductions in floral diversity. Conversion to pine forests is less severe and may result in species mix changes and either positive or negative changes in floral production. Areas upstream and downstream from impoundments are estimated to be unaffected. Near-term implications of channelization are expected to be drastic reductions in flora, with long-term effects uncertain. Vegetation under levees is at first destroyed, but the effect on land influenced by the levee may be either positive or negative.

Overall Impact

Unweighted tallies of the effects of disruptive activities on sensory experience through various bottomland hardwood forest characteristics indicate that conversion of bottomland forests to rice, soybeans, pine plantations, aquaculture, or on-site impoundments produces negative impacts. Channelization also has negative implications on sensory experience; upstream and downstream impacts are neutral. Only levees may have a positive effect on sensory experience.

Recreational Experience

Outdoor activities are an important and growing component of private recreation, and preservation of remaining bottomland hardwood forests for recreational use may be economically viable as well as ecologically sound. The challenge lies in correctly valuing recreational experience associated with bottomland hardwood forests, because much of the recreational value is not established through consumer demand in well-functioning markets.

This section of the CRE Workgroup report estimates how characteristics of flow velocity, water depth minimum, stream sinuosity, physical access, contiguity and size, fauna and flora, and water quality affect the recreational use of bottomland hardwood forests. Furthermore, the impacts of various disruptive activities on bottomland hardwood forest characteristics and recreational experience are analyzed (Table 5). Bottomland hardwood forest characteristics chosen to analyze the recreational experience lean heavily toward water-based recreation. This may be appropriate, because water plays such an important role in bottomland hardwood ecosystems.

Water Flow Velocity

Recreational use of waterways for boating requires a minimum flow velocity that varies with the type of watercraft being used. Once this threshold velocity is reached, the quality of the recreational experience remains fairly constant for a substantial range of flow velocities and then deteriorates as velocities become too high for safe navigation by various

Table 5. Recreational experience as affected by the impact of disruptive activities on bottomland hardwood forest characteristics.

Bottomland hardwood forest characteristic	Relationship of characteristic to use[1]	Activity[2]								
		R	S	I-U	I-0	I-D	C	L	P	A
Water flow velocity	-	-	-	NA	0	0	+	0-	NA	-
Water depth minimum	+-	-	-	0	NA	+	+	0	0	NA
Stream pattern complexity	+-	-	-	0	-	+	-	0	0	NA
Physical access	+-	-+	-+	+	-	+	+	+	+	NA
Contiguity and size	+-	-+	-+	-	-	-	-	-	NA	-
Water quality	+	-	-	0	-+	-+	-+	0	NA	-
Flora and fauna	+	-+	-+	-	-	-	-+	0-	+-	-
Overall impact on recreational experience		-	-	-	-	+	-+	0	-	-

1 See footnote 1 in Table 4.
2 See footnote 2 in Table 4.

recreational watercraft (Hydra 1978, Smardon 1988b). Conversion of bottomland forests to rice, soybeans, or aquaculture withdraws on-site land from recreational boating and is likely to reduce flows in adjacent streams. Water velocities on the stream side of a levee may increase somewhat; velocities on the protected side of a levee are likely to decrease. Stream velocities at on-site impoundments obviously decrease, but the effect on recreational use was considered inconsequential. More consistent velocities downstream from the impoundment were also considered to have no effect on use.

Streams and wet areas of bottomland hardwood forests are not normally converted to pine plantations, so the stream velocity indicator is not applicable for that activity. Water velocity upstream from impoundments was also considered not applicable. Channelization, however, can provide increased stream flow velocity and improved access by larger recreational watercraft.

Water Depth Minimum

Shallow waters physically and psychologically inhibit recreational boating and may preclude access to bottomland forests by boat. Water depth required for access, launching, and navigation depends upon the type of watercraft used (Hydra 1978, Sargent and Berke 1979).

Water depth is eliminated by conversion of bottomland hardwood forests to rice or soybean farming. It is not applicable to pine plantation activities because streams and wet areas cannot be converted to pine plantations successfully. While conversion to aquaculture and impoundments may increase water depth in some cases, these land uses are so different from bottomland hardwood forests that varying water depths in this way was not considered applicable to a bottomland hardwood recreational experience. Minimum depths of water upstream from impoundments are not affected, nor are water depths on the landward side of levees. Streams inside levees have sufficient minimum water depths for a greater part of the year than they had before levee construction. Channelization also increases water depth minimums, ensuring a positive relationship between channelization and recreational experience acting through water depth changes.

Stream Pattern Complexity (Sinuosity)

For a scene to evoke a positive visual response it must be both legible and complex. Legibility is related to an individual's ability to adapt or understand a visual experience; complexity is related to variation and surprise. Visual complexity has a positive effect on visual preference until the scene is no longer legible. Finally, as complexity continues to increase, cognitive overload can result in feelings of chaos and confusion. Interaction of these two attributes is an important determinant of recreational experience as a traveler encounters a series of views, or "sequential glimpses," down a stream corridor (Lee 1983).

Conversion to rice, soybeans, aquaculture, or on-site impoundments eliminates most stream segments and straightens others. Channelization also has a strongly negative effect on recreational experience through reduced sinuosity of streams. Streams below impoundments have more stable flows and are somewhat less complex, while upstream areas are not affected by an impoundment. Levees are assumed to have no effect on recreational experience through changes in stream sinuosity.

Physical Access

The collective enjoyment of people who visit recreational sites increases throughout substantial ranges of access potential. If access becomes so convenient that crowding of visitors becomes extreme, the relationship between further improvements in physical access and recreational experience may become negative.

On-site impoundments destroy bottomland hardwood forests for recreation, as does conversion to aquaculture. Bottomland forests upstream and downstream of impoundments, however, may offer improved recreational experiences because the impoundments may become new launching sites. Channelization and levees also improve recreation through increased access by boats. Land access is improved by conversion to pine plantations, rice, and soybeans. Access by water, however, is decreased when bottomland forests are converted to rice or soybeans, resulting in potential reductions in water-based recreational experience.

Contiguity and Size

Recreational experience is positively correlated with wetland size up to a point. The relationship becomes negative as effective access becomes less practical. The size of a wetland can be extended effectively through contiguity between waterbodies providing boat passageways.

Water impoundments destroy bottomland forests on-site and interrupt stream contiguity between upstream and downstream locations, effecting a negative impact on recreational experience. Channelization affects recreational experience negatively by segmenting and dividing bottomland hardwood wetlands, as do levees that break continuity and reduce the size of bottomland hardwood areas.

Conversion of bottomland hardwood forests to rice, soybeans, or aquaculture reduces tract size and interrupts contiguity, but conversion to pine plantations has no effect because plots are still timbered, and natural wetlands are not converted.

Water Quality

Algal blooms may be sufficiently unsightly to discourage recreational use, and chemical concentrations or high bacterial counts may pose a direct health hazard (Kooyoomijian and Clesceri 1973). Once safe thresholds of water quality have been reached and no visual, tactile, or olfactory annoyances are noticed, further improvements in water quality go unnoticed and do not change recreational experience.

Conversion of bottomland hardwood forests to rice, soybeans, or aquaculture may impair water quality through all production phases because of increased sediment loading and chemical concentrations. Water quality upstream from impoundments is unchanged, while on-site impoundments and downstream impoundments affect water quality and recreational experience in uncertain ways. Chemicals and sediments may be trapped by the impoundment, improving water quality downstream. On the other hand, algal blooms, bacteria counts, and water temperature may be increased in the impoundment, decreasing water quality and recreational experience both in the impoundment and below.

Channelization decreases water quality and thereby recreational experience during and shortly after development, but is likely to improve water quality and recreation thereafter by reducing stagnation. Levees were estimated to have no long-term effect on water quality as was conversion to pine plantations, because the land is still in timber. Short-term (construction period) effects of these two activities are negative because of increased sedimentation from construction work.

Flora and Fauna

Recreational experience varies directly with (is positively related to) abundance of flora and fauna. On-site impoundments and conversion to aquaculture produce drastic changes in plants and animals and have a negative effect on recreation. Communities downstream from impoundments change because of reductions in backwater flooding, with a resulting negative impact on recreational experience. Upstream and downstream locations are isolated from each other, affecting recreation negatively through adverse impacts on large mammals.

Conversion to rice, soybeans, and pine plantations drastically changes vegetation, but may either enhance or impair habitat, depending upon species. This may result in either positive or negative effects on recreational experience. Channelization was estimated to have a negative impact on recreation through its short-term effects on flora and fauna and long-term species shifts that could be either positive or negative. Levees were judged to reduce flora and fauna and, thereby, recreational experience on the land side; flora and fauna are left largely unchanged on the stream side of the levee.

Overall Impact

All except three of the activities considered at the workshop were estimated to produce negative effects on bottomland hardwood recreational experience. Conversion of bottomland hardwood ecosystems to rice or soybeans was estimated to have the strongest negative implications for recreational experience. Impoundments and conversions to aquaculture were judged to be less damaging to recreation, but were also estimated to have negative impacts.

Levees and pine plantations were judged to have no net effect on recreational experience in bottomland hardwood areas when the positive implications from some characteristics are balanced against the negative and neutral impacts expected from other characteristics.

Only channelization may impact recreational experience in a positive way. Positive effects through enhanced waterflow velocity, greater water depth minimums (depending on management), and improved access slightly outnumber the clearly negative impacts associated with impaired stream pattern complexity and reduced contiguity and size. Impacts on water quality, flora, and fauna may be either positive or negative depending upon which species are being considered.

Information Storage

The information storage use of bottomland hardwood ecosystems relates to various historic attributes which include those that arise from information contained in the complex, living ecosystem; those that are of cultural significance; and those that have documented potential for future scientific, educational, and cultural use. This function includes, by example, natural history. Explorations in the mid-Mississippi Valley reveal the use of animals such as turtles, fish, waterfowl, beaver (*Castor canadensis*), and raccoon (*Procyon lotor*) by early native Americans (Smith 1975). Cultural history also includes bottomland hardwood ecosystems as a source of skins for clothing. Along the southern coast of Carolina, traders worked on agreements with Indians for obtaining fur resources from the floodplains and swamps of the river systems in Georgia, Alabama, and Mississippi (Niering 1979). There is also cultural history associated with logging of timber in southern river swamps.

Cultural and historical significance by association refers to bottomland hardwood ecosystems that inspired or were the actual basis for art, literature, and music (Fritzell 1979). Wetland art related to cultural history is discussed in Elman (1972). Famous paintings of bottomland hardwood forests include R. W. Meeker's (1829-1889) painting of a river hardwood forest draped in Spanish moss (*Bayou Teche*) and *Muddy Alligators* by John Singer Sargent (1856-1926). The literary heritage of wetlands in general is well treated by Fritzell (1979). Specific pieces of literature that relate to wetlands can be found in the works of Twain, Rolvaag, Faulker, Hemingway, and Welty. Particular literary works that may relate to bottomland hardwood ecosystems include Faulkner's *The Bear* and Twain's *Huckleberry Finn*. The musical heritage of bottomland hardwood forests can be heard in such works as Ferde Grofe's *Mississippi Suite* and Frederick Delivo's *Florida Suite*.

Current and future potential informational uses of bottomland hardwood forests include research on hydrology, biological productivity, nutrient cycling, diversity, and spatial patterns. The value of a particular site may be related to existing documentation of previous research. Specialized historical research, such as palynology and archaeology, can be conducted only on sites with appropriate historical materials. Even more specialized are those sites that have rare or endangered species (e.g., the endangered Florida royal palm [*Roystonea elata*], an associate in cypress swamps).

One of the most promising potential uses of bottomland hardwood forests is educational. As Niering (1979:517) reminds us by example "...small alligators in floating water lettuce from the boardwalk in National Audubon Society's Corkscrew Cypress Swamp..." can be very educational. Certain sites have high educational potential, especially if there are many species or features in close proximity to each other or if they are close to educational institutions (Smardon 1975, 1978).

Presence of Historic/Archaeological Sites

The existence of historic/archaeological sites bears a strong positive relationship to information storage use (Coles 1984). No actions can be taken to "create" these sites, but the sites can be preserved. Many sites where artifacts, such as arrowheads, are abundant are located in higher bottomland hardwood locations. Unfortunately, areas with high probabilities for containing archaeological sites are also excellent sites for "borrowing" fill and construction materials.

Conversion of bottomland hardwood forests to rice, soybeans, aquaculture, or on-site impoundments risks severe disturbance or covering of historic/archaeological sites, thereby impairing information storage (Table 6). Locations upstream and downstream from impoundments are unaffected. Pine plantations may result either in the discovery and potential protection of historic/archaeological sites or their complete destruction if they are not discovered when bottomland hardwood forests are converted to pines.

Channelization has a high probability of total disruption of historic/archaeological sites during construction. Levees are also likely to be totally destructive at the dredge or fill location. However, bottomland hardwood forests protected from flooding by the levee may preserve historic/archaeological sites, while bottomland hardwood forests on the water side of a levee are likely to have deeper, higher velocity water that may damage historic/archaeological sites.

Representative Ecosystems

Information storage potential is decreased or lost when ecosystems that are representative of natural bottomland hardwood forests are converted to rice, soybeans, aquaculture, or on-site impoundments. Bottomland hardwood forests upstream and downstream of an

Table 6. Information storage as affected by the impact of disruptive activities on bottomland hardwood forest characteristics.

Bottomland hardwood forest characteristic	Relationship of characteristic to use[1]	Activity[2]								
		R	S	I-U	I-O	I-D	C	L	P	A
Historic/ archaeological	+	-	-	0	-	0	-	--+	0-	-
Representative ecosystems	+	-	-	0	-	0	-	0--	-	-
Relict stands	+	-	-	0	-	0	-	0--	-	-
Existing documentation	+	NA	NA	NA	NA	NA	NA	NA	NA	NA
Overall impact on information storage		-	0	-	0	-	-	-	-	-

1 See footnote 1 in Table 4.
2 See footnote 2 in Table 4.

impoundment are not likely to be changed. Conversion to pine plantations destroys the representative ecosystems characteristic and replaces it with a monoculture. Channelization destroys some representative bottomland hardwood ecosystems and segments other stands, affecting the function negatively. Bottomland hardwood forests directly impacted by levees are destroyed as representative ecosystems, but land on the unprotected side of the levee remains unchanged because it continues to be flooded periodically. Bottomland hardwood forests protected by the levee are either unaffected or impacted negatively, depending upon the amount of change in the water regime caused by the levee.

Relict Stands

Relict stands either currently exist or they will not exist during the time period relevant to this analysis. When they do exist, these stands have a very positive relationship to information storage. They are affected by activities exactly as representative systems are affected, for identical reasons, except that when relict stands are discovered during pine plantation conversion, they are more likely to be preserved.

Existing Documentation

Existing documentation consists of photographic materials, paintings, and literature that describe bottomland forests for current visitors, vicarious observers, or intergenerational transfer. Obviously, existing documentation is unaffected by any activity impacting bottomlands. However, bottomland hardwood forests provide the opportunity for further documentation in the future. Any activity that interferes with natural bottomland hardwood ecosystems has a negative effect on these future documentation opportunities.

Overall Impact

No activity considered during this analysis was estimated to have a net positive effect on information storage through characteristics identified and analyzed by the workgroups. In fact, only one positive relationship was found in the entire analysis (Table 6). Conversion of bottomland hardwood forests to rice, soybeans, aquaculture, or on-site impoundments, as well as channelization, were all judged to have negative effects on the information storage that can be provided by bottomland hardwood ecosystems.

Renewable Harvesting

While it is recognized that crop production, aquaculture, and silvicultural plantations are important sources of food and fiber, the consideration of renewable harvesting will focus on the harvesting uses that depend upon bottomland hardwood forest functions and related characteristics. As discussed earlier, crop production, aquaculture, and silvicultural plantations are viewed as activities that require substantial disruption of forest functions, thus affecting characteristics such as presence of food, finfish, shellfish, wildlife, and fiber plant species.

Table 7 gives a summary of the impacts of these forest-modifying activities on renewable harvesting uses of bottomland hardwood forests.

Presence of Food Plant Species

The presence of food species is positively correlated with renewable harvesting uses (Niering and Palmisano 1979, Reinhold and Hardisky 1979). Conversion of bottomland hardwood forests to rice or soybeans dramatically increases the total amount of food that will be produced on a parcel. However, the effect of the new landuse will be to decrease the presence of naturally occurring food species on the parcel. As a result, the effect is given as both plus and minus in Table 7. Levees have a positive effect because they improve access to foods. Conversion to pine plantations may have either a positive or negative impact depending upon the species being harvested.

Aquaculture and on-site impoundments place land under water and eliminate production of harvestable vegetation. Upstream and downstream production potential are likely to be largely unchanged. Channelization reduces renewable harvest through loss or displacement caused by the channel itself and by making some land more wet and other land more dry.

Presence of Finfish and Shellfish

The presence of finfish and shellfish is positively correlated with food harvesting (Gosselink et al. 1974, Batie and Wilson 1978, Niering and Palmisano 1979). Conversion of bottomland hardwood forests to rice or soybean production eliminates the above two groups of animals. Channelization is likely to reduce finfish and shellfish abundance also, especially in the short-term. The degree of impairment in the long-term depends upon management practices that may maintain or restore bank shading, instream riffles, pools, and segments of old channels.

On-site impoundments and conversions to aquaculture were estimated to increase shellfish and finfish production and thereby improve the renewable harvest function of the site. Stream reaches or locations downstream of impoundments are not likely to be significantly impacted. Access by spawning fish and shellfish to upstream reaches can be impaired.

Table 7. Renewable harvesting as affected by the impact of disruptive activities on bottomland hardwood forest characteristics.

Bottomland hardwood forest characteristic	Relationship of characteristic to use[1]	Activity[2] R	S	I-U	I-0	I-D	C	L	P	A
Presence of food species	+	-+	-+	0	-	0	-	+	+-	-
Presence of finfish, shellfish	+	-	-	0	+	0	-	+	NA	+
Presence of wildlife	+	-+	-+	0	+-	0	-	+	+-	+-
Presence of fiber species	+	-	-	0	-	0	+-	+	+-	-
Overall impact on renewable harvesting		-	-	0	-	0	-	+	+-	-

1 See footnote 1 in Table 4.
2 See footnote 2 in Table 4.

Conversion to pine plantations does not occur on lands that are sufficiently wet to be of significant importance in finfish or shellfish production; this change in landuse is therefore not applicable. Levees were estimated to increase harvest of finfish and shellfish through improved access.

Presence of Wildlife

The presence of wildlife is positively related to renewable harvesting (Chabreck 1979, Thomas et al. 1979). Wildlife abundance and harvesting may be affected either positively or negatively by rice and soybean production, on-site impoundments, conversion to pine plantations, or conversion to aquaculture, depending upon the particular wildlife species being harvested.

Habitats may be drastically changed by any of these activities, but food supplies can also be increased for some species. Impoundments and aquaculture may also improve habitat for some waterfowl. Channelization disrupts habitat in the short-term and segments ranges in the long-term. Only levees have a clearly positive impact, through improved access, on renewable harvesting. Streams below or above impoundments were assumed not to change significantly from the effects of the impoundment.

Presence of Fiber Species

Renewable harvesting is a positive function of the presence of fiber species (Niering and Palmisano 1979, Reinhold and Hardisky 1979). Again, conversion of bottomland hardwood forests to pine plantations has a strong positive effect on the total amount of fiber harvested because yields per hectare are improved. However, the type of wood harvested is substantially different. As a result the impact is given as both plus and minus in Table 7. Levees can increase fiber harvest through improved access, while channelization is likely to reduce fiber harvest because of losses resulting from channelizing and clearing.

Conversion of bottomland hardwood forests to rice, soybeans, on-site impoundments, or aquaculture destroys fiber production capabilities. It was assumed that downstream and upstream areas are unchanged by impoundments.

Overall Impacts

Levees have a clearly positive effect on renewable harvest by improving access. Conversion to pine plantations has a modestly positive net effect on renewable harvest uses. All other activities have strong negative effects, with the exception that impacts downstream of an impoundment were assumed to be negligible.

Overview of Impacts

The CRE Workgroup, following a thorough review of all activities and their impact on the four uses of bottomland hardwood forests, made composite judgments on the overall impacts. These are summarized in

Table 8. All except one of the impacts of the activities were judged negative or neutral. Only levee construction was viewed as having possible net positive impacts on the renewable harvest function.

Table 8. Summary of the net impacts of various disruptive activities on the four uses of interest.[1]

	Functions			
Activities	Sensory Experience	Recreation Experience	Information Storage	Renewable Harvesting
Conversion to rice, soybeans	-	-	-	-
Conversion to pine plantations	-	0	-	-
Impoundments, aquaculture	- on site 0 upstream 0 downstream	- on-site - upstream - downstream	- on-site 0 upstream 0 downstream	- on-site 0 upstream 0 downstream
Channelization	-	0	-	-
Levee construction	0	0	-	+

1 The impact shown in each cell is a composite judgement of the CRE Workgroup based on individual weighting of the positive, negative, and neutral effects of the various activities on the characteristics. A plus indicates that the use is enhanced by the activity. A minus indicates that the use is impaired. A zero indicates no effect.

CONCLUSIONS

Systematic information is needed for the public decision-making process dealing with alteration of bottomland hardwood forests to represent public values adequately. Descriptions of forest functions and linkages between various forest-using activities, especially between those that alter the forest and those that utilize the forest in its natural state, are also needed. In addition, capability to estimate the demand for various bottomland hardwood functions and associated products and services is an important element in expressing benefits and detriments of actions in common terms. The above analysis is but a subset of the information needed in the decision process. The CRE Workgroup developed the following suggestions for improving the available information:

1) Develop a generic appraisal process that includes several levels of valuation, ranging from market prices and simulated prices to intangible or nonmonetary valuations. These valuations may be organized into national and local accounts such as those used in the principles and guidelines developed by the U.S. Water Resources Council (1983). Kiker and Lynne (1989) have recently provided some features of such an approach.
2) Establish a hierarchy that identifies the relative importance of various bottomland hardwood forest characteristics to the many uses of the forest. The level of quantification required and the order of collecting information in suggestion 1) will depend upon these designations.
3) Collect primary data and review literature for the bottomland hardwood systems of interest. This work should include human values, functions, and proposed activities in bottomland hardwoods. Human values can be divided into the components shown in Figure 2 for analytical purposes. The depth of study and resources allocated to the study should be determined by the perceived importance of the specific case. Existing information should be used to make the appraisals whenever possible. The analysis and reference materials (including data sources and models) should be accessible and understandable to regulatory and higher level decision-makers in addition to the analysts who are responsible for developing them. This "generic" information should include sufficient quantification and be of sufficient scientific credibility to establish its utility in the regulatory review process. It must be current, relevant, and testable in terms of reliability, validity, and sensitivity.

REFERENCES CITED

Batie, S. S., and J. R. Wilson. 1978. Economic values attributable to Virginia's coastal wetlands as inputs in oyster production. So. J. of Ag. Econ. 10(1):160-118.

Chabreck, R. H. 1979. Wildlife harvest in wetlands of the United States. Pages 618-631 *in* P. E. Greeson, J. R. Clark and J. E. Clark, eds. Wetlands functions and values: the state of our understanding. American Water Resources Association, Minneapolis, Minnesota.

Coles, J. M. 1984. The archeology of wetlands. Edinburgh University Press, Edinburgh.

Elman, R., ed. 1972. The great American shooting prints. Knopf, New York.

Fritzell, P. A. 1979. American wetlands as cultural symbol: places of wetlands in American culture. Pages 523-534 *in* P. E. Greeson, J. R. Clark, and J. E. Clark, eds. Wetland functions and values: the state of our understanding. American Water Resources Association, Minneapolis, Minnesota.

Gosselink, J. G., E. P. Odum, and R. M. Pope. 1974. The value of the tidal marsh. Center for Wetland Resources, Louisiana State University, Baton Rouge, Louisiana. LSU-SG-74-03.

Hydra, R. 1978. Methods of assessing instream flows for recreation. U.S. Fish and Wildlife Service, Fort Collins, Colorado. FWS/OBS/-78/34.

Kiker, C. F., and G. D. Lynne. 1989. Can MAMA help? Multiple alternative/multiple attribute evaluation of wetland use. Pages 25-38 *in* J. Luzar and S. Henning, eds. Alternative perspectives on wetland valuation and use. Proceedings of a regional workshop. Southern Rural Development Center, Mississippi State University, Mississippi. SRDC Pub. no. 120.

Kooyoomijian, J. K., and N. L. Clesceri. 1973. Perception of water quality by selected respondent groupings in inland-based recreational environments. Water Resour. Bull. 10(4).

Lee, M. S. 1983. Assessing visual preference for Louisiana river landscapes. Pages 43-63 *in* R. C. Smardon, ed. The future of wetlands: assessing visual-cultural values. Allanheld, Osum and Co., Totowa, New Jersey.

Newby, F. L. 1971. Understanding the visual resources. Pages 68-72 *in* E. H. Larson, ed. The Forest Recreation Symposium. State University of New York, College of Forestry, Syracuse and U.S. Department of Agriculture, Forest Service, Northeast Forest Experiment Station, Upper Darby, Pennsylvania.

Niering, W. A. 1979. Our wetland heritage: historic, artistic, and future perspectives. Pages 505-522 *in* P. E. Greeson, J. R. Clark, and J. E. Clark, eds. Wetlands functions and values: the state of our understanding. American Water Resources Association, Minneapolis, Minnesota.

__________, and A. W. Palmisano. 1979. Use values: harvest and heritage. Pages 100-12 *in* J. R. Clark and J. E. Clark, eds. Scientists report: the National Symposium on Wetlands. National Wetlands Technical Council, Washington, D.C.

Reinhold, R. J., and M. A. Hardisky. 1979. Nonconsumptive use values of wetlands. Pages 558-564 *in* P. E. Greeson, J. R. Clark and J. E. Clark, eds. Wetland functions and values: the state of our understanding. American Water Resources Association, Minneapolis, Minnesota.

Sargent, F. O., and P. R. Berke. 1979. Planning undeveloped lakeshore: a case study on Lake Champlain, Ferrisberg, Vermont. Water Resour. Bull. 15(3):826-83.

Shabman, L. A., and S. S. Batie. 1988. Socioeconomic value of wetlands: literature review, 1970-1985. U.S. Army Engineer Waterways Experiment Station, Vicksburg, Mississippi. Technical Report Y-88.

Smardon, R. C. 1975. Assessing visual-cultural values of inland wetlands in Massachusetts. Pages 289-318 *in* E. H. Zube, R. O. Brush, and J. Gy. Fabos, eds. Landscape assessment: value, perceptions, and resources. Dowden, Hutchinson, and Ross, Stroudsburg, Pennsylvania.

__________. 1978. Visual-cultural values of wetlands. Pages 535-544 *in* P. E. Greeson, J. R. Clark, and J. E. Clark, eds. Wetland functions and values: the state of our understanding. American Water Resources Association, Minneapolis, Minnesota.

__________. 1983. State of the art in assessing visual-cultural values. Pages 5-16 *in* R. C. Smardon, ed. The future of wetlands: assessing visual-cultural values. Allanheld, Osum and Co., Totowa, New Jersey.

__________. 1988a. Visual-cultural assessment and wetland evaluation. Pages 103-114 *in* D. Hook et al., eds. The ecology and management of wetlands: vol. 2. Management use and value of wetlands. Timber Press, Portland, Oregon.

Smardon, R. C. 1988b. Water recreation in North America. Landscape and Urban Plan. 16:127-143.

Smith, B. D. 1975. Middle Mississippi explorations of animal populations. Museum of Anthropology, University of Michigan. Anthropological Paper No. 57.

Thomas, M., B. C. Liu, and A. Randall. 1979. Economic aspects of wildlife habitat and wetlands. Midwest Research Institute, Kansas City, Missouri.

U.S. Water Resources Council. 1983. Economic and environmental principles and guidelines for water and related land resources implementation studies. Superintendent of Documents, U.S. Government Printing Office, Washington, D.C.

19. THE REGULATION AND MANAGEMENT OF BOTTOMLAND HARDWOOD FOREST WETLANDS: IMPLICATIONS OF THE EPA-SPONSORED WORKSHOPS

James G. Gosselink
Marine Sciences Department, Louisiana State University
Baton Rouge, LA 70803

Lyndon C. Lee
L.C. Lee & Associates, Inc.
3016 W. Elmore Street, Seattle WA 98199

Thomas A. Muir
Division of Endangered Species and Habitat Conservation
U.S. Fish and Wildlife Service
Washington, D.C. 20240

ABSTRACT

Forested bottomlands of the southeast are complex ecosystems whose physical, chemical and biotic functional processes interact to perform valuable services to society. These processes operate at three scales - local, ecosystem, and landscape - and new properties emerge at each higher level in this hierarchy. Wetland regulation under Section 404 of the Clean Water Act is focused primarily at the local level, often ignoring the emergent properties. As a result human activities that lead to cumulative impacts are not adequately addressed. Effective management must (1) focus on the landscape, and (2) anticipate and plan, rather than react. Regulatory authority for wetland and landscape planning is provided in the Clean Water Act Advance Identification program, the Water Resources Planning Act of 1965, and the Watershed Protection and Flood Prevention Act of 1954. In order to control cumulative impacts and

---

Ecological Processes and Cumulative Impacts: Illustrated by Bottomland Hardwood Wetland Ecosystems. Edited by James G. Gosselink, Lyndon C. Lee, and Thomas A. Muir. © 1990 by Lewis Publishers, Inc. Chelsea, MI 48118. Printed in USA.

conserve natural resources, the opportunities available in these statutes must be gradually incorporated into the framework by which wetlands are regulated.

INTRODUCTION - WORKSHOP OBJECTIVES

Earlier chapters in this book, especially the workgroup summaries (Chapters 11 to 18), addressed the following general questions. (1) What key ecological structures and processes characterize forested bottomland hardwood ecosystems? (2) Is the zonal concept (Clark and Benforado 1981) useful as an organizing principle? (3) What are the major human activities in river floodplains, and how do they affect bottomland hardwood forest ecosystems? (4) From a practical point of view can the Wetland Evaluation Technique (WET) developed by Adamus and Stockwell (1983), be refined for bottomland hardwood forest ecosystems? (5) What are the implications of (1), (2) and (3) for management of cumulative impacts in forested floodplains? In this chapter we summarize the answers to these questions as formulated by the participants in the EPA-sponsored bottomland hardwood forest workshops. We emphasize ecological concepts that require modification of the regulatory procedures now in operation. We discuss management implications of these concepts, as synthesized from workshop deliberations and from our own experience. Finally, we discuss how these concepts can be incorporated into existing regulatory and non-regulatory programs.

FUNCTIONAL PROCESSES AND HUMAN IMPACTS IN BOTTOMLAND HARDWOOD FOREST ECOSYSTEMS

Chapters 2-16 and 18 summarize the roles of physical (water, nutrients, soils and substrates), biotic (vegetation, fish, shellfish, and wildlife), and socioeconomic components and processes in bottomland hardwood forest ecosystems, and their interactions. The Ecosystem Workgroup report (Chapter 17) integrates the information about these specific processes and components, to show how the parts of the system interact as a whole.

Throughout this book emphasis has been placed on processes and components of forested bottomland ecosystems that are valued by humans. These "services" are many, as perusal of the workgroup reports show. They can be summarized into four groups: hydrologic services (especially the moderation of severe floods through peak stage and current velocity reduction; water quality services (especially the reduction and transformation of stream nutrient and suspended sediment concentrations); production of harvestable commodities (e.g., timber, waterfowl, and game), and provision of habitat (to maintain balanced indigenous populations of biota); and non-consumptive recreational use (for such amenities as bird watching and boating). Physical, chemical, and biological characteristics of bottomland forests control these societal services. The number of different, although related, ecosystem characteristics in Table 1 is an indication of the complexity of the bottomland hardwood forest ecosystems we seek to manage.

Human activities influence forested bottomland ecosystem services by modifying the characteristics listed in Table 1. Table 2 lists the seven most common activities. They represent different combinations of 5 generic effects: (1) forest clearing, (2) draining, (3) impoundment, (4) stream channel modification, and (5) isolation of bottomland forests from their hydrologic connections with adjacent streams (e.g., by levee construction). By considering the influence of an activity on each characteristic (see workgroup reports), it is possible to evaluate the overall impact of that activity on the services performed by bottomland hardwood forest ecosystems (Table 3). All seven activities have generally adverse impacts. This is because all seven disrupt the natural vegetation structure and hydrologic regime that are crucial to maintaining the performance of the services.

The direction of an impact (Table 3) is put into perspective by consideration of three additional factors (1) areal extent, (2) intensity, and (3) relative permanence. This is illustrated in Figure 1, which groups activities that commonly occur in bottomland forest ecosystems. Large public flood control and navigation projects that involve levees, channelization and reservoir impoundments affect large areas (directly or indirectly). They are permanent (that is, reversible only at great expense), and they result in severe modification of bottomland hardwood forest ecosystems. Mining is the most intense activity because it completely changes the processes controlling these ecosystems; however, the total area affected is small. Dry land farming practices that require clearing

Table 1. Characteristics that control bottomland forest services, as identified by the workgroups (Roelle 1987).

Characteristics	Workgroup(s)[1]
Position and Dimensions	
Width/lenth ratio	H
Sinuosity	H, F
Size of tract	WQ, W
Proximity to stream	WQ
Proximity to flyway corridors	W
Proximity to Gulf coast	W
Proximity to mainline levees or floodways	W
Stream pattern complexity	CRE
Physical access (may also require ownership)	CRE
Topography	
Ground surface roughness	H, WQ
Elevation	H
Slope	H, WQ, F
Soils	
Infiltration rate	H
Permeability	H
Organic matter content	WQ
Clay content	WQ
Hydrology	
Downstream flood stage/standard flood	EP
Hydrologic detention time	EP
Waterflow velocity	CRE
Water depth minimum	CRE
Topography and hydrology	
Surface area of active floodplain	H
Detention storage	H
Internal drainage	H
Extent of natural ponding	WQ
Duration of flooding	WQ
Frequency of flooding	WQ
Areal/stage relationship	F
Access to permanent water	F
Shallowly flooded area	W
Access/transport vectors	EP
% of freely flooded area in watershed	EP

Table 1. Continued

Characteristics	Workgroup(s)[1]
<u>General cover maps</u>	
Natural floodplain vegetation	F
Forested watercourses	W
Open water	W
Area of contiguous forest	W
Vegetation corridors connecting BLH forests and uplands	W
% of regional landscape	EP
Effective size	EP
Landscape mosaic	EP
% of watershed in BLH forests	EP
Buffer configuration of BLH forests	EP
Linear continuity	EP
Contiguity and size	CRE
<u>Detailed cover maps</u>	
Interspersion of swamps and higher ground	W
Breadth	EP
Number and area of community types and zones	EP
Edge between community types	EP
Vegetative diversity	CRE
<u>Site visits</u>	
Vegetation cover (density)	H
Debris	H
Soil saturation	H
Density of understory	WQ, W
Density of overstory	WQ
Snags and instream cover	F
Sedimentation	F
Presence of weed seeds, fruits, aquatic tubers	W
Presence of tree cavities	W
Presence of large, unflooded, basal cavities	W
Presence of browse, forbes, grasses	W
Structural diversity of vegetation (within stand)	W, EP
Mast production and diversity	W, EP
Native species richness	EP
Nesting and escape cover	EP
Presence of food species	CRE
Presence of finfish, shellfish	CRE
Presence of wildlife	CRE
Presence of fiber species	CRE
Visual access	CRE
Animal species diversity	CRE
Flora	CRE
Fauna	CRE

Table 1. Concluded

Characteristics	Workgroup(s)[1]
Detailed investigations	
Evapotranspiration	H
Hydraulic gradient	H
Sediment input from stream	WQ
Sediment input from uplands	WQ
Upland runoff	WQ
Presence of toxics	F
Dissolved oxygen content	F
Net primary production	EP
Organic matter processing	EP
Water quality	CRE
Other	
Soil disturbance	WQ
Historical/archeological	CRE
Representative ecosystem	CRE
Relict stand	CRE
Existing documentation	CRE

[1]H = Hydrology, WQ = Water Quality, F = Fisheries, W = Wildlife, EP = Ecosystem Processes, CRE = Cultural/Recreational/Economic.

Table 2. Dominant human activities in bottomlands, grouped by their generic effect on bottomland forest ecosystems.

	Activity	Symbol	Generic Effect on Bottomland Hardwood Ecosystems	
Agriculture/forestry	Soybean cultivation	S	Draining	Clearing
	Pine plantation	P		
	Rice cultivation	R	Impoundment	
	Aquaculture	A		
Flood control navigation	Impoundment			
	upstream	I-U		
	onsite	I-O		
	downstream	I-D		
	Channelization	C	Channel works	
	Levee construction	L	Isolation/constriction	

Table 3. Summary of the impacts of various human activities on the services identified by the workgroups (Roelle 1987).[1]

Function	R	S	I-U	I-O	I-D	C	L	P	A
Hydrology									
Flood storage capacity	-	+	0	-	0	+	-	+0	-
Velocity reduction	-	-	0	NA	0	-	-	-+0	-
Ground water discharge modication	-	+	0	-	+	-	-	+	-
Water Quality[2]									
Sediment retention	-	-			-	-	-	-	-
Erosion control	-	-			0+	-	-	0	-0
Nutrient retention and transformation	-	-			-	-	-	-	-
Contaminant retention and transformation	-	-			-	-	-	-	-
Fisheries									
Finfish habitat	-	-	-	-	-	-	-	-	-
Shellfish habitat	-	-	0	-	-	-	-	-	-
Wildlife									
Wood duck habitat	-	-		-		-	-	-	-
Wintering dabbling duck habitat	-	-		-		-	-	-	-
Black bear habitat	-	-		-		-	-	-	-
White-tailed deer habitat	-	-		-		-	-	?	-
Migrating passerine habitat	-	-		-		-	-	-	-
Ecosystem Processes[2]									
Maintenance of natural biotic diversity, cumulative level	-	-		-		-	-	-	-
Maintenance of natural biotic diversity site-specific level	-	-		-		-+	-+	-	-
Food chain support	-	-		-		-	-	-	-
Streamflow mediation	-	0		+		-	-	0	0+
Transforming and filtering	-	-		-+		-	-+	-	-0
Cultural/Recreational/Economic									
Sensory experience	-	-	0	-	0	-	+	-	-
Recreation experience	-	-	-	-	-	+	0	0	-
Information storage	-	-	0	-	0	-	-	-	-
Renewable harvesting	-	-	-	-	-	-	+	+	-

[1]A plus = enhancement, a minus impairment, a zero no impact. NA = not applicable. R = conversion of site to rice, S = conversion to soybeans, I-U = impact upstream of poundment, I-O = impacts of on-site impoundment (except Ecosystem Processes Workgroup where I-O = impoundment impacts in general), I-D = impacts downstream from impoundment, C = stream channelization adjacent to site, L = levee construction on-site, P = site conversion to pine plantation, A = site conversion to aquaculture.
[2]Workgroup did not summarize impacts. Entries are author's interpretation.

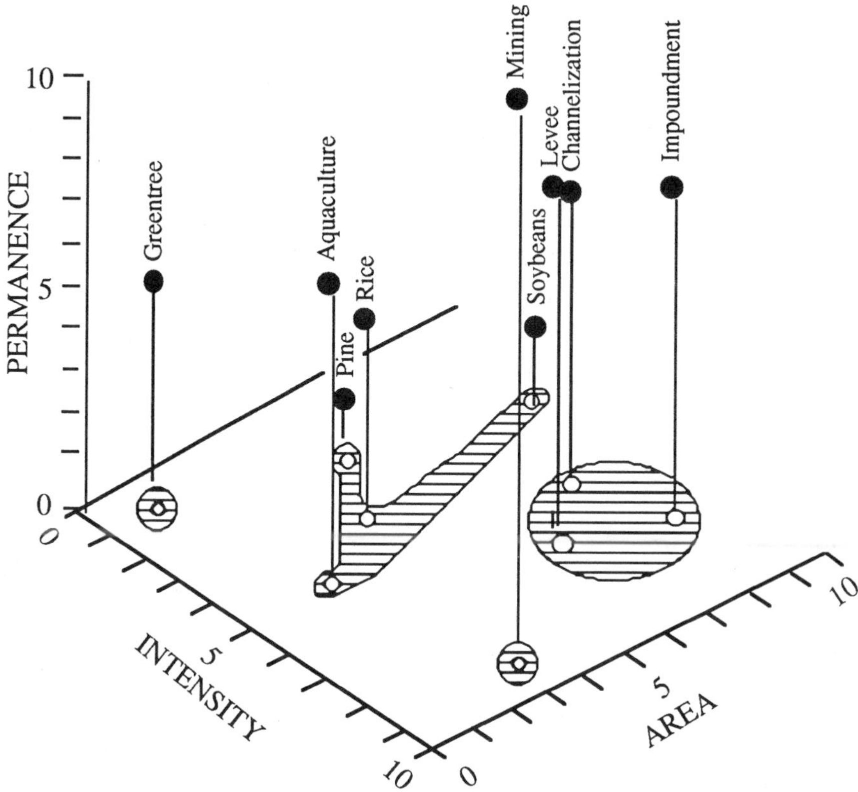

Figure 1. Simple ordination of activities in bottomlands by their cumulative impacts (from Lee and Gosselink 1988, p.596; Copyright © by Springer-Verlag New York Inc., reprinted with permission).

(e.g. soybean culture) are reversible (albiet, on a scale of 100 years or more for full replacement of functions), but they affect more bottomland area than any other activity. Cultivation of other commodities (pine trees, aquaculture products, rice) is a more permanent activity than soybean farming because of the construction of impoundments. However, these activities occupy less area. Greentree reservoirs, timber harvest (not shown) and hunting (not shown) were considered among the least intrusive of common activities, which, when carried out using best management practices, have minimal effect on the long term integrity of bottomland forest ecosystem structure and function.

THE CONCEPT OF HIERARCHY IN BOTTOMLAND HARDWOOD FOREST SYSTEM ORGANIZATION, AND ITS MANAGEMENT IMPLICATIONS

Three generalizations from the workshop discussions concern (1) the importance of hydrology in bottomland hardwood forest ecosystems, (2) the role of vegetation, and (3) the notion of a natural hierarchy of spatial and temporal scales in considerations of the ecology of bottomland forests and their responses to human impacts. The first two generalizations have been discussed in detail elsewhere (Goseslink et al. Chapter 11, Conner et al. Chapter 14, Johnson and McCormick 1979, Wharton et al. 1982, Clark and Benforado 1981, Brinson et al. 1981), and will not be considered further here. In this section we consider in some detail the notion of ecosystem hierarchy. The topic has recently received considerable attention within an emerging discipline of landscape ecology (Forman and Godron 1986, Turner 1989) and as applied in conservation biology (Soulé and Wilcox 1980). But the practical implications for wetland regulation through Section 404 of the Clean Water Act have scarcely been addressed.

Bottomland hardwood forests can be characterized in terms of a temporal and spatial hierarchy. Aspects of this hierarchy are shown in Figures 2 and 3. Both figures demonstrate that temporal scales generally increase with the size of the structures and processes considered. This is true regardless of whether one is concerned with vegetation patterns, ecosystem processes, or human disturbance regimes.

In the bottomland hardwood forest workshops, three hierarchical levels of bottomland forest system organization were clearly distinguished. The local site is characterized by analysis of species, species groups, and individual processes. At this level, typical questions concerned the relation of flood hydrographs to local flooding; population

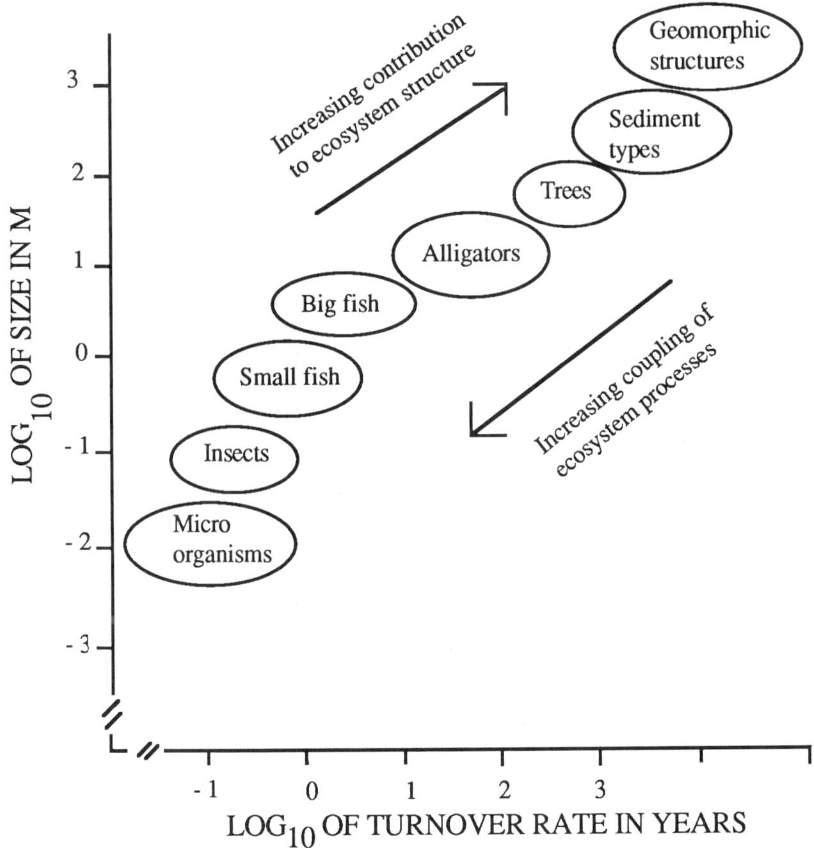

Figure 2. The generalized relationship between structural size and temporal scale of associated processes (Gosselink et al. Chapter 17)..

size of a cavity-nesting species and their relationship to the density and distribution of tree snags; or the movement of water across a floodplain as controlled by forest litter, soil permeability, and nutrient dynamics.

Ecosystems provide the next higher level of the conceptual hierarchy of bottomland forest systems organization. At the ecosystem level, individual sites are viewed in a context that combines all the major variables and interactions of the first (site-specific) level into an integrated,

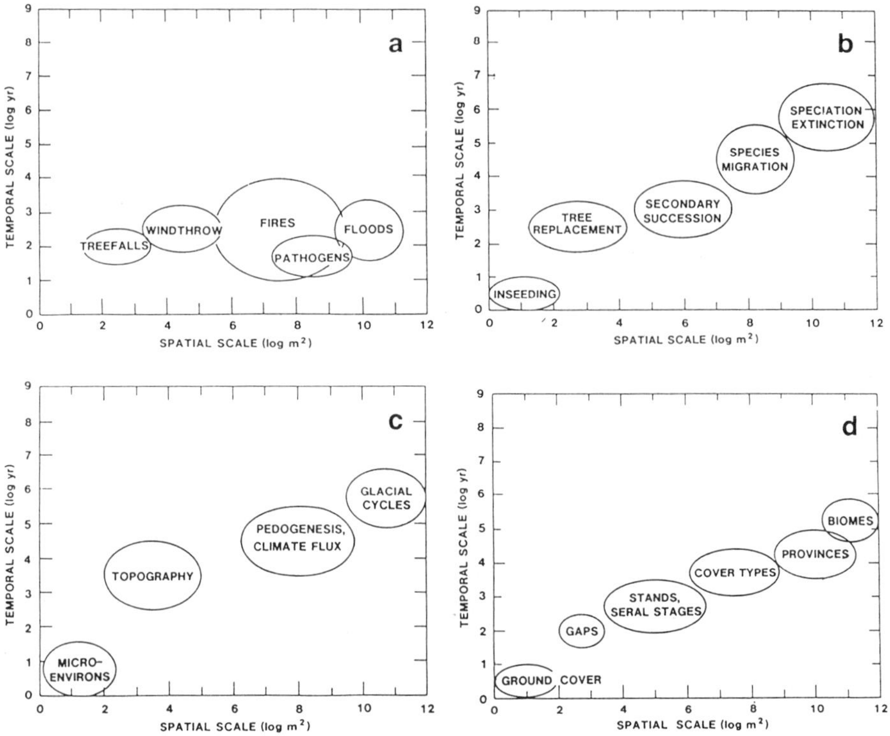

Figure 3. **(a) Disturbance regimes, (b) forest processes, (c) environmental constraints, and (d) vegetation patterns, viewed in the context of space-time domains (from Urban et al. 1987, page 120; Copyright © 1987 by the American Institute of Biological Sciences, reprinted with permission).**

functional unit. Typically, ecosystems ecology deals with such concepts as external, physical, forcing functions; nutrient cycling; trophic structure; and community diversity. These concepts have little meaning at the lower level of the hierarchy. Bottomland forest ecosystem characterization enables one to estimate the "cascade" effect of human activities; for example, to trace the impact of a hydrologic modification to the forest vegetation and from there through the forest's trophic levels.

A final level of bottomland forest systems organization is that of the landscape. While there is considerable overlap between concepts concerning ecosystems and landscapes (see Gosselink et al. Chapter 17), the latter term is usually reserved for large heterogeneous areas composed of several ecosystems that are spatially and temporally linked and that function as an integrated unit. The importance of spatial patterns in landscapes is explicitly recognized. Landscapes have their own properties, related to large areas and to long time frames. For example, a river is a complex system of many parts that function together as an integrated whole. Headwaters, upland slopes, floodplains, terraces, and river channels are all spatially and structurally integrated and interrelated. River systems cannot be fully understood by a study of these individual parts; their interrelationship is a fundamental property of the system.

The concept of a spatial and temporal hierarchy of bottomland forest systems organization has a number of implications that influence our understanding of these systems and the way we manage them.

1) <u>Management of individual processes or species generally ignores the integrated nature of bottomland hardwood forest systems</u>. As a result, related processes are ignored, and the resulting management actions often have unforeseen and undesirable side effects. For example, flood control projects are too often designed to minimize on-site flooding, with little regard to impacts on water quality and biota, or even on resulting downstream flooding and/or secondary effects to the biological integrity of downstream reaches. Since bottomland flooding regimes determine the quality of most of the valued services of bottomland forest systems, such a focus maximizes flood control at the expense of other services.

There has been considerable effort devoted to the development of evaluation systems for rating bottomland hardwood forests and other wetland ecosystems. Evaluation protocols such as the Habitat Evaluation Procedure (HEP, U.S. Fish and Wildlife Service 1980), the Golet Evaluation System (Golet and Larson 1974) and the Wetland Evaluation Technique (WET, Adamus et al. 1987) were developed with the idea of rating the services of a specific wetland site, or of comparing the relative value of services provided by alternative sites. They all focus on the first level of bottomland forest systems organization, site-specific processes.

If models are correctly bounded, the WET and Golet procedures deals qualitatively with second level (ecosystem) processes such as primary production, flood reduction, and water quality protection. No current rating procedure adequately addresses third level (landscape) processes. Indeed the context value of a specific site (that is, its context in the landscape) can only be evaluated if that context is assumed to persist through time. This presumption is unwarranted in the absense of an implemented natural resource management plan for the entire watershed landscape. For example, the value of a riparian site for flood abatement changes if other flood detention areas are isolated from the river by levees, if a dam is constructed across the river upstream, or if land clearing increases runoff.

2) Bottomland hardwood forest systems operate as integrated functional units. This statement is related to the previous point. Rather than managing to maximize one service at the expense of other services, evaluation and management procedures should recognize the integrated functioning of bottomland hardwood forest systems, and manage to optimize for the greatest combined value of all services. At a minimum, management decisions should be based on an explicit recognition of the balance between gains in one service and the loss of others. An excellent example of the integrated functioning of bottomland hardwood forest systems is the functional role of the elevation gradient found at many sites. As elevation changes across a floodplain, flooding frequency, depth, and duration also change. As a result, soil characteristics and vegetation patterns also change in relatively predictable ways (see Chapters 12 and 14). Biotic and geochemical processes related to soil saturation, inundation, and accessibility (primary productivity, nutrient recycling, decomposition, fish reproduction) also change predictably across this gradient (Gosselink et al. 1981). One issue discussed at length in the first workshop was whether classifying bottomland hardwood forest sites according to flooding (and related vegetation and soil) zones was a useful management devise (see Clark and Benforado 1981). Despite considerable evidence that ecological processes change predictably across the gradient, and despite consensus about the utility of the zonal concept for organizing and communicating some types of information about bottomland forest systems, workshop participants rejected the idea of using zones as a management tool precisely because the concept ignored

the integrated nature of the bottomland ecosystems along complex environmental gradients. Repeatedly individuals and workgroups emphasized that the functions of each bottomland "zone" (as defined in Chapters 12 and 14) changed as water levels changed. The land-water interface moves up and down the zones with changes in river stage, and as this occurs the ecological processes associated with that interface move. Thus, while the idea of rating zones (for example, giving highest priority for preservation to the annually flooding zone, and progressively lower priority to higher zones), seemed to be a reasonable regulatory innovation, it ignored points repeatedly made in the workshops: (a) zones are interdependent, that is, the quantity and quality of water, materials, and biota in a particular zone depend on changes that occur as these items cross and interact with other portions of the floodplain ecosystem; (b) functional processes move among zones, depending on flooding depth; and (c) different zones perform different (or very similar) services and support different (or very similar) flora and fauna at different seasons, depending on flooding.

3) <u>The regulatory focus on an individual site ignores the context of that site in the landscape.</u> The ecological value of a site depends in part on its position in the landscape. For example, construction in a bottomland hardwood forest site upstream from a reservoir could threaten an urban water supply. Or, clearing a bottomland hardwood forest stand near a city could have a larger adverse effect on hunters, than clearing a similar stand many miles away from an urban area. Although the concept of landscape context is broadly recognized, from a wetlands regulatory perspective it has been difficult to address. Current §404 regulations consider ecosystem processes, but generally the focus has been on site-specific functions and services. Early evaluation approaches and procedures dealt strictly with site-specific conditions but not the landscape contexts in which wetland sites exist. More recently, WET (Adamus et al. 1987) began to broaden the focus of functional assessments by evaluating the <u>opportunity</u> and <u>significance</u> of a particular wetland process which may occur in a wetland, but landscape contextual considerations and their influence on the functional attributes of a particular wetland at both site-specific and landscape scales are narrowly interpreted.

4) <u>Important ecological processes occur at landscape scales.</u> Landscapes, as integrated ecological units, are characterized by processes

that have meaning only at landscape scales. Fortunately, these processes can often be measured at specific points in the landscape, the measurement reflecting the functional integrity of the whole system. For example, the hydrograph at a specific point on a river is a function of precipitation over the upstream portions of the watershed, land use on the watershed, the frequency and intensity of upstream human activities that affect water flows, watershed slope, and soil characteristics. Similarly, water quality at a specific site on a river, reflects upstream loading, which in turn is controlled by all the above parameters. Large, continuous tracts of bottomland hardwood forest were identified by both the wildlife and ecosystem processes workgroups as landscape structures with functions and attributes that transcend the average size of individual §404 permit sites. Large forested bottomland tracts consist of a diverse array of subsystems that have the potential to support a full set of species with narrow habitat requirements, especially those species adapted to forest interiors. These tracts maintain spatially heterogeneous patch-dynamic processes whereby shade-tolerant and shade-intolerant species can occur in all life stages. They enhance genetic interchange and diversity, which are especially important to maintain viable populations of large, terrrestrial organisms. They also maintain faunal assemblages that have community and trophic integrity.

5) A site-specific focus cannot deal adequately with cumulative effects. Cumulative effects are, by their nature, landscape-level processes. They can be characterized as occuring in five general categories: time and space crowded, synergistic, indirect, and nibbling (Beanlands and Duinket 1983). The additive impacts (which result from nibbling), occur as a result of human activities on many different sites, which taken together have significant effects on the structural and functional integrity of the environment. Since evaluation of the environment of a single site through assessment techniques such as HEP or WET considers only the impact of modifying that site, it cannot and does not consider what happens to the whole landscape when many different sites are modified. Thus, almost by definition, a site-specific focus does not consider the landscape, and therefore ignores cumulative effects.

RECOMMENDATIONS FOR BOTTOMLAND HARDWOOD FOREST MANAGEMENT

Hierarchical considerations, as discussed above, lead to a number of recommendations for management of forested bottomland wetlands.

1) Regulation and management procedures must focus on the landscape as well as on site-specific impacts. Section 404 of the Clean Water Act, and attendant regulations and guidelines (33 CFR § 230, 40 CFR § 320-30), provide for regulation of dredge and fill activities in Waters of the United States, including wetlands. Section 404 does not regulate development of upland sites. Nevertheless, because uplands, wetlands and waterbodies (within river basins or hydrologic units), function as an integrated whole, effective management of bottomland hardwood forest wetlands requires that upstream and upslope ecosystems be managed as part of an integrated landscape. As discussed above, landscape management is necessary to: (a) maintain landscape-level ecological properties; (b) maintain the functional integrity of individual sites within the landscape, since these sites are influenced by what occurs around them; and (c) control cumulative impacts.

2) A landscape focus requires preplanning. As part of the federal Section 404 permit process, judgments about human actions in wetlands are made by permit specialists in response to individual permit requests prepared by project proponents. If regulatory actions are to maintain and restore the physical, chemical, and biological integrity of the nations's waters, as required in the Clean Water Act, they must be made in the context of plans for the entire landscape. This follows from previous arguments: if decisions are made on individual sites from site information alone, there can be no effective management of the landscape. And if the landscape degrades because of the cumulative effects of many site-specific decisions, the individual sites will also degrade, because they are part of the landscape. This implies that decisions about sites should be made only in the context of landscape plans. Such plans must be formulated and articulated prior to local decisions (Gosselink and Lee 1989, Lee and Gosselink, 1988). Although site evaluations are primarily reactive, landscape evaluations must be anticipatory, setting the conditions for site development.

3) The planning process requires an iterative sequence of ecological assessment, goal-setting, and planning for spatially discrete implementation. Assessment of the ecological condition of a landscape unit is a necessary precursor to its management. Assessment provides information about the condition and potential of the landscape system, from which goals and plans can be formulated (Gosselink et al. Chapter 17, Gosselink and Lee 1989). Such assessments should focus on landscape-scale processes, not details of individual sites. (Tables 6 and 7 in Chapter 17 list such processes and suggest indices by which to measure them.) Goal-setting forces all interested parties to reach consensus about the desired future of the total resource, not selected aspects of it. Finally, planning is the process of implementing the goals through prioritized actions at specific locations within the assessment unit. Within this context, permit decisions should be based on the "direction" of the impact of the proposed action with respect to the goals (Lee and Gosselink 1988). Generally, permits would be approved if cumulatively they move the landscape system toward stipulated goals. Permits would be denied if proposed projects move the landscape system away from approved goals. This kind of process requires monitoring of the system and provision of an "institutional memory" so that actions are recorded as they occur, and implementation strategies can be revised as the landscape system approaches the goals.

RECENT DEVELOPMENTS IN WETLANDS REGULATION

The bottomland hardwood forest workshops were held during a time when both the scientific and regulatory communities were becoming increasingly concerned with the issue of cumulative impacts. Since then, several workshops and/or symposia had cumulative impacts as their major focus and a number of articles on the subject were published (Bedford and Preston 1988a, CEARC 1986, Gosselink and Lee 1989, Williamson and Hamilton 1989). One of the workshops, held by the U.S. Environmental Protection Agency (EPA) in January 1987, was initiated specifically to develop more systematic cumulative impact assessment methods within a consistent conceptual framework. In the Preface to the published report of the workshop, Bedford and Preston (1988a) pointed to the "fundamental incongruity" that confronts those who regulate or study wetland ecosystems. "The scale at which they observe human impacts on wetland resources to be accumulating is far greater than the scale at which they ask questions or make decisions. Entire wetland landscapes have been altered inadvertently through the cumulative effects of numerous localized individual actions. Insights gained through research conducted at one site and on one process cannot provide straightforward answers about the consequences of multiple interacting processes operating at the scale of watersheds and landscapes" (p. 561).

In subsequent articles in this EPA workshop publication this theme is repeated many times. In their summary article Bedford and Preston (1988b) stated: "The primary conclusion to be drawn from the articles and the workshop is that improving the scientific basis for regulation will not come merely from acquiring more information on more variables. It will come from recognizing that a perceptual shift in temporal, spatial, and organizational scale is overdue. The shift in scale will dictate different -- not necessarily more -- variables to be measured in future wetland research and considered in wetland regulation" (p. 752).

The excellent articles resulting from the EPA workshop focus primarily on documentation of cumulative impacts to hydrology, water quality and life support functions of wetland ecosystems; and on conceptual approaches to measuring these impacts. The emphasis was on how to "scale up" research to make it consistent with the scale of management problems. The workshop did not address how, given appropriate technical information, cumulative impacts can be managed.

In contrast, the recommendations of the National Wetland Policy Forum, which met during 1987 and 1988, focused on action programs to solve a number of wetland problems, including the cumulative loss of wetlands. The Forum was an EPA-commissioned effort to develop national policy recommendations by individuals representing a cross-section of groups most interested in wetland issues. Participants included representatives of environmental organizations, academic interests, developers, farmers, foresters, and local, state and federal agencies. In its final report (Conservation Foundation 1988) the Forum recommended a national goal of "no overall net loss of the nation's remaining wetlands base in the short term, and an increase in the quantity and quality of the nation's wetlands resource base in the long term" (p.3).

One of a number of specific recommendations for implementing the no net loss goal is advance planning. "A major element in these reforms is increased emphasis on advance planning to guide our wetlands protection and management programs. The Forum recommends all states undertake the preparation of State Wetlands Conservation Plans to provide a basis for all subsequent acquisition, regulation, and other wetlands protection and management activities. These efforts, which should reflect local land-use plans and other societal values, should result in the nation's wetlands programs anticipating needs and problems rather than merely reacting to them" (Conservation Foundation 1988, p. 4).

The no net loss policy, if implemented, would elevate the status of wetland mitigation, since presumably dredge and fill permits that involved "unavoidable losses" would impose conditions for wetland creation and/or restoration to offset the losses. Currently, mitigation is carried out without much regard for whether created wetlands replace the services of the developed site in a landscape context (Larson 1987, Kusler and Kentula 1989). Federal guidelines state a preference for on-site, in-kind (i.e., local) replacement of wetlands that are impaired, but it has been extremely difficult for regulators and wetland scientists to address the question of functional equivalence, especially as related to position and function of wetlands within watersheds. Advance planning, which establishes a blue print for the future of a landscape unit, could identify key sites for wetland creation and/or restoration, that would enhance overall landscape integrity (Gosselink et al. 1990).

REGULATORY STRATEGIES TO IMPLEMENT PLANNING WITH A LANDSCAPE FOCUS

Wetland Regulation under Section 404 of the Clean Water Act

In the United States, the principle regulatory tool to protect wetlands is Section 404 of the Clean Water Act, and associated U.S. Army Corps of Engineers (ACE) and EPA regulations and guidelines (33 CFR §§ 320-330; 40 CFR §230). Aside from initial uncertainties about the regulatory intent of the Act and the incremental clarification of the regulations through successive court challenges (Anonymous 1988), the chief hindrance to effective regulation is probably the incompatibility of cumulative impact issues with site by site permit review. By definition, a site-specific focus cannot limit cumulative impacts, except in the most restrictive case of denying all §404 permit requests. That is, in the absence of landscape-level planning, there can be no effective mechanism to decide how much development is enough, or to allow development of one site but not another, if the site evaluations are equivalent. Under current administration of §404, the development of one site often sets a precedent for the approval of all similar permit requests (or types of projects), since the criteria for permit approval are usually the same regardless of landscape context. This inevitably leads to an "all or none" syndrome, in which environmental activists are pitted against development interests over isolated issues and local sites, rather than focusing on site-specific project impacts in the context of an overall plan.

Inability of the §404 program to manage cumulative impacts is compounded by the issuance of nationwide permits, particularly Nationwide Permit 26 (33 CFR 330.5 (a)(26)). This general permit, which allows up to 10 acres of fill to be placed into headwater wetlands (wetlands located in headwater reaches of tributary streams with a mean annual flow of less than 5 ft^3 sec^{-1}) and in isolated wetlands (wetlands that are not bordering, neighboring, or contiguous to Waters of the United States), effectively exempts about 7 million ha of wetlands from regulation (Anonymous 1988), and may have led to significant cumulative impacts to water quality, hydrology, and food web support functions throughout the range of forested wetlands in the Southeastern and Mid-Atlantic states . Headwaters are critically important because they form the primary contact between terrestrial and aquatic ecosystems in landscapes that support bottomland hardwood wetlands. They also exist as the headward-most member of bottomland hardwood wetland continua.

Originally, the Nationwide Permit program was designed to reduce paperwork and agency response time to permit requests that were repetitive (e.g., Nationwide Permit 3 - repair and replacement activities), which had potentially minor impacts (e.g., Nationwide Permit 1 - aids to navigation), or which potentially impacted relatively small areas (e.g., Nationwide Permit 26 - headwaters, isolated or intermittent waters). Dredge and fill activities permitted under Nationwide Permit 26 usually proceed without requirements for mitigation. Thus, leniency in Nationwide Permit 26 permitting of projects proposed for headward extensions of watersheds is, by definition, bound to result in incremental, unmitigated cumulative impacts. This situation is exacerbated by the fact that incremental "nibbling" of headwater and isolated wetlands is common for three major reasons: (1) unscrupulous practices on the part of project proponents, who have been known to design projects so that individual parcels (in development projects) are less than 10 acres (e.g., 100 acre tracts split into multiple parcels of 9.9 acre); (2) high personnel turn-over rate within regulatory agencies, which leads to failure to track the cumulative effects of Nationwide Permit 26 permitting over time within a given watershed; (3) the slow pace of incremental wetland development (from an institutional viewpoint) that can occur over many years. Even with the best institutional memory, the cumulative effects of Nationwide Permit 26 activities are difficult to except after years of incremental change in a watershed.

Long Range Opportunities for a Landscape Focus and Planning

In the four years since the EPA-sponsored bottomland hardwood forest workshops, the idea that planning is a necessary precondition of cumulative impact management has gradually gained recognition. For example, the report of the National Wetlands Policy Forum states "To be effective, the nation's wetlands protection and management programs must anticipate rather than react.....They should consider the whole, not just the individual parts. In short, the programs should be based on comprehensive planning for wetlands protection and management...." (Conservation Foundation 1988:19). In addition, planning, which sets the conditions for management of the wetland resource, has been distinguished from regulation, which is one (albiet probably the strongest) of a number of tools available to implement planning. These tools run the gamut from programs of persuasion, offering incentives and disincentives, through regulatory prohibition, to acquisition. A difficulty in both planning and implementation is that they can involve many different agencies at federal, state, and local levels, as well as public and private interests, whose individual missions and interests must be merged to achieve a consensus about the future of the resource and to implement the plan. Each agency works within relatively restrictive regulations and guidelines developed as the best way to satisfy its mission. In this context it is appropriate to ask two questions about our institutional capability to address cumulative impacts. First, what agency, if any, has either the responsibility or the authority to initiate comprehensive planning? Second, within the existing regulatory programs of EPA (as the nation's lead environmental protection agency), what flexibility exists for comprehensive planning?

Federal Authority for Comprehensive Natural Resource Planning

In the United States no federal agency, or even group of agencies, has either the mission or the explicit authority to undertake comprehensive natural resource planning at landscape scales. This is in part because of the nature of the political system in this country. Generally, the legislative branch responds to particular issues with legislation tailored to address that issue. Legislation, therefore, is usually focused on a single resource (e.g., water quality, rare and endangered species), and agencies with responsibility for implementing the legislation are constrained by that focus. They do not have a mandate for managing the landscape to optimize all resources. A recent report on the nation's floodplain management activities (Johnston 1989), states (with reference to natural floodplain values) "....federal, state and local programs to manage these natural values are often not focused on the floodplain, but on the particular resource or activity. For example, programs have been developed to protect water quality, but they are not focused on managing floodplains for water quality protection...... Floodplain management and/or protection of natural floodplain values is typically not an explicit program objective" (pages 7-20 to 7-21).

Two reasons for the reluctance of federal legislators to address comprehensive planning are: (1) the strong tradition in the United States against land-use planning, which is seen as abrogating rights of individual ownership of land, and (2) the historical rights of states and local jurisdictions in these matters (i.e., environmental federalism). As a result, federal environmental legislation has generally avoided any hint of landuse planning. Rather, it has attempted to protect and enhance the public resources of air, water and federally owned or managed lands. Any restrictions on the use of private lands have been enacted to protect these public resources, and only after bitter battles in the Congress and in the courts. The requirement for a permit for wetland development, as required under § 404 of the Clean Water Act, is an excellent example of this type of legislation. The Act's purpose is to maintain and restore the physical, chemical, and biological integrity of the nation's waters, which include wetlands (40 CFR 328.3 (a) 1-7). Thus, regulation of dredge and fill activities in privately owned wetlands is justified in the Act because wetlands have been shown to function in several ways that directly influence the maintenance of water quality. In contrast, land management practices in uplands, which can also strongly affect downslope water quality, are not usually regulated. (We refer here to such practices as land clearing, not to the production of point source effluents, which are regulated.)

Despite these political considerations, several Acts passed by the Congress during the late 60's and 70's do provide possibilities for natural resource planning on landscape scales. The 1970 National Environmental Policy Act (NEPA, 42 U.S.C. 4371 *et seq.*) broadly recognized natural environmental values and incorporated them into federal decision-making for large projects. Although the requirement to address cumulative impacts is specific in NEPA, in practice the Environmental Impact Statement is largely reactive. Environmental Impact Statement review is not usually viewed as a comprehensive planning process, and thus has never effectively addressed cumulative impacts.

An earlier Act, the Water Resources Planning Act of 1965 (WRPA), has greater potential for comprehensive planning. Although it was passed to address the widespread problem of flood losses, in combination with NEPA it has the potential to encourage the development of comprehensive resource management plans. The WRPA created the Water Resources Council, which was charged with (1) assessing the adequacy of basin plans and establishing principles and standards for federal participation in river basin planning; and (2) creation, operation and termination of interstate government river basin planning commissions. The Principles and Standards issued by the Water Resources Council (1973) established three "accounts" by which proposed actions were to be evaluated; (1) a

"National Economic Development" account, which, as implied, evaluated the economic utility of the project; (2) an "Environmental Quality" account, and (3) an account called "Social Well Being". This document was a major attempt to describe a procedure for standardized basin-level planning that addressed multiple management objectives including both economic development and environmental quality (Field 1979)

The Water Resources Council was active in promoting the establishment of River Basin Commissions. Several were established in the late 60's and early 70's, and were required to prepare comprehensive, coordinated plans for their region or basin.

The ambitious programs set forth by the WRPA did not fare well. In 1982, funding was withdrawn from the Water Resources Council and River Basin Commissions. The Water Resources Council was disbanded, and the River Basin Commissions gradually closed their operations. The 1983 revision of "Principles and Standards" dropped the "Environmental Quality" and "Social Well-Being" accounts. As a result the "National Economic Development" account provides the sole basis for project justification (Interagency Task Force on Floodplain Management 1986). In the absence of the Water Resources Council, the WRPA and Executive Orders related to floodplain and wetland management (Executive Orders 11988 and 11990) have been coordinated by an Interagency Task Force on Floodplain Management (created in 1975 to develop a Unified National Program for Floodplain Management, and formerly chaired by a Water Resources Council representative). While this task force has sponsored several important initiatives it does not consider itself a policy group, and does not have the influence of the Water Resources Council.

Another Act that has potential for comprehensive natural resource management is the Watershed Protection and Flood Prevention Act of 1954, administered by the Soil Conservation Service. This Act authorized the Soil Conservation Service to participate in comprehensive watershed management projects in cooperation with states and their subdivisions. Eligible projects are limited to watersheds of less than 250,000 acres (Stembridge, undated). Although the goals of the Act were water and erosion management, it is currently being used in a much broader context, for example as a mechanism for wetlands protection.

Generally, responsibility for land-use planning is reserved for the states and local authorities. State and local governments are in a stronger position than the federal government to institute comprehensive planning (1) through their authority to initiate zoning and other land use restrictions, and (2) because they are closer to the local problems and thus able to tailor plans to local conditions. Therefore, federal legislation, such as the WRPA and the CWA, does not preempt this authority, but through a system of incentives encourages state assumption. For example, the CWA encourages states (through grants for compliance) to assume responsibility for regulation of local water quality, including (in the 1987 revisions), the development of plans for control of non-point pollution sources. The report of the National Wetlands Policy Forum reflects this trend. "The Forum recommends that: state and local governments and regional agencies, with the support and cooperation of the relevant federal agencies, undertake wetlands planning to achieve the goal of no net loss, and that Congress allocate adequate funds to assist with these efforts" (Conservation Foundation 1988, p. 19).

In summary, there is at present little Congressional, executive, or federal agency encouragement or authority for comprehensive multiple objective landscape planning, of the kind needed to manage cumulative impacts. In some respects (for example, with the Water Resources Council and the WRPA) there has been a retreat, since the 1970's, in the executive branch's willingness to deal with comprehensive planning. Some agencies are working to institute such planning within the authority of existing statutes and regulations. However the nature of these statutes often limits the scope of planning efforts. Nevertheless, some existing statutes provide mechanisms to address cumulative impacts. In the following section we examine EPA's opportunities for cumulative impact management through its mandate for wetland protection.

EPA Opportunities for Landscape Planning and Cumulative Impact Management

Advance Identification. Section 404 of the Clean Water Act requires EPA and the ACE to regulate dredge and fill activities in wetlands. Nationwide the ACE has authority to evalute and issue permits for dredge and fill activities, but EPA has authority to veto ACE decisions under §404(c) of the CWA. The procedure, as described in regulations issued by the ACE and EPA (33 CFR §§ 320-330; 40 CFR §230) is largely reactive (that is permit evaluations are usually initiated in response to an individual application). Because the §404 permit process focuses on individual permit sites it fails to address landscape contextual problems, and thus provides no mechanism to manage cumulative impacts.

EPA, however, has authority for planning in the Advance Identification program, as described in Section 230.80 of the 404(b)(1) Guidelines (40 CFR Part 230). The ACE has a similar authority under their "Special Area Management Plans". Under §230.80 EPA and the permitting authority (ACE or an approved State agency), act jointly to identify wetlands and other waters of the United States as possible future disposal sites or areas generally unsuitable for disposal site specification. The results are informational, not regulatory. Any person may still submit an application for a §404 permit regardless of the designation of the site. The results of the advance identification process simply put the applicant on notice about the relative probability of obtaining a permit. Thus this program is used to improve the public's understanding of wetlands and to provide a degree of predictability to the regulated community that does not exist within the context of individual permit reviews. It also provides EPA regional offices opportunities to coordinate more effectively in local or state planning processes. Documents developed in the advance identification process contain wetlands assessment and other data that are potentially useful in other wetland protection activities such as the development of local zoning and regulations, the identification of valuable sites for purchase or zoning easements, and for public education efforts.

Although the §404 (b)(1) guidelines are specific as to the purpose of Advance Identifications (i.e., designation of wetlands as unsuitable for disposal site specification) EPA's Advance Identification guidelines (Office of Wetland Protection 1989) clearly encourage a broad perspective on ecosystem protection and management. For example, the program can be used in combination with other portions of the §404 (b)(1) guidelines to address large geographically-based issues. Advanced identifications can be focused to reduce or reverse regional trends of wetland losses, and/or for assessing cumulative and secondary impacts within designated boundaries (§230.80 (g) & (h)). The preferred units for advanced identification program are watersheds or ecosystems because components of these systems functionally relate to one another. Broad participation by all agencies and groups with interests in the geographic area is usually encouraged, beginning early in the advance identification process (Office of Wetland Protection 1989). Thus the advance identification program may be used as a means to initiate comprehensive planning, even though EPA's authority under Section 404 is limited to jurisdictional wetlands.

The potential of Advance Identifications for use as a planning mechanism for cumulative impact management is just beginning to be realized, with the recent issuance of guidelines by the EPA Office of Wetland Protection. The number of advance identification programs initiated across the U.S. has increased rapidly in the past three years, and several have been completed. They vary in size from several to over 100,000 ha. Documents prepared for these programs are limited to the authority under §230.80 to identify wetlands unsuitable for disposal. It is not clear, therefore, to what extent they are being used for more comprehensive planning. One difficulty is that no agency has authority to expand the advance identification program into a comprehensive multiple objective project that can result in the development of natural resource protection goals and bring to bear many different regulatory and non-regulatory tools to implement the goals. Some broader authority is needed to accomplish this. It could be located at the state level, but most states are not organized to handle programs that cut across federal and state agency lines. The Advance Identification guidelines (Office of Wetland Protection 1989) provide a broad perspective, and specific recommendations for comprehensive planning, but their effective implementation appears to require broader authority and more focused support from EPA.

Interim Opportunities for Cumulative Impact Management. As EPA and cooperating federal, state and private sector groups gain more experience with and more confidence in Advance Identifications, we can expect better recognition and management of cumulative impacts in wetlands. In the short term, the §404(b)(1) requirements for alternatives analysis, impact assessment and impact minimization can provide opportunities to improve the permit review process and specifically to address cumulative impacts. However, there are distinct administrative, conceptual, and practical limits regarding the extent to which cumulative impact assessments can be effectively incorporated into individual §404 permit reviews. These limits are defined mostly by agency time and by staffing constraints. As discussed above, agency personnel are significantly influenced by lack of clear definition of "goals" for management of an individual permit site in the context of the landscape in which it occurs, by lack of "institutional memory" on the part of regulatory agencies, by lack of familiarity with the rapidly evolving field of landscape ecology, and by the unavailability of landscape-scale data in a usable and accessible form.

Goals set the context for assessment of cumulative effects and their perceived impacts on the condition of wetlands. Even though the §404(b)(1) Guidelines direct federal agencies to review "cumulative effects" (230.11(g)) and "secondary effects" (230.11(h)) in the context of individual permit reviews, only broad direction is given in the §404(b)(1) Guidelines, the ACE Regulations (33 CFR 320-330) and from the Clean Water Act itself regarding standards or criteria for review of cumulative and secondary effects. Thus, while the agencies are required by their guidelines to review cumulative and secondary effects, they currently have no specific standards for doing so. In addition, while most agency managers and staff are relatively well trained in basic biology, jurisdictional delineation of wetlands, and the administrative or regulatory details of individual §404 permit review, they are relatively unfamiliar with the new and rapidly evolving field of landscape ecology. This lack of command of the subject matter often leads to review of §404 permits in the context of what is familiar to agency staff and management (i.e., site-specific review), and not necessarily what is most technically valid (i.e., site specific reviews combined with landscape-level reviews).

In practice, the importance of cumulative and secondary effects has been recognized by relatively innovative regional offices of the EPA, ACE and U.S. Fish and Wildlife Service. Such recognition is reflected in documents (such as letters of denial or written recommendations for project revisions) that address cumulative and secondary effects. However, most of these agency comments are developed in reaction to an individual permit request and without substantive technical support in the form of data and/or credible models that document cumulative or secondary effects. The lack of sound technical substantiation of such impacts is open to successful challenge by individuals with knowledge of current methods and data.

The main point to emphasize is that assessment of cumulative impacts needs to be carefully and slowly implemented, and incorporated into the standard procedures used by regulatory agencies during their reviews of wetland-related projects. Implementation should be incremental, along with "business-as-usual" considerations of individual permit reviews. Because introduction of landscape-level reviews into the §404 process is experimental, agencies need to develop and maintain a dual track approach to permit review. When possible, data from cumulative impact assessment efforts initiated under regional Advance Identifications or Special Area Management Plans should be used to substantiate claims of cumulative or secondary effects related to individual permit reviews (e.g., EPA-Region Three Recommended Section 404c determination for the Ware Creek, VA Project, USEPA, Philadelphia, PA, 1989). Simultaneously, agencies need to devote increased time and resources to Advance Identifications and/or Special Area Management Plans that rely principally on landscape-level analyses that focus on management of cumulative impacts and anticipatory approaches to permit reviews (Gosselink and Lee 1989).

Another effective tool in cumulative impact management is the "layering" of State §401(Water Quality) authority over individual as well as nationwide permit reviews. This approach has proven to be an effective tool consistent with the goals of environmental federalism as articulated by the current (Bush) administration. For example, states with §401 permitting authority can effectively veto isuance of individual §404 or Nationwide Permit 26 permits through denial of state wetlands permits or §401 certification.

A final suggestion regarding current approaches for dealing with cumulative effects relates to two major concerns articulated throughout the bottomland hardwood forest workshops by agency personnel: (1) the need for a unified approach to the definition of wetlands under federal jurisdiction, and (2) the need to deal consistently with agricultural and silvicultural exemptions in conformity with §404(f)(1) and (f)(2).

With regard to development of a consistent, unified approach for definition of wetlands, the "Federal Manual for Identification and Delineation of Jurisdictional Wetlands", issued January 10, 1989 (Federal Interagency Committee 1989), promises to go a long way towards shifting the focus of §404 review from questions concerning jurisdiction to more substantive issues. This is a positive step for the national wetlands protection process. It has particular relevance for dealing with cumulative impacts because agreement on wetland jurisdictional boundaries allows participants in the §404 process to devote more time and effort to deal with identification of project purpose, siting alternatives outside jurisdictional wetlands, minimization of impacts, and mitigation of unavoidable impacts, as required in the §404 Guidelines. With the recent emphases on landscape ecology and "no net loss" of wetlands, such efforts are bound to incorporate landscape-level thinking into the planning processes associated with individual permit reviews and with Advance Identifications and Special Area Management Plans.

Section 404(f)(1) & (2) of the Clean Water Act address agricultural, silvicultural and ranching exemptions for "on-going and established" operations (40 CFR 232.3). The new wetlands delineation manual clarifies regulatory jurisdiction over these exemptions. Combined with the Swampbuster provisions of the Food Security Act of 1985, consistent application and enforcement of §404(f)(1) & (2) can work to maintain bottomland hardwood forest patch size and structural integrity, help to limit non-point source inputs to Waters of the United States, and maintain reasonably natural hydrologic connections with functioning (even though farmed or forested) wetland ecosystems.

The National Wetlands Policy Forum (Conservation Foundation 1988) identified a number of pressing wetlands issues that need to be addressed. The Forum suggested that while some of the present regulatory deficiencies can be addressed by modifying present regulations, others probably require new legislation at both federal and state levels. Recognizing the slow pace of major legislative initiatives we have, in this chapter, focused on the possibilities for improving wetlands protection within the present statutory and regulatory framework. Summarizing these possibilities, a three-pronged approach, operating within current regulatory constraints, can produce significant improvements in management of cumulative impacts in bottomland hardwood forests.

1) Gradually incorporate landscape ecology principles and cumulative impact evaluation into current wetland permit reviews that are completed under federal or similar regional, state, or municipal authorities;

2) Expand the use of advance planning authority under §230.80 of the EPA §404 Guidelines, and the ACE Special Area Management Program, to address landscape level planning and cumulative impacts assessment;

(3) Continue to explore ways to use the comprehensive planning authority of the Water Resources Planning Act of 1965 and the Watershed Protection and Flood Prevention Act of 1954, to incorporate comprehensive planning at landscape scales. This approach integrates the multiple missions of the many federal, state, and local authorities and private interests that have a stake in wetlands. In the final analysis only comprehensive planning can lead to effective management of cumulative impacts, and effective protection of forested bottomland resources.

LITERATURE CITED

Adamus, P. R., and L. T. Stockwell. 1983. A method for wetland functional assessment. U.S. Army Waterways Experiment Station, Vicksburg, MS. Tech. Rep. D-77-223. 72 p.

______________, E. J. Clairain, Jr., R. D. Smith, and R. E. Young. 1987. Wetland evaluation technique (WET); Vol. II: Methodology. Operational Draft Technical Report Y-87-__. U.S. Army Engineer Waterways Experiment Station, Vicksburg, MS.

Anonymous. 1988. A guide to federal wetlands protection under Section 404 of the Clean Water Act. Natural Resources Law Institute, Lewis and Clark Law School, Portland, OR. Anadromous Fish Law Memo 46: 1-34.

Beanlands, G. E., and P. N. Duinker. 1983. An ecological framework for environmental impact assessment in Canada. Institute for Resource and Environmental Studies, Dalhousie Univ., Halifax, and Federal Environmental Assessment Review Office, Hull, Quebec.

Bedford, B. L., and E. M. Preston, eds. 1988a. Cumulative effects on landscape systems of wetlands. Environmental Management 12 (5): 561-775.

______________________________.. 1988b. Conclusions. Pages 751-771 in B. L. Bedford and E. M. Preston, eds. 1988. Cumulative effects on landscape systems of wetlands. Environmental Management 12 (5): 561-775.

Brinson, M. M., R. L. Swift, R. C. Plantico, and J. S. Barclay. 1981. Riparian ecosystems: their ecology and status. U.S. Fish Wildl. Serv., Biol. Serv. Prog., Washington, D.C. FWS/OBS-81-17. 151 p.

CEARC (Canadian Environmental Assessment Research Council) and NRC (United States National Research Council. 1986. Cumulative environmental effects: a binational perspective. Minister of Supply and Services Canada. 174 pp.

Clark, J. B., and J. Benforado, eds. 1981. Wetlands of bottomland hardwood forests. Elsevier Sci. Publ. Co., N.Y. 401 pp.

Conservation Foundation. 1988. Protecting America's wetlands: an action agenda. The final report of the National Wetlands Policy Forum. Washington, D.C. 69 pp.

Federal Interagency Committee for Wetlands Delineation. 1989. Federal manual for identifying and delineating jurisdictional wetlands. U.S. Army Corps of Engineers, U.S. Environmental Protection Agency, U.S. Fish Wildl. Serv., and U.S.D.A. Soil conservation Service. Washington, D.C. Cooperative Technical Publication. 76 pp. & Appendices.

Field Associates, R. M. 1979. Regional water resources planning: a review of level B study impacts. Prepared for the U.S. Water Resources Council, Washington, D.C.

Forman, R. T. T., and M. Godron. 1986. Landscape ecology. J. Wiley and Sons, New York. 619 pp.

Golet, F. C., and J. S. Larson. 1974. Classification of wetlands in the glaciated Northeast. U. S. Fish Wildl. Serv., Washington, D.C. Resources Publ. 116. 56 pp.

Gosselink, J. G., S. E. Bayley, W. H. Conner, and R. E. Turner. 1981. Ecological factors in the determination of riparian wetland boundaries. Pages 197-219 *in* J. B. Clark and J. Benforado, eds. Wetlands of Bottomland Hardwood Forests. Elsevier Sci. Publ. Co., N.Y. 401 pp.

______________, G. P. Shaffer, L. C. Lee, D. Burdick, D. Childers, N. Taylor, S. Hamilton, R. Boumans, D. Cushman, S. Fields, M. Koch and J. Visser. 1990. Can we manage cumulative impacts? Testing a landscape approach in a forested wetland watershed. Bioscience (in press).

Gosselink, J. G., and L. C. Lee. 1989. Cumulative impact assessment in bottomland hardwood forests. Wetlands 9:83-174.

Interagency Task Force on Floodplain Management. 1986. A unified national program for floodplain management. Federal Emergency Management Agency, Washington, D.C. FEMA 100/March 1986.

Johnson, R. R., and J. F. McCormick. 1979. Strategies for protection and management of floodplain wetlands and other riparian ecosystems. U. S. Forest Service, U. S. Dept. Agriculture, Washington, D.C. General Tech. Rep. WO-12.

Johnston Associates, L. R. 1989. Status report on the nation's floodplain management activity. An interim report. Prepared for the Interagency Task Force on Floodplain Management, Washington, D.C.

Kusler, J. A., and M. E. Kentula. 1989. Wetland creation and restoration: the status of the science. Vols. I and II. U. S. Environmental Protection Agency, Corvallis, OR. EPA/600/3-89/038.

Larson, J. S. 1987. Increasing our wetland resources. Proc. Nat. Wildl. Fed. Symp., 5 Oct 87. Washington, D.C.

Lee, L. C., and J. G. Gosselink. 1988. Cumulative impacts on wetlands: linking scientific assessments and regulatory alternatives. Environmental Management 12 (5): 591-602.

Office of Wetland Protection. 1989. Use of Advance Identification authorities under Section 404 of the Clean Water Act. Guidance to EPA regional offices. U.S. Environmental Protection Agency, Office of Water, Washington, D.C. IM-89-2.

Roelle, J. E. 1987. Workshop summary. Pages 156-171 *in* Roelle, J. E., G. T. Auble, D. B. Hamilton, G. C. Horak, R. L Johnson, and C. A. Segelquist, eds. Results of a workshop concerning impacts of various activities on the functions of bottomland hardwoods. U.S. Fish Wildl. Serv., National Ecology Center, Fort Collins, CO. NEC-87/15. 171 pp.

Soulé, M. E., and B. A. Wilcox, eds. 1980. Conservation biology: an evolutionary-ecological perspective. Sinauer Associates, Sunderland, MA. 305 pp.

Stembridge, J. E., Jr. undated. The role of Conservation Districts in urban floodplain management. Coast Environment Resources Institute, OR. Working Paper No. 4.

Turner, M. G. 1989. Landscape ecology: the effect of pattern on process. Annual Rev. Ecol. Systematics 20: 171-197.

U.S. Fish and Wildlife Service. 1980. 101 ESM -Habitat as a basis for environmental assessment. 102 ESM - Habitat evaluation procedures. Division of Ecological Services, Washington, D.C.

Urban, D. L., R. V. O'Neill, and H. H. Shugart, Jr. 1987. Landscape ecology. Bioscience 37 (2): 119-127.

Wharton, C. H., W. M. Kitchens, E. C. Pendleton, and T. W. Sipe. 1982. The ecology of bottomland hardwood swamps of the Southeast: a community profile. U. S. Fish Wildl. Serv., Biol. Serv. Prog., Washington, D.C. FWS/OBS-81/37.

Williamson, S.C., and K. Hamilton. 1989. Anotated bibliography of ecological cumulative impacts assessment. U.S. Fish. Wildl. Serv., Washington, D.C. Biol. Rep. 89(11). 80 pp.

Water Resources Council. 1973. Principles and standards for planning water and related land resources. Federal Register 38 (174): 10 Sep 1973, 24778, with 2 appendices.

APPENDIX A. SOURCES OF INFORMATION, MAPPING, AND CLASSIFICATION, FOR BOTTOMLAND HARDWOOD ECOSYSTEMS.

I. SATELLITE IMAGERY AND AERIAL PHOTOGRAPHY

Satellite imagery and aerial photography provide a high resolution perspective and repeat coverage capability of earth scenes. Imagery such as this can be used to refine and update existing map information on a wide range of disciplines.

GEOLOGY
- Geologic formations
- Cultural features
- Mapping

FORESTRY
- Stand density
- Disease, flood, and fire damage
- Vegetation assemblages
- Recovery and regeneration over time from human and natural disturbances

WATER AND ENVIRONMENT
- Hydrologic transport processes
- Sediment plumes and plume boundaries
- Landscape stability
- Flood coverage and damage
- Wildlife habitat mapping

URBAN PLANNING AND CARTOGRAPHY
- Vegetation types and boundaries
- Roads
- Demographic changes
- Map revisions

AGRICULTURE
- Crop type and coverage
- Drainage patterns
- Rate of conversion to agricultural land
- Boundaries
- Surface water on fields

A. SATELLITE IMAGERY

SOURCE	PRODUCT
EROS Data Center U.S. Geological Survey User Services Section Sioux Falls, S.D. 57198 (605) 594-6511 Information and ordering EROS is now owned by EOSAT Earth Observation Satellite Company EOSAT 8201 Corporate Drive Metroplex II Suite 450 Landover, MD 20785 (301) 552-0500 Presently, continue ordering and information from EROS.	Landsat (1972-present) satellite Multispectral Scanner (MSS) and Thematic Mapper (TM) imagery. Available in various bands (black and white to thermal infrared). Coverage of one earth scene of 185 x 185 km is repeated every 18 days at 30 meter resolution. Imagery can be acquired on film, color hardcopy, or computer compatible tapes (CCT).
Spot Image Corporation 1897 Preston White Drive Reston, VA 22091-4326 (703) 620-2200 TELEX: 4993073	Spot satellite imagery is available in black and white (10 meter resolution) and records color in three spectral bands (20 meter resolution). Coverage of one given point is repeated every 26 days. A full scene format is 60 x 60 km. Imagery can be acquired on film, paper, or computer compatible tapes at several processed levels (A-C). Spot imagery is also available in stereoscope pairs of images.

B. AERIAL PHOTOGRAPHY

SOURCE	PRODUCT
National Cartographic Information Center U.S. Geological Survey National Space Technology Laboratories NSTL Station, MS 59529 (601) 688-3544 FTS 494-3544	National High Altitude Photography Program (black and white and color infrared photography). Imagery available from 1978 to present. Total U.S. coverage.

SOURCE	PRODUCT
EROS Data Center U. S. Geological Survey Sioux Falls, S.D. 57198 (605) 594-6511 FTS 784-7151	U.S. Geological Survey medium altitude photography (black and white). Imagery available 1950 to present, and in some cases 1940 to present. Total U.S. coverage.
NCIC National Headquarters National Cartographic Information Center U.S. Geological Survey 507 National Center Reston, VA 22092 (703) 860-6045 FTS 928-6045	
Agriculture Stabilization & Conservation Service (ASCS) U.S. Department of Agriculture Aerial Photography Field Service P. O. Box 30010 222 West 2300 South Salt Lake City, Utah (801) 524-5856	Low and medium altitude black and white aerial photography (1950 to present). Excellent national coverage through many years.
Soil Conservation Service Remote Sensing Specialist P. O. Box 2890 Washington, D. C. 20013 (202) 447-4452	Low and medium altitude black and white and color infrared aerial photography. Phone or write for inventory of coverage.
U.S. Department of Agriculture Photography Division O.G.P.A. Office of Information 4407 South Building Washington, D. C. 20250 (202) 447-2791	Low and medium altitude black and white and color infrared aerial photography. Photography catalog listed by subject.

SOURCE	PRODUCT
National Ocean Service National Oceanic and Atmospheric Administration Photogrammetry Branch, N/CG2314 Nautical Charting Division Rockville, MD 20852 (301) 443-8601	Low and medium altitude black and white and color infrared aerial photography. Coastal areas only.
U.S. Forest Service USDA P. O. Box 2417 Washington, D. C. 20013 (202) 235-8071	Low and medium altitude black and white and color infrared aerial photography of the Forest Service holdings.
Photographic Repository for the Forest Service is the ASCS	
U.S. Environmental Protection Agency Photogrammetry EMSL-LV AMS P. O. Box 15027 Las Vegas, NV 89114 (702) 798-2100	Coverage based on project requests. Low and medium altitude color infrared photography. Phone or write for inventory of coverage.
Markhurd Aerial Surveys 345 Pennsylvania Ave., S. Minneapolis, MN 55426 (612) 545-2583	Private firm with good national coverage in high altitude black and white and color infrared aerial photography.

OTHER SOURCES OF AERIAL PHOTOGRAPHY

U.S. Army Corps of Engineers

State Department of Transportation

State Department of Natural Resources

State Department of Planning

OTHER SOURCES OF AERIAL PHOTOGRAPHY (Continued)

State Forestry Service

State Water Management Service

Regional Utility Companies

Regional Forestry Companies

Regional Coal Companies

Regional Geological and Petrochemical Companies

State Universities with Forestry, Geology, Geography and Remote Sensing Departments

II. MAP PRODUCTS

SOURCE	PRODUCT
U.S. Geological Survey Eastern Distribution Branch 1200 South Eads Street Arlington, VA 22202 (areas east of Mississippi River)	1:24,000 topographic maps 1:50,000 scale mapping 1:100,000 scale quadrangle maps 1:250,000 scale quadrangle maps
Western Distribution Branch U.S. Geological Survey Box 25286, Federal Center Denver, CO 80225 (areas west of Mississippi River)	1:1,000,000 scale maps 1:2,000,000 scale sectional maps State Map Series U.S. Base Maps Land Use/Land Cover Maps

SOURCE	PRODUCT
National Mapping Program Department of the Interior U.S. Geological Survey Reston, VA 22092	Coastal Ecology Inventory Maps Hydrologic Unit Maps of States and U.S. Geologic Map of States and U.S. Digital Elevation Models

Further Information

National Headquarters National Cartographic Information Center U. S. Geological Survey 507 National Center Reston, VA 22092 (703) 860-6045 FTS 928-6045 Regional Offices Eastern Mapping Center-NCIC U.S. Geological Survey 536 National Center Reston, VA 22092 (703) 860-6336 FTS 928-6336 Mid-Continent Mapping Center-NCIC U.S. Geological Survey 1400 Independence Road Rolla, MO 65401 (314) 341-0851 FTS 277-0851 National Cartographic Information Center U.S. Geological Survey National Space Technology Laboratories NSTL Station, MS 39529 (601) 688-3544 FTS 494-3544	7.5 and 15 minute topographic maps Digital line graphs and digital elevation model data Orthophotoquad 7.5 minute maps National high altitude photography Detailed mapping information

SOURCE	PRODUCT
Further Information (Continued) Rocky Mountain Mapping Center-NCIC U.S. Geological Survey Box 25046, Stop 504 Federal Center Denver, CO 80225 (303) 236-5829 FTS 776-5829 Western Mapping Center-NCIC U.S. Geological Survey 345 Middlefield Road Menlo Park, CA 94025 (415) 323-8111, ext. 2427 FTS 467-2427 Alaska Office-NCIC U.S. Geological Survey Skyline Building 218 E. Street Anchorage, AK 99501 (907) 271-4148 FTS 271-4159	
National Wetlands Inventory U.S. Fish & Wildlife Service Order From: National Cartographic Information Center U.S. Geological Survey 507 National Center Reston, VA 22092 (703) 860-6045 (703) 860-6167	1:24,000; 1:62,500; and 1:63,360 scale wetlands inventory maps
National Wetlands Inventory U.S. Fish and Wildlife Service Dade Building Suite 217 9620 Executive Center Dr., N St. Petersburg, FL 33702 (813) 893-3624	Ongoing wetlands inventory

SOURCE	PRODUCT
Further Information (Continued)	
National Coastal Ecosystems Team U.S. Fish and Wildlife Service Information Transfer Specialist NASA Slidell Computer Complex 1010 Gause Blvd. Slidell, LA 70458 (504) 646-7310 FTS 8-680-7310	Community profiles and limited maps
U.S. Forest Service USDA Southern Region Geometronics 1720 Peachtree Rd., NW Atlanta, GA 30367 (404) 881-3986	Maps of all forest service timber holdings
U.S. Forest Service Southeastern Experimental Station 200 Weaver Blvd. Asheville, N.C. 28804 (704) 259-0758	Inventory survey of all forested lands in the southeast
U. S. Forest Service Southern Experimental Station T-10210 U.S. Postal Service Bldg. 701 Loyola Avenue New Orleans, LA 70113 (504) 589-6712	Inventory survey of all forested lands in the south and southwest
Soil Conservation Service Soils Division P. O. Box 2890 Washington, D.C. 20013 (202) 382-1819	National soil maps and updates on state and county soil survey coverage
State Soil Conservation Service State Soil Scientist State of Interest County of Interest	County soil maps of each state are available through the respective state

III. INFORMATION ON BOTTOMLAND HARDWOOD RESEARCH DISCIPLINES

A. CLIMATE

- NATIONAL CLIMATIC DATA CENTER (NCDC)
  National Oceanic and Atmospheric Administration
  Asheville, N.C. 28801
  (704) 259-0682

The National Climatic Data Center: (NCDC) acquires, processes, archives, analyzes, and disseminates global climatological data; develops analytical and descriptive products to meet user requirements; and provides facilities for the World Data Center-A (Meteorology). It is the collection center and custodian of all United States weather records and the largest climatic center in the world.

Retrospective weather satellite images (from 1960 to present) are available from NCDC's Satellite Data Services Division, World Weather Building, Room 100, Washington, D. C. 20233.

- OFFICE OF SATELLITE DATA PROCESSING AND DISTRIBUTION
  E/ER2
  NOAA/NESDIS
  Washington, D.C. 20233
  (202) 634-7722

The Office of Satellite Data Processing and Distribution directs the operation of NESDIS central ground facilities and field stations. It processes and distributes current weather satellite data and derived products to the National Weather Service and other domestic and foreign users. Direct readout systems on NOAA's weather satellites enable users to receive, on relatively low-cost equipment, images and weather charts directly from the satellites.

B. HYDROLOGY

- OFFICE OF WATER DATA COORDINATION (OWDC)
  U.S. Geological Survey
  417 National Center
  Reston, VA 22092

The OWDC is the focal point for interagency coordination of ongoing and planned water-data acquisition activities of all Federal agencies and many non-Federal organizations. The "National Handbook of Recommended Methods for Water-Data Acquisition," indexes to the "Catalog of Information on Water Data," and other publications are available from OWDC.

- NATIONAL WATER DATA EXCHANGE (NAWDEX)
U.S. Geological Survey
421 National Center
Reston, VA 22092

NAWDEX maintains a computerized data system that identifies sources of water data and indexes information on the water data available from the sources. The NAWDEX Program Office and local Assistance Centers assist data users in locating sources of water data, identifying sites at which data have been collected, and obtaining specific data.

- HYDROLOGIC INFORMATION UNIT
U.S. Geological Survey
420 National Center
Reston,VA 22092

Questions about water resources in general and about the water resources of specific areas of the United States can be directed to the Hydrologic Information Unit. This office will also answer inquiries about the availability of reports of water-resources investigations.

- NATIONAL OCEANOGRAPHIC DATA CENTER (NODC)
NOAA
Washington, D.C. 20235
(202) 634-7232

NODC acquires, processes, archives, analyzes, and disseminates global oceanographic data; develops analytical and descriptive products to meet user requirements; and provides facilities for the World Data Center-A (Oceanography). It was the first NODC established and houses the world's largest usable collection of marine data.

- U.S. FOREST SERVICE
Southern Region
1720 Peachtree Rd., NW
Atlanta, GA 30367
(404) 881-3986

Forest Service has hydrological information on experimental watersheds (e.g. Coweeta, North Carolina) and other forestry service holdings regarding rainfall amount, periodicity, intensity and base and peak waterflow data that may be transposable to areas of interest.

- FEDERAL AND PRIVATE UTILITIES WITH DAM SYSTEMS

Many federal utilities (Tennessee Valley Authority, Bonneville Power, Washington) and private utilities maintain excellent records of rainfall, base and peak flows of streams, and rivers servicing dam networks.

- STATE WATER MANAGEMENT AND WATER RESOURCES AGENCIES

These agencies, especially in states prone to flooding or requiring management of water flow, maintain excellent records of rainfall and flow discharges.

C. VEGETATION

- U.S. Fish and Wildlife Service
  Regional Director, Research
  18th + C St, NW
  Washington, D.C. 20240
  (202) 343-6394

- National Coastal Ecosystems Team
  Information Transfer Specialist
  U.S. Fish and Wildlife Service
  NASA-Slidell Computer Complex
  1010 Gause Blvd.
  Slidell, LA 70458
  (504) 646-7310
  FTS 8-680-7310

Both of these offices are engaged in wetlands evaluation of coastal and interior regions. They are an excellent source of information and publications regarding wetlands and bottomland hardwoods research. The regional offices of Fish and Wildlife Service are also a good source for information concerning programs in that area. The addresses and telephone numbers for all the regions are available from the Washington office.

- National Wetlands Inventory (NWI)
  U.S. Fish and Wildlife Service
  Dade Building Suite 217
  9620 Executive Center Drive, North
  St. Petersburg, FL 33702
  (813) 893-3624

A useful source of information for categorizing wetlands under the Fish and Wildlife's classification system. The NWI is currently developing an inventory of coastal and interior wetlands, their location and boundary delineation, for use by other agencies and the private sector.

- U.S. Forest Service
  Southeastern Experimental Station
  200 Weaver Blvd.
  Asheville, N.C. 28804
  (704) 259-0759

- U.S. Forest Service
  Southern Experimental Station
  T-10210
  U.S. Postal Service Bldg.
  701 Loyola Avenue
  New Orleans, LA 70113
  (504) 589-7013

The Forest Service Experimental Stations usually have complete inventory surveys of all forested areas in the south and southeast. These surveys are performed at various times, therefore inventory material can range from current to quite dated.

- State Forestry or Natural Resources Departments

These departments usually have information on forest resources which can range from excellent to poor. These surveys may also only inventory commercial or commercially available timber stands. Forested wetlands may be totally omitted. Some states are even converting to computer digitized vegetation maps.

- Private Timber Industries

These companies may or may not have inventoried bottomland hardwoods in their area. Usually all landholdings are surveyed and inventoried as to forest type, stand density, age, accessibility, etc.

D. SOILS

- Soil Conservation Service (SCS)
  Soils Division
  P. O. Box 2890
  Washington, D.C. 20013
  (202) 382-1819

The SCS can provide listings, addresses, and inventories of state soil conservation service. In addition, the SCS can give soil survey updates on which counties have and have not been done in each state.

- State Soil Conservation Service Based in State

The repository of all county soil survey maps.

- State Universities

State universities with forestry, soils, and agricultural departments or private universities with these respective disciplines may have additional information on forest soils in particular. Many universities conduct ongoing research in forests and their soil properties.

E. GEOLOGY

- Geologic Inquiries Group
  U.S. Geological Survey
  907 National Center
  Reston, VA 22092

Questions about all aspects of geology can be directed to the Geologic Inquires Group. This office answers questions on topics such as earthquakes, energy and mineral resources, the geology of specific areas, and geologic maps and mapping. Information on geologic map indexes for the 50 states is also available.

- Geophysical Data Center (NGDC)
  National Oceanic and Atmospheric Administration
  Boulder, CO 80303

NGDC acquires, processes, archives, analyzes, and disseminates global solid Earth and marine geophysical data, as well as ionospheric, solar, and other space environment data; develops analytical, climatological, and descriptive products to meet user requirements; and provides facilities for World Data Center-A (Glaciology, Solar Terrestrial Physics, Solid Earth Geophysics, and Marine Geology and Geophysics).

- State Geological Survey

These agencies will have the state, county, and 1:24,000 topographic quadrangle geologic maps. Indexes of the topographic quadrangles for the topographic state are usually available from them.

- Universities

Universities with geology departments can have extensive information and data on certain areas in the state, unique and/or troublesome geologic formations, or special areas of geological expertise.

- Private Geological Companies

These companies may have extensive information on the geology of a state, county or local area. In particular, these companies also perform extensive geological data bases for EPA Environmental Impact Statements (EIS). Particularly in developing urban areas with adjacent wetlands voluminous amounts of information can be found from EIS documents.

IV. ENVIRONMENTAL DATA BASE DIRECTORY

- National Environmental Data Referral Service (NEDRES)
  Program Office
  Assessment and Information Services Center
  NOAA/NESDIS (E/AIX3)
  3300 Whitehaven Street, N.W.
  Washington, D.C. 20235
  (202) 634-7722
  FTS 634-7722

NEDRES is designed to provide convenient, economical, and efficient access to widely scattered environmental data. In development since 1980, NEDRES is

- a publicly available service which identifies the existence, location, characteristics, and availability conditions of environmental data sets.
- a national network of federal, state, and private organizations cooperating to improve access to environmental data for anyone who needs it.

INDEX

Rivers and Harbors Act of 1899, 232, 605